Interdisciplinary Applied Mathematics

Problems in engineering, computational science, and the physical and biological sciences are using increasingly sophisticated mathematical techniques. Thus, the bridge between the mathematical sciences and other disciplines is heavily traveled. The correspondingly increased dialog between the disciplines has led to the establishment of the series: *Interdisciplinary Applied Mathematics*.

The purpose of this series is to meet the current and future needs for the interaction between various science and technology areas on the one hand and mathematics on the other. This is done, firstly, by encouraging the ways that mathematics may be applied in traditional areas, as well as point towards new and innovative areas of applications; and, secondly, by encouraging other scientific disciplines to engage in a dialog with mathematicians outlining their problems to both access new methods and suggest innovative developments within mathematics itself.

The series will consist of monographs and high-level texts from researchers working on the interplay between mathematics and other fields of science and technology.

Interdisciplinary Applied Mathematics

For further volumes:
http://www.springer.com/series/1390

Zohar Yosibash

Singularities in Elliptic Boundary Value Problems and Elasticity and Their Connection with Failure Initiation

Springer

Zohar Yosibash
Department of Mechanical Engineering
Ben-Gurion University of the Negev
PO Box 653
84105 Beer-Sheva
Israel

ISSN 0939-6047
ISBN 978-1-4614-1507-7 e-ISBN 978-1-4614-1508-4
DOI 10.1007/978-1-4614-1508-4
Springer New York Dordrecht Heidelberg London

Library of Congress Control Number: 2011940836

Mathematics Subject Classification (2010): 35B40, 35B65, 35C20, 35J15, 35J25, 35J52, 35Q74, 47A75, 65N30, 74A45, 74F05, 74G70, 74R10, 74S05, 80M10

Printed on acid-free paper

Springer is part of Springer Science+Business Media (www.springer.com)

To my wife, Gila, and our children,
Royee, Omer, and Inbar

Preface

Things in life break, and as my son used to say after being asked why he broke one of his toys, *"It happens."* This monograph is mainly aimed at providing mathematical insight into why "it happens," especially when brittle materials are of interest. We are interested also in investigating whether "nature is acquainted with the mathematical solution," i.e., does the experimental evidence correspond to the mathematical predictions?

We are motivated by the theory of fracture mechanics, which has matured over the past half century and is able nowadays to predict failure incidents in mechanical components due to an existing *crack*. The classical approach to fracture mechanics is based on a simplified postulate, namely the correlation of a parameter characterizing the linear elastic solution in a neighborhood of the crack tip to experimental observations. It is well known that the linear elastic solution is singular at the crack tip, i.e., its gradient (associated with the stress field) tends to infinity. Thus, from an engineering viewpoint, the linear elastic solution is meaningless in the close vicinity of the crack tip, because of evident nonlinear effects such as large strains and plastic deformations.

Nevertheless, when the nonlinear behavior is confined entirely to some small region inside an elastic solution, then it can be determined through the solution of the linear elastic problem. Consequently, experimental observations on failure initiation and propagation in the neighborhood of a crack tip have been shown to correlate well with the linear elastic solution in many engineering applications.

Although attracting much attention, a crack tip is only a special, and rather simple case of singular points. In a solid body, singular solutions occur at reentrant corners, where material properties abruptly change along a free surface; at interior points where three or more zones of different materials intersect; or where an abrupt change in boundary conditions occurs. In the introduction we show some examples of the aforementioned singularities in "two-dimensional" domains.

From the mathematical viewpoint, the linear elastic solution in the vicinity of any of the above cases has the same characteristics as the solution in the neighborhood of a crack tip. Thus, an unavoidable question comes to mind: *Can one predict failure initiation at the singular points based on parameters of the elastic solution?*

The answer to this question is of major engineering importance due to its broad applicability to failures in electronic devices, composite materials and metallic structures. As in linear elasticity, the solution to heat-conduction problems has similar behavior near singularities, and the coupled thermo elastic response is crucial in understanding failure-initiation events in electronic components.

The first step toward a satisfactory answer is the capability to reliably compute the singular solution and/or functionals associated with it in the neighborhood of any singularity. This is one of the main motivations in writing this monograph. We also wanted to gather as many *explicit* mathematical results as possible on the linear elastic and heat-conduction solutions in the neighborhood of singular points, and present these in engineering terminology for practical usage. This means that we will rigorously treat the mathematical formulations from an engineering viewpoint. We present numerical algorithms for the computation of singular solutions in anisotropic materials and multi material interfaces, and advocate for the proper interpretation of the results in engineering practice, so that these can be correlated to experimental observations.

In the third part of the book, three-dimensional domains and singularities associated with edges and vertices are addressed. These have been mostly neglected in the mathematical analysis due to the tedious required treatment. In the past ten years, major achievements have been realized in the mathematical description of the singular solution in the vicinity of 3-D edges, with new insights into these realistic 3-D solutions. These are summarized herein together with new numerical methods for the extraction of so-called edge stress intensity functions and their relevance to fracture initiation. We also derive exact solutions in the vicinity of vertex singularities and extend the numerical methods for the computation of these solutions when analytical methods become too complex to be applied.

I have tried to make this book introductory in nature and as much as possible self-contained, and much effort has been invested to make the text uniform in its form and notation. Nevertheless, some preliminary knowledge of the finite element method is advised (see, e.g., [178]) but not mandatory, because we use the method for the solution of example problems (a short chapter is devoted to finite element fundamentals). It is aimed at the postgraduate level and to practitioners (engineers and applied mathematicians) who are working in the field of failure initiation and propagation. *Many examples of engineering relevance are provided and solved in detail.* We apologize to authors of relevant works that have not been cited; this is the result of my ignorance rather than my judgment.

The book is divided into fourteen chapters, each containing several sections. Most of it (the first nine chapters) addresses two-dimensional domains, where only singular *points* exist. The thermo elastic system and the feasibility of using the eigen pairs and GSIFs for predicting failure initiation in brittle material in engineering practice are addressed. Several failure laws for two-dimensional domains with V-notches and multi material interfaces are presented, and their validity is examined by comparison to experimental observation. A sufficient simple and reliable condition for predicting failure initiation (crack formation) in micron-level electronic devices, involving singular points, is still a topic of active research and interest, and we address it herein. Three-dimensional problems are addressed in the next five chapters, discussing the singular solution decomposition into edge, vertex, and edge-vertex singular solutions. I conclude with circular edges in 3-D domains and some remarks on open questions.

I have the pleasure of thanking many of my colleagues and friends who have assisted in various ways toward the successful completion of this manuscript and with whom I have had the privilege to collaborate over the past two decades: Prof. Barna Szabó (Washington University, St. Louis, MO, USA) for the motivation to write the monograph (he is a coauthor of papers based on which Chapters 3-6 are developed), Profs. Monique Dauge and Martin Costabel (University of Rennes 1, Rennes, France) for stimulating discussions and acute contributions to the understanding of edge flux/stress intensity functions (parts of Chapters 10, 13, and 14 are based on joint papers), Prof. George Karniadakis (Brown University, Providence, RI, USA) for the connection to the publisher and the encouragement to write the book. The first five chapters of the monograph were composed for the special course "Singularities in elliptic problems and their treatment by high-order finite element methods" taught in the Division of Applied Mathematics at Brown University in spring 2003 while I was on a sabbatical stay in Prof. Karniadakis's group. Many thanks are also extended to Prof. Dominique Leguillon (University of Paris 6, Paris, France) for inspiring discussions on failure laws and singularities, Prof. Ernst Rank (Technical University of Munich, Munich, Germany) for many interesting and stimulating discussions on $p-$finite element methods. I would like to thank Profs. Sue Brenner (Louisiana State University, Baton Rouge, LA, USA), Ivo Babuška (University of Texas, Austin, TX USA); and Christoph Schwab (ETH, Zurich, Switzerland) for interesting discussions on a variety of topics associated with singularities, and Dr. Tatianna Zaltzman (Sapir College, Sderot, Israel) for her help with vertex singularities (she is a coauthor on a paper based on which Chapter 12 is developed). Thanks are extended to some of my graduate students who read parts of the manuscript and provided me with their comments and insights, and especially to Dr. Netta Omer; the chapters discussing edge flux/stress intensity functions are based her doctoral dissertation, and Mr. Samuel Shannon - the last chapter is based on his MSc dissertation.

I gratefully acknowledge the permission granted by all the publishers to quote from my material previously published by them in various journals. Part of the material in this monograph is reproduced by permission of Elsevier, Wiley, and Springer publishers.

I would like to acknowledge the sponsorship of the research work reported in this book by the Air Force Office of Scientific Research, the Israel Ministry of Absorption - Center for Science Absorption, Israel Ministry of Industry and Commerce under 0.25μ Consortium Grant and the Israel Science Foundation.

Finally I would like to thank my family, Gila, Royee, Omer, and Inbar, for their understanding and patience during the writing of this book.

Beer-Sheva, Israel Zohar Yosibash

Contents

List of Main Symbols

$\underset{\approx}{\mathbf{a}}$	Denotes a tensor.
$\mathbf{a}$	Denotes a vector.
[a]	Denotes a matrix.
$f_{,x}$	Denotes $\frac{\partial f}{\partial x}$.
$f'(x)$	Denotes $\frac{df}{dx}$.
$\mathcal{E}(\Omega)$	The energy space of functions over the domain Ω. A function belongs to $\mathcal{E}(\Omega)$ if it has final "strain energy."
$\mathcal{E}_c(\Omega)$	The complementary energy space of fluxes/stresses over the domain Ω. A flux vector/stress tensor belongs to $\mathcal{E}_c(\Omega)$ if it satisfies the heat equation/equilibrium equation.
$\mathcal{E}_{ij}$	Edge between vertices V_i and V_j in a 3-D domain.
$\mathcal{G}$	The energy release rate (ERR).
$\mathcal{G}_c$	Fracture energy, also known as critical energy release rate (ERR).
$\mathcal{U}$	Strain energy within an elastic domain.
$\mathcal{B}(\tau, \chi)$	The bilinear form of the weak formulation.
$\mathcal{F}(\chi)$	The linear form of the weak formulation.
α, β	Elliptical coordinates.
α_i	The ith singular exponent (ith eigenvalue). The ith singular scalar solution is $\tau_i = r^{\alpha_i} s_i^+(\theta)$.
γ_i	The ith singular exponent (ith eigenvalue) associated with a vertex singularity. The ith singular scalar solution is $\tau_i = \rho^{\gamma_i} s_i^+(\theta, \varphi)$.
∂_β	Derivative operator $\frac{\partial}{\partial x_\beta}$.
$\underset{\approx}{\boldsymbol{\varepsilon}}$	The strain tensor.
Γ_1, Γ_2	Boundaries intersecting at the singular point.
Γ_R, Γ_{R^*}	Circular boundary around the singular point having a radius of R (resp. R^*).

$\Phi_i(\xi, \eta)$	The ith shape function over the standard finite element.
$\Phi_i^{(\alpha_j)}, \boldsymbol{\Phi}_i^{(\alpha_j)}$	The edge heat conduction/elasticity eigenfunction (for $i = 0$) primal or shadow function (for $i \geq 1$) associated with the α_j eigenvalue. $\boldsymbol{\Phi}_i^{(\alpha_j)}(r, \theta) = r^{\alpha_j + i} \boldsymbol{\varphi}_i(\theta)$.
$\Psi_i^{(\alpha_j)}, \boldsymbol{\Psi}_i^{(\alpha_j)}$	The edge dual heat conduction/elasticity eigenfunction (for $i = 0$) or dual shadow function (for $i \geq 1$) associated with the α_j eigenvalue. $\boldsymbol{\Psi}_i^{(\alpha_j)}(r, \theta) = r^{-\alpha_j + i} \boldsymbol{\psi}_i(\theta)$.
κ	Kolosov constant: $(3-\nu)/(1+\nu)$ for plane-stress, $(3-4\nu)$ for plane-strain.
λ	One of the two Lamé constants.
μ	Shear modulus $E/(2(1+\nu))$ (one of the two Lamé constants). Also the normalized crack length associated with Leguillon's failure criteria at the rounded V-notch tip (μ_0 is normalized crack length for ℓ_0).
ν	Poisson ratio.
ω	Rigid V-notch angle.
Ω	2-D or 3-D domain of interest.
$\partial\Omega$	The boundary of Ω.
ρ	V-notch tip radius, or the radius vector of the spherical coordinate system.
$\boldsymbol{\sigma}$	The elastic stress vector $(\sigma_{11}, \sigma_{22}, \sigma_{33}, \sigma_{23}, \sigma_{13}, \sigma_{12})^T$.
$\tilde{\boldsymbol{\sigma}}$	The elastic stress vector expressed in cylindrical/spherical coordinates, $(\sigma_{rr}, \sigma_{\theta\theta}, \sigma_{zz}, \sigma_{\theta z}, \sigma_{rz}, \sigma_{r\theta})^T$, or $(\sigma_{rr}, \sigma_{\theta\theta}, \sigma_{\phi\phi}, \sigma_{\theta\phi}, \sigma_{r\phi}, \sigma_{r\theta})^T$.
$\underset{\approx}{\boldsymbol{\sigma}}$	The stress tensor.
σ_c	Tensile strength.
$\tau(\boldsymbol{x})$	Temperature field - the solution to the heat conduction equation (scalar elliptic equation).
θ	Polar coordinate. In some chapters it is measured from one of the V-notch/crack edge and in others from the bisector of the solid angle.
(ξ, η)	The coordinates of the standard finite element, $-1 \leq \xi, \eta \leq 1$.
A_i	The ith generalized flux/stress intensity factor or function (for edges).
$A_{\mathrm{Ic}}^{\mathrm{blunt}}$	Critical mode I GSIF for rounded V-notches.
$B_m(x_3)$, $JB_m(x_3)$	Extraction polynomial and the Jacobi extraction polynomial of order m, that depends on the coordinate x_3 along the edge.
$[D]$	Differential operator. In 3-D $[D]^T = \begin{bmatrix} \partial_1 & 0 & 0 & 0 & \partial_3 & \partial_2 \\ 0 & \partial_2 & 0 & \partial_3 & 0 & \partial_1 \\ 0 & 0 & \partial_3 & \partial_2 & \partial_1 & 0 \end{bmatrix}$ and in 2-D $[D]^T = \begin{bmatrix} \partial_1 & 0 & \partial_2 \\ 0 & \partial_2 & \partial_1 \end{bmatrix}$.

$[D^{(r,\theta)}]$	$[D]$ operator in cylindrical coordinates.
e or $\boldsymbol{e}$	Error between the exact and FE solutions. $e = \tau - \tau_{FE}$, $\boldsymbol{e} = \boldsymbol{u} - \boldsymbol{u}_{FE}$.
E	Young's modulus.
$[E]$	Elastic material matrix with elements denoted by E_{ij}.
H_{11}	A function that associates the small virtual crack increment at the V-notch tip and ERR - it depends on the V-notch tip geometry and boundary conditions ΔH_{11} is the change in H_{11} between cracked and uncracked rounded notch tip.
$[k]$, k_{ij}	The thermal conductivity matrix and the coefficient of thermal conductivity in the x_i and x_j directions.
$[K]$	The stiffness matrix with elements denoted by K_{ij}.
k_c	Critical material-dependent parameter at failure initiation at a V-notch tip $k_c = A_1 S_{\theta\theta}^{(1)}(0)$.
K_I, K_{II}	Mode I and mode II stress intensity factors for cracks ($K_I = \sqrt{2\pi} A_1$, $K_{II} = \sqrt{2\pi} A_2$).
K_{Ic}	Fracture toughness.
$K_m^{(\alpha_i)}[B]$	Quasidual singular function.
$[M_R]$	The mass matrix associated with Γ_R edge with $(M_R)_{ij}$ its i, j element.
ℓ_0	Characteristic length.
$\boldsymbol{n}$	Outer normal unit vector to the surface $(n_1, n_2, n_3)^T$.
n_g	Gauss quadrature order.
N	Number of degrees of freedom (DOFs), also number of terms in the singular asymptotic expansion.
r, θ	Cylindrical coordinates.
$s_i^+(\theta)$	The ith angular part of the primal singular function (ith eigenfunction) of the temperature/displacement.
$s_i^-(\theta)$	The ith angular part of the dual singular function (ith dual eigenfunction) of the temperature/displacement.
$\boldsymbol{S}_i^+(\theta)$	The ith angular part of the primal eigenstress tensor.
$\boldsymbol{S}_i^-(\theta)$	The ith angular part of the dual eigenstress tensor.
$\boldsymbol{t}$	Tangential unit vector to the surface $(t_1, t_2, t_3)^T$.
T	T-stress in the vicinity of a crack tip.
$\boldsymbol{T}$	Traction vector to the surface $T_i = \sigma_{ji} n_j$.
$\boldsymbol{q}(\boldsymbol{x})$	The flux vector. It is connected to the heat conduction solution by $\boldsymbol{q} = (q_1, q_1, q_3)^T = [k]\nabla\tau$.
$\boldsymbol{u}$	The elastic displacements (solution of the Navier-Lamé elasticity system) $(u_1, u_2, u_3)^T$.
$\tilde{\boldsymbol{u}}$	The elastic solution (displacements) expressed in cylindrical/spherical coordinates: $(u_r, u_\theta, u_z)^T$, or $(u_\rho, u_\theta, u_\phi)^T$.
V_i	A vertex in a 3-D domain.
$\boldsymbol{x}$	Cartesian coordinates $(x_1, x_2, x_3)^T$.

Chapter 1
Introduction

The point of departure is the motivation to write this monograph, and the assumptions under which linear theories predict well failure initiation and propagation effects. Thereafter, a layout of the book is provided, after which a rather simplified model problem presents the notation adopted.

The main goal of this book is to provide a unified approach for the analysis of singular points, both analytically and numerically, and the subsequent use of the computed data in engineering practice for predicting and eventually preventing failures in structural mechanics. We also summarize recent new insights on the solutions of realistic three-dimensional domains in the vicinity of singular edges and vertices. We strive to provide a rigorous mathematical framework for singularities in two- and three-dimensional domains in a systematic and simple manner. We then turn to numerical methods, specifically high-order finite element analysis, and summarize advanced methods for the computation of the necessary mathematical quantities for realistic problems too complex to be tackled analytically. Failure criteria based on the generated data are being proposed and supported by experimental observations.

1.1 What Is It All About?

During the last two decades, several books on singular solutions of elliptic boundary value problems have been published, among them [49, 72, 73, 97, 98, 109, 123, 127]. A comprehensive, rigorous, and up-to-date mathematical treatment of corner singularities and analytic regularity for linear elliptic systems is about to be published in a new monograph [45], which may serve as a reference to more mathematically oriented readers. Singularities of elliptic equations in polyhedra domains are rigorously covered from the mathematical viewpoint in a recent book [117]. These books provide an excellent mathematical foundation on singular solutions of linear elliptic boundary value problems. However, most of them require highly mathematical

Z. Yosibash, *Singularities in Elliptic Boundary Value Problems and Elasticity and Their Connection with Failure Initiation*, Interdisciplinary Applied Mathematics 37,
DOI 10.1007/978-1-4614-1508-4_1, © Springer Science+Business Media, LLC 2012

proficiency and are not aimed at practical applications to failure initiation and propagation in real-life structures (except [109]). At the same time, high-order finite element methods (FEMs), namely the p- and hp-versions of the FEM, were developed and proved to be very efficient for approximating the solutions of elliptic boundary value problems with singular points on the boundary [9–11]. The use of these p-FEMs together with new extraction methods enables the computation of special singular solutions [12, 13, 177] elegantly and very efficiently suitable for use in engineering practice. Furthermore, three-dimensional explicit solutions for edge and vertex singularities are seldom provided, and their connection to two dimensional approximation is not well documented. Because of a growing demand for efficient and reliable means for predicting and eventually preventing failure initiation and propagation in multi-chip modules (MCM), electronic packages, and composite materials subjected to mechanical and thermal loads, there is a need to clearly address these singular solutions and utilize them in engineering practice. Thermal, elastic, and thermoelastic problems associated with large-scale integrated circuits, electronic packaging, and composites increase in complexity and importance. These components are assemblages of dissimilar materials with different thermal and mechanical properties. The mismatch of the physical properties causes flux and stress intensification at the corners of interfaces and can lead to mechanical failures. For example, in a conference paper on electronic components [119] the following was stated: *"The catastrophic effects of the residual stresses in electronic devices has been very well documented..."* However, no appropriate solutions are available yet: *"Most of these analyses though, have been based on elementary strength of materials concepts such as beam theory and proved inadequate to predict the shear stress magnitude at material interfaces."*

The traditional finite element analysis of stresses is also considered inadequate [81]: *"Since the stress and displacement fields near a bonding edge show singularity behavior, the adhesive strength evaluation method, using maximum stresses calculated by a numerical stress analysis, such as the finite element method, is generally not valid."*

These material interfaces, as well as crack tips, are called singular points because the temperature fluxes are infinite in the linear theory of steady-state heat conduction, and so are the stresses in the linear theory of elasticity. For example, typical singular points where failures initiate and propagate in an electronic device are illustrated in Figure 1.1.

Typical cracks can be observed by sectioning a VLSI device followed by a scanning electron microscope inspection, as shown in Figure 1.2. As observed, the failure initiates at the vertex of a reentrant V-notch.

It has been known for several decades that in large metallic structures, cracks may cause catastrophic failures. One of the recent and well-documented events of a structural failure in a civil airplane is the Aloha Airlines Flight 243 accident on April 28, 1988. A section of the upper fuselage was torn away in flight at 24,000 ft in a Boeing B-777-200 due to cracks originating in multiple places around riveted holes; see Figure 1.3. The airplane had flown 89,680 flights over its 19-year lifetime. Aircraft bulkheads can also break due to fatigue cracks, as did the F-16 bulkhead shown in Figure 1.4. There are many other examples of failed structures

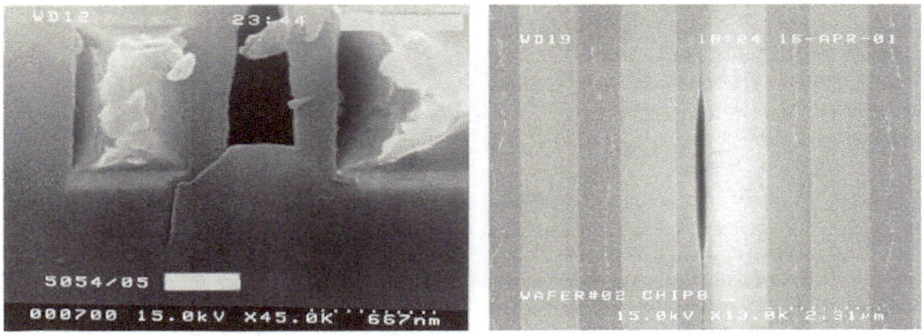

Fig. 1.1 Typical sites of failure initiation in an electronic device.

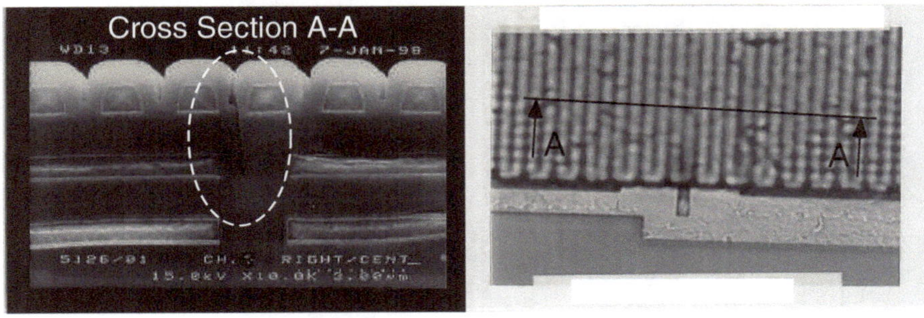

Fig. 1.2 Cracks in the passivation layer of a VLSI device: on the right a top view of the wafer, on the left a scanning electron microscope image showing the cross-section.

Fig. 1.3 The Aloha Airline Boeing 777 immediately after landing, April 1988.

Fig. 1.4 A broken bulkhead from a F-16 aircraft due to a small surface crack.

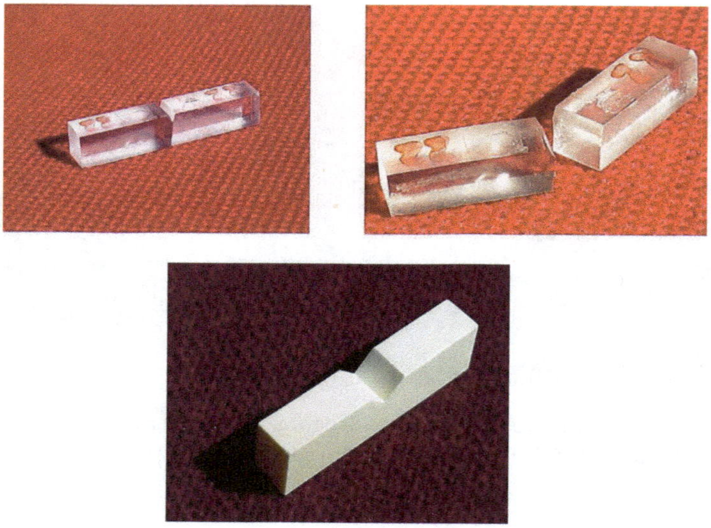

Fig. 1.5 PMMA and Alumina-7%Zirconia specimens that break due to failures starting at V-notch tips.

such as those shown in Figure 1.5; where the failure starts at the V-notch tip in a PMMA polymer, or an Alumina-7%Zirconia ceramic. Such failures and their possible prediction will be discussed in this monograph.

New approaches to predicting the initiation and extension of delaminations in plastic-encapsulated LSI (large scale integrated circuit) devices, for example, are based on the computation of certain functionals, called the generalized flux/stress intensity factors (GFIFs/GSIFs); the strength of the stress singularity; and in thermoelastic problems, the thermal stress intensity factors (TSIFs). These are defined in the sequel, and they apply to many types of singular points, such as reentrant corners, abrupt change in boundary conditions, multimaterial interfaces, and at an internal intersection point of several materials. We show in Figure 1.6 some examples of the aforementioned singularities in two-dimensional domains.

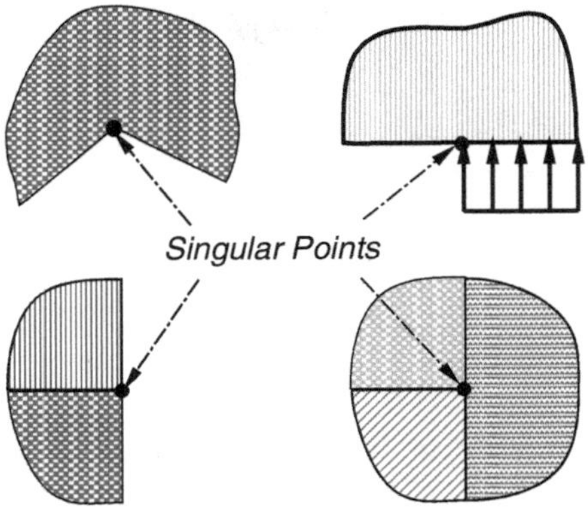

Fig. 1.6 Typical singular points in two-dimensional domains.

Some may claim that failure initiation and crack propagation are inherently nonlinear processes, and the linear elastic solution may not be of practical application. However, when the nonlinear behavior is confined entirely to some small region inside an elastic solution, then it can be determined through the solution of the linear elastic problem. Consequently, experimental observations on failure initiation and propagation in the neighborhood of a crack tip have been shown to correlate well with the linear elastic solution in many engineering applications. An overview of the mechanical problems in electronic devices supports the new trend [85]. *"The author has organized the research committee on the mechanical problems in electron devices, which consists of the members from Japanese universities and private industries. The committee examined the research results on the mechanical problems in electron devices... The intensity and the order of the stress singularity are the main parameters to determine the (failure) criterion..."*

The approach of correlating the GFIFs/GSIFs or TSIFs (determined through an elastic analysis) to experimental observations for establishing failure laws seems to be *the right approach*, as shown by several recent publications [58, 59, 77, 81, 143, 206]. Quoting from [81], for example: *"... in the case of plastic encapsulated LSI (Large Scale Integrated Circuit) devices, the thermal-expansion mismatch of utilized materials causes thermal stresses... these thermal stresses could cause serious reliability problems, such as interface debonding, resin cracking... A new method for evaluating adhesive strength was developed which uses two-stress-singularity parameters... this method was applied to estimate delamination behavior of plastic encapsulated LSI models, and these estimated results coincide well with the observed results using scanning acoustic tomography".*

1.2 Principles and Assumptions

It is assumed throughout the monograph that the principles of continuum mechanics remain valid everywhere within the body. Let us describe the various assumptions shown experimentally to be valid for brittle materials, on the basis of a two-

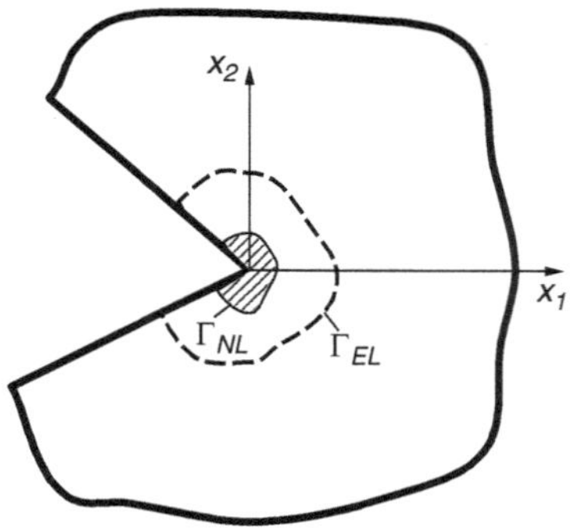

Fig. 1.7 Definition of Γ_{NL} and Γ_{EL}.

dimensional domain containing a singular point. Let $\boldsymbol{u}_{\text{NL}} = \{u_1, u_2\}_{\text{NL}}$ be the displacement vector (in x_1 and x_2 directions) that is the solution to the fully solid mechanics nonlinear problem. It is expected that failure initiation will depend on $\boldsymbol{u}_{\text{NL}}$, or some functionals computed from it, in the strongly nonlinear region of the singular point bounded by a boundary Γ_{NL}, as shown in Figure 1.7. This region is called the *process zone*. Let Γ_{EL} be a curve outside of Γ_{NL}, with $\boldsymbol{u}_{\text{NL}}|_{\Gamma_{\text{EL}}}$ the trace of $\boldsymbol{u}_{\text{NL}}$ on this curve. Denoting the solution of the linear elastic problem by $\boldsymbol{u}_{\text{EL}}$, then the following reasonable assumptions hold for brittle materials:

Assumption 1.1 *Inside of Γ_{EL} the error $\boldsymbol{u}_{\text{NL}}|_{\Gamma_{\text{EL}}} - \boldsymbol{u}_{\text{EL}}|_{\Gamma_{\text{EL}}}$ is so small that conclusions based on $\boldsymbol{u}_{\text{EL}}|_{\Gamma_{\text{EL}}}$ are sufficiently close to conclusions based on $\boldsymbol{u}_{\text{EL}}|_{\Gamma_{\text{NL}}}$ for practical purposes.*

This assumption is valid whenever the nonlinear behavior is confined entirely to some small region inside Γ_{EL} (a typical situation for brittle metals and ceramic materials). Assumption 1.1 leads to the important conclusion that failure initiation, which depends on the solution of the nonlinear problem inside of Γ_{NL}, can be determined through a solution of the linear elastic problem, even though all basic assumptions of the linear theory may be violated inside Γ_{NL}. Consequently, failure initiation in the neighborhood of a singular point can be predicted on the basis of the theory of linear elasticity.

Assumption 1.2 *There exists a physical principle that establishes the relationship between crack initiation and the stress field on the basis of information obtained from the linear solution $\boldsymbol{u}_{\text{EL}}$ only.*

The theory of linear elastic fracture mechanics, having been used successfully in engineering practice for over half a century, is a typical application of Assumption 1.2, where not the total elastic solution is of interest, but a specific parameter characterizing its behavior in the vicinity of the singular point. In general, the linear solution $\boldsymbol{u}_{\text{EL}}$ is not known, and only an approximation to it, obtained by finite element methods, for example, and denoted by $\boldsymbol{u}_{\text{FE}}$ is known. Therefore the following assumption is necessary:

Assumption 1.3 *There exists a norm* $\| \bullet \|$ *such that when* $\|\boldsymbol{u}_{\mathrm{EL}} - \boldsymbol{u}_{\mathrm{FE}}\|$ *is sufficiently small, then the physical principle of Assumption 1.2 is not sensitive to replacement of* $\boldsymbol{u}_{\mathrm{EL}}$ *with* $\boldsymbol{u}_{\mathrm{FE}}$.

Of course, the specific norm is expected to depend on the physical principle of Assumption 1.2, which is material-dependent.

Based on these assumptions, linear elastic computations can be used for prediction of failure initiation and propagation even though failure processes are nonlinear in nature. There are two essential elements of failure initiation analysis:

1. A hypothesis concerning the relationship between certain parameters of the stress/strain field and observed failure initiation or crack propagation events.
2. Convincing experimental confirmation that the hypothesis holds independently of variations in geometric attributes, loading, and constraints.

It would not be sensible to perform failure initiation analysis unless a detailed understanding of $\boldsymbol{u}_{\mathrm{EL}}$ is achieved, and an accurate estimate of $\boldsymbol{u}_{\mathrm{FE}}$ is obtained. Thus it is our aim in this book to explore the solution in the vicinity of singularities and its approximation by FE methods.

1.3 Layout

The book is divided into fourteen chapters, each containing several sections. The first nine chapters address two-dimensional domains, where only singular *points* exist. Thermoelastic singularities, failure laws and their application for predicting failure initiation in electronic devices are presented in Chapters 7–9. We then proceed to three-dimensional problems addressed in Chapters 10–13. We conclude with circular 3-D edges and remarks on open questions.

In the introduction the notation and problems of interest are presented. We formulate mathematically the problems of heat conduction and linear elasticity in two and three dimensions and present the general functional representation of the singular solutions. Based on the simple Laplace equation, we derive explicitly the singular solution in the vicinity of a reentrant corner. Chapter 2 provides a short introduction to the finite element method (FEM), especially the p-version of the FEM. The singular solutions have a strong impact on the rates of convergence of the finite element approximations: thus these are discussed also. Chapters 3 and 4 are devoted to two-dimensional heat conduction singular solutions. Basic ideas are presented and computation of so-called eigenpairs by the modified Steklov weak formulation is performed in Chapter 3. The modified Steklov weak eigenproblem is derived for a general scalar elliptic equation representing heat conduction in anisotropic domains and multimaterial interfaces. In the case of an isotropic domain, the weak eigenproblem is simplified and corresponds to the Laplace equation, for which the explicit solution has been given in the introduction. In Chapter 4 we present a method for the computation of the generalized flux intensity factors (GFIFs) by two methods: the dual weak formulation and the eigenpairs, and the dual singular functions followed by many examples. Chapters 5 and 6 address the linear elastic singular solutions in a two-dimensional domain. We discuss some unique features of the elastic system, as complex eigenpairs (giving rise to oscillatory solutions in the vicinity of the singular point), as well as power-logarithmic stress singularities. In Chapter 5 we derive analytically the asymptotic representation of the displacement and stress field in an isotropic material containing a V-notch or crack, and address cracks at a bimaterial interface where complex eigenpairs exist. Thereafter we formulate the modified Steklov weak form for the computation of eigenpairs for cases in which analytical methods are too complex to be applied. Chapter 6 is dedicated to the extraction of generalized stress intensity factors (GSIFs) from FE solution by the contour integral method (CIM) and the complementary energy method (CEM). Again, computation of GSIFs by the CIM and CEM for realistic engineering problems are provided. We then proceed to the problem of thermoelasticity and its singular solution in Chapter 7. This problem is solved in a decoupled way, first obtaining the temperature distribution, with its singular behavior, which is thereafter imposed as a thermal loading on the elastic domain. Due to the thermal loading an inhomogeneous elastic problem is obtained, giving rise to thermal generalized stress intensity factors (TGSIFs), which are computed by a sequence of solutions. TGSIFs for multimaterial interface problems and crack tips are provided to demonstrate the method's performance. In Chapter 8 we discuss the various possible interpretations of the elastic singular solutions and their correlation to failure criteria for mechanical components. We propose some extensions of fracture mechanics failure laws to multimaterial interfaces and general two-dimensional singular points. Several available failure criteria have been tested against experimental observations provided herein, leading to a good correlation. The application in engineering practice of a new failure criterion for preventing thermoelastic failures in an electronic device at the manufacturing process is illustrated and demonstrated by experimental observations in Chapter 9.

The remainder of the text is devoted to three-dimensional domains, where edge singularities, vertex singularities, and vertex-edge singularities are evident. After a short explanation of the three different singularity types, based on the Laplace equation, we consider the decomposition of the solution in the neighborhood of a straight edge in Chapter 10 for general heat conduction equations, first treating the Laplace equation. Here we emphasize the difference between two-dimensional and three-dimensional edge singularities. We introduce in Chapter 11 the quasidual function method for extracting edge flux intensity functions (EFIFs), which may be viewed as an extension of the dual singular function method in 2-D domains. This accurate and efficient method provides the *polynomial representation* of the EFIFs along the edge and is implemented as a postsolution operation in conjunction with p-FEMs. Vertex singularities for the Laplace equation are investigated in Chapter 12. An exact solution for the case of axisymmetric conical points is discussed first, followed by a numerical method that is an extension of the modified Steklov method for 3-D vertex singularities. Chapter 13 is devoted to the computation of edge eigenpairs for linear elasticity, and the various methods for extracting the edge stress intensity functions along a given straight edge. Numerical examples are presented. Finally, we provide in Chapter 14 some recent results on circular singular edges in three-dimensional domains, and a review of open problems in this field from the mathematical, conceptual viewpoint, as well as from an engineering viewpoint.

Many appendices include topics that are connected with the mainstream topic of the monograph but are not essential for understanding the methods and ideas discussed. In Appendix A we provide the definition of norms and function spaces, which play an important role in FEMs. The exact solution to scalar elliptic problems in two-dimensional anisotropic domains (and multimaterial domains) is derived in Appendix B, and in Appendix C we discuss the asymptotic solution near circular edges in two-dimensional domains intersecting at a singular point when the Laplace equation is of interest. We show that the eigenvalues of the heat conduction problem in a two-dimensional domain are real numbers in Appendix D. The path-independent integral and the orthogonality of the eigenfunctions for general heat-conduction problems in 2-D are derived and proved in Appendix E. Finally, we discuss the energy release rate (ERR) method for cracks in appendix F and its connection to the J-integral and the extraction of stress-intensity-factors.

1.4 A Model Problem

For illustrating the basic characteristics of the solution of a typical elliptic partial differential equation in the vicinity of a singular point, we address herein the simplified Laplace equation (physically it describes the steady-state heat conduction problem in an isotropic material) over a two-dimensional domain denoted by Ω. The boundary $\partial\Omega$ consists of two straight lines, denoted by Γ_1 and Γ_2, which intersect at the singular point P, creating a reentrant corner with a solid angle of ω radians. We attach a Cartesian coordinate system x_1, x_2 to P, with r and θ being the polar coordinates; see Figure 1.8. We consider the Laplace equation over Ω, with Dirichlet boundary conditions on its boundary $\partial\Omega$:

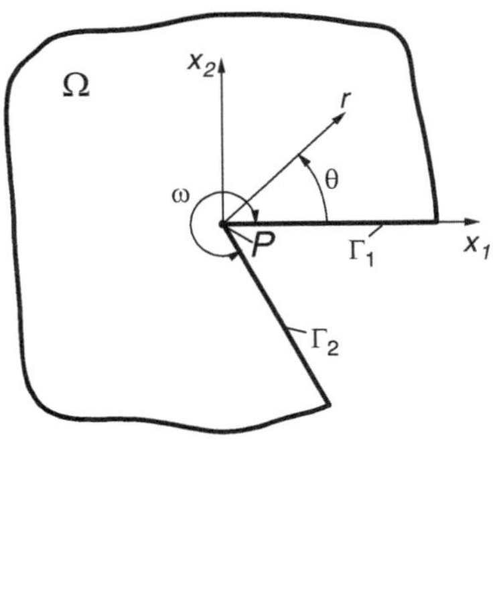

Fig. 1.8 Two-dimensional domain with a reentrant corner, notation.

$$\nabla^2 \tau \overset{\text{def}}{=} \frac{\partial^2 \tau}{\partial r^2} + \frac{1}{r}\frac{\partial \tau}{\partial r} + \frac{1}{r^2}\frac{\partial^2 \tau}{\partial \theta^2} = 0 \qquad \text{in } \Omega, \tag{1.1}$$

$$\tau = \hat{\tau} \qquad \text{on } \partial\Omega - \Gamma_1 - \Gamma_2, \tag{1.2}$$

$$\tau = 0 \qquad \text{on } \Gamma_1 \cup \Gamma_2, \tag{1.3}$$

where $\hat{\tau}$ is a prescribed function on a part of the boundary. *Throughout the book, unless otherwise explicitly stated, we assume that in the close vicinity of the singular point P, homogeneous boundary conditions are applied.* The solution in the vicinity of the singular point is of interest, and because in this subdomain the partial differential equation and boundary conditions are homogeneous, one seeks a homogeneous solution by separation of variables of the form:

$$\tau_H = R(r)s(\theta). \tag{1.4}$$

Substituting (1.4) in (1.1), one obtains, after elementary algebraic manipulations,

$$r^2 \frac{R''(r)}{R(r)} + r\frac{R'(r)}{R(r)} = -\frac{s''(\theta)}{s(\theta)} = \alpha^2. \tag{1.5}$$

Here primes represent differentiation with respect to r and θ for $R(r)$ and $s(\theta)$ respectively. The constant is positive, denoted by α^2; otherwise, $s(\theta)$ is an exponential function, which cannot possibly satisfy the homogeneous boundary conditions on $\theta = 0, \omega$. The function $R(r)$ satisfies the Euler ordinary differential equation of second order, whose solution is of the form

$$R(r) = \begin{cases} \text{for } \alpha \neq 0, & ar^{\alpha} + br^{-\alpha}, \\ \text{for } \alpha = 0, & a + b\log r, \end{cases} \tag{1.6}$$

where a and b are generic constants.

The function $s(\theta)$ satisfies

$$\alpha^2 s(\theta) + s''(\theta) = 0, \quad 0 \geq \theta \geq \omega. \tag{1.7}$$

One may observe that for both positive and negative α, the equations that determine $s(\theta)$ are identical. The solution to (1.7) is

$$s(\theta) = \begin{cases} \text{for } \alpha \neq 0, & c\cos(\alpha\theta) + d\sin(\alpha\theta), \\ \text{for } \alpha = 0, & c + d\theta, \end{cases} \tag{1.8}$$

where c and d are generic constants.

The homogeneous Dirichlet boundary conditions (1.3) imply on $s(\theta)$ the following boundary conditions

$$\tau(r, \theta = 0) = \tau(r, \theta = \omega) = 0 \quad \Rightarrow \quad s(0) = s(\omega) = 0. \tag{1.9}$$

Applying the boundary conditions on (1.8), one may observe that the solution associated with $\alpha = 0$ is the trivial solution, and we are left with

$$s(\omega) = d \sin(\alpha\omega) = 0 \Longrightarrow \alpha_i = \pm i\pi/\omega, \quad i \in \mathbb{Z}. \tag{1.10}$$

Both positive and negative α_i values satisfy (1.1). We denote by $s^+(\theta)$ the functions associated with the positive value of α, and by $s^-(\theta)$ those associated with the negative value of α. Although for the Laplace equation $s^+(\theta) \equiv s^-(\theta)$ ((1.7) is quadratic in α), for a general scalar elliptic equation, and for the elasticity system this is no longer the case.

The restriction $\alpha_i \geq 0$ is imposed because of "physical" reasoning, since τ at $r = 0$ should be finite, so that we deal with solutions belonging to the Sobolev space $H^1(\Omega)$ (see Appendix A). The negative values are nevertheless of interest for other mathematical manipulations and for describing the "far field," as will be discussed in the sequel. In view of (1.10) and (1.6), the solution to (1.1) admits the expansion

$$\tau_H = \sum_{i=1}^{\infty} A_i r^{\alpha_i} s_i^+(\theta), \quad \text{with} \quad s_i^+(\theta) = \sin(\alpha_i \theta), \quad \alpha_i = i\pi/\omega,$$

$$\Downarrow$$

$$\tau_H = \sum_{i=1}^{\infty} A_i r^{i\pi/\omega} \sin\left(\tfrac{i\pi}{\omega}\theta\right), \tag{1.11}$$

α_i and $s_i(\theta)$ are called *eigenpairs*, and these are determined uniformly by the geometry and boundary conditions in the neighborhood of the singular point. We also define the "primal" eigenfunction and "dual" eigenfunction to be the two functions corresponding to the same positive and negative eigenvalues $s_i^+(\theta)$ and $s_i^-(\theta)$, respectively.

The series (1.11) is an *asymptotic series*, i.e., taking a finite number of terms, say N, the series is more and more accurate as r becomes smaller, but for a given r, the series might even diverge as N gets larger and larger. The series coefficients A_i can be bounded in the vicinity of the singular point for $r < R$ (see, for example, [138, Chapter 2]):

$$|A_i| < C\sqrt{i}R^{-\alpha_i}, \tag{1.12}$$

where C is a generic constant, and R represents the largest radius in the vicinity of the singular point where (1.11) still holds.

It has been shown that all α_i in (1.10) are real numbers for the Laplace equation. It can be shown also that all eigenpairs of scalar isotropic domains (either open domains or multimaterial interfaces) are real. Mantic et al. [115] proved that all eigenpairs of scalar anisotropic domains (multimaterial interfaces or single material) in open domains are real. For *periodic*, anisotropic multimaterial interfaces, however, complex eigen-pairs can appear (see Appendix D). For the linear elastic problem, complex eigenpairs are known to exist, and their interpretation will be discussed in the sequel.

Notice that if $\alpha_i < 1$, the corresponding i^{th} term in the expansion (1.11) for the flux vector $\boldsymbol{q}_H = \mathbf{grad}\tau_H \stackrel{\text{def}}{=} \left(\frac{\partial \tau_H}{\partial x_1}, \frac{\partial \tau_H}{\partial x_2}\right)$ is unbounded as $r \to 0$.

Proposition 1.1 *We say that τ_H is singular at* 0 *if $\boldsymbol{q}_H = \mathbf{grad}\tau_H$ tends to infinity as $r \to 0$. The solution τ_H in* (1.11) *is therefore singular at* 0 *if $\omega > \pi$.*

Problem 1.1. Show that $\boldsymbol{q}_H = \mathbf{grad}\tau_H$ for τ_H given by (1.11) can be represented as $\boldsymbol{q}_H = \sum_{i=1}^{\infty} A_i r^{\alpha_i - 1} (\boldsymbol{q}_H)_i^+(\theta)$. Find the vectors $(\boldsymbol{q}_H)_i^+(\theta)$ explicitly, and show that for $\omega > \pi$, the first term in the series for $\boldsymbol{q}_H$ is singular.

Hint: $\frac{\partial}{\partial x_1} = \cos\theta \frac{\partial}{\partial r} - \frac{1}{r}\sin\theta \frac{\partial}{\partial \theta}$, $\frac{\partial}{\partial x_2} = \sin\theta \frac{\partial}{\partial r} + \frac{1}{r}\cos\theta \frac{\partial}{\partial \theta}$.

Proposition 1.1 is slightly ambiguous, because A_1 may be zero, and although $\omega > \pi$ the solution will not tend to infinity. The coefficients A_i are so far undetermined, and depend on the boundary conditions away from the singular point and the right hand side of the Laplace equation if it exists. We can think of the coefficients A_i of these terms as analogous to the stress intensity factors of elasticity in linear elastic fracture mechanics (this topic will be addressed in detail in the sequel). We generalize this terminology, and refer to all coefficients A_i, whether or not the corresponding flux terms are singular, as *generalized flux intensity factors* (GFIFs) .

To be more precise in mathematical terms, it is necessary to discuss the *regularity of the solution* of the Laplace equation $\nabla^2 \tau = f$ (more details can be found in [109, Chapter 4] and references therein). For the Laplace equation with a "smooth" (say $f \in H^k$) right-hand side, homogeneous Dirichlet BCs *and a smooth boundary without kinks,* the solution is as smooth as the right-hand side allows: $f \in H^k \Rightarrow \tau \in H^{k+2}$. This is called the "shift theorem." In case $\partial\Omega$ has corner points, then the shift theorem no longer holds, even if the right-hand side is in C^∞. In this case, however, the solution may be decomposed into a singular part and a regular remainder, which can again be as smooth as the right-hand side allows:

$$\tau = \sum_{i=1}^{N} A_i r^{\alpha_i} s_i^+(\theta) + \tau^{\text{reg}}$$

where $\tau^{\text{reg}} \in H^{1+q}(\Omega)$ and $q > \alpha_N$ depends on N.

Following the regularity concept, and because the series (1.11) may not converge in general and should be understood as an asymptotic series, we are interested in a finite number of terms N in the solution of (1.1):

$$\left(\tau_H - \sum_{i=1}^{N} A_i r^{\alpha_i} s_i^+(\theta)\right) \stackrel{r \to 0}{\longrightarrow} 0.$$

A very similar analysis can be performed for two other boundary conditions:

$$\begin{cases} \text{Neumann B.C.s} & \frac{\partial \tau}{\partial n} = \mathbf{grad}\tau \cdot \mathbf{n} = 0 \quad \text{on } \Gamma_1 \cup \Gamma_2, \\ \text{Newton B.C.s} & \tau = 0 \quad \text{on } \Gamma_1 \quad \frac{\partial \tau}{\partial n} = 0 \quad \text{on } \Gamma_2, \end{cases} \tag{1.13}$$

where $\mathbf{n}$ is the normal outward unit vector to the boundary. The eigenpairs for the above boundary conditions may be easily computed and are explicitly given:

$$\alpha_i = \begin{cases} i\pi/\omega & \text{for Neumann B.C.s,} \\ (2i-1)\pi/2\omega & \text{for Newton B.C.s,} \end{cases} \tag{1.14}$$

$$s_i^+(\theta) = s_i^-(\theta) = \begin{cases} \cos(\alpha_i\theta) & \text{for Neumann B.C.s,} \\ \cos(\alpha_i\theta) & \text{for Newton B.C.s.} \end{cases} \tag{1.15}$$

Problem 1.2. Construct the asymptotic expansion for τ_H in the neighborhood of a vertex subject to Neumann and Newton boundary conditions (obtain the results in equations (1.14) and (1.15)).

The GFIFs cannot be determined in general analytically and therefore special numerical methods have been developed. One of the most efficient methods is based on a path-independent integral, introduced next.

1.4.1 A Path-Independent Integral

Several interesting mathematical properties can be derived for elliptic second order PDEs, one of which is a path independent integral (along an arbitrary curve starting on Γ_1 and terminating anywhere on Γ_2), derived herein as an example for the Laplace equation (see [12, 27]). For ease of presentation we choose the path as an arc of radius R, centered at the singular point of interest. Consider the shaded subdomain in the vicinity of the singular point shown in Figure 1.9. The shaded domain is a part of Ω bounded by $R^* \leq r \leq R$. Multiplying the Laplace equation (1.1) by a function χ and integrating over the shaded subdomain, one obtains

$$\iint \nabla^2 \tau \chi \, r dr d\theta = 0. \tag{1.16}$$

Use Green's theorem, and note that along the arcs of the shaded domain $\frac{\partial \bullet}{\partial n} = \pm \frac{\partial \bullet}{\partial r}$, (1.16) becomes

$$\iint \tau \nabla^2 \chi \, r dr d\theta + \int_{\Gamma_1 \cup \Gamma_2} \left[\chi \frac{\partial \tau}{\partial n} - \tau \frac{\partial \chi}{\partial n} \right] d\Gamma + \int_0^\omega \left[\chi \frac{\partial \tau}{\partial r} - \tau \frac{\partial \chi}{\partial r} \right]_R R d\theta - \int_0^\omega \left[\chi \frac{\partial \tau}{\partial r} - \tau \frac{\partial \chi}{\partial r} \right]_{R^*} R^* d\theta = 0. \tag{1.17}$$

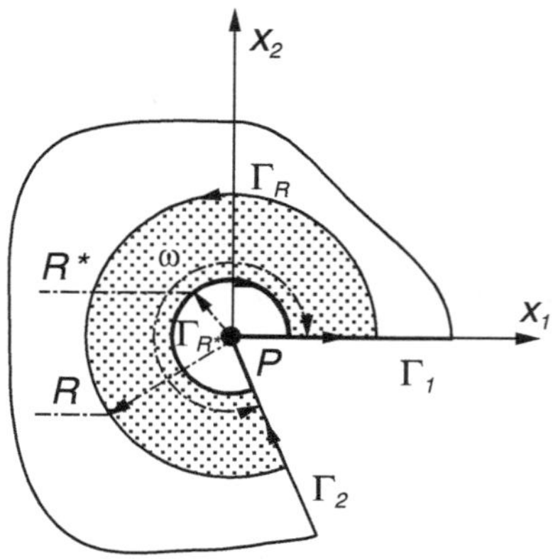

Fig. 1.9 Sub-domain in the vicinity of reentrant corner.

If χ is chosen so as to satisfy the same equation as τ, namely $\nabla^2 \chi = 0$, then the first term vanishes. The second term vanishes because we assume either homogeneous Dirichlet or Neumann boundary conditions on $\Gamma_1 \cup \Gamma_2$ (both for χ and τ), so that (1.17) becomes

$$I_\Gamma(\tau, \chi) \stackrel{\text{def}}{=} \int_0^\omega \left[\tau \frac{\partial \chi}{\partial r} - \chi \frac{\partial \tau}{\partial r} \right]_R R d\theta = \int_0^\omega \left[\tau \frac{\partial \chi}{\partial r} - \chi \frac{\partial \tau}{\partial r} \right]_{R*} R^* d\theta. \quad (1.18)$$

The integral $I_\Gamma(\tau, \chi)$ is path-independent because its value is the same for any R.

1.4.2 Orthogonality of the "Primal" and "Dual" Eigenfunctions

Choose $\tau = r^{\alpha_i} s_i^+(\theta)$ and $\chi = r^{-\alpha_j} s_j^-(\theta)$ (no summation) that satisfy (1.1) and the homogeneous boundary conditions, so both can be inserted in (1.18)

$$\int_0^\omega \left[R^{\alpha_i} s_i^+(\theta)(-\alpha_j) R^{-\alpha_j - 1} s_j^-(\theta) - R^{-\alpha_j} s_j^-(\theta) \alpha_i R^{\alpha_i - 1} s_i^+(\theta) \right] R d\theta$$

$$= \int_0^\omega \Big[(R^*)^{\alpha_i} s_i^+(\theta)(-\alpha_j)(R^*)^{-\alpha_j - 1} s_j^-(\theta)$$

$$- (R^*)^{-\alpha_j} s_j^-(\theta) \alpha_i (R^*)^{\alpha_i - 1} s_i^+(\theta) \Big] R^* d\theta,$$

$$\text{no summation on } i \text{ and } j, \quad (1.19)$$

which in turn equals

$$(\alpha_j + \alpha_i)\left[R^{\alpha_i-\alpha_j} - (R^*)^{\alpha_i-\alpha_j}\right]\int_0^\omega s_i^+(\theta)s_j^-(\theta)d\theta = 0,$$

$$\text{no summation on } i \text{ and } j. \tag{1.20}$$

One may choose any R and R^*. Then if $i \neq j$, in order for (1.20) to hold, one obtains

$$\int_0^\omega s_i^+(\theta)s_j^-(\theta)d\theta = 0 \quad \text{if } i \neq j. \tag{1.21}$$

This shows that the "primal" and "dual" eigenfunctions are orthogonal with respect to a path integral along an arc starting at Γ_1 and terminating at Γ_2 (this orthogonal property holds for any path starting at Γ_1 and terminating at Γ_2). The dual eigenfunctions in conjunction with the path-independent integral are used in Chapter 4 for the extraction of GFIFs from FE solutions.

1.4.3 Particular Solutions

In addition to the homogeneous part of the solution in (1.1), a logarithmic type of singularities may exist [111] if a right-hand-side term is considered, i.e., solutions to the Poisson equation are considered

$$\nabla^2\tau = f(x_1, x_2) \qquad \text{in } \Omega. \tag{1.22}$$

For simplicity, we assume that $f \in C^\infty$ in the vicinity of P, i.e., the function f is infinitely many times differentiable, so we may expand f in a series of the form

$$f(r,\theta) = \sum_{i=0}^\infty r^i f_i(\theta). \tag{1.23}$$

The Poisson equation admits a solution that is a combination of a homogeneous part (1.11) and a particular solution τ_P. By the shift theorem, the particular solution for the cases $i + 2 \neq \alpha_j\ \forall i, j$ is

$$\tau_P = \sum_{i=0}^\infty r^{i+2} F_i(\theta). \tag{1.24}$$

Otherwise, for each α_j that satisfies $i + 2 = \alpha_j$, the particular solution τ_P will contain a term of the form

$$r^{i+2}\left[c_j \ln(r) s_j(\theta) + F_i(\theta)\right]. \tag{1.25}$$

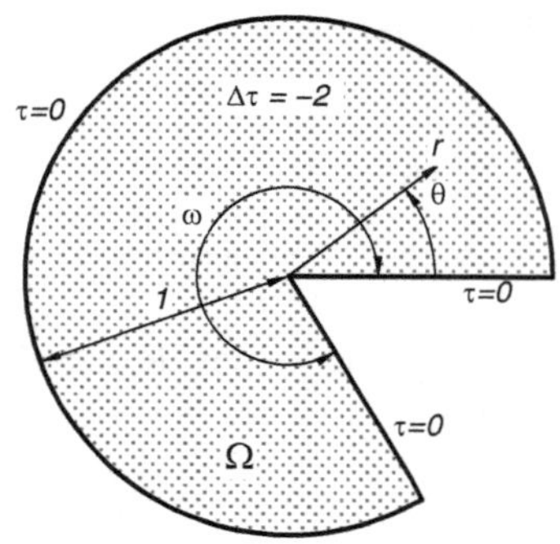

Fig. 1.10 Cross-section of a rod with a reentrant corner.

Let us demonstrate the overall solution by considering a simple example, the Saint Venant torsion problem, formulated in terms of Prandtl's stress potential, denoted also by τ, as follows[1] (see e.g., [167, Chapter 35]):

$$\nabla^2 \tau = -2 \qquad \text{in } \Omega, \tag{1.26}$$

$$\tau = 0 \qquad \text{on } \partial\Omega. \tag{1.27}$$

Consider a long rod with a cross-section in the shape of a circular sector of radius 1, shown in Figure 1.10. The solution to this problem, $\tau = \tau_H + \tau_P$, is [121]

$$\begin{aligned}\tau_H(r,\theta) &= \sum_{i=1,3,5\ldots} \frac{8}{i\pi[4-(i\pi/\omega)^2]} r^{i\pi/\omega} \sin(i\theta\pi/\omega) \\ &= A_1 r^{\pi/\omega} \sin(\theta\pi/\omega) + A_3 r^{3\pi/\omega} \sin(3\theta\pi/\omega) + \mathcal{O}(r^{5\pi/\omega}),\end{aligned} \tag{1.28}$$

$$\tau_P(r,\theta) = r^2 \sum_{i=1,3,5\ldots} \frac{8}{i\pi[4-(i\pi/\omega)^2]} \sin(i\theta\pi/\omega), \quad \omega \neq \pi/2, 3\pi/2, \tag{1.29}$$

$$\tau_P(r,\theta) = r^2 \frac{-2}{\pi}\left[\frac{\pi}{4} + \ln(r)\cos(2\theta) - \theta\sin(2\theta)\right], \quad \omega = \pi/2, \tag{1.30}$$

$$\tau_P(r,\theta) = r^2 \frac{-2}{3\pi}\left[\frac{3\pi}{4} - \ln(r)\cos(2\theta) + \theta\sin(2\theta)\right], \quad \omega = 3\pi/2. \tag{1.31}$$

In general, the scalar elliptic problems in two dimensions in the vicinity of a singular point allow the following expansion of the solution [47, 95]

---

[1] The shear stresses are related to the stress potential via

$$\sigma_{13} = \mu k \frac{\partial \tau}{\partial x_2}, \qquad \sigma_{23} = -\mu k \frac{\partial \tau}{\partial x_1},$$

where μ is the shear modulus, and k is the angle of twist per unit length of the rod.

$$\tau = \sum_{i=1}^{\infty} \sum_{j=1}^{J} \sum_{k=1}^{K} A_{ijk} r^{\alpha_i + j} \ln^k(r) s_{ijk}(\theta), \tag{1.32}$$

where K might be 1 for integer eigenvalues (and a right-hand-side term exists), and J differs from 0 if Γ_1 and/or Γ_2 are curvilinear arcs. The singular term due to the curvature (with $i = 1, j = 1$) may be more singular than the second term (with $i = 2, j = 0$). This happens if

$$\alpha_1 + 1 < \alpha_2, \quad \text{i.e., } \omega < \pi.$$

Remark 1.1 *The homogeneous Laplace equation with homogeneous boundary conditions does not admit logarithmic terms in the expansion* (1.32).

1.4.4 Curved Boundaries Intersecting at the Singular Point

The cases treated in this monograph assume straight boundaries intersecting at the singular point. For curved boundaries, an asymptotic analysis can be performed that shows that the leading term of the singular solution is as if the boundaries were straight. However more terms are involved in the description of the singular solution. In Appendix C, the analysis for deriving the singular solution (1.33) for a simple case of two circular arcs intersecting at an angle of $\omega = 3\pi/2$ following the steps in [195] is provided. This particular case provides explicitly the structure of the singular asymptotic series

$$\tau = \sum_{i} \sum_{j=1,2,\cdots} A_{ij} r^{\alpha_i + j} s_{ij}(\theta). \tag{1.33}$$

1.5 The Heat Conduction Problem: Notation

Following the brief explanation of a typical case for which the solution of the Laplace equation is singular in the vicinity of a reentrant corner, let us define the mathematical problem of heat conduction, which will be rigourosly treated in a two-dimensional setting in Chapter 3. In general, Ω is a three-dimensional domain with a boundary denoted by $\partial\Omega$. Unless otherwise specifically stated, a Cartesian coordinate system is used with coordinates x_1, x_2, and x_3, and the summation notation is adopted. The partial derivative with respect to x_i is denoted by

$$\partial_i \overset{\text{def}}{=} \frac{\partial}{\partial x_i}$$

and the divergence operator on a vector is the usual

$$\mathbf{div}\, \boldsymbol{v} \stackrel{\text{def}}{=} \partial_1 v_1 + \partial_2 v_2 + \partial_3 v_3 = \partial_i v_i .$$

We seek for the heat conduction a solution (the temperature function) denoted by τ in Ω satisfying the Fourier steady-state heat conduction equation

$$-\,\mathbf{div}\,([k]\mathbf{grad}\tau) = Q \quad \text{in } \Omega, \tag{1.34}$$

which in summation notation can be expressed as

$$-\,\partial_i \left(k_{ij} \partial_j \tau\right) = Q \quad \text{in } \Omega, \tag{1.35}$$

with indices $i, j = 1, 2, 3$. The 3×3 symmetric matrix $[k]$ consists of the heat conduction coefficients

$$[k] = \begin{bmatrix} k_{11} & k_{12} & k_{13} \\ & k_{22} & k_{23} \\ & & k_{33} \end{bmatrix} .$$

In the vicinity of an edge (or a singular point in two-dimensional domains), k_{ij} are assumed to be constants (or piecewise constants in multimaterial interface problems), or at most functions of the angle θ (and independent of r in the vicinity of the singularities). In isotropic materials ($k_{ij} = k\delta_{ij}$) (1.34) becomes the Poisson equation. The matrix $[k]$ satisfies the ellipticity condition

$$\boldsymbol{x}^T [k] \boldsymbol{x} \geq \gamma \boldsymbol{x}^T \boldsymbol{x} \quad \forall \boldsymbol{x} \in \Omega, \tag{1.36}$$

where γ is a generic positive constant called the ellipticity constant. Here Q is the heat generation term per unit of volume. We consider two sorts of boundary conditions: Dirichlet boundary conditions ($\tau = \hat{\tau}$, $\hat{\tau}$ being a given function) on a part of the domain boundary, denoted by $\partial\Omega_D$; and Neumann boundary conditions $(([k]\mathbf{grad}\tau) \cdot \mathbf{n} = \hat{q}_n)$ on the other part of the domain boundary, denoted by $\partial\Omega_N$. Homogeneous boundary conditions are assumed on the surface in the vicinity of the the singular edge or vertex.

The heat conduction equation in its classical form (1.34) can be given a weak formulation. Multiplying (1.34) by a test function χ, then integrating by parts and using the boundary conditions, one obtains the weak heat conduction formulation (for details the reader is referred to [178]):

$$\text{Seek} \qquad \tau \in \mathcal{E}(\Omega),\ \tau = \hat{\tau} \text{ on } \partial\Omega_D \qquad \text{such that}$$

$$\mathcal{B}(\tau, \chi) = \mathcal{F}(\chi) \qquad \forall \chi \in \mathcal{E}_o(\Omega), \tag{1.37}$$

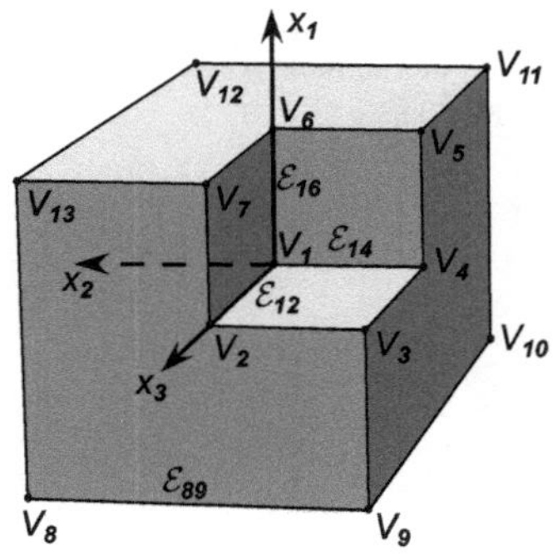

Fig. 1.11 Typical three-dimensional singularities.

where the bilinear form is

$$\mathcal{B}(\tau,\chi) = \int_\Omega k_{ij}\ \partial_i\chi\ \partial_j\tau\ d\boldsymbol{x}, \tag{1.38}$$

and the linear form is

$$\mathcal{F}(\chi) = \int_{\partial\Omega_N} \hat{q}_n\chi d\Gamma + \int_\Omega Q\chi\ d\boldsymbol{x}. \tag{1.39}$$

The energy spaces $\mathcal{E}$ and $\mathcal{E}_o$ are defined in Appendix A together with their connection to the Sobolev spaces H^1 and H^1_o.

In a three-dimensional domain, the singular solution of (1.34) or (1.37) is decomposed into three different forms, depending on whether it is in the neighborhood of an edge, a vertex, or an intersection of the edge and the vertex. Mathematical details on the decomposition can be found, e.g., in [6,15,40,49,73,76] and the references therein. A representative three-dimensional polyhedral domain containing typical 3-D singularities is shown in Figure 1.11. Vertex singularities arise in the neighborhood of the vertices V_i, and edge singularities arise in the neighborhood of the edges $\mathcal{E}_{ij}$. Close to the vertex/edge intersection, vertex-edge singularities arise.

Before treating three-dimensional singularities, we simplify the problem and address the heat conduction problem over a two-dimensional domain. One may consider the two-dimensional problem as a restriction to edge singularity on a plane perpendicular to the edge, as shown in Figure 1.11. Over this two-dimensional domain we consider in Chapters 3-4 the solution to (1.35), except that the indices are $i,j = 1,2$. If $k_{ij} \neq 0$ for $i \neq j$, and if a multimaterial interface exists (i.e., k_{ij} is piecewise constant in the neighborhood of a singular point), as shown in Figure 1.12, neither the eigenpairs nor the GFIFs are known analytically, and numerical methods for their determination are needed. The modified Steklov method for the determination of eigenpairs associated with the heat conduction

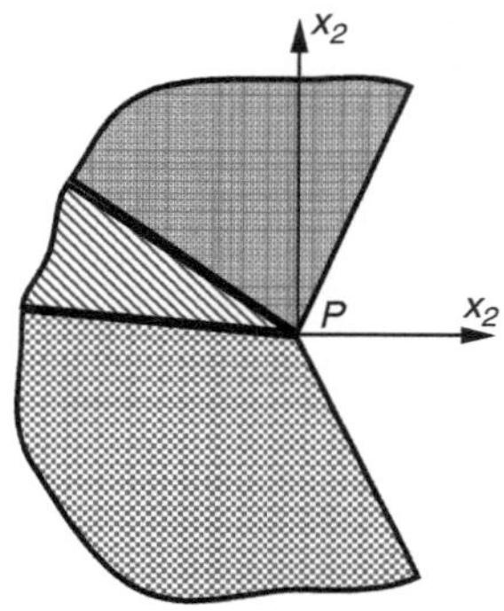

Fig. 1.12 Two-dimensional multimaterial interface singularity.

problem over a two-dimensional domain is described in detail in Chapter 3, and the computation of the GFIFs by dual weak formulation is described in Chapter 4.

1.6 The Linear Elasticity Problem: Notation

The three-dimensional problem of linear elasticity consists of an elliptic system of three second-order partial differential equations for three components u_1, u_2, u_3 of displacement (in a Cartesian coordinate system). We denote by $\boldsymbol{u}$ the displacement vector. The components of the linear strain (second-order) tensor are related to the displacements by

$$\varepsilon_{ij}(\boldsymbol{u}) = \frac{1}{2}\left(\partial_j u_i + \partial_i u_j\right) \stackrel{\text{def}}{=} \frac{1}{2}\left(u_{i,j} + u_{j,i}\right).$$

We also define the engineering shear strain

$$\gamma_{ij} = 2\varepsilon_{ij}, \quad i \neq j,$$

and we use the Voigt notation[2] to define the strain vector

$$\boldsymbol{\varepsilon} \stackrel{\text{def}}{=} (\varepsilon_{11}\ \varepsilon_{22}\ \varepsilon_{33}\ \gamma_{23}\ \gamma_{13}\ \gamma_{12})^T = [D]\boldsymbol{u}, \tag{1.40}$$

[2] Named after Woldemar Voigt (September 2, 1850 - December 13, 1919), a German physicist who also introduced (among many other important things such as the Lorentz transformation) the word tensor in its current meaning in 1899.

where $[D]$ is the differential operator

$$[D] \stackrel{\text{def}}{=} \begin{bmatrix} \partial_1 & 0 & 0 \\ 0 & \partial_2 & 0 \\ 0 & 0 & \partial_3 \\ 0 & \partial_3 & \partial_2 \\ \partial_3 & 0 & \partial_1 \\ \partial_2 & \partial_1 & 0 \end{bmatrix}. \tag{1.41}$$

Throughout the book the stress tensor will be given in either its tensor or vector form:

$$\underset{\approx}{\boldsymbol{\sigma}} \stackrel{\text{def}}{=} \begin{bmatrix} \sigma_{11} & \sigma_{12} & \sigma_{13} \\ \sigma_{12} & \sigma_{22} & \sigma_{23} \\ \sigma_{13} & \sigma_{23} & \sigma_{33} \end{bmatrix} \quad \text{or} \quad \boldsymbol{\sigma} \stackrel{\text{def}}{=} (\sigma_{11}\ \sigma_{22}\ \sigma_{33}\ \sigma_{23}\ \sigma_{13}\ \sigma_{12})^T .$$

The elastic constitutive law (Hooke's law) connects the stress vector with the strain/displacement vectors:

$$\boldsymbol{\sigma} = [E]\boldsymbol{\varepsilon} = [E][D]\boldsymbol{u}. \tag{1.42}$$

In general, $[E]$ is a symmetric positive definite matrix with 21 independent entries. For orthotropic materials, there are nine independent entries, and for isotropic materials, the symmetric matrix $[E]$ is given by

$$[E] = \frac{E}{(1+\nu)(1-2\nu)} \begin{bmatrix} (1-\nu) & \nu & \nu & 0 & 0 & 0 \\ & (1-\nu) & \nu & 0 & 0 & 0 \\ & & (1-\nu) & 0 & 0 & 0 \\ & & & (1-2\nu)/2 & 0 & 0 \\ & & & & (1-2\nu)/2 & 0 \\ & & & & & (1-2\nu)/2 \end{bmatrix},$$

where E and ν are engineering notations, Young's modulus and the Poisson ratio, respectively. Most mathematical publications use instead the Lamé constants λ and μ, connected to E and ν by

$$E = \frac{\mu(3\lambda + 2\mu)}{\lambda + \mu}, \quad \nu = \frac{\lambda}{2(\lambda + \mu)}, \quad \mu = \frac{E}{2(1+\nu)}, \quad \lambda = \frac{\nu E}{(1+\nu)(1-2\nu)}.$$

In tensorial notation, Hooke's law for isotropic materials is also given by

$$\sigma_{ij} = C_{ijkl}\epsilon_{kl}, \quad C_{ijkl} = \lambda\delta_{ij}\delta_{kl} + \mu\left(\delta_{ik}\delta_{jl} + \delta_{il}\delta_{jk}\right), \quad i,j,k,l = 1,2,3. \tag{1.43}$$

The stress tensor satisfies the equilibrium equations at every point, resulting in a system of three second-order partial differential equations:

$$-\partial_i \sigma_{ji}(\boldsymbol{u}) = f_j \quad \text{in } \Omega, \qquad j = 1,2,3, \tag{1.44}$$

where $\boldsymbol{f} = (f_1\ f_2\ f_3)^T$ is the body force per unit volume, and the notation $\sigma_{ji}(\boldsymbol{u})$ implies that stress components are expressed in terms of the three displacement functions.

For an isotropic material, (1.44) can be explicitly written in terms of displacements; these are known as the Navier-Lamé (in Cartesian coordinates) equations[3]:

$$-\mu\nabla^2\boldsymbol{u} - (\lambda + \mu)\,\mathbf{grad}\,\mathbf{div}\,\boldsymbol{u} = \boldsymbol{f} \quad \text{in } \Omega. \tag{1.45}$$

On the boundaries of the domain, two kinds of boundary conditions may be prescribed:

- Dirichlet (displacements) boundary conditions;

$$u_i = \hat{u}_i \quad \text{on } (\partial\Omega_D)_i, \tag{1.46}$$

  where i might be a normal or tangential component of the displacement vector. We denote all parts of the boundary on which displacement boundary conditions are imposed by $\cup_i(\partial\Omega_D)_i$.
- Neumann (traction) boundary conditions;

$$T_i = \sigma_{ji}(\boldsymbol{u})n_j = \hat{T}_i \quad \text{on } \partial\Omega_N, \tag{1.47}$$

  where i here denotes the components of the traction vector.

In elasticity one may have on the same boundary a combination of the two boundary conditions, i.e., the normal component of the displacement field may be zero, and the two tangential tractions may be zero; this case is an idealization of a contact surface with free sliding. Homogeneous boundary conditions are assumed in the vicinity of the singular point.

Remark 1.2 *It is sometimes more convenient to use the displacements and stresses in a cylindrical coordinate system. To distinguish these from those in a Cartesian coordinate system, we denote them as follows:*

[3]The elasticity system in a general curvilinear coordinate system in terms of "dilatations" and "rotations" is provided in [113].

$$\tilde{\boldsymbol{u}} \overset{\text{def}}{=} \begin{Bmatrix} u_r \\ u_\theta \\ u_z \end{Bmatrix}, \qquad \tilde{\boldsymbol{\varepsilon}} \overset{\text{def}}{=} \begin{Bmatrix} \varepsilon_{rr} \\ \varepsilon_{\theta\theta} \\ \varepsilon_{zz} \\ \gamma_{\theta z} \\ \gamma_{rz} \\ \gamma_{r\theta} \end{Bmatrix}, \qquad \tilde{\boldsymbol{\sigma}} \overset{\text{def}}{=} \begin{Bmatrix} \sigma_{rr} \\ \sigma_{\theta\theta} \\ \sigma_{zz} \\ \sigma_{\theta z} \\ \sigma_{rz} \\ \sigma_{r\theta} \end{Bmatrix}. \tag{1.48}$$

The kinematic relations (displacement-strain relations) in cylindrical coordinates are

$$\varepsilon_{rr} = \partial_r u_r, \quad \varepsilon_{\theta\theta} = \tfrac{1}{r}\partial_\theta u_\theta + \tfrac{u_r}{r}, \quad \varepsilon_{zz} = \partial_z u_z, \tag{1.49}$$
$$\gamma_{\theta z} = \partial_z u_\theta + \tfrac{1}{r}\partial_\theta u_z, \quad \gamma_{rz} = \partial_r u_z + \partial_z u_r, \quad \gamma_{r\theta} = \tfrac{1}{r}\partial_\theta u_r + \partial_\theta u_r - \tfrac{u_\theta}{r},$$

whereas the constitutive relation (Hooke's law) is identical to (1.43), where the Cartesian stress and strain tensors are replaced by the cylindrical ones.

The elasticity system can be brought to a weak formulation by multiplying (1.44) by a displacement test function vector $\boldsymbol{v}$, then integrating over the domain and using Green's theorem (see details in [178, Chapter 5]):

$$\text{Seek} \quad \boldsymbol{u} \in \mathcal{E}(\Omega), \ \boldsymbol{u} = \hat{\boldsymbol{u}} \text{ on } \cup_i (\partial\Omega_D)_i \quad \text{such that}$$

$$\mathcal{B}(\boldsymbol{u}, \boldsymbol{v}) = \mathcal{F}(\boldsymbol{v}) \qquad \forall \boldsymbol{v} \in \mathcal{E}_o(\Omega), \tag{1.50}$$

where the bilinear form is

$$\mathcal{B}(\boldsymbol{u}, \boldsymbol{v}) = \int_\Omega ([D]\boldsymbol{v})^T [E][D]\boldsymbol{u}\, d\boldsymbol{x}, \tag{1.51}$$

and the linear form is

$$\mathcal{F}(\boldsymbol{v}) = \int_{\partial\Omega_N} \hat{\boldsymbol{T}}^T \boldsymbol{v}\, d\Gamma + \int_\Omega \boldsymbol{f}^T \boldsymbol{v}\, d\boldsymbol{x}. \tag{1.52}$$

The energy spaces for elasticity, $\mathcal{E}$ and $\mathcal{E}_o$, are defined in Appendix A.

Planar Elasticity

Situations exist (for isotropic materials) for which the elasticity problem can be solved over a two-dimensional domain. These are called *plane stress* and *plane strain* situations (for more details the reader is referred to [167]). The only difference between them is Hooke's law, i.e., the material matrix:

$$[E] = \begin{cases} \frac{E}{1-\nu^2} \begin{bmatrix} 1 & \nu & 0 \\ & 1 & 0 \\ & & (1-\nu)/2 \end{bmatrix} & \text{plane stress,} \\ \frac{E}{(1+\nu)(1-2\nu)} \begin{bmatrix} (1-\nu) & \nu & 0 \\ & (1-\nu) & 0 \\ & & (1-2\nu)/2 \end{bmatrix} & \text{plane strain.} \end{cases} \quad (1.53)$$

In tensorial notation, Hooke's law for isotropic materials (1.43) reads

$$\sigma_{\beta\gamma} = C_{\beta\gamma\zeta\varsigma}\epsilon_{\zeta\varsigma}, \quad C_{\beta\gamma\zeta\varsigma} = \tilde{\lambda}\delta_{\beta\gamma}\delta_{\zeta\varsigma} + \mu\left(\delta_{\beta\zeta}\delta_{\gamma\varsigma} + \delta_{\beta\varsigma}\delta_{\gamma\zeta}\right), \quad \beta, \gamma, \zeta, \varsigma = 1, 2, \quad (1.54)$$

with

$$\tilde{\lambda} = \begin{cases} \lambda & \text{plane strain,} \\ 2\lambda\mu/(\lambda + 2\mu) & \text{plane stress.} \end{cases}$$

In 2-D elasticity, only two displacement functions, u_1 and u_2, are sought, and the stress/strain vectors consist of three unknown entries with subscripts 11, 22, and 12 e.g., $\boldsymbol{\varepsilon} = (\varepsilon_{11}\ \varepsilon_{22}\ \gamma_{12})^T$.

The operator matrix $[D]$ for plane elasticity is

$$[D] \stackrel{\text{def}}{=} \begin{bmatrix} \partial_1 & 0 & 0 \\ 0 & \partial_2 & 0 \\ \partial_2 & \partial_1 & 0 \end{bmatrix}. \quad (1.55)$$

In Sections 5.1–5.2 we show that in the vicinity of a singular point in an isotropic two-dimensional domain, the elastic solution admits the asymptotic expansion

$$\boldsymbol{u} = \begin{pmatrix} u_1 \\ u_2 \end{pmatrix} = \sum_{i=1}^{\infty} A_i r^{\alpha_i} \boldsymbol{s}_i^+(\theta), \qquad \boldsymbol{s}_i^+(\theta) = \begin{pmatrix} s_1^+(\theta) \\ s_2^+(\theta) \end{pmatrix}_i. \quad (1.56)$$

and the corresponding stress tensor is given by:

$$\boldsymbol{\sigma} = \begin{Bmatrix} \sigma_{11} \\ \sigma_{22} \\ \sigma_{12} \end{Bmatrix} = \sum_{i=1}^{\infty} A_i r^{\alpha_i - 1} \boldsymbol{S}_i^+(\theta), \qquad \boldsymbol{S}_i^+(\theta) = \begin{Bmatrix} S_{11}^+(\theta) \\ S_{22}^+(\theta) \\ S_{12}^+(\theta) \end{Bmatrix}_i. \quad (1.57)$$

For a general two-dimensional singular point, the asymptotic series (1.56) is more complicated:

$$u = \begin{pmatrix} u_1 \\ u_2 \end{pmatrix} = \sum_{i=1}^{\infty} \sum_{j=1}^{J} \sum_{k=1}^{K} A_{ijk} r^{\alpha_i + j} \ln^k(r) \boldsymbol{s}^+_{ijk}(\theta), \qquad \boldsymbol{s}^+_{ijk}(\theta) = \begin{pmatrix} s_1^+(\theta) \\ s_2^+(\theta) \end{pmatrix}_{ijk}. \tag{1.58}$$

The similarity between the elastic singular expansion (1.58) and that of the heat conduction singular expansion (1.32) is notable. For elasticity, however, complex eigenpairs may (and frequently do) exist, and a multiple eigenvalue may exist with different eigenvectors (as in the case of two-dimensional crack tips, where $\alpha_1 = \alpha_2 = 1/2$). There is the possibility that a multiple eigenvalue exists with a lower number of corresponding eigenvectors (the algebraic multiplicity is higher than the geometric multiplicity). This case is associated with the special cases in which K is greater than zero, and power-logarithmic singularities are evident (see details in [28, 52, 136]). When tractions are applied in the close vicinity of the singular point, power-logarithmic singularities may exist as well (the reader is referred to [37]). J differs from 0 if Γ_1 and/or Γ_2 are curvilinear arcs.

Chapter 2
An Introduction to the p- and hp-Versions of the Finite Element Method

The various methods described in this monograph for the computation of eigenpairs and GFIF/GSIFs cannot in general be carried out by means of analytical techniques; thus they require the use of numerical methods. We use the $p-$version of the FE method as the machinery for obtaining the required quantities, therefore, this chapter provides a brief introduction to these methods, their main features and characteristics. Readers interested in the mathematical aspects of $p-$ and $hp-$FEMs are encouraged to consult Schwab's book [157] whereas details on the applicative aspects of $p-$ and spectral FE methods are well documented in Karniadakis and Scherwin's book [92] and Szabó and Babuška's book [178].

2.1 The Weak Formulation

Let us consider the strong (or classical) formulation of the heat conduction equation in a two dimensional domain as the departing point, which is similar to (1.35) with Greek indices $\alpha, \beta = 1, 2$:

$$- \partial_\alpha \left(k_{\alpha\beta} \partial_\beta \tau \right) = Q \quad \text{in } \Omega \in \mathbb{R}^2. \tag{2.1}$$

For ease of explanation let us assume that homogeneous Dirichlet boundary conditions are specified on a part of the boundary $\partial\Omega_D$, and Neumann boundary conditions on the rest of the boundary $\partial\Omega_N = \partial\Omega / \partial\Omega_D$:

$$\tau = 0 \quad \text{on } \partial\Omega_D, \tag{2.2}$$

$$([k]\mathbf{grad}\tau) \cdot \mathbf{n} = \hat{q}_n \quad \text{on } \partial\Omega_N. \tag{2.3}$$

The Dirichlet boundary condition (2.2) is called an "essential" boundary condition, whereas the Neumann boundary condition (2.3) is called a "natural" boundary

Z. Yosibash, *Singularities in Elliptic Boundary Value Problems and Elasticity and Their Connection with Failure Initiation*, Interdisciplinary Applied Mathematics 37, DOI 10.1007/978-1-4614-1508-4_2, © Springer Science+Business Media, LLC 2012

condition. Multiplying (2.1) by a function $\chi(x_1, x_2)$, chosen to satisfy the "essential" boundary conditions, i.e., $\chi = 0$ on $\partial\Omega_D$, and integrating over Ω, one obtains

$$-\int_\Omega \chi\, \partial_\alpha \left(k_{\alpha\beta}\partial_\beta\tau\right) d\Omega = \int_\Omega \chi\, Q\, d\Omega. \tag{2.4}$$

Applying Green's theorem to the left-hand side of (2.4), one obtains

$$-\int_\Omega \chi\, \partial_\alpha \left(k_{\alpha\beta}\partial_\beta\tau\right)\, d\Omega = -\int_{\partial\Omega} \chi\, k_{\alpha\beta}\partial_\beta\tau n_\alpha\, d\Gamma + \int_\Omega k_{\alpha\beta}\, \partial_\beta\tau\, \partial_\alpha\chi\, d\Omega. \tag{2.5}$$

Because χ is zero on $\partial\Omega_D$, the boundary integral in (2.5) reduces to the boundary $\partial\Omega_N$, on which we can use (2.3). Thus the left-hand side of (2.4) becomes

$$-\int_{\partial\Omega_N} \chi\hat{q}_n\, d\Gamma + \int_\Omega k_{\alpha\beta}\, \partial_\beta\tau\, \partial_\alpha\chi\, d\Omega. \tag{2.6}$$

Inserting (2.6) into (2.4), one obtains

$$\int_\Omega k_{\alpha\beta}\, \partial_\beta\tau\, \partial_\alpha\chi\, d\Omega = \int_\Omega \chi\, Q\, d\Omega + \int_{\partial\Omega_N} \chi\hat{q}_n\, d\Gamma. \tag{2.7}$$

What has been obtained in (2.7) is the weak formulation already presented in (1.37) in a 3-D setting, where the left-hand side is denoted the bilinear form $\mathcal{B}(\tau, \chi)$ (see (1.38)), and the right-hand side is denoted by the linear form $\mathcal{F}(\chi)$ (see (1.39)). In summary, the weak formulation for this example problem is

$$\text{Seek } \tau_{\text{EX}} \in \mathcal{E}_o(\Omega) \text{ such that } \quad \mathcal{B}(\tau_{\text{EX}}, \chi) = \mathcal{F}(\chi) \qquad \forall\chi \in \mathcal{E}_o(\Omega), \tag{2.8}$$

with $\mathcal{E}_o(\Omega) = \{\tau(x_1, x_2) \mid \tau \in \mathcal{E}(\Omega),\ \tau = 0 \text{ on } \partial\Omega_D\}$ and $\mathcal{E}(\Omega)$ is defined in Appendix A.

It can be shown that the weak formulation is equivalent to the principle of minimum potential energy (see, for example, [178]). Define the potential energy (a functional) by

$$\Pi(\tau) \overset{\text{def}}{=} \frac{1}{2}\mathcal{B}(\tau, \tau) - \mathcal{F}(\tau). \tag{2.9}$$

The principle of minimum potential energy states that the exact solution is the one that gives the potential energy a minimum value:

$$\Pi_{\text{EX}} \overset{\text{def}}{=} \Pi(\tau_{\text{EX}}) = \min_{\chi\in\mathcal{E}(\Omega)} \Pi(\chi). \tag{2.10}$$

Of course, if essential boundary conditions are prescribed, the energy space has to be adjusted accordingly. The main role of the potential energy principle is its use in estimating the error of the approximated FE solution, and in understanding the concept of monotonic convergence when using hierarchical spaces (which will be discussed in the next subsection).

2.2 Discretization

As formulated, $\mathcal{E}_o(\Omega)$ is an infinitely large space, so that in order to solve (2.8), one needs to find a function within an infinite number of functions. This is, of course, not practically possible. Instead, and *this is the major step where the discretization errors are introduced*, one may consider a finite-dimensional subspace, i.e., in $\mathcal{E}_o^N(\Omega) \subset \mathcal{E}_o(\Omega)$, such that $\dim(\mathcal{E}_o^N) = N$. So instead of solving (2.8), we are seeking an approximate solution, denoted by $\tau_{\text{FE}} \in \mathcal{E}_o^N(\Omega)$:

$$\text{Seek } \tau_{\text{FE}} \in \mathcal{E}_o^N(\Omega), \text{ such that} \qquad \mathcal{B}(\tau_{\text{FE}}, \chi) = \mathcal{F}(\chi) \qquad \forall \chi \in \mathcal{E}_o^N(\Omega). \quad (2.11)$$

One important property is that from all functions in $\mathcal{E}_o^N(\Omega)$, the function τ_{FE} that satisfies (2.11) is the closest to τ_{EX} when measured in the *energy norm*:

$$\min_{\tau \in \mathcal{E}_o^N(\Omega)} \|\tau_{\text{EX}} - \tau\|_{\mathcal{E}_o^N(\Omega)} = \|\tau_{\text{EX}} - \tau_{\text{FE}}\|_{\mathcal{E}_o^N(\Omega)}, \qquad (2.12)$$

and furthermore, the numerical error $e(\boldsymbol{x}) \stackrel{\text{def}}{=} \tau_{\text{EX}} - \tau_{\text{FE}}$ is orthogonal to the space $\mathcal{E}_o^N(\Omega)$. The important question that arises is, how can one estimate the numerical error $e(\boldsymbol{x})$? The answer to this question requires the definitions of hierarchical spaces and extensions.

Consider, for example, a set of hierarchical subspaces (see a graphical interpretation of hierarchical subspaces as opposed to non-hierarchical subspaces in Figure 2.1) $\mathcal{E}_o^{N_1}(\Omega) \subset \mathcal{E}_o^{N_2}(\Omega) \subset \cdots \subset \mathcal{E}_o(\Omega)$, having the property that

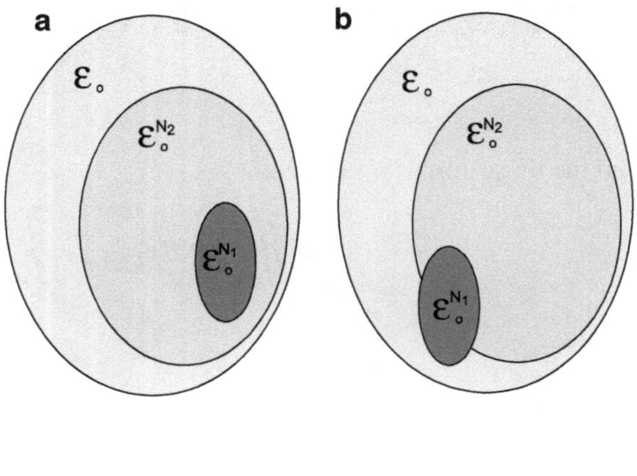

Fig. 2.1 (**a**) Hierarchical as opposed to (**b**) non-hierarchical subspaces.

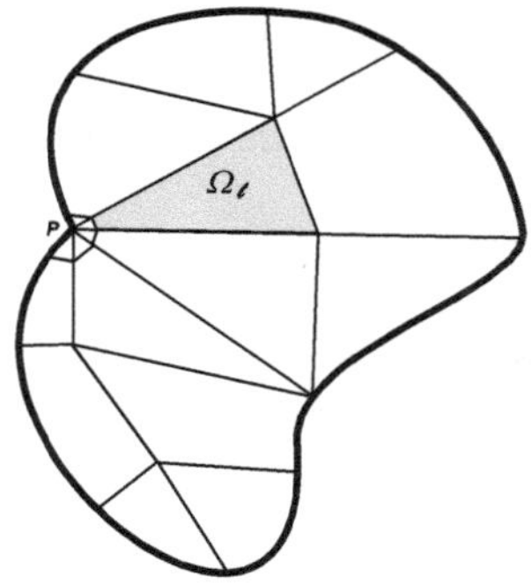

Fig. 2.2 A typical 2-D FE p-mesh.

$N_1 < N_2 < \cdots$. Then by (2.12), $\|\tau_{EX} - \tau_{FE_2}\|_{\mathcal{E}_o(\Omega)} \leq \|\tau_{EX} - \tau_{FE_1}\|_{\mathcal{E}_o(\Omega)}$. Of course, as the space $\mathcal{E}_o^N(\Omega)$ is enriched by more and more functions, then the approximated solutions become closer to the exact solution. The systematic enrichments of the subspaces are called *extensions*, and we will elaborate on a special extension procedure, called p- and hp-extensions.

In order to solve (2.8), we partition the domain over which the integration has to be performed into quadrilateral or triangular subdomains, called *elements*. The collection of these elements is called the FE "mesh." Each element is denoted by Ω_ℓ, so that $\cup\Omega_\ell = \Omega$; see Figure 2.2 as an example of a typical FE p-mesh. Having the mesh, the integral over the entire domain can be replaced by the sum of integrals over each subdomain (element), so that the weak form in (2.8) can now be stated as

$$\text{Seek } \tau_{FE} \in \mathcal{E}_o^N(\Omega), \text{ such that } \quad \sum_\ell \mathcal{B}_\ell(\tau_{FE}, \chi) = \sum_\ell \mathcal{F}_\ell(\chi) \qquad \forall \chi \in \mathcal{E}_o^N(\Omega), \tag{2.13}$$

where $\mathcal{B}_\ell(\tau, \chi)$, for example, is the bilinear form over the element Ω_ℓ,

$$\mathcal{B}_\ell(\tau, \chi) = \int_{\Omega_\ell} k_{\alpha\beta}\, \partial_\beta \tau\, \partial_\alpha \chi\, d\Omega = \iint_{\Omega_\ell} \begin{pmatrix} \frac{\partial \chi}{\partial x_1} & \frac{\partial \chi}{\partial x_2} \end{pmatrix} \begin{bmatrix} k_{11} & k_{12} \\ k_{12} & k_{22} \end{bmatrix} \begin{Bmatrix} \frac{\partial \tau}{\partial x_1} \\ \\ \frac{\partial \tau}{\partial x_2} \end{Bmatrix} dx_1 dx_2, \tag{2.14}$$

and the linear form for the element is

$$\mathcal{F}_\ell(\chi) = \iint_{\Omega_\ell} \chi Q\, dx_1 dx_2 + \int_{\partial\Omega_{N\ell}} \chi \hat{q}_n\, d\Gamma. \tag{2.15}$$

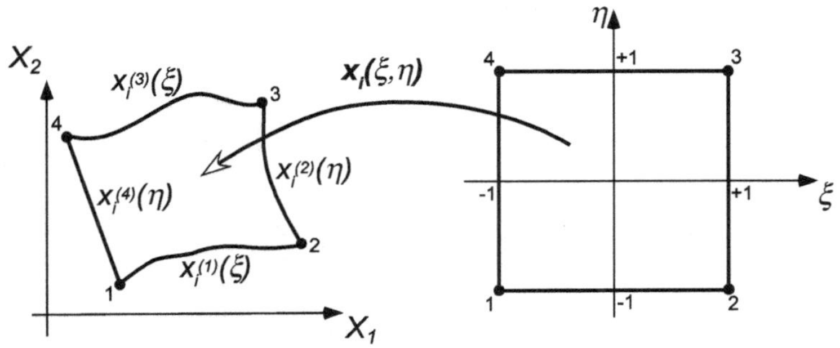

Fig. 2.3 Blending mapping from the standard element to the "physical" element.

2.2.1 Blending Functions, the Element Stiffness Matrix and Element Load Vector

Integrating over different-shaped quadrilaterals and triangles is a complicated procedure. To overcome this difficulty, a standard element is introduced. Assume that one needs to evaluate (2.14) over the quadrilateral domain shown in the left side of Figure 2.3, where the sides of the element have parametric representations. In view of Figure 2.3, we denote by $x_i^{(j)}(t)$, $-1 \le t = \xi, \eta \le 1$, the parametric representation of the curved edge j and by $X_i^{(j)}$ the coordinates of vertex j. With these definitions and using the "blending function method"[68], it is possible to "map" the standard quadrilateral element $\Omega_{st}^{Q} = \{(\xi,\eta) \mid -1 \le \xi \le 1, \ -1 \le \eta \le 1\}$ into any quadrilateral element with curved boundaries Ω_{ℓ}^{Q}:

$$\begin{aligned} x_i(\xi,\eta) &= \frac{1}{2}(1-\eta)x_i^{(1)}(\xi) + \frac{1}{2}(1+\xi)x_i^{(2)}(\eta) + \frac{1}{2}(1+\eta)x_i^{(3)}(\xi) \\ &+\frac{1}{2}(1-\xi)x_i^{(4)}(\eta) \\ &-\frac{1}{4}(1-\xi)(1-\eta)X_i^{(1)} - \frac{1}{4}(1+\xi)(1-\eta)X_i^{(2)} \\ &-\frac{1}{4}(1+\xi)(1+\eta)X_i^{(3)} - \frac{1}{4}(1-\xi)(1+\eta)X_i^{(4)}, \quad i = 1,2. \end{aligned} \tag{2.16}$$

With the aid of the blending functions, one has

$$\left\{ \begin{matrix} \frac{\partial}{\partial \xi} \\ \\ \frac{\partial}{\partial \eta} \end{matrix} \right\} = [J] \left\{ \begin{matrix} \frac{\partial}{\partial x_1} \\ \\ \frac{\partial}{\partial x_2} \end{matrix} \right\}, \quad \text{with} \quad [J] = \begin{bmatrix} \frac{\partial x_1}{\partial \xi} & \frac{\partial x_2}{\partial \xi} \\ \frac{\partial x_1}{\partial \eta} & \frac{\partial x_2}{\partial \eta} \end{bmatrix},$$

so the bilinear form for element ℓ is

$$\mathcal{B}_\ell(\tau,\chi) = \iint_{-1}^{1} \left(\frac{\partial\chi}{\partial\xi}\ \frac{\partial\chi}{\partial\eta}\right) \left([J]^{-1}\right)^T [k] \left([J]^{-1}\right) \begin{Bmatrix} \frac{\partial\tau}{\partial\xi} \\ \\ \frac{\partial\tau}{\partial\eta} \end{Bmatrix} |J| d\xi d\eta. \tag{2.17}$$

Similarly, if for example the boundary of the domain coincides with the element edge 2-3 (see Figure 2.3), then the linear form for element ℓ is

$$\mathcal{F}_\ell(\chi) = \iint_{-1}^{1} \chi Q\ |J| d\xi d\eta + \int_{-1}^{1} (\chi\hat{q}_n)\,|_{\xi=1} d\eta. \tag{2.18}$$

2.2.2 *The Finite Element Space*

There are three ways of increasing the FE space, i.e., there are three different extension possibilities:

1. *h-Extension:* Refining the FE mesh (i.e., adding more elements), while keeping over each element a basis consisting of a given number of functions.
2. *p-Extension:* Keeping the FE mesh fixed and increasing the number of basis functions over each element.
3. *hp-Extension:* Changing the mesh and the number of basis functions over individual or all elements.

A necessary condition for a function to be in $\mathcal{E}(\Omega)$ is that it be C^0 continuous, i.e., continuity across elements' boundaries is maintained. Instead of defining a basis function over the entire Ω, we define a set of element basis functions, so that on combining all together, they provide a C^0 continuous overall function. Since the weak formulation has been split into a sum over all elements, and furthermore, all integrations are performed over the standard element, it is only natural to define a basis function for approximating both the test and trial functions on the standard element.

Let the trial function τ and the test function χ be expressed in terms of an elemental basis functions $\Phi_i(\xi,\eta)$ (spanning a finite-dimensional subspace) in the standard element

$$\tau(\xi,\eta) = \sum_{i=1}^{DOF} b_i^{(\ell)} \Phi_i(\xi,\eta) = \boldsymbol{\Phi}^T \boldsymbol{b}^{(\ell)} \qquad \chi(\xi,\eta) = \sum_{i=1}^{DOF} c_i^{(\ell)} \Phi_i(\xi,\eta) = (\boldsymbol{c}^{(\ell)})^T \boldsymbol{\Phi}, \tag{2.19}$$

where $b_i^{(\ell)}$ and $c_i^{(\ell)}$ are the amplitudes of the basis functions in element ℓ, and Φ_i are products of integrals of Legendre polynomials in ξ and η. Substituting (2.19) into (2.17), one obtains an expression for the unconstrained elemental stiffness matrix $[K^{(\ell)}]$ associated with $\mathcal{B}_\ell$;

$$\mathcal{B}_\ell(\tau,\chi) = (\boldsymbol{c}^{(\ell)})^T [K^{(\ell)}] \boldsymbol{b}^{(\ell)}, \tag{2.20}$$

and the entries of $[K^{(\ell)}]$ are computed by

$$K_{ij}^{(\ell)} = \iint_{-1}^{1} \left(\frac{\partial \Phi_i}{\partial \xi} \; \frac{\partial \Phi_i}{\partial \eta} \right) \left([J]^{-1}\right)^T [k] \left([J]^{-1}\right) \begin{Bmatrix} \frac{\partial \Phi_j}{\partial \xi} \\ \frac{\partial \Phi_j}{\partial \eta} \end{Bmatrix} |J| d\xi d\eta. \tag{2.21}$$

Substituting (2.19) into (2.18), one obtains an expression for the unconstrained load vector $\boldsymbol{r}^{(\ell)}$ associated with $\mathcal{F}_\ell$;

$$\mathcal{F}_\ell(\chi) = (\mathbf{c}^{(\ell)})^T \boldsymbol{r}^{(\ell)}, \tag{2.22}$$

and the entries of $\boldsymbol{r}^{(\ell)}$ are computed by

$$r_i^{(\ell)} = \iint_{-1}^{1} \Phi_i Q \; |J| d\xi d\eta + \int_{-1}^{1} (\Phi_i \hat{q}_n) \,|_{\xi=1} d\eta. \tag{2.23}$$

Hierarchic Basis (Shape) Functions for Quadrilateral Elements

There are many possibilities for choosing a basis of functions to span the space $\mathcal{E}^N$. Usually it is constructed by specially chosen polynomials based on Legendre or Jacobi polynomials (see for example [32,91,178]). Here we present shape functions for the classical h-version of the FEMs and a family of hierarchical shape functions over quadrilateral elements for the p-version of the FEM, as described in [178], based on the Legendre polynomials. This basis function is extendable to triangular elements also (details are provided in [91, 178]).

Conventional Parabolic (second-order) h-Space

Serendipity (8-nodes)

Vertex

$$\Phi_1(\xi,\eta) = -\frac{1}{4}(1-\xi)(1-\eta)(1+\xi+\eta)$$

$$\Phi_2(\xi,\eta) = -\frac{1}{4}(1+\xi)(1-\eta)(1-\xi+\eta)$$

$$\Phi_3(\xi,\eta) = -\frac{1}{4}(1+\xi)(1+\eta)(1-\xi-\eta)$$

$$\Phi_4(\xi,\eta) = -\frac{1}{4}(1-\xi)(1+\eta)(1+\xi-\eta)$$

Product (9-nodes)

Vertex

$$\Phi_1(\xi,\eta) = \frac{1}{4}\xi\eta(\xi-1)(\eta-1)$$

$$\Phi_2(\xi,\eta) = \frac{1}{4}\xi\eta(\xi+1)(\eta-1)$$

$$\Phi_3(\xi,\eta) = \frac{1}{4}\xi\eta(\xi+1)(\eta+1)$$

$$\Phi_4(\xi,\eta) = \frac{1}{4}\xi\eta(\xi-1)(\eta+1)$$

Edge

$$\Phi_5(\xi,\eta) = \frac{1}{2}\left(1-\xi^2\right)(1-\eta)$$

$$\Phi_6(\xi,\eta) = \frac{1}{2}(1+\xi)\left(1-\eta^2\right)$$

$$\Phi_7(\xi,\eta) = \frac{1}{2}\left(1-\xi^2\right)(1+\eta)$$

$$\Phi_8(\xi,\eta) = \frac{1}{2}(1-\xi)\left(1-\eta^2\right)$$

Edge

$$\Phi_5(\xi,\eta) = \frac{1}{2}\left(1-\xi^2\right)\eta\,(\eta-1)$$

$$\Phi_6(\xi,\eta) = \frac{1}{2}\left(1-\eta^2\right)(1+\xi)$$

$$\Phi_7(\xi,\eta) = \frac{1}{2}\left(1-\xi^2\right)(1+\eta)$$

$$\Phi_8(\xi,\eta) = \frac{1}{2}(1-\eta)\,\xi\,(\xi-1)$$

Face

$$\Phi_9(\xi,\eta) = \left(1-\xi^2\right)\left(1-\eta^2\right)$$

Hierarchical (p-version) Trunk Space

Vertex

$$\Phi_1(\xi,\eta) = \frac{1}{4}(1-\xi)(1-\eta)$$

$$\Phi_2(\xi,\eta) = \frac{1}{4}(1+\xi)(1-\eta)$$

$$\Phi_3(\xi,\eta) = \frac{1}{4}(1+\xi)(1+\eta)$$

$$\Phi_4(\xi,\eta) = \frac{1}{4}(1-\xi)(1+\eta)$$

Edge

$$\Phi_5(\xi,\eta) = -\sqrt{\frac{3}{32}}\left(1-\xi^2\right)(1-\eta)$$

$$\Phi_6(\xi,\eta) = -\sqrt{\frac{3}{32}}(1+\xi)\left(1-\eta^2\right)$$

$$\Phi_7(\xi,\eta) = -\sqrt{\frac{3}{32}}\left(1-\xi^2\right)(1+\eta)$$

$$\Phi_8(\xi,\eta) = -\sqrt{\frac{3}{32}}(1-\xi)\left(1-\eta^2\right)$$

Edge (cont.)

$$\Phi_9(\xi,\eta) = -\sqrt{\frac{5}{32}}\,\xi\left(1-\xi^2\right)(1-\eta)$$

$$\Phi_{10}(\xi,\eta) = -\sqrt{\frac{5}{32}}\,\eta\,(1+\xi)\left(1-\eta^2\right)$$

$$\Phi_{11}(\xi,\eta) = -\sqrt{\frac{5}{32}}\,\xi\left(1-\xi^2\right)(1+\eta)$$

$$\Phi_{12}(\xi,\eta) = -\sqrt{\frac{5}{32}}\,\eta\,(1-\xi)\left(1-\eta^2\right)$$

$$\Phi_{13}(\xi,\eta) = \sqrt{\frac{7}{512}}\left(1-\xi^2\right)\left(1-5\xi^2\right)(1-\eta)$$

$$\Phi_{14}(\xi,\eta) = \sqrt{\frac{7}{512}}(1+\xi)\left(1-\eta^2\right)\left(1-5\eta^2\right)$$

$$\Phi_{15}(\xi,\eta) = \sqrt{\frac{7}{512}}\left(1-\xi^2\right)\left(1-5\xi^2\right)(1+\eta)$$

$$\Phi_{16}(\xi,\eta) = \sqrt{\frac{7}{512}}(1-\xi)\left(1-\eta^2\right)\left(1-5\eta^2\right)$$

Face

$$\Phi_{17}(\xi,\eta) = \frac{3}{8}\left(1-\xi^2\right)\left(1-\eta^2\right)$$

$$\vdots$$

The specific vertex, edge or face number i with which a shape function Φ_i is associated is shown in Figure 2.4. A graphical representation of the hierarchical truth space shape functions is shown in Figure 2.5.

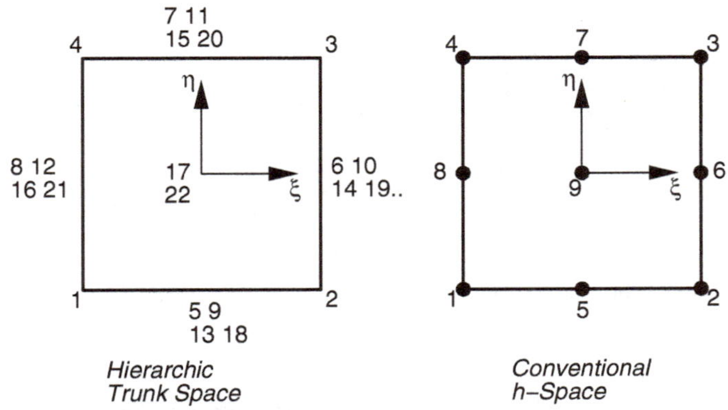

Fig. 2.4 Standard element and notation of shape functions.

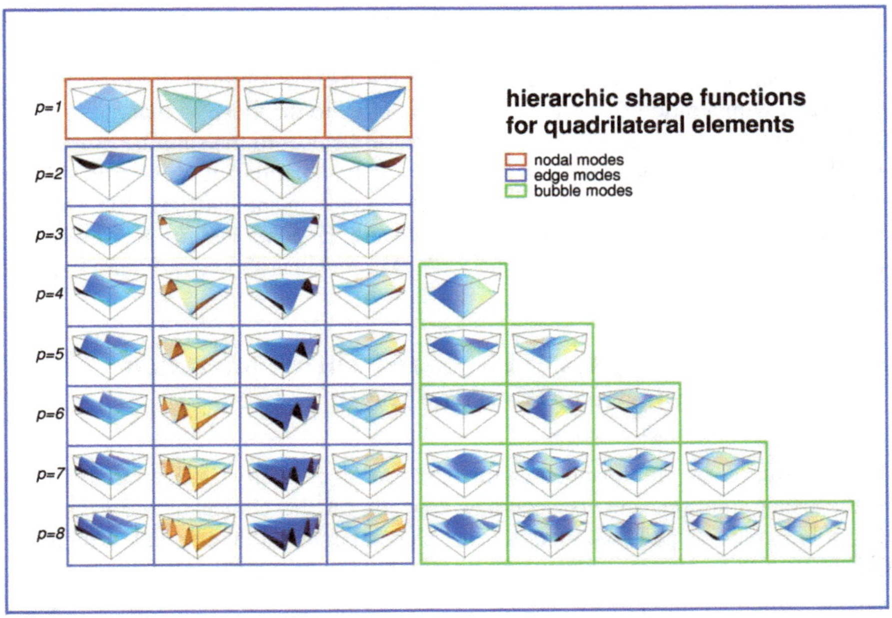

Fig. 2.5 Trunk space hierarchic shape functions over quadrilaterals (from prof. Ernst Rank of TUM-Germany).

For a given mesh and $p-$ level, the global stiffness matrix and load vector are obtained by an assembly procedure of the elemental stiffness matrices and load vectors, so the weak form (2.13) becomes

$$c^T[K]b = c^T r, \quad \forall c, \quad \Rightarrow \quad [K]b = r. \tag{2.24}$$

The solution of (2.24) determines b, thus defines the finite element solution τ_{FE} for a given discretization.

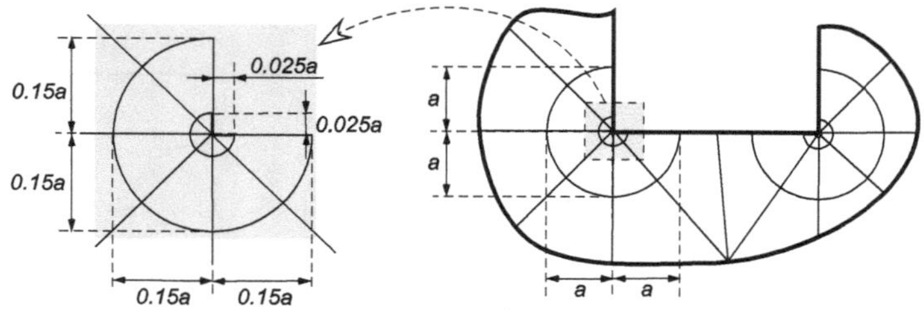

Fig. 2.6 An example of a mesh design with geometric mesh refinement in the vicinity of singular points.

2.2.3 *Mesh Design for an Optimal Convergence Rate*

For domains with singular points, there exists an optimal design of the discretization in the neighborhood of the singularity: the finite elements should be created so that their sizes decrease in geometric progression towards the singular point, and the polynomial degree over the elements decreases. The optimal *geometric mesh refinement* with a ratio $(\sqrt{2}-1)^2 \approx 0.17$ is applied to the mesh so that $h_{i+1}/h_i = 0.17$, where i increases as the nodes are closer and closer to the singular point. The grading factor is independent of the strength of the singularity, and applicable to both scalar and vector elliptic problems (heat conduction and elasticity). In practice a geometric grading with a factor 0.15 is used, and the generated meshes are called geometric graded meshes. An example for 2-D domains is shown in Figure 2.6.

2.3 Convergence Rates of FEMs and Their Connection to the Regularity of the Exact Solution

The FE solution (τ_{FE} for heat conduction and $\boldsymbol{u}_{FE}$ for elasticity) is an approximation to the exact solution, and its accuracy depends on the choice of the FE mesh, the polynomial degree assigned to the elements, and the mapping functions. Quantifying this error in energy norm is as important as the FE solution itself, and thus we provide estimates to $\|e\|_{\mathcal{E}(\Omega)} \overset{\text{def}}{=} \|\tau_{EX} - \tau_{FE}\|_{\mathcal{E}(\Omega)}$ for heat conduction, or $\|\boldsymbol{e}\|_{\mathcal{E}(\Omega)} \overset{\text{def}}{=} \|\boldsymbol{u}_{EX} - \boldsymbol{u}_{FE}\|_{\mathcal{E}(\Omega)}$ for elasticity. The error estimates are presented as error bounds, and are expressed in terms of h, a characteristic length of the largest element in the domain, and p, the polynomial degree of the test and trial functions. Because both h and p are associated with the number of degrees of freedom[1] N, the error bounds are expressed as

$$\|e\|_{\mathcal{E}(\Omega)} \leq C h^n p^m \approx C f(N), \tag{2.25}$$

[1]The connection between h, p, and N depends on the mesh and the dimension of the problem:

where C, n, m are generic constants independent of the discretization parameters and $f(N)$ is a decreasing function of N (see details in [29, Chapter II7] and [14]). The rate at which $f(N) \overset{N\to\infty}{\longrightarrow} 0$ depends on the "regularity" (sometimes also called in the engineering community the "smoothness") of the exact solution (as will be precisely defined in the sequel) in addition to h and p and the FE-extension method (either the h- p- or hp-version).

Definition 2.1. For simplicity consider the heat conduction problem with $k_{ij} = \delta_{ij}$, and instead of the energy norm we use the H^1 norm (these norms are equivalent). Then we say that the bilinear weak form (1.37) has H^s **regularity** if the solution to

$$(\tau, \chi)_{H^1(\Omega)} = (Q, \chi)_{L^2(\Omega)} \qquad \forall \chi \in H^1(\Omega)$$

belongs to $H^s(\Omega)$, i.e., $\tau \in H^s(\Omega)$, for every $Q \in H^{s-2}$ and there exists a constant $c(s, \Omega)$ such that

$$\|\tau\|_{H^s(\Omega)} \leq c(s, \Omega) \|Q\|_{H^{s-2}(\Omega)}.$$

Roughly speaking, the more regular the solution, the less its value and first derivatives change over a given short distance in the domain. Following [176, 178], we may differentiate the solutions of elliptic PDEs based on their regularity into three categories:

- Category A: The exact solution is analytic on each element including on the boundary, $\boldsymbol{u}, \tau \in \mathcal{C}^\infty(\Omega)$.
- Category B: The exact solution is analytic on each element including on the boundary, except at some vertices at which nodes are located (and edges in 3-D domains). *The regularity of the exact solution is determined by the smallest eigenvalue that characterizes the most singular solution in the domain:* $\alpha \overset{\text{def}}{=} \min_i \alpha_i$.
- Category C: The exact solution is neither in category A nor in category B.

A very brief description of the mathematical steps followed to obtain estimates such as (2.25) from various finite element methods (for the heat conduction problem for example) are as follows:

- First bound the error between the function τ and its interpolant $I_h\tau$ in a given norm H^k, in terms of $\|\tau\|_{H^t}$, where $t > k$. That is, provide estimates

$$\|\tau - I_h\tau\|_{H^k(\Omega)} \leq \|\tau\|_{H^t(\Omega)}.$$

$$N = \begin{cases} Ch^{-1}p & \text{for 1-D domains,} \\ Ch^{-2}p^2 & \text{for uniform or radical mesh in 2-D domains,} \\ Ck\rho_o p^2 & \text{for geometric mesh in 2-D domains,} \end{cases} \tag{2.26}$$

where k represents the number of layers and C is a generic constant.

- Second, use regularity theorems (shift theorems), the simplest of which says: *Let the elliptic bilinear form have sufficiently smooth coefficient functions. Then if Ω is convex, the Dirichlet problem is H^2 regular (see definition below). If Ω has a C^s boundary with $s \geq 2$, then the Dirichlet problem is H^s regular.*
- A combination of the first and second steps enables one to bound the interpolation error in terms of a constant depending on the input data (left-hand side of equation and BCs), shape of the boundary, and material coefficients $[k]$.
- The last step is the use of Cea's lemma, bounding the finite element error by the interpolation error:

$$\|\tau - \tau_{\text{FE}}\|_{H^1(\Omega)} \leq \frac{C}{\gamma} \inf_{\chi \in S_h} \|\tau - \chi\|_{H^1(\Omega)}.$$

2.3.1 Algebraic and Exponential Rates of Convergence

The convergence rates represent the speed at which $f(N) \to 0$ as $N \to \infty$. They depend on the FE extension method and the regularity of the exact solution. The two different functions describing $f(N)$ for $N \gg 1$ (termed the asymptotic range) are

$$\text{Algebraic Rate:} \quad \left.\begin{matrix} \|e\|_{\mathcal{E}(\Omega)} \\ \|e\|_{\mathcal{E}(\Omega)} \end{matrix}\right\} \leq \frac{k}{N^{\beta}}, \tag{2.27}$$

$$\text{Exponential Rate:} \quad \left.\begin{matrix} \|e\|_{\mathcal{E}(\Omega)} \\ \|e\|_{\mathcal{E}(\Omega)} \end{matrix}\right\} \leq \frac{k}{\exp(\gamma N^{\phi})}; \tag{2.28}$$

$k, \gamma > 0$ are independent of N. The notion of "algebraic rate" of convergence is due to the straight line on a log-log scale obtained by applying the log operator to the convergence estimate (2.27):

$$\log \|e\|_{\mathcal{E}(\Omega)} \leq \log \left(\frac{k}{N^{\beta}} \right). \tag{2.29}$$

For large N the inequality becomes "almost equal," so that (2.29) reads

$$\log \|e\|_{\mathcal{E}(\Omega)} \approx \log k - \beta \log N, \tag{2.30}$$

which is a straight line with slope of $-\beta$, called the "convergence rate."

Depending on the version of the FE method and the regularity of the exact solution, it is possible to estimate the rates of convergence for elliptic problems in 2-D and 3-D as summarized in Tables 2.1 and 2.2 (from [176]).

Table 2.1 Asymptotic rates of convergence in energy norm, two-dimensions.

	Type of Extension		
Category	h	p	hp
A	Algebraic $\beta = p/2$	Exponential $\phi \geq 1/2$	Exponential $\phi \geq 1/2$
B	Algebraic (see Note 1) $\beta = \frac{1}{2}\min(p, \alpha)$	Algebraic $\beta = \alpha$	Exponential $\phi \geq 1/3$
C	Algebraic $\beta > 0$	Algebraic $\beta > 0$	See Note 2

Table 2.2 Asymptotic rates of convergence in energy norm, three-dimensions.

	Type of Extension		
Category	h	p	hp
A	Algebraic $\beta = p/3$	Exponential $\phi \geq 1/3$	Exponential $\phi \geq 1/3$
B	See Note 3		Exponential $\phi \geq 1/5$
C	Algebraic $\beta > 0$	Algebraic $\beta > 0$	See Note 2

Note 1: Uniform or quasiuniform meshes are assumed. The maximum possible value of β obtainable with optimal (adaptively determined) meshes is $p/2$.

Note 2: When the exact solution has a recognizable structure, then nearly exponential convergence rates can be obtained with hp-adaptive schemes [11].

Note 3: The characterization of smoothness in 3-D is much more difficult than in 2-D. Nevertheless, as in the 2-D case, the rate of p-convergence is twice the rate of oh-convergence when quasiuniform meshes are used.

The error in energy norm may be computed by (see [178, p. 69])

$$\left.\begin{matrix} \|e\|_{\mathcal{E}} \\ \|\boldsymbol{e}\|_{\mathcal{E}} \end{matrix}\right\} = \left.\begin{matrix} \sqrt{\frac{1}{2}\mathcal{B}(e,e)} \\ \sqrt{\frac{1}{2}\mathcal{B}(\boldsymbol{e},\boldsymbol{e})} \end{matrix}\right\} = \sqrt{\Pi_{\mathrm{FE}} - \Pi_{\mathrm{EX}}}. \tag{2.31}$$

Although the exact solution is unknown, it is possible to obtain very sharp estimates for the error in energy norm using three consecutive FE solutions with increasing hierarchical spaces having $1 << N_1 < N_2 < N_3$. Combine (2.31) and (2.27) to obtain

$$\Pi_{\mathrm{FE}} - \Pi_{\mathrm{EX}} \approx \frac{k^2}{N^{2\beta}}. \tag{2.32}$$

Table 2.3 Exponential convergence rates in energy norm for $hp-$extensions for 3-D problems having different types of singularities using optimal meshes [75].

extension	Regular	Edge	Edge-Vertex	Vertex
hp	$\phi = 1/3$	$\phi = 1/4$	$\phi = 1/5$	$\phi = 1/4$

The three unknowns $\Pi_{\text{EX}}, k, \beta$ can be computed if we express (2.32) for the three consecutive FE solutions:

$$\Pi_{\text{FE}_i} - \Pi_{\text{EX}} \approx \frac{k^2}{N_i^{2\beta}}, \qquad i = 1, 2, 3. \tag{2.33}$$

These three equations may be solved and one obtains an implicit equation for determining Π_{EX}:

$$\frac{\Pi_{\text{EX}} - \Pi_{\text{FE}_3}}{\Pi_{\text{EX}} - \Pi_{\text{FE}_2}} \approx \left(\frac{\Pi_{\text{EX}} - \Pi_{\text{FE}_2}}{\Pi_{\text{EX}} - \Pi_{\text{FE}_1}}\right)^{\frac{\log N_2 - \log N_1}{\log N_3 - \log N_2}}. \tag{2.34}$$

Problem 2.1. Use (2.33) for three consecutive FE solutions to obtain (2.34).

Once Π_{EX} is determined, the error of each FE solution can be estimated by (2.31). These error estimates are progressively better as N increases.

Some Remarks

Before we present some numerical examples, let us provide some remarks on $hp-$extensions and optimal convergence rates (see Table 2.3 and [14]):

- The optimal convergence rate is achieved by taking a geometric graded mesh with a factor 0.17, and a linear-degree vector $\boldsymbol{p}$ such that $p_j = [(2\alpha) - 1)j] + 1$, where j indicates the element at the jth layer away from the singular point.
- For $p-$extensions over a geometric mesh with uniform $p = [(2\alpha) - 1)n]$, where n is the number of layers around the singular point, one still obtains an exponential convergence rate, but the exponent ϕ reduces by a factor of about 0.7.
- It is interesting to note that for the h-extensions, the geometric refinement is not optimal, but the radical one is [14]. Using a radical mesh, when refining further and further one obtains the envelope of the h-p version if the polynomial level on every element is increased linearly as the number of elements is increased.

2.3.1.1 Numerical Examples

The examples discussed in this section were constructed so that the exact solutions are known and used to demonstrate the convergence rates of the $h-$ and $p-$extensions.

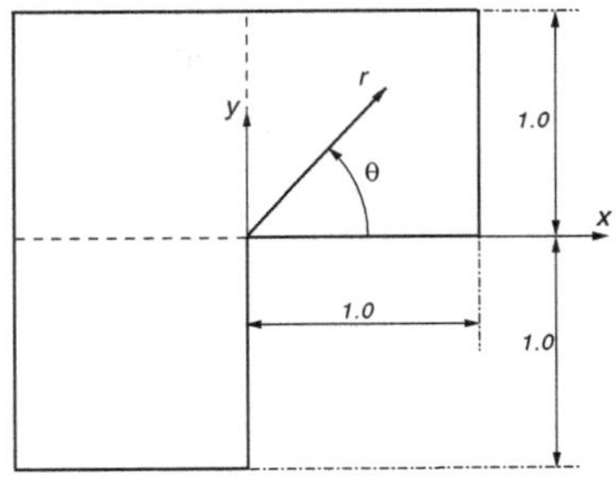

Fig. 2.7 The L-shaped domain (with the coarsest FE mesh having 3 elements).

The Laplace equation over a L-shaped 2-D domain

First, we consider the heat conduction equation in an isotropic material (Laplace equation) $\nabla^2\tau = 0$ over the L-shaped domain in Figure 2.7 (containing a $3\pi/2$ corner) with homogeneous Neumann boundary conditions on the boundaries intersecting at $(x, y) = (0, 0)$, having an exact solution (see (1.11–1.14))

$$\tau_{\text{EX}} = A_1 r^{2/3} \cos\left(\frac{2\theta}{3}\right). \tag{2.35}$$

On the domain's boundaries $x = \pm 1$ and $y = \pm 1$, Neumann boundary conditions according to (2.35) are prescribed:

$$\begin{Bmatrix} \frac{\partial\tau}{\partial x} \\ \frac{\partial\tau}{\partial x} \end{Bmatrix} = \begin{Bmatrix} \frac{\partial\tau}{\partial r}\frac{\partial r}{\partial x} + \frac{\partial\tau}{\partial\theta}\frac{\partial\theta}{\partial x} \\ \frac{\partial\tau}{\partial r}\frac{\partial r}{\partial y} + \frac{\partial\tau}{\partial\theta}\frac{\partial\theta}{\partial y} \end{Bmatrix} = \begin{Bmatrix} \frac{\partial\tau}{\partial r}\cos\theta - \frac{\partial\tau}{\partial\theta}\frac{\sin\theta}{r} \\ \frac{\partial\tau}{\partial r}\sin\theta + \frac{\partial\tau}{\partial\theta}\frac{\cos\theta}{r} \end{Bmatrix} = \frac{2}{3}A_1 r^{-1/3} \begin{Bmatrix} \cos\frac{\theta}{3} \\ \sin\frac{\theta}{3} \end{Bmatrix}. \tag{2.36}$$

The exact potential energy is

$$\Pi_{\text{EX}} = \frac{1}{2}\mathcal{B}(\tau,\tau) - \mathcal{F}(\tau) = -\frac{1}{2}\mathcal{B}(\tau,\tau) = -\frac{1}{2}\iint_{\Omega}\left[\left(\frac{\partial\tau}{\partial x}\right)^2 + \left(\frac{\partial\tau}{\partial y}\right)^2\right] dxdy = -0.91811318807 A_1^2. \tag{2.37}$$

Problem 2.2. Use (2.35) and (2.36) to obtain (2.37).

For $h-$extensions with $p = 1$ or $p = 2$ on a uniform mesh, an algebraic convergence is expected with a convergence rate of $-\alpha/2 = -1/3$, whereas a $p-$extension is expected to converge with a convergence rate of $-\alpha = -2/3$. If $p-$extensions on a geometrical graded mesh are performed, an exponential convergence rate is realized for low p-levels, and as the error decreases, the convergence rate may deteriorate to an algebraic rate. In Figure 2.8 we demonstrate numerically the convergence rates that closely match the anticipated theoretical rates for the aforementioned problem with $A_1 = 1$.

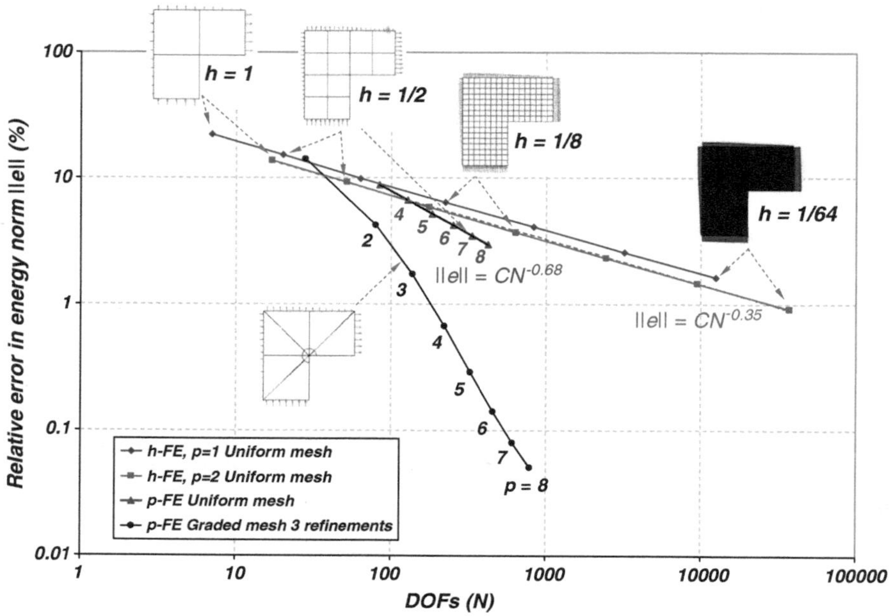

Fig. 2.8 Convergence rates. Heat conduction (Laplace equation).

Elasticity Problem over a L-shaped 2-D Domain

Here we consider the elasticity system over the same L-shaped domain as in the previous example problem rotated by 90° as shown in Figure 6.7 under plane-strain conditions. On the boundaries of the domain, tractions that correspond to the exact "Mode I" stress field in (6.52-6.54) are prescribed, where $A_1 = 1$ and $A_2 = 0$ with $\alpha_1 = 0.5444837368$, and $Q_1 = 0.543075597$ are constants determined so that the solution satisfies the equilibrium equations and the traction-free boundary conditions on the reentrant edges. A Young's modulus $E = 1$ and Poisson ratio $\nu = 0.3$ are chosen so that the exact potential energy for this example problem is given by

$$\Pi_{\text{EX}} = -\frac{1}{2}\mathcal{B}(\boldsymbol{u}, \boldsymbol{u}) = -4.15454423. \tag{2.38}$$

On the 12-quadrilateral uniform mesh ($h = 1/2$) we obtain FE solutions for p ranging from 1 to 8. In addition, we also used h extensions on a sequence of uniformly refined meshes so that h ranges from 1 to $1/16$ and fixed $p = 1$ or 2. A strongly graded mesh having three layers of elements with a geometric progression of 0.15 and p ranging from 1 to 8 was also considered. The four convergence paths are shown in Figure 2.9. The algebraic convergence rate of $\approx \alpha = -0.54$ is observed for $p-$extensions over a uniform mesh, and $\approx \alpha/2 = -0.27$ is observed for $h-$extensions. For $p-$extensions over the geometrically graded mesh, a pre-asymptotic exponential convergence rate is realized.

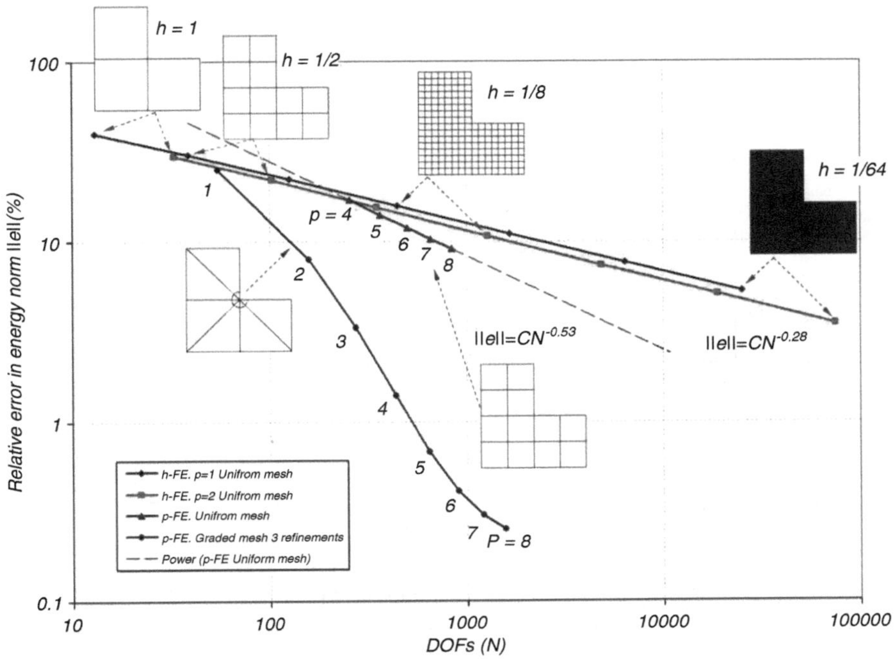

Fig. 2.9 Convergence rates. Elasticity problem over a L-shaped domain.

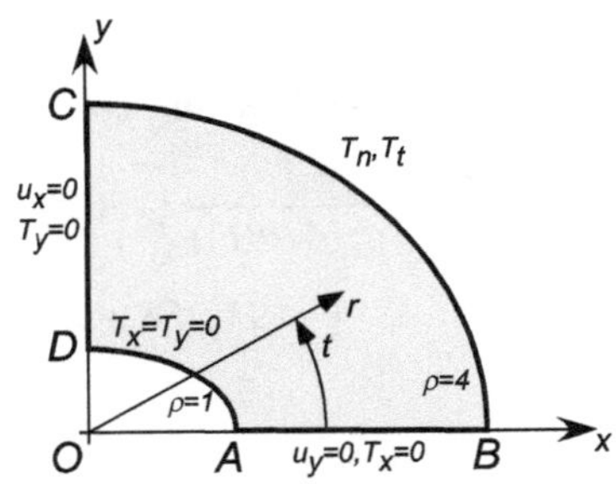

Fig. 2.10 Elliptical domain and boundary conditions for the elasticity problem with an analytical solution.

Elasticity Problem in a 2-D Domain Having an Elliptical/Circular Hole

For problems having an analytical solution (the regularity is as high as required) $p-$Extensions are expected to converge exponentially. To demonstrate this convergence pattern we consider a 2-D plate bounded by two ellipses and the x-y axes shown in Figure 2.10. The inner elliptical boundary AD has major axis $1 + m$ and minor axis $1 - m$, whereas the outer one, BC, has major axis $4 + m/4$ and minor axis $4 - m/4$. One may observe that for $m = 0$ one obtains concentric circles, and as $m \to 1$, the inner ellipse tends to a crack. On the inner elliptical boundary

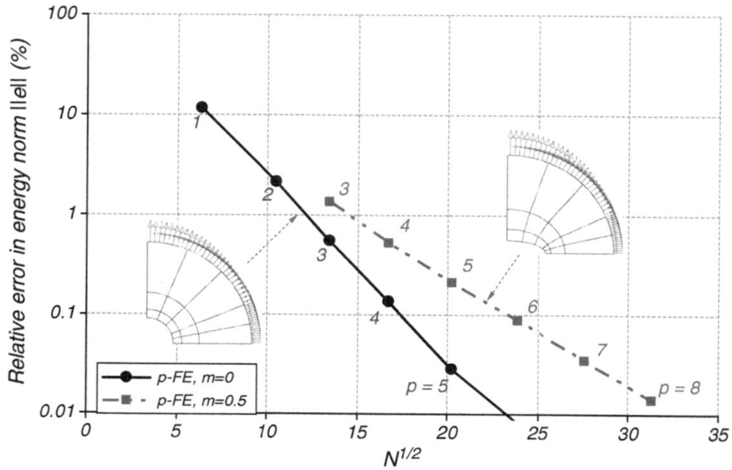

Fig. 2.11 Convergence rates. Elasticity problems with analytical solutions.

AD traction-free conditions are prescribed; on the two straight boundaries AB and DC symmetry boundary conditions are prescribed and on the elliptical boundary BC normal and tangential tractions are prescribed according to (see details in [209])

$$T_t = \frac{1}{g^2(4,\theta(t))}\{\sin 2\theta(t)[480(1+m)(16-m^2)+255(m^2+256)] \\ -4080m\sin 4\theta(t)\}, \tag{2.39}$$

$$T_n = \frac{1}{g^2(4,\theta(t))}\{15(256-2m-m^2)(16-m^2)+\cos 2\theta(t) \\ \left[-193(256+m^2)-512(m^2+1)\right]+7200m\cos^2 2\theta(t)\}, \tag{2.40}$$

where

$$g(4,\theta(t)) = 256-32m\cos 2\theta(t)+m^2, \quad \text{and} \quad \theta(t) = \arctan\left\{\left(\frac{16+m}{16-m}\right)\tan(t)\right\}.$$

For an isotropic material under the assumption of plane stress with $E = 1$ and $\nu = 0.3$, the following values for the potential energy for the two different m's are obtained: $\Pi_{\text{EX}}(m = 0) = -26.33892955$ and $\Pi_{\text{EX}}(m = 0.5) = -27.08611104$. By varying m from 0 to 0.5, we obtain stress concentration factors at point A that range from 3 to 7, respectively, but the solution is still analytic. A p–FE analysis was performed on a mesh of 15 quadrilateral elements by increasing the polynomial order from 1 to 8, where an exponential convergence rate is expected:

$$\|e\|_{\mathcal{E}} \leq \frac{k}{\exp(\gamma N^{\phi})}.$$

Applying the log operator to the convergence estimate (2.28), and since the problem is in Category A, then according to Table 2.2,

$$\log \|\boldsymbol{e}\|_{\mathcal{E}} \leq \log k - \gamma N^{1/2} \log(\exp). \tag{2.41}$$

For large N the inequality becomes "almost equal" and we plot in Figure 2.11 the log of the relative error in energy norm as a function of $N^{1/2}$. As expected, a straight line is obtained both for the domain with $m = 0$ and for $m = 0.5$.

Having described and demonstrated the high convergence rates of the $p-$version of the FEM, we apply it in the next chapters for the computation of the eigenpairs to extract GFIFs and GSIFs.

Chapter 3
Eigenpair Computation for Two-Dimensional Heat Conduction Singularities

Here we present a numerical procedure based on the p-version of the finite element method for computing efficiently and reliably approximations to the eigenpairs associated with two-dimensional heat conduction singular points. The proposed method, called the "modified Steklov method," is general, that is, applicable to singularities associated with corners, nonisotropic multimaterial interfaces, and abrupt changes in boundary conditions. We introduce the method and demonstrate some of its main characteristics on several numerical examples.

3.1 Overview of Methods for Computing Eigenpairs

In Chapter 1 we showed that in the vicinity of a singular point the solution to the heat conduction problem is represented by a series of eigenpairs α_i and $s_i^+(\theta)$, see (1.32). For general singular points, as cracks in anisotropic multimaterial interfaces and V-notches in composite materials, only numerical approximations to the eigenpairs are available.

Most of the research performed in the past concentrated on solutions to problems corresponding to isotropic linear elastic materials or the Laplace equation. In this case, the eigenpairs can be computed analytically [189]. For example, in [150] the eigenpairs for cracks along the interface of two dissimilar isotropic materials are explicitly given, and in [174] explicit eigenfunctions for a crack along the interface of anisotropic materials are provided as well. It is easier to obtain explicit *eigenvalues* than eigenfunctions (see, for example, [197] and [54] for cases of up to three subdomains). Complete exact solutions are restricted to rather simple geometries.

For general singular points, that is, singularities associated with corners, non-isotropic multimaterial interfaces, and abrupt changes in boundary conditions, the eigenfunctions and very often the eigenvalues as well cannot be computed explicitly and have to be computed by numerical methods.

Z. Yosibash, *Singularities in Elliptic Boundary Value Problems and Elasticity and Their Connection with Failure Initiation*, Interdisciplinary Applied Mathematics 37, DOI 10.1007/978-1-4614-1508-4_3, © Springer Science+Business Media, LLC 2012

Several numerical methods have been presented for the computation of the eigenvalues during the last decade, but only scant attention has been given to the computation of the eigenfunctions and their connection to the GFIFs/GSIFs.

- Leguillon and Sanchez-Palencia [109] proposed the matrix and the determinant methods based on the h-version of the FEM to compute the eigenvalues and the corresponding eigenfunctions.
- In [138, 139], Papadakis and Babuška propose a method by which the eigenvalues are determined using the matrix method followed by an iterative approach and the "shooting method," while the eigenfunctions are evaluated using numerical integration based on the Runge-Kutta-Fehlberg method. This enables the calculation of the eigenpairs for anisotropic inhomogeneous materials with several types of boundary conditions. A detailed discussion on the evaluation of the eigenvalues, which involve the solution of a nonlinear eigenvalue problem, is presented. Both methods reduce the 2-D problem to a 1-D ordinary differential equation.
- Barsoum in [21] and [20] proposed the iterative finite element method for the computation of the *first smallest* eigenpair. This method, however, is less efficient, robust, and general than the matrix and the determinant methods.
- Gu and Belytschko [74] proposed a method based on a stress function and the interpolation of displacements (based on the finite element method) for the computation of the eigenvalues, without discussing the computation of the eigenfunctions.
- The Steklov method, first mentioned in [100] and used for vertex singularities of the Laplace operator in [15] and for elastic edge singularities in [6], has been modified by Yosibash and Szabo [200, 210]. This modified method is called the *modified Steklov method*. It makes it possible to compute reliably eigenpairs resulting from singularities due to corners, abrupt changes in material properties, and boundary conditions. The modified Steklov method, used in conjunction with the p-version of the finite element method, is robust and efficient, and the computed eigenpairs are shown to converge strongly and accurately.
- Costabel and Dauge [41] proposed a determinant method, that is somewhat different from that of [109], and more accurate and efficient. This method provides the first $4n$ eigenpairs, where n is the number of materials in the angular direction of the singular point. In [44], a semianalytic method for the computation of eigenpairs for general two-dimensional singular points or 3-D edges in linear elasticity is presented. It is based on the construction of a matrix of a lower dimension, whose determinant is zero for the sought eigen-values. Except during the preliminary step at which the computation of the roots of a certain polynomial are done numerically, the construction of the matrix is analytic. Therefore this method is very accurate and fast.
- Pageau et al. [135] (and the references to their work therein) propose a finite element method based on the h-version FEM and the formulation of Yamada and Okumura for the determination of the eigenpairs. These methods involve the solution of a quadratic eigenproblem. The method is less efficient and accurate when compared to the others.

- Ying and Katz [198, 199] proposed an explicit closed-form expression for the eigenequation in the case of a trimaterial isotropic edge. However, explicit expressions become extremely cumbersome for anisotropic multimaterial interfaces, and even for isotropic multimaterial interfaces with more than three materials.
- Mantic et al. [115] use the "transfer matrix" for the computation of eigenpairs, especially for multimaterial corner problems. They obtained explicit forms of eigenequations for evaluation of the singularity exponent in the case of multimaterial corners.

3.2 Formulation of the Modified Steklov Eigenproblem

We address the heat-conduction problem in a two-dimensional domain in the vicinity of a singular point P as shown in Figure 3.1. Zooming in on the vicinity of P, we let the heat-conduction coefficients be θ-dependent, which enables us to consider multimaterial interfaces in the close vicinity of P. Although $k_{ij}(\theta)$ represent the heat conduction in the x_1, x_2 directions, they may change in the θ direction. We may consider a zoomed subdomain close to P because the eigenpairs depend only on the geometry, material properties, and boundary conditions in P's vicinity. Therefore, we define the subdomain Ω_R ("modified Steklov domain") as follows (see Figure 3.1):

$$\Omega_R = \{\boldsymbol{x} \mid \boldsymbol{x} \in \Omega \cap \{R^* \leq r \leq R\}\}. \tag{3.1}$$

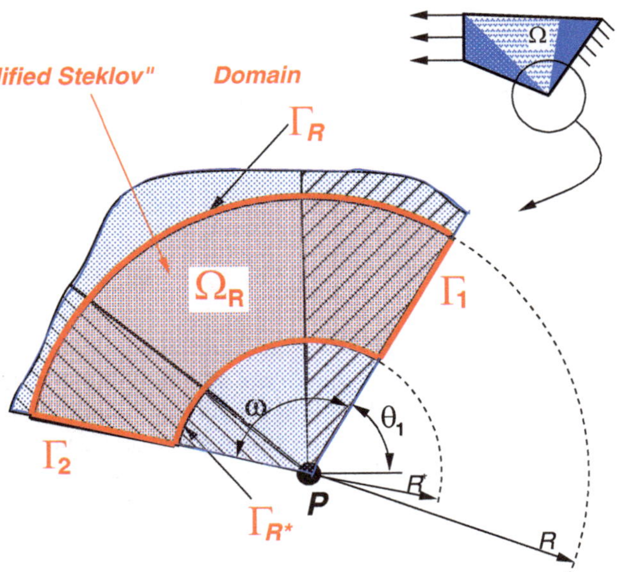

Fig. 3.1 The "modified Steklov" domain and notation.

One has to solve over Ω_R the equation

$$\partial_\beta \left(k_{\beta\gamma}(\theta)\partial_\gamma \tau\right) = 0, \quad \text{in } \Omega_R \qquad \beta, \gamma = 1, 2, \tag{3.2}$$

subject to either homogeneous Dirichlet or Neumann boundary conditions on Γ_1 and Γ_2 (or a combination of these). We first consider homogeneous Neumann boundary conditions on Γ_1 and Γ_2 for simplicity of presentation, assuming that in the neighborhood of P, the right-hand-side heat source vanishes ($Q = 0$ in (1.34))

$$\partial_\beta \left(k_{\beta\gamma}(\theta)\partial_\gamma \tau\right) = 0 \quad \text{in } \Omega_R, \tag{3.3}$$

$$k_{\beta\gamma}(\theta)\partial_\beta \tau \, n_\gamma = 0 \quad \text{on } \Gamma_1 \cup \Gamma_2, \quad \text{i.e. } \theta = \theta_1, (\theta_1 + \omega), \tag{3.4}$$

where $\boldsymbol{n}$ is the outward normal vector. This problem is not well posed, because no boundary conditions are specified (yet) on the artificial circular boundaries Γ_{R^*} and Γ_R. To create these boundary conditions, let us assume that α and $s(\theta)$ are an eigenpair satisfying (3.3) and the boundary conditions (3.4) on Γ_1 and Γ_2. Thus $\tau = r^\alpha s(\theta)$ is a solution, and on the circular boundary Γ_R it satisfies

$$\frac{\partial \tau}{\partial n} = \frac{\partial \tau}{\partial r} = \alpha r^{\alpha-1} s(\theta)|_{r=R} = \frac{\alpha}{R}\tau \quad \text{on } \Gamma_R. \tag{3.5}$$

Similarly, we have on Γ_{R^*},

$$\frac{\partial \tau}{\partial n} = -\frac{\partial \tau}{\partial r} = -\alpha r^{\alpha-1} s(\theta)|_{r=R^*} = -\frac{\alpha}{R^*}\tau \quad \text{on } \Gamma_{R^*}. \tag{3.6}$$

We may now summarize the eigenproblem to be solved:

$$\partial_\beta \left(k_{\beta\gamma}(\theta)\partial_\gamma \tau\right) = 0 \quad \text{in } \Omega_R, \tag{3.7}$$

$$k_{\beta\gamma}(\theta)\partial_\beta \tau n_\gamma = 0 \quad \theta = \theta_1, (\theta_1 + \omega), \tag{3.8}$$

$$\frac{\partial \tau}{\partial r} = \frac{\alpha}{R}\tau \quad \text{on } \Gamma_R, \quad \text{i.e., } r = R, \tag{3.9}$$

$$\frac{\partial \tau}{\partial r} = \frac{\alpha}{R^*}\tau \quad \text{on } \Gamma_{R^*}, \quad \text{i.e., } r = R^*, \tag{3.10}$$

Equations (3.7)–(3.10) constitute the "generalized mixed problem of Steklov type," first considered by Steklov for the Laplace equation in 1902 [169]. Note that one has to find an eigenvalue, appearing here in the boundary condition, and an associated eigenfunction. There exists a complete discrete set of eigen-pairs to the above problem that is sought, and for the Laplace equation these are real. In Appendix D we prove that the eigen-pairs of the general heat conduction problem, with constant coefficients and homogeneous boundary conditions in the vicinity of the singular point, are real.

An analytic solution of the "classical" Steklov eigenproblem, (3.7)–(3.10), cannot be obtained in general. Thus the weak formulation has to be used, based on which the finite element method will be applied. We follow the same steps as in Chapter 2, first multiplying (3.7) by a test function χ and integrating over the domain Ω_R, then applying Green's theorem to obtain (2.7), where $Q = 0$. Because we consider homogeneous Neumann boundary condition on Γ_1 and Γ_2, all that remains from the right-hand side of (2.7) is

$$\int_{\Gamma_R} \chi\, k_{\beta\gamma}\, \partial_\beta \tau\, n_\gamma\, d\Gamma + \int_{\Gamma_{R^*}} \chi\, k_{\beta\gamma}\, \partial_\beta \tau\, n_\gamma\, d\Gamma. \tag{3.11}$$

We concentrate our discussion on the first term in (3.11), because the second term is treated in exactly the same manner only that R is replaced by R^* and the direction of integration is reversed. Note that $(n_1, n_2) \equiv (\cos\theta, \sin\theta)$, so that the first term in (3.11) becomes

$$\int_{\Gamma_R} [(k_{11}\partial_1\tau + k_{12}\partial_2\tau)\cos\theta + (k_{21}\partial_1\tau + k_{22}\partial_2\tau)\sin\theta]\, \chi d\Gamma.$$

Furthermore, expressing the derivatives with respect to x_1, x_2 by derivatives with respect to r, θ one obtains

$$\int_{\Gamma_R} \left\{ \left[k_{11}\left(\cos\theta \frac{\partial\tau}{\partial r} - \frac{1}{r}\sin\theta \frac{\partial\tau}{\partial\theta} \right) + k_{12}\left(\sin\theta \frac{\partial\tau}{\partial r} + \frac{1}{r}\cos\theta \frac{\partial\tau}{\partial\theta} \right) \right] \cos\theta \right\} \chi d\Gamma$$
$$+ \int_{\Gamma_R} \left\{ \left[k_{21}\left(\cos\theta \frac{\partial\tau}{\partial r} - \frac{1}{r}\sin\theta \frac{\partial\tau}{\partial\theta} \right) + k_{22}\left(\sin\theta \frac{\partial\tau}{\partial r} + \frac{1}{r}\cos\theta \frac{\partial\tau}{\partial\theta} \right) \right] \sin\theta \right\} \chi d\Gamma.$$

Note that $k_{21} = k_{12}$ and that $d\Gamma = R d\theta$ on Γ_R and use the Steklov-type boundary condition (3.9) to obtain for the first term on the right-hand side

$$\alpha \int_{\theta_1}^{\theta_1+\omega} (k_{11}\cos^2\theta + k_{12}\sin 2\theta + k_{22}\sin^2\theta)\tau\chi d\theta$$
$$+ \int_{\theta_1}^{\theta_1+\omega} [(k_{22} - k_{11})\sin\theta\cos\theta + k_{12}\cos 2\theta] \frac{\partial\tau}{\partial\theta} \chi d\theta. \tag{3.12}$$

The second term in (3.11), corresponding to the boundary $r = R^*$, is treated similarly. Substituting (3.12) and the term corresponding to the boundary $r = R^*$ in (2.7), then collecting terms that are α independent of the left-hand side, we obtain the "modified Steklov weak eigenproblem" for homogeneous Neumann BCs corresponding to (3.7)–(3.10):

$$\text{Seek } \alpha \in \mathbb{C}, \quad 0 \neq \tau \in \mathcal{E}(\Omega_R) \text{ such that} \tag{3.13}$$
$$\mathcal{B}(\tau,\chi) - [\mathcal{N}_{R^*}(\tau,\chi) + \mathcal{N}_R(\tau,\chi)] = \alpha\, [\mathcal{M}_{R^*}(\tau,\chi) + \mathcal{M}_R(\tau,\chi)], \ \forall \chi \in \mathcal{E}(\Omega_R),$$

where

$$\mathcal{M}_R(\tau,\chi) \stackrel{\text{def}}{=} \int_{\theta_1}^{\theta_1+\omega} (k_{11}(\theta)\cos^2\theta + k_{12}(\theta)\sin 2\theta + k_{22}(\theta)\sin^2\theta)\,[\tau\chi]_R\,d\theta, \tag{3.14}$$

$$\mathcal{N}_R(\tau,\chi) \stackrel{\text{def}}{=} \int_{\theta_1}^{\theta_1+\omega} [(k_{22}(\theta) - k_{11}(\theta))\sin\theta\cos\theta + k_{12}(\theta)\cos 2\theta]\left[\frac{\partial\tau}{\partial\theta}\chi\right]_R d\theta, \tag{3.15}$$

and $\mathcal{M}_{R^*}$, $\mathcal{N}_{R^*}$ are exactly the same as (3.14), (3.15) only that the expressions are to be evaluated at R^*.

There are important and interesting properties that arise when one examins the weak eigenproblem (3.14).

Remark 3.1. *For isotropic homogeneous materials $k_{11} = k_{22} = k$; $k_{12} = 0$, and thus $\mathcal{N}_R = \mathcal{N}_{R^*} \equiv 0$ and (3.15) involves only symmetric terms, i.e., eigenpairs are real. This case is associated with the Laplace equation.*

Remark 3.2. *For the general heat-conduction equation the bilinear forms $\mathcal{N}_R$ and $\mathcal{N}_{R^*}$ are asymmetric with respect to τ and χ. This may give rise to complex eigenvalues and eigenvectors for anisotropic domains. However, the eigenpairs are always real unless an internal singular point within a multimaterial interface with anisotropic materials is considered (see [115]). In the numerical algorithm we always seek $\alpha \in \mathbb{C}$.*

Remark 3.3. *The weak form has some unusual properties:*

(a) The expressions $\mathcal{N}_R$ (respectively $\mathcal{N}_{R^*}$) involve the derivative of τ along a curve. However, if one seeks $\tau \in \mathcal{E}(\Omega_R)$, then the trace of its derivatives on Γ_R (resp. Γ_{R^*}), which is nothing more than $\frac{\partial\tau}{\partial\theta}$, is not well defined on its boundary since $\left(\frac{\partial\tau}{\partial\theta}\right) \notin L^2(\Gamma_R)$ (resp. $L^2(\Gamma_R^*)$). A heuristic way to bypass this difficulty is to use the following argument. We are really interested in $\tau = r^\alpha s(\theta)$ such that $s(\theta)$ is piecewise analytic, so that $\frac{\partial\tau}{\partial\theta} \in L^2(\Gamma_R)$ (resp. $L^2(\Gamma_R^*)$). This means that we restrict the space in which τ lies to be $\mathcal{E}(\Omega_R) \cap L^2(\Gamma_R) \cap L^2(\Gamma_R^*)$. In the finite element method we design the mesh such that the element boundaries coincide with Γ_R and Γ_{R^*}. Thus we automatically satisfy the new restriction on the space by choosing only polynomials along Γ_R and Γ_{R^*}.

(b) It is necessary to show that any eigenpair that is the solution of the weak form, is of the form $r^\alpha s(\theta)$, and is also the eigenpair of the Steklov strong formulation.

(c) The weak form (3.13) is not a variational principle in the sense that a minimum principle is equivalent to it (this is because the bilinear forms $\mathcal{N}_R$ and $\mathcal{N}_{R^*}$ are asymmetric). Thus the problem formulated in the weak form loses its self-adjoint property, and the "minimax principle" does not hold. Therefore, nonmonotonic convergence of the approximated eigenpairs is expected even when the sequence of finite element spaces S_i ($i = 1, 2, \ldots$) is hierarchic,

that is $S_i \subset S_{i+1}$. Indeed, as we show by numerical examples, the approximated eigenvalues are sometimes smaller than the exact ones and sometimes larger.

It is natural to expect that by enlarging the finite element space, the eigenpairs will approach the exact values. To prove convergence, it is necessary to prove consistency and stability. Consistency is ensured, since the trial space is dense in H^1, and the approximated eigenfunction converges to any function in H^1 in the H^1 norm. The stability proof, which requires that the numeric operators be uniformly invertible, is more difficult. However, the general proof of convergence rate of non-self-adjoint problems given in [8] is useful for our purposes because it establishes that the Steklov problem does converge.

Remark 3.4. *The weak form* (3.13) *does not exclude the existence of negative eigenpairs. This is because solutions of the form $r^{-\alpha}s^{-}(\theta)$ do belong to the space $\mathcal{E}(\Omega_R)$. Therefore both the positive and negative eigenpairs are obtainable. It is interesting to note that for the Laplace equation if α is an eigenvalue, with a corresponding eigenfunction called $s^{+}(\theta)$, then $-\alpha$ is also an eigenvalue with an associated eigenfunction $s^{-}(\theta) \equiv s^{+}(\theta)$. For a general heat conduction problem, $-\alpha$ is an eigenvalue, but the associated eigenfunction $s^{-}(\theta)$ is different from $s^{+}(\theta)$. The negative eigenpairs are usually used for the computation of the GFIFs by a powerful method called the "dual singular function method" (see for example [12, 27]). Thus, the capability of computing $s^{-}(\theta)$ by the modified Steklov method may be appreciated for post-processing operation for extracting GFIFs.*

Remark 3.5. *The domain Ω_R does not include singular points, hence no special refinements of the finite element mesh is required. Furthermore, Ω_R is small in size so that very few finite elements are needed in using the FEM.*

A similar method to the one presented here has been addressed in [15, 100]. However, these approaches have not excluded the vertex P from the domain of interest. Thus the performance of the previous methods, called the "classical" Steklov weak eigenproblem (this is the historical reason that the present method is called a "modified" Steklov method), is considerably inferior to the one presented here. For a comparison of the methods the reader is referred to [210].

3.2.1 Homogeneous Dirichlet Boundary Conditions

Homogeneous Dirichlet boundary conditions may be applied on Γ_1 and Γ_2 instead of (3.4):

$$\tau(\theta = \theta_1) = 0 \quad \text{and} \quad \tau(\theta = \theta_1 + \omega) = 0. \tag{3.16}$$

In this case, the only difference in the weak modified Steklov eigenproblem (3.13) is the restriction of the space $\mathcal{E}(\Omega_R)$ to

$$\mathcal{E}_o(\Omega_R) = \{\tau | \tau \in \mathcal{E}(\Omega_R), \text{ and } \tau(\theta_1) = \tau(\theta_1 + \omega) = 0\}, \tag{3.17}$$

and the modified Steklov eigen-problem is identical to (3.13) with $\mathcal{E}(\Omega_R)$ replaced by $\mathcal{E}_o(\Omega_R)$.

3.2.2 *The Modified Steklov Eigen-problem for the Laplace Equation with Homogeneous Neumann BCs*

The modified Stekov weak eigen-problem is considerably simplified in case $k_{11}(\theta) = k_{22}(\theta) = \text{const} \stackrel{\text{def}}{=} k$ and $k_{12} = 0$. When these conditions are satisfied, we say that the material is isotropic, and the strong formulation is simply the Laplace equation. In this case the expressions for $\mathcal{N}_R$ and $\mathcal{N}_{R^*}$ are identically zero, and the "modified Steklov weak eigenproblem" becomes

$$\text{Seek } \alpha \in \mathbb{R}, \quad 0 \neq \tau \in H^1(\Omega_R) \text{ such that}$$
$$\mathcal{B}(\tau, \chi) = \alpha \left[\mathcal{M}_{R^*}(\tau, \chi) + \mathcal{M}_R(\tau, \chi)\right], \; \forall \chi \in H^1(\Omega_R), \tag{3.18}$$

where

$$\mathcal{B}(\tau, \chi) = \int_R^{R^*} \int_{\theta_1}^{\theta_1 + \omega} \partial_\beta \chi \, \partial_\beta \tau \, r dr d\theta, \tag{3.19}$$

$$\mathcal{M}_R(\tau, \chi) \stackrel{\text{def}}{=} \int_{\theta_1}^{\theta_1 + \omega} [\tau \chi]_R \, d\theta, \tag{3.20}$$

and $\mathcal{M}_{R^*}$, is exactly the same as (3.20), only that the expressions is to be evaluated at R^*. One may observe that all forms in (3.18) are symmetric; thus only real eigenpairs are obtained.

3.3 Numerical Solution of the Modified Steklov Weak Eigenproblem by p-FEMs

The weak eigenproblem (3.13) may be formulated in matrix representation and solved by means of the p-FEMs. In Ω_R the exact solution is analytic, thus p-extension is preferable because of its exponential rate of convergence. We replace the infinite space $\mathcal{E}(\Omega)$ by a finite subspace of dimension DOF, consisting of piecewise polynomials (the finite element space). The domain Ω_R is divided into

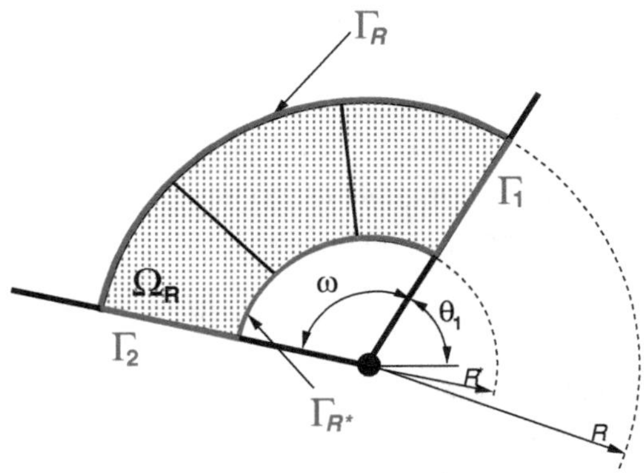

Fig. 3.2 Finite element mesh over the artificial Ω_R subdomain.

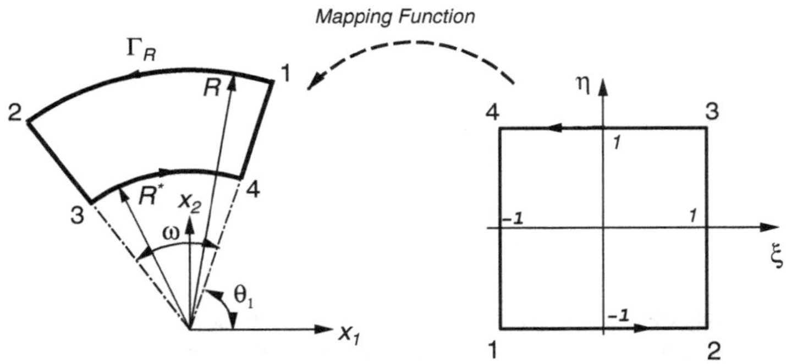

Fig. 3.3 The mapping of the standard element to the "real" element.

finite elements through a meshing process in the θ direction only. Let us divide the domain shown in Figure 3.1 into three finite elements as illustrated in Figure 3.2. These elements are mapped from a standard element in the ξ, η plane such that $-1 \le \xi \le 1$, $-1 \le \eta \le 1$, by an appropriate mapping function $x_1 = x_1(\xi, \eta)$, $x_2 = x_2(\xi, \eta)$. Let us consider a single element that is mapped from the standard element as shown in Figure 3.3. The mapping is given by the following blending functions:

$$x_1(\xi, \eta) = \cos\left(\frac{\omega}{2}\xi + \frac{\omega + 2\theta_1}{2}\right)\left[R\frac{1-\eta}{2} + R^*\frac{1+\eta}{2}\right], \tag{3.21}$$

$$x_2(\xi, \eta) = \sin\left(\frac{\omega}{2}\xi + \frac{\omega + 2\theta_1}{2}\right)\left[R\frac{1-\eta}{2} + R^*\frac{1+\eta}{2}\right]. \tag{3.22}$$

The polynomial basis and trial functions are defined on the standard element. Let the temperature function τ and the trial function χ be expressed in terms of the basis functions $\Phi_i(\xi, \eta)$ given in Chapter 2:

$$\tau(\xi,\eta) = \sum_{i=1}^{N} b_i \Phi_i(\xi,\eta) = \mathbf{\Phi}^T \mathbf{b}, \qquad \chi(\xi,\eta) = \sum_{i=1}^{N} c_i \Phi_i(\xi,\eta) = \mathbf{c}^T \mathbf{\Phi}, \quad (3.23)$$

where b_i and c_i are the amplitudes of the basis functions, and Φ_i are the shape functions. Substituting (3.23) into the expression for $\mathcal{B}(\tau,\chi)$ given by (2.14), one obtains the unconstrained stiffness matrix $[K]$ associated with $\mathcal{B}$, given by (2.21):

$$\mathcal{B}(\tau,\chi) = \mathbf{c}^T [K] \mathbf{b}. \quad (3.24)$$

Let us consider now the computation of the matrices associated with $\mathcal{M}_R$ and $\mathcal{N}_R$ presented in (3.14)–(3.15):

$$\mathcal{M}_R(\tau,\chi) = \mathbf{c}^T [M_R] \mathbf{b}, \qquad \mathcal{N}_R(\tau,\chi) = \mathbf{c}^T [N_R] \mathbf{b}. \quad (3.25)$$

One may observe, that many of the shape functions Φ_i are zero on the side 1-2 of the standard element, so that many of the entries of the matrices $[M_R]$ and $[N_R]$ are zero. On this side $-1 \le \xi \le 1$, whereas $\eta = -1$, and only the angle θ is a function of ξ as follows:

$$\theta = \frac{\omega}{2}\xi + \frac{\omega + 2\theta_1}{2}.$$

The only entries of the matrices that are nonzero are associated with the shape functions Φ_1, Φ_2, Φ_5, Φ_9, and so on. We compute only those that are nonzero (i.e. $i, j = 1, 2, 5, 9, \ldots$):

$$(M_R)_{ij} = \frac{\omega}{2} \int_{-1}^{1} \Big(k_{11}(\theta)\cos^2\theta + k_{12}(\theta)\sin 2\theta \quad (3.26)$$
$$+ k_{22}(\theta)\sin^2\theta \Big) \Phi_i(\xi,\eta=-1)\Phi_j(\xi,\eta=-1)\, d\xi,$$

$$(N_R)_{ij} = \int_{-1}^{1} [(k_{22}(\theta) - k_{11}(\theta))\sin\theta\cos\theta + k_{12}(\theta)\cos 2\theta] \left[\frac{\partial\Phi_i}{\partial\xi}\Phi_j\right]_{\eta=-1} d\xi. \quad (3.27)$$

Note that the entries of the matrices $[N_{R*}]$ and $[M_{R*}]$ have the same values as those of $[N_R]$ and $[M_R]$, but of opposite sign. This is because the shape functions on Γ_R and Γ_{R*} are the same, and so is the mapping to the standard plane, except that the integration is from 1 to -1. The entries of $[N_{R*}]$ and $[M_{R*}]$ that are not zero, are associated with $i, j = 3, 4, 7, 11, \ldots$.

Usually more than one element is required over Ω_R, each having element matrices $[K]$, $[M_R]$, $[M_{R*}]$, $[N_R]$, and $[N_{R*}]$. These are assembled (all element matrices $[K]$ are assembled to form the global $[K]$ matrix, etc). If we denote by

$\mathbf{b}_{tot}$ the set of all coefficients, these associated with Γ_R we denote by $\mathbf{b}_R$, and those associated with Γ_{R*} we denote by $\mathbf{b}_{R*}$, the eigenproblem to be solved is

$$[K]\boldsymbol{b}_{tot} - ([N_R]\boldsymbol{b}_R + [N_{R*}]\boldsymbol{b}_{R*}) = \alpha\,([M_R]\boldsymbol{b}_R + [M_{R*}]\boldsymbol{b}_{R*})\,. \tag{3.28}$$

We assemble the left-hand part of (3.28), denote the whole matrix (which becomes nonsymmetric) by $[\tilde{K}]$, and assemble the right-hand part which is denoted by $[M_{R\cup R*}]$:

$$[\tilde{K}]\boldsymbol{b}_{tot} = \alpha[M_{R\cup R*}]\boldsymbol{b}_{R\cup R*}, \tag{3.29}$$

where $\boldsymbol{b}_{R\cup R*} = \boldsymbol{b}_R \cup \boldsymbol{b}_{R*}$. The vector that represents the total number of nodal values in Ω_R may be divided into two vectors such that one contains the coefficients $\boldsymbol{b}_{R\cup R*}$, and the other contains the remaining coefficients: $\boldsymbol{b}_{tot}^T = \{\boldsymbol{b}_{R\cup R*}^T, \boldsymbol{b}_{in}^T\}$. By partitioning $[\tilde{K}]$, we can write the eigenproblem (3.29) in the form

$$\begin{bmatrix} [K_{R\cup R*}] - ([N_R] + [N_{R*}]) & [K_{R-in}] \\ [K_{in-R}] & [K_{in}] \end{bmatrix} \begin{Bmatrix} \boldsymbol{b}_{R\cup R*} \\ \boldsymbol{b}_{in} \end{Bmatrix} = \alpha \begin{bmatrix} [M_{R\cup R*}] & [0] \\ [0] & [0] \end{bmatrix} \begin{Bmatrix} \boldsymbol{b}_{R\cup R*} \\ \boldsymbol{b}_{in} \end{Bmatrix}. \tag{3.30}$$

Equation (3.30) can be used to eliminate $\boldsymbol{b}_{in}$ by static condensation as follows. The system of equations (3.30) can be partitioned into two systems:

$$\{[K_{R\cup R*}] - ([N_R] + [N_{R*}])\}\,\boldsymbol{b}_{R\cup R*} + [K_{R-in}]\boldsymbol{b}_{in} = \alpha[M_{R\cup R*}]\boldsymbol{b}_{R\cup R*}, \tag{3.31}$$

$$[K_{in-R}]\boldsymbol{b}_{R\cup R*} + [K_{in}]\boldsymbol{b}_{in} = 0. \tag{3.32}$$

It is possible to eliminate the vector $\mathbf{b}_{in}$ by expressing it in terms of $\boldsymbol{b}_{R\cup R*}$, obtaining from (3.32)

$$\boldsymbol{b}_{in} = -[K_{in}]^{-1}[K_{in-R}]\boldsymbol{b}_{R\cup R*}. \tag{3.33}$$

Substituting $\boldsymbol{b}_{in}$ from (3.33) in (3.31), one obtains the reduced eigenproblem

$$[K_S]\boldsymbol{b}_{R\cup R*} = \alpha[M_{R\cup R*}]\boldsymbol{b}_{R\cup R*}, \tag{3.34}$$

where

$$[K_S] = ([K_{R\cup R*}] - ([N_R] + [N_{R*}])) - [K_{R-in}][K_{in}]^{-1}[K_{in-R}]. \tag{3.35}$$

For the solution of the eigenproblem (3.34), it is important to note that $[K_S]$ is, in general, a full matrix. However, since the order of the matrices is relatively small, the solution (using Cholesky factorization to compute $[K_{in}]^{-1}$) is inexpensive.

Using any available routine for a generalized eigenvalue problem,[1] we may solve (3.34), obtaining the eigenvalues α_i and their corresponding eigenvectors.

The entries of the above matrices are computed by numerical integration based on 14 point Gauss quadrature. Numerical experiments have shown that satisfactory accuracy is achieved with 14 integration points. The difference in the numerically approximated eigenvalues, when using 13 and 14 Gauss integration points, is significant only after the 14th digit. The computations were done with 16 significant digits.

3.4 Examples on the Performance of the Modified Steklov Method

In this section we provide a detailed numerical example for computing eigenpairs using the modified Steklov method, followed by several other examples consisting of singularities in "anisotropic" domains, multimaterial interfaces, and internal points at the intersection of different materials.

3.4.1 A Detailed Simple Example

We provide a simple example problem solved in detail to illustrate the solution procedure. Let us consider a domain with a $\pi/2$ corner as shown in Figure 3.4. The Laplace equation is considered over the domain (heat-transfer problem in an isotropic domain), with Neumann boundary conditions along the x_1- and x_2-axes. We compute an approximation to the smallest eigenpair by the modified Steklov weak eigenformulation, using the FEM with the trunk space and polynomial degree $p = 2$ (8 DOFs).

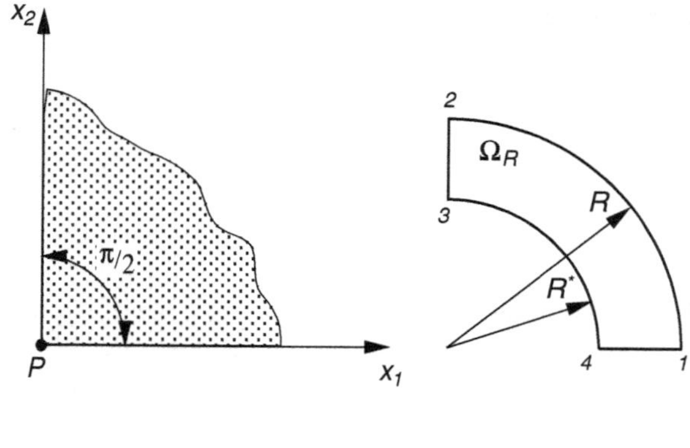

Fig. 3.4 Example problem 3.4.1: $\pi/2$ corner, and the corresponding Ω_R subdomain.

[1] In our studies the LAPACK library was used [4].

Computation of the stiffness matrix $[K]$:

We use one finite element as shown in Figure 3.4. The blended mapping function from the standard element to the element of interest is

$$x_1(\xi,\eta) = \cos\left(\frac{\pi}{4}\xi + \frac{\pi}{4}\right)\left[\frac{1-\eta}{2}R + \frac{1+\eta}{2}R^*\right], \tag{3.36}$$

$$x_2(\xi,\eta) = \sin\left(\frac{\pi}{4}\xi + \frac{\pi}{4}\right)\left[\frac{1-\eta}{2}R + \frac{1+\eta}{2}R^*\right], \tag{3.37}$$

so that

$$[J] = \begin{bmatrix} \frac{-\pi}{4}\sin\left(\frac{\pi}{4}\xi + \frac{\pi}{4}\right)\left(\frac{1-\eta}{2}R + \frac{1+\eta}{2}R^*\right) & \frac{\pi}{4}\cos\left(\frac{\pi}{4}\xi + \frac{\pi}{4}\right)\left(\frac{1-\eta}{2}R + \frac{1+\eta}{2}R^*\right) \\ \frac{1}{2}\cos\left(\frac{\pi}{4}\xi + \frac{\pi}{4}\right)(-R + R^*) & \frac{1}{2}\sin\left(\frac{\pi}{4}\xi + \frac{\pi}{4}\right)(-R + R^*) \end{bmatrix},$$

$$|J| = \frac{\pi}{8}(R - R^*)\left[\frac{1-\eta}{2}R + \frac{1+\eta}{2}R^*\right], \tag{3.38}$$

and

$$[J]^{-1} = \frac{1}{|J|}\begin{bmatrix} \frac{1}{2}\sin\left(\frac{\pi}{4}\xi + \frac{\pi}{4}\right)(-R + R^*) & \frac{-\pi}{4}\cos\left(\frac{\pi}{4}\xi + \frac{\pi}{4}\right)\left(\frac{1-\eta}{2}R + \frac{1+\eta}{2}R^*\right) \\ \frac{-1}{2}\cos\left(\frac{\pi}{4}\xi + \frac{\pi}{4}\right)(-R + R^*) & \frac{-\pi}{4}\sin\left(\frac{\pi}{4}\xi + \frac{\pi}{4}\right)\left(\frac{1-\eta}{2}R + \frac{1+\eta}{2}R^*\right) \end{bmatrix}. \tag{3.39}$$

For the Laplace equation $[k]$ is reduced to the identity matrix. Substituting (3.39) into (2.21), the i, j entry in the stiffness matrix becomes

$$\begin{aligned} K_{ij} &= \iint_{-1}^{1} \left(\frac{\partial\Phi_i}{\partial\xi}, \frac{\partial\Phi_i}{\partial\eta}\right)^T \frac{1}{|J|} \begin{bmatrix} \frac{1}{4}(R^* - R)^2 & 0 \\ 0 & \frac{\pi^2}{16}\left(\frac{1-\eta}{2}R + \frac{1+\eta}{2}R^*\right)^2 \end{bmatrix} \begin{Bmatrix} \frac{\partial\Phi_j}{\partial\xi} \\ \frac{\partial\Phi_j}{\partial\eta} \end{Bmatrix} d\xi d\eta \\ &= \iint_{-1}^{1} \frac{\partial\Phi_i}{\partial\xi}\frac{\partial\Phi_j}{\partial\xi}\frac{(R^* - R)^2}{4|J|} + \frac{\partial\Phi_i}{\partial\eta}\frac{\partial\Phi_j}{\partial\eta}\frac{\pi^2}{16|J|} \\ &\quad \left(\frac{1-\eta}{2}R + \frac{1+\eta}{2}R^*\right)^2 d\xi d\eta. \end{aligned} \tag{3.40}$$

We choose in all cases $R = 1$, and in this particular case $R^* = 0.9$ (the eigenproblem is virtually independent of R^* for $R^* > 1/2$). Let us explicitly compute the value of K_{11}:

$$K_{11} = \iint_{-1}^{1} \frac{1}{|J|} \left\{ \left(\frac{\partial \Phi_1}{\partial \xi} \right)^2 (0.05)^2 \right.$$
$$\left. + \left(\frac{\partial \Phi_1}{\partial \eta} \right)^2 \frac{\pi^2}{16} \left(\frac{1-\eta}{2} + \frac{1+\eta}{2} 0.9 \right)^2 \right\} d\xi d\eta$$
$$= \frac{160}{\pi} \iint_{-1}^{1} \frac{1}{1.9 - 0.1\eta} \left\{ \left(\frac{1-\eta}{4} \right)^2 (0.05)^2 \right.$$
$$\left. + \left(\frac{1-\xi}{4} \right)^2 \frac{\pi^2}{16} \left(\frac{1-\eta}{2} + 0.9 \frac{1+\eta}{2} \right)^2 \right\} d\xi d\eta$$
$$= 4.99596. \tag{3.41}$$

Similarly, we may compute all entries of $[K]$:

$$[K] = \begin{bmatrix} 4.9960 & 2.4653 & -2.4983 & -4.9630 & -3.0461 & 0.1204 & 3.0461 & 0.2002 \\ & 4.9960 & -4.9630 & -2.4983 & -3.0461 & 0.2002 & 3.0461 & 0.1204 \\ & & 4.9971 & 2.4641 & 3.0461 & -0.2276 & -3.0461 & -0.0930 \\ & & & 4.9971 & 3.0461 & -0.0930 & -3.0461 & -0.2276 \\ & & & & 3.0281 & -0.1309 & -2.9622 & -0.1309 \\ & & & & & 9.9618 & 0.1309 & 4.9608 \\ & & & & & & 3.0304 & 0.1309 \\ & & & & & & & 9.9618 \end{bmatrix}. \tag{3.42}$$

Computation of the Matrices $[M_R]$ and $[M_{R^*}]$

For the Laplace equation the entries of the matrix $[M_R]$ can be computed based on the simplified equation (3.26), with $k_{11} = k_{22} = 1$, and $k_{12} = 0$:

$$(M_R)_{ij} = \frac{\pi}{4} \int_{-1}^{1} \Phi_i(\xi, \eta = -1) \Phi_j(\xi, \eta = -1)\, d\xi. \tag{3.43}$$

Substituting the functions Φ_1, Φ_2, and Φ_5 (all others are zero on the side 1-2 on the quadrilateral element), we obtain

$$[M_R]\begin{Bmatrix} b_1 \\ b_2 \\ b_5 \end{Bmatrix} = \frac{\pi}{4}\begin{bmatrix} \frac{2}{3} & \frac{1}{3} & \frac{-1}{\sqrt{6}} \\ & \frac{2}{3} & \frac{-1}{\sqrt{6}} \\ & & \frac{2}{5} \end{bmatrix}\begin{Bmatrix} b_1 \\ b_2 \\ b_5 \end{Bmatrix}. \tag{3.44}$$

The matrix $[M_{R^*}]$ is similar to $[M_R]$, with oposite signs, and it multiplies a vector of different coefficients:

$$[M_{R^*}]\begin{Bmatrix} b_3 \\ b_4 \\ b_7 \end{Bmatrix} = \frac{\pi}{4}\begin{bmatrix} \frac{-2}{3} & \frac{-1}{3} & \frac{1}{\sqrt{6}} \\ & \frac{-2}{3} & \frac{1}{\sqrt{6}} \\ & & \frac{-2}{5} \end{bmatrix}\begin{Bmatrix} b_3 \\ b_4 \\ b_7 \end{Bmatrix}. \tag{3.45}$$

We may now assemble the right-hand side of (3.29):

$$[M_R]\begin{Bmatrix} b_1 \\ b_2 \\ b_5 \end{Bmatrix} + [M_{R^*}]\begin{Bmatrix} b_3 \\ b_4 \\ b_7 \end{Bmatrix} = \frac{\pi}{4}\begin{bmatrix} \frac{2}{3} & \frac{1}{3} & \frac{-1}{\sqrt{6}} & 0 & 0 & 0 \\ & \frac{2}{3} & \frac{-1}{\sqrt{6}} & 0 & 0 & 0 \\ & & \frac{2}{5} & 0 & 0 & 0 \\ & & & \frac{-2}{3} & \frac{-1}{3} & \frac{1}{\sqrt{6}} \\ & & & & \frac{-2}{3} & \frac{1}{\sqrt{6}} \\ & & & & & \frac{-2}{5} \end{bmatrix}\begin{Bmatrix} b_1 \\ b_2 \\ b_5 \\ b_3 \\ b_4 \\ b_7 \end{Bmatrix}. \tag{3.46}$$

One needs to rearrange rows and collumns in the above matrix to bring it to the form in (3.30):

$$\begin{bmatrix} \underbrace{\frac{\pi}{4}\begin{bmatrix} \frac{2}{3} & \frac{1}{3} & 0 & 0 & \frac{-1}{\sqrt{6}} & 0 \\ & \frac{2}{3} & 0 & 0 & \frac{-1}{\sqrt{6}} & 0 \\ & & \frac{-2}{3} & \frac{-1}{3} & 0 & \frac{1}{\sqrt{6}} \\ & & & \frac{-2}{3} & 0 & \frac{1}{\sqrt{6}} \\ & & & & \frac{2}{5} & 0 \\ & & & & & \frac{-2}{5} \end{bmatrix}}_{[M_{R\cup R^*}]} & \begin{bmatrix} 0 & 0 \\ 0 & 0 \\ 0 & 0 \\ 0 & 0 \\ 0 & 0 \\ 0 & 0 \end{bmatrix} \\ & \begin{bmatrix} 0 & 0 \\ & 0 \end{bmatrix} \end{bmatrix}\begin{Bmatrix} b_1 \\ b_2 \\ b_3 \\ b_4 \\ b_5 \\ b_7 \\ b_6 \\ b_8 \end{Bmatrix}. \tag{3.47}$$

For the Laplace equation the matrices $[N_R]$ and $[N_{R^*}]$ vanish.

Static Condensation and Eigenvalues

In order to perform static condensation on the stiffness matrix, we have to rearrange the rows and columns so that $[K]$ will multiply the unknown vector $(b_1, b_2, b_3, b_4, b_5, b_7, b_6, b_8)^T$. This rearrangement produces the matrix

$$\left[\begin{array}{cc}\underbrace{\begin{bmatrix} 4.9960 & 2.4653 & -2.4983 & -4.9630 & -3.0461 & 3.0461 \\ & 4.9960 & -4.9630 & -2.4983 & -3.0461 & 3.0461 \\ & & 4.9971 & 2.4641 & 3.0461 & -3.0461 \\ & & & 4.9971 & 3.0461 & -3.0461 \\ & & & & 3.0281 & -2.9622 \\ & & & & & 3.0304 \end{bmatrix}}_{[K_{R\cup R^*}]} & \underbrace{\begin{bmatrix} 0.1204 & 0.2002 \\ 0.2002 & 0.1204 \\ -0.2276 & -0.0930 \\ -0.0930 & -0.2276 \\ -0.1309 & -0.1309 \\ 0.1309 & 0.1309 \end{bmatrix}}_{[K_{R-in}]} \\ & \underbrace{\begin{bmatrix} 9.9618 & 4.9608 \\ & 9.9618 \end{bmatrix}}_{[K_{in}]} \end{array}\right]. \tag{3.48}$$

Using (3.48), we may condense the matrix $[K]$ according to (3.35), obtaining a symmetric matrix (because $[K]$ itself is symmetric):

$$[K_S] = [K_{R\cup R^*}] - [K_{R-in}][K_{in}]^{-1}[K_{R-in}]^T$$
$$= \begin{bmatrix} 4.99188 & 2.46251 & -2.4959 & -4.9585 & -3.04324 & 3.04324 \\ & 4.99188 & -4.9585 & -2.4959 & -3.04324 & 3.04324 \\ & & 4.99188 & 2.46251 & 3.04324 & -3.04324 \\ & & & 4.99188 & 3.04324 & -3.04324 \\ & & & & 3.02576 & -2.95987 \\ & & & & & 3.02812 \end{bmatrix}. \tag{3.49}$$

The last step left is to solve the generalized eigenproblem (3.34). Because $[M_{R\cup R^*}]$ is nonsingular, we can transform the generalized eigenproblem to a regular one by multiplying (3.34) by $[M_{R\cup R^*}]^{-1}$:

$$[M_{R\cup R^*}]^{-1}[K_S]\mathbf{b}_{R\cup R^*} = \alpha \mathbf{b}_{R\cup R^*}. \tag{3.50}$$

Solving (3.50), one obtains the following six approximate eigenvalues:

$$10^{-7}, 10^{-7}, 2.205, 4.966, -2.205, -5.001.$$

We may observe that there are two ≈ 0 eigenvalues, corresponding to so-called rigid body motion (this always occurs for homogeneous Neumann boundary conditions on both boundaries Γ_1 and Γ_2). These eigenvalues are associated with constant-value eigenvectors, and are of no interest. The smallest approximate eigenvalue $(\alpha_1)_{p=2} = 2.205$, which has a 10.05% relative error compared to the exact value of 2 (for the Laplace equation the approximate eigenvalues are always larger than the exact ones, as can be shown using the Rayleigh-quotient).

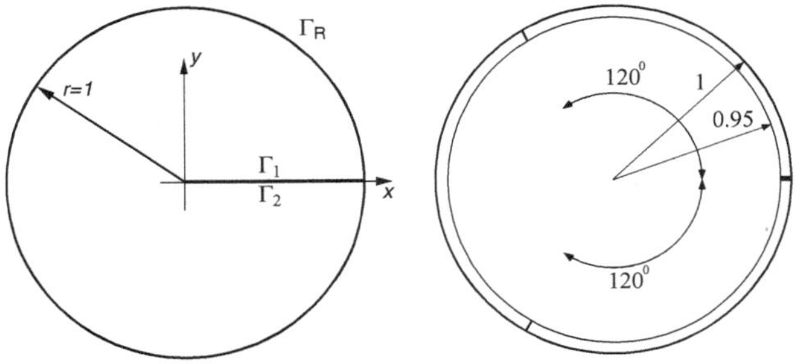

Fig. 3.5 A crack with homogeneous Newton BCs in a circular domain and the FE mesh used for the computation of the eigenpairs.

If we increase the polynomial degree over the element, we obtain the following values for $(\alpha_1)_{p=i}$:

$$((\alpha_1)_{p=3},\ (\alpha_1)_{p=4},\ (\alpha_1)_{p=5},\ (\alpha_1)_{p=6},\ (\alpha_1)_{p=7},\ (\alpha_1)_{p=8})$$
$$= (2.000568859638720, 2.000568859628970, 2.000000278341443,$$
$$2.000000278341425, 2.000000000037642, 2.000000000037600).$$

As observed, we may obtain for polynomial degree $p = 8$ the first eigenvalue with a relative error of $1.85 \times 10^{-9}\%$. Not only the positive eigenvalues are obtained, but also the negative ones, and their corresponding eigenvectors. The second eigenvalue ($\alpha_2 = 4$) has $3 \times 10^{-6}\%$ relative error at $p = 8$.

It is important to realize that as we increase the polynomial degree, the size of the condensed matrix $[K_S]$ increases by 2 for each order of the polynomial degree, although the overall number of basis functions $\Phi_i(\xi, \eta)$ may increase considerably. For example, at $p = 8$ one has 47 shape functions, but the size of the matrix $[K_S]$ is 18×18.

3.4.2 A Crack with Homogeneous Newton BCs (Laplace Equation)

Let Ω be the unit circle slit along the positive x-axis with Γ_1 the upper face of the slit, Γ_2 the lower face of the slit, and Γ_R the circular portion of the boundary of Ω as shown on the left of Figure 3.5. We consider in this unit circle slit the problem

$$\nabla^2 \tau = 0 \qquad \text{in } \Omega, \tag{3.51}$$

$$\tau = 0 \ \text{ on } \Gamma_1, \qquad \tfrac{\partial \tau}{\partial \theta} = 0 \ \text{ on } \Gamma_2, \qquad \frac{\partial \tau}{\partial r} = y \ \text{ on } \Gamma_R.$$

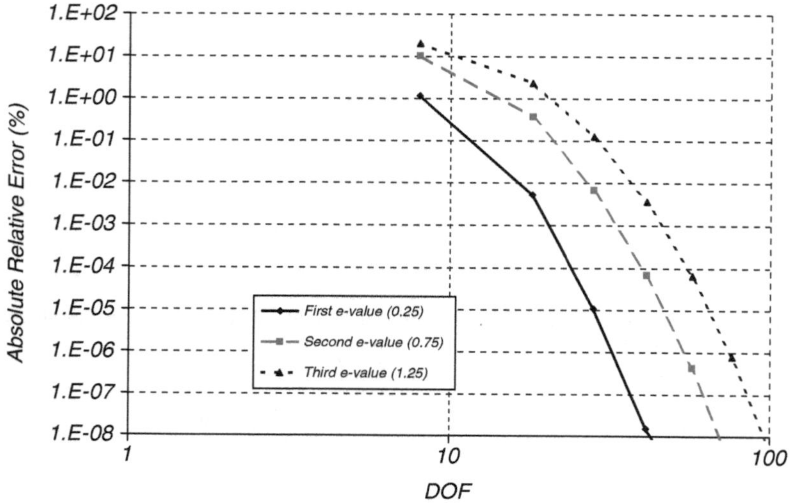

Fig. 3.6 Relative error (%) of the first three eigenpairs.

The solution to this problem, accurate up to the sixth significant digit, is given in [12]:

$$\tau(r,\theta) = -1.35812r^{1/4}\sin(\theta/4) + 0.970087r^{3/4}\sin(3\theta/4) - 0.452707r^{5/4}\sin(5\theta/4) + O(r^{7/4}), \quad (3.52)$$

and $\mathcal{B}(\tau,\tau) \stackrel{\text{def}}{=} \int_\Omega \Delta\tau \cdot \Delta\tau \, d\Omega = 4.52707$.

We first study the convergence of the eigenvalues computed by the modified Steklov method, and show that the value of R^* has a minor influence on the accuracy of the results if chosen in the range $0.5 \leq R^* \leq 0.95$.

First we compute the eigenpairs using three finite elements having 120° each, with $R = 1$ and $R^* = 0.95$, as shown in Figure 3.5. The convergence of the first three computed eigen-values as the p-level over each element is increased from $p = 1$ up to $p = 6$ is demonstrated in Figure 3.6. The absolute relative error as a percentage is computed as $100 \times \text{abs}[\frac{(\alpha_i)_{\text{FE}} - (\alpha_i)_{\text{EX}}}{(\alpha_i)_{\text{EX}}}]$. This example demonstrates the efficiency and accuracy of the modified Steklov method in computing the eigenpairs. Eigenvalues have relative errors of less than $10^{-8}\%$ with fewer than 100 DOFs.

Next, we keep the polynomial degree fixed $p = 8$ over the three elements, and extract the first three eigenpairs taking R^* to the range between $R^* = 0.3$ and $R^* = 0.99$. We plot the absolute relative error in the first three eigenvalues as a function of R^* in Figure 3.7. One may notice that if R^* is chosen in the range $0.5 \leq R^* \leq 0.95$, excellent results are obtained. Even if chosen out of this range, still the computed values are very accurate. We use in all computations a value of R^* in the specified range.

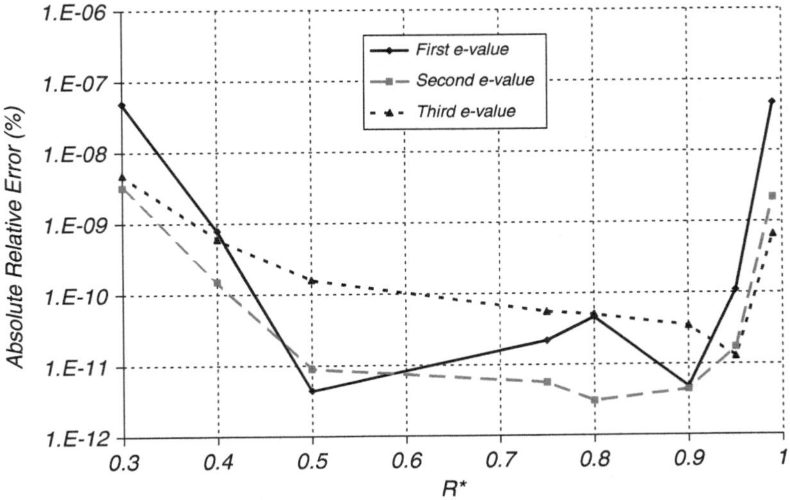

Fig. 3.7 Relative error (%) of the first three eigenpairs as a function of R^*.

3.4.3 A V-Notch in an Anisotropic Material with Homogeneous Neumann BCs.

Consider the heat conduction problem in an anisotropic material governed by the equation

$$4\frac{\partial^2 \tau}{\partial x_1^2} + \frac{\partial^2 \tau}{\partial x_2^2} = 0, \tag{3.53}$$

prescribed over a domain Ω whose boundary consists of a reentrant corner of 90° generated by two edges, Γ_1 and Γ_2. On the two edges Γ_1 and Γ_2, which meet at the origin of the coordinate system, flux-free boundary conditions (3.4) are applied and $\tau = 0$ is specified at $(0,0)$.

The solution τ is (the derivation is provided in Appendix B)

$$\tau = \sum_{i=1}^{\infty} A_i r^{\frac{2i}{3}} 2^{\frac{-2i}{3}} \left(1 + 3\sin^2\theta\right)^{i/3} \cos\left[\frac{2i}{3}\arctan(2\tan\theta)\right], \tag{3.54}$$

where r and θ are polar coordinates centered on the reentrant corner such that $\theta = 0$ coincides with the Γ_1 boundary. The first term in the expansion (3.54) for $\nabla\tau$ is unbounded as $r \to 0$.

Let Ω be the unit circle sector shown in Figure 3.8. The circular boundary of the domain, Γ_R, is loaded by a flux boundary condition that corresponds to the first symmetric eigenfunction of the asymptotic expansion of τ:

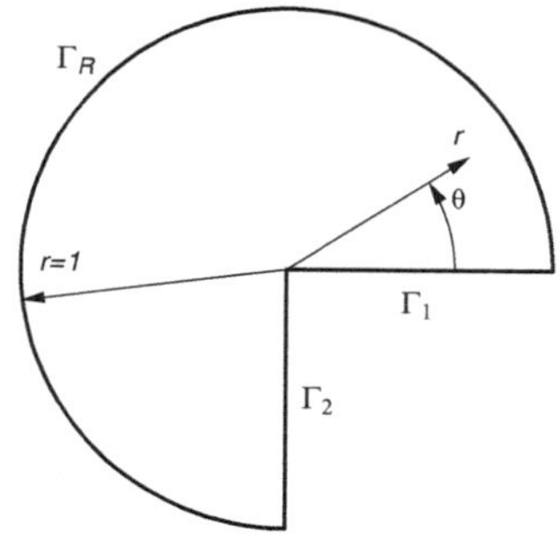

Fig. 3.8 Domain for the anisotropic heat transfer problem with Neumann homogeneous boundary conditions.

$$\begin{aligned}\frac{\partial\tau}{\partial r} \stackrel{\text{def}}{=} q_r &= (k_{11}\cos^2\theta + k_{22}\sin^2\theta)\frac{\partial\tau}{\partial r} + \frac{1}{2}\sin 2\theta(k_{22}-k_{11})\left(\frac{1}{r}\frac{\partial\tau}{\partial\theta}\right)\\ &= A_1 r^{-1/3}[2(1+3\sin^2\theta)]^{-2/3}\left\{2\sin 2\theta\sin\left[\frac{2}{3}\arctan(2\tan\theta)\right]\right.\\ &\left.+\left[\frac{2}{3}(1+3\cos^2\theta)(1+3\sin^2\theta)-\frac{3}{2}\sin^2 2\theta\right]\cos\left[\frac{2}{3}\arctan(2\tan\theta)\right]\right\}.\end{aligned} \tag{3.55}$$

On the other two boundaries flux-free boundary conditions are applied. The GFIF A_1 is arbitrarily selected to be $A_1 = 1$, while the others are $A_i = 0$, $i = 2, 3, \ldots, \infty$. The exact solution to this problem is given by the first term with $i = 1$ in (3.54).

We use three finite elements having 90° each, with $R = 1$ and $R^* = 0.95$. The convergence of the first three computed eigenvalues as the p-level over each element is increased from $p = 1$ up to $p = 8$ is demonstrated in Figure 3.9, showing the absolute relative error as a percentage versus the number of degrees of freedom. We also show in Figure 3.10 the first three eigenfunctions $s_1^+(\theta), s_2^+(\theta), s_3^+(\theta)$ and their dual counterparts $s_1^-(\theta), s_2^-(\theta), s_3^-(\theta)$ computed at $p = 8$. For the anisotropic material the primal eigenfunctions $s_i^+(\theta)$ are different from to the dual ones $s_i^-(\theta)$.

3.4.4 *An Internal Singular Point at the Interface of Two Materials*

Let $\Omega = \{(r,\theta) : r \le 2,\ 0 \le \theta \le 2\pi\}$ and let Ω_i, $i = 1, 2$, be the two subdomains of Ω occupying the sectors $0 \le \theta \le \pi/2$ and $\pi/2 \le \theta \le 2\pi$. See Figure 3.11. Continuity of the function and the fluxes is assumed at the materials' interface.

We first consider the problem with two isotropic materials, where the eigenpairs are all real:

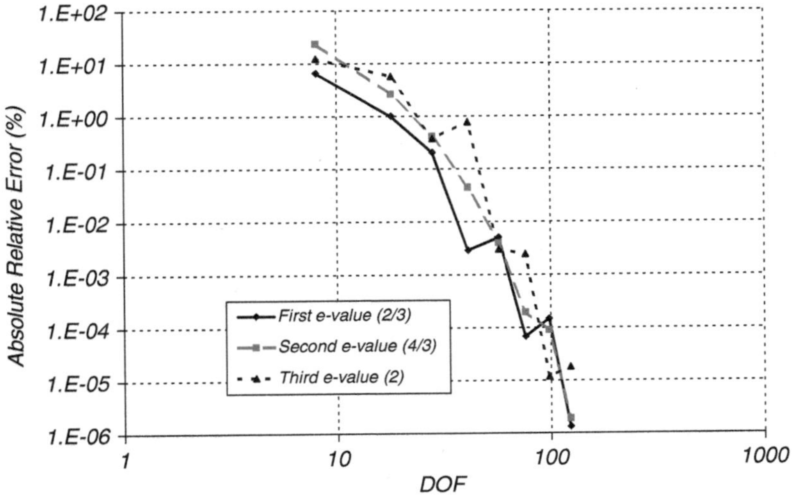

Fig. 3.9 Absolute relative error (%) of the first three eigenpairs.

$$k^{(i)}\nabla^2\tau = 0 \qquad \text{in } \Omega_i, \quad i = 1, 2, \tag{3.56}$$

with the boundary conditions

$$\frac{\partial\tau}{\partial r} = k^{(i)}\left[\alpha_1 r^{\alpha_1-1}s_1^+(\theta) + \alpha_2 r^{\alpha_2-1}s_2^+(\theta)\right] \qquad \text{on } \Gamma_i = \partial\Omega_i,\ i = 1, 2. \tag{3.57}$$

$$\begin{aligned} k^{(1)} &= 10 \text{ and} & k^{(2)} &= 1, \\ \alpha_1 &= 0.731691779 \text{ and} & \alpha_2 &= 1.268308221, \end{aligned} \tag{3.58}$$

$$s_1^+(\theta) = \begin{cases} \cos[(1-a)\theta] + c_1\sin[(1-a)\theta] & 0 \le \theta \le \pi/2, \\ c_1\cos[(1-a)\theta] + c_2c_3\sin[(1-a)\theta] & \pi/2 \le \theta \le 2\pi, \end{cases} \tag{3.59}$$

$$s_2^+(\theta) = \begin{cases} \cos[(1+a)\theta] - c_3\sin[(1+a)\theta] & 0 \le \theta \le \pi/2, \\ c_1\cos[(1+a)\theta] - c_2c_3\sin[(1+a)\theta] & \pi/2 \le \theta \le 2\pi, \end{cases} \tag{3.60}$$

$c_1 = 6.31818181818182$, $c_2 = -2.68181818181818$, $c_3 = 0.64757612580273$, and $a = 0.26830822130025$.

Then the unique solution (up to an additive constant) to this interface problem is given in [132]:

$$\tau(r,\theta) = A_1 r^{\alpha_1}s_1^+(\theta) + A_2 r^{\alpha_2}s_2^+(\theta), \tag{3.61}$$

where $A_1 = A_2 = 1$.

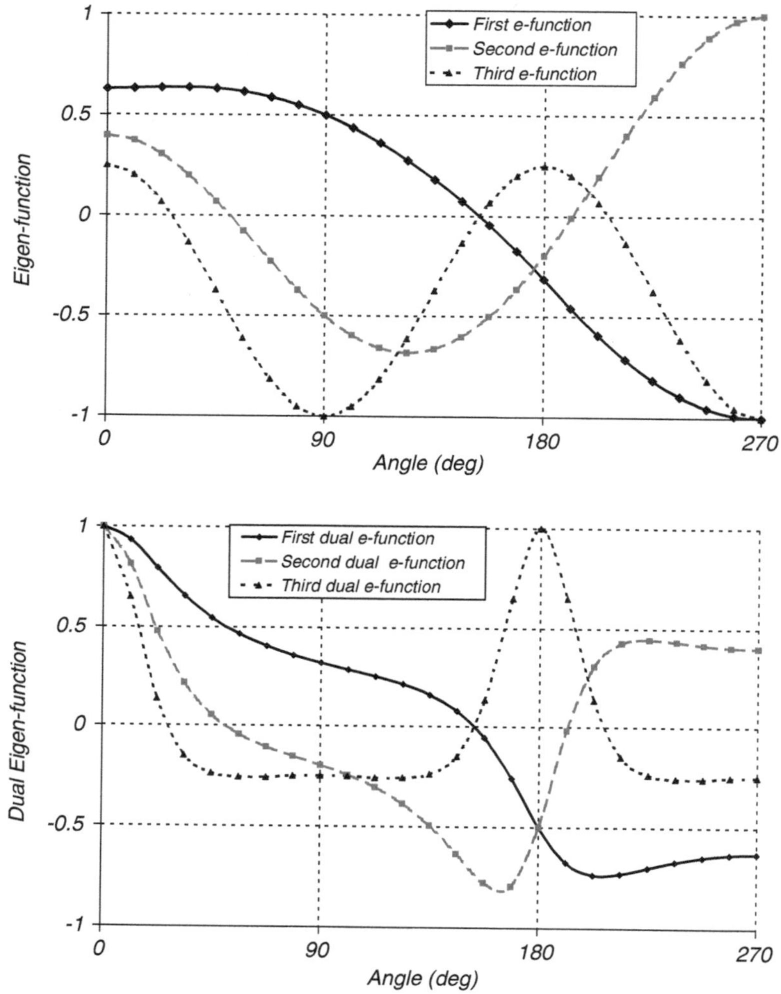

Fig. 3.10 The first three eigenfunctions and dual eigenfunctions at $p = 8$.

The performance of the modified Steklov method is demonstrated in Table 3.1, where we report the relative error of the first and second computed eigenvalues, using a four element mesh.

As a second example, we consider the same domain as shown in Figure 3.11, except that this time, one of the materials is anisotropic, namely

$$\frac{\partial^2 \tau}{\partial x_1^2} + \frac{\partial^2 \tau}{\partial x_2^2} = 0 \qquad \text{in } \Omega_1,$$

$$10\frac{\partial^2 \tau}{\partial x_1^2} + 0.1\frac{\partial^2 \tau}{\partial x_2^2} = 0 \qquad \text{in } \Omega_2. \tag{3.62}$$

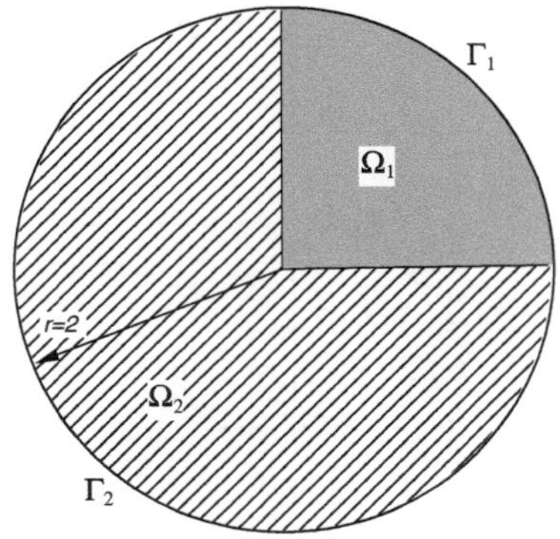

Fig. 3.11 Internal interface with two materials.

Table 3.1 Relative error (%) in first two eigenvalues ($\alpha_1^{EX} = 0.731691779, \alpha_2^{EX} = 1.268308221$).

	p = 1	p = 2	p = 3	p = 4	p = 5	p = 6	p = 7	p = 8
DOF	8	18	28	41	57	76	98	123
$e_{\alpha_1}(\%)$	10.32	0.377	0.0069	7.0*e*-5	4.3*e*-7	1.0*e*-9	3.0*e*-11	1.0*e*-10
$e_{\alpha_2}(\%)$	−3.94	0.909	0.0270	4.4*e*-4	4.5*e*-6	7.9*e*-8	1.0*e*-10	1.0*e*-10

Table 3.2 Real and imaginary parts of the first eigen-value for the bi-material anisotropic internal singular point.

p-	DOF	$\Re\alpha_1$	$\Im\alpha_1$
1	30	0.9210204825	0.2354645754
2	73	0.8943243330	0.3172441170
3	116	0.8792150895	0.3247298610
4	173	0.8806907064	0.3235360557
5	244	0.8816130960	0.3230501821
6	329	0.8816518948	0.3230476439
7	428	0.8816075676	0.3230757155
8	541	0.8815999758	0.3230801526
[115]	∞	0.8816020381	0.3230787589

This case has been shown in [115] to have complex eigenpairs. Using the modified Steklov method, we compute the first eigenvalue on a mesh consisting 14 elements, 3 in Ω_1 and 11 in Ω_2. We present the real and imaginary parts of the first computed eigenvalue in Table 3.2. The absolute relative error as a percentage versus the number of degrees of freedom is shown in Figure 3.12. Again, even for cases in which complex eigenvalues appear, the modified Steklov method provides good results and captures accurately the existence of complex eigenpairs.

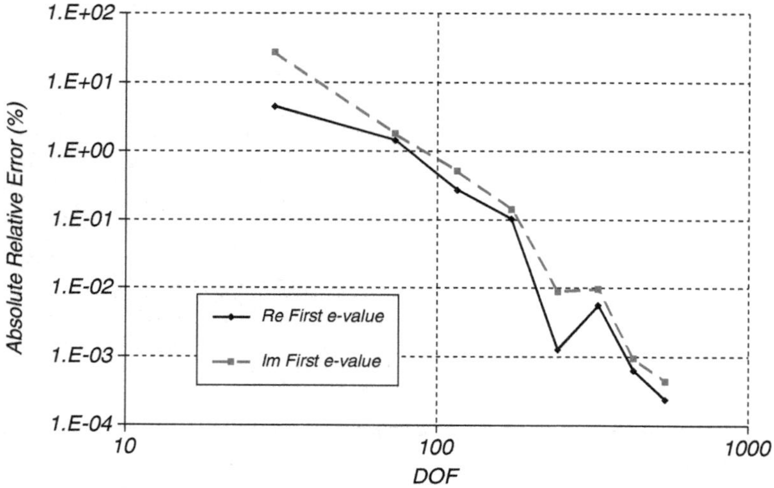

Fig. 3.12 Absolute relative error (%) of the real and imaginary parts of the first eigenvalue for the bimaterial anisotropic internal singular point.

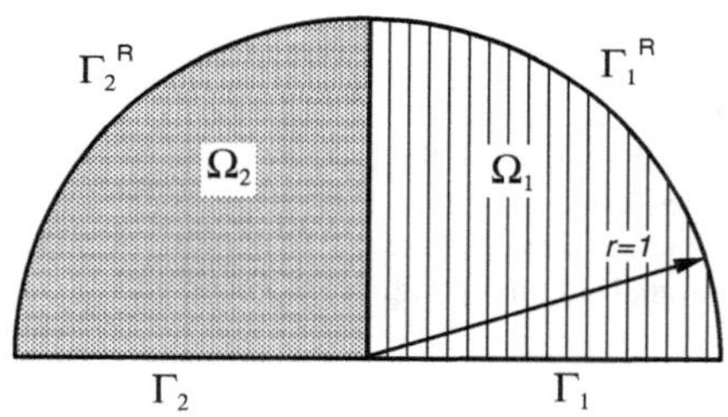

Fig. 3.13 Domain for the two anisotropic flux-free bimaterial scalar problem.

3.4.5 An Anisotropic Flux-Free Bimaterial Interface

The solution to the scalar elliptic problem associated with an anisotropic bimaterial interface with flux-free boundary conditions on the free edge in a neighborhood of a singular point is given in Appendix B.

The domain under consideration is defined by $\Omega = \{(x, y) : (x^2 + y^2) \le 1 \cap y > 0\}$, and let Ω_i be the two subdomains of Ω occupying the sectors $0 \le \theta \le \pi/2$, $\pi/2 \le \theta \le \pi$. See Figure 3.13.

The equations to be solved in each subdomain are

$$-k_{\beta\gamma}^{(i)} \frac{\partial^2 \tau}{\partial x_\beta \partial x_\gamma} = 0 \qquad \text{in } \Omega_i, \quad i = 1, 2, \tag{3.63}$$

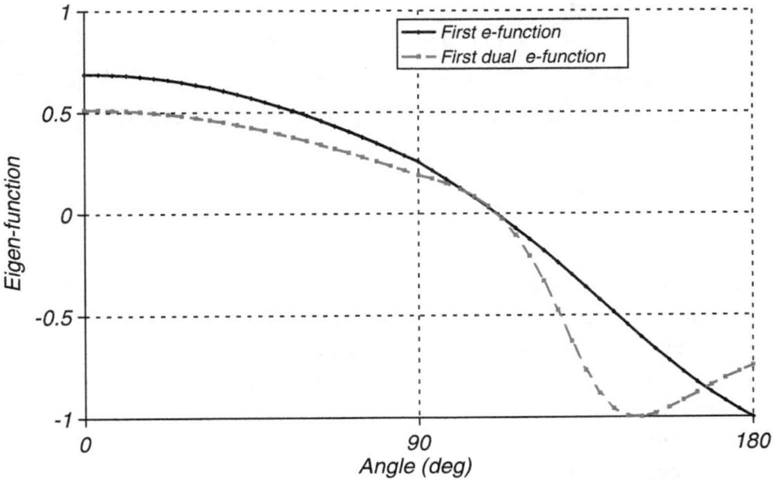

Fig. 3.14 The first primal and dual eigenfunctions for the bimaterial anisotropic interface as computed at $p = 8$ using six elements.

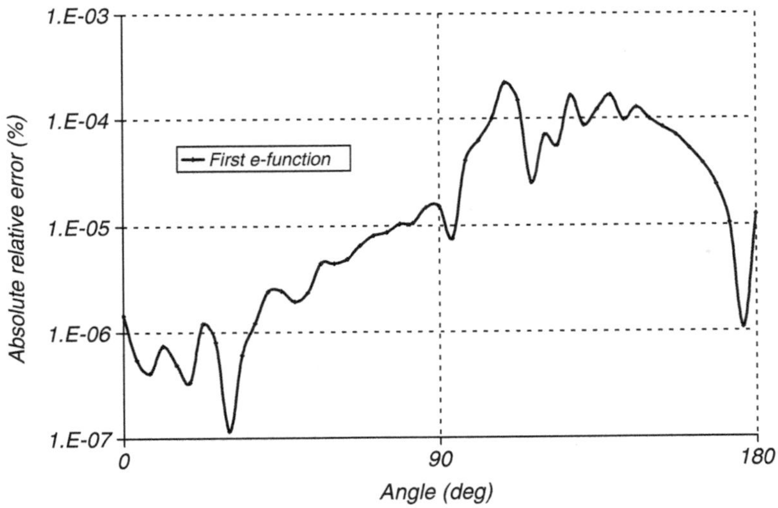

Fig. 3.15 Absolute relative error (%) of the first eigenfunction for the bimaterial anisotropic interface.

having the following heat conduction coefficients:

$$k_{11}^{(1)} = k_{11}^{(2)} = k_{22}^{(1)} = k_{22}^{(2)} = 1.0,$$

$$k_{12}^{(1)} = 0.0, \quad k_{12}^{(2)} = -0.75,$$

The flux-free boundary conditions (3.4) are applied on the straight edges intersecting in the singular point, and continuity of the function and flux across the interface boundary are maintained, i.e.m τ is continuous and $k_{\beta\gamma}^{(1)}\partial_\beta\tau n_\gamma = k_{\beta\gamma}^{(2)}\partial_\beta\tau n_\gamma$. Taking the boundary condition on the circular boundaries as

$$\frac{\partial\tau}{\partial r} = \frac{\partial\tau_1^{(i)}}{\partial r}, \qquad i = 1, 2, \text{ on the circular edges } \Gamma_i^R, \tag{3.64}$$

where $\tau_1^{(i)}$ is given by (B.41) and (B.43) (evaluated at $r = 1$), the exact solution in each subdomain is given by (B.41) and (B.43).

Using a finite element mesh having six elements, we computed the eigenpairs for $p = 1$ to $p = 8$. At $p = 6$ with 145 DOFs we obtain the eigenvalues correct up to the seventh digit as shown in the analytical solution. We present the first primal and dual eigenfunctions computed at $p = 8$ in Figure 3.14, and the absolute relative error as a percentage in Figure 3.15.

After computing the eigenpairs the next task is the extraction of GFIFs, discussed in the next chapter.

Chapter 4
GFIFs Computation for Two-Dimensional Heat Conduction Problems

Having computed the eigenpairs associated with a 2-D singular point, the next task is the computation of the coefficients of the series expansion A_i's, called for the heat conduction equation "generalized flux intensity functions" (GFIFs). The eigenpairs may be viewed as characterizing the straining modes, and their amplitudes (the GFIFs) quantify the amount of "energy" residing in particular straining modes. For this reason, failure theories directly or indirectly involve the GFIFs. As a simple example, consider a solution for which all eigenpairs are given. Although the first eigenvalue may be very small, if the corresponding GFIF is zero, the solution does not manifest this singular behavior.

Many methods exist for the computation of GFIFs, mainly associated with *cracks*, from finite element solutions. For example, the J-integral method, the energy release rate method, the stiffness derivative method, the dual singular function method (also known as the contour integral method CIM), the cutoff function method (CFM), the singular superelement method, etc. See references [12, 27, 31, 177] and the references therein. Most of the methods, however, are applicable to crack singularities in isotropic materials only and do not provide any desired number of stress intensity factors.

One of the most efficient ways for extracting the GFIFs in a superconvergent manner is by an indirect extraction procedure using the dual singular function [12, 27]. These efficient procedures use specially constructed extraction functions (the dual eigenpairs), and will be presented in the next section.

4.1 Computing GFIFs Using the Dual Singular Function Method

Consider the scalar elliptic equation

$$-\partial_\beta \left(k_{\beta\gamma}\partial_\gamma \tau\right) = 0$$

Z. Yosibash, *Singularities in Elliptic Boundary Value Problems and Elasticity and Their Connection with Failure Initiation*, Interdisciplinary Applied Mathematics 37, DOI 10.1007/978-1-4614-1508-4_4, © Springer Science+Business Media, LLC 2012

over the domain shown in Figure 4.1, for example, and assume that the boundary conditions are homogeneous in a close vicinity of the singularity. The point of departure in deriving the dual singular function method is the path-independent integral defined in Appendix E in (E.18), over the arc Γ_R:

$$I_{\Gamma_R} = \int_{\theta_1}^{\theta_1+\omega} \left\{ \left(\frac{\partial \tau}{\partial r}\chi - \tau \frac{\partial \chi}{\partial r} \right) \left(k_{11}\cos^2\theta + k_{12}\sin 2\theta + k_{22}\sin^2\theta\right) + \frac{1}{r}\left(\frac{\partial \tau}{\partial \theta}\chi - \tau \frac{\partial \chi}{\partial \theta} \right) \left[\frac{(k_{22}-k_{11})}{2}\sin 2\theta + k_{12}\cos 2\theta \right] \right\}_{r=R} R d\theta. \quad (4.1)$$

If the ith GFIF (A_i) is of interest, we choose as the "extraction function" χ_i the ith dual eigenpair multiplied by a constant to be specified in the sequel:

$$\chi_i = c_i r^{-\alpha_i} s_i^-(\theta). \quad (4.2)$$

Of course, the chosen function χ_i, being constructed using the dual eigenpair, satisfies the differential equation and the homogeneous boundary conditions, thus is justified to be used in the path independent integral. Substituting (4.2) in (4.1) yields

$$I_{\Gamma_R} = c_i R^{-\alpha_i} \int_{\theta_1}^{\theta_1+\omega} \left\{ \left(r\frac{\partial \tau}{\partial r} + \alpha_i \tau \right) \left(k_{11}\cos^2\theta + k_{12}\sin 2\theta + k_{22}\sin^2\theta\right) s_i^- + \left(s_i^- \frac{\partial \tau}{\partial \theta} - \tau (s_i^-)' \right) \left[\frac{(k_{22}-k_{11})}{2}\sin 2\theta + k_{12}\cos 2\theta \right] \right\}_{r=R} d\theta,$$

no summation on i (4.3)

If the solution τ is known along a circular path around the singular point, it can be substituted in (4.3) to obtain the value of I_{Γ_R}. On the other hand, in the vicinity of the singular point, the solution can be represented also as $\tau = \sum_j A_j r^{\alpha_j} s_j^+(\theta)$. Inserting the series solution in (4.3), one obtains

$$I_{\Gamma_R} = c_i \sum_j A_j R^{\alpha_j - \alpha_i} \int_{\theta_1}^{\theta_1+\omega} \left\{ (\alpha_i + \alpha_j) s_j^+ s_i^- \left(k_{11}\cos^2\theta + k_{12}\sin 2\theta + k_{22}\sin^2\theta\right) + \left(s_i^-(s_j^+)' - s_j^+(s_i^-)' \right) \left[\frac{(k_{22}-k_{11})}{2}\sin 2\theta + k_{12}\cos 2\theta \right] \right\} d\theta,$$

no summation on i. (4.4)

Because of the orthogonality of the primal and dual eigen-pairs (see (E.21)), all integrals in the sum for which $j \neq i$ vanish, so that (4.4) becomes:

$$I_{\Gamma_R} = A_i c_i \int_{\theta_1}^{\theta_1+\omega} \Big\{ 2\alpha_i s_i^+ s_i^- \left(k_{11}\cos^2\theta + k_{12}\sin 2\theta + k_{22}\sin^2\theta\right) + \left(s_i^-(s_i^+)' - s_i^+(s_i^-)'\right)\left[\frac{(k_{22}-k_{11})}{2}\sin 2\theta + k_{12}\cos 2\theta\right]\Big\}\, d\theta,$$
$$\text{no summation on } i. \tag{4.5}$$

In view of (4.5), we are at the stage to provide the expression for the constant c_i, which can be computed by the ith primal and dual eigenpairs, and the coefficients of heat conduction

$$c_i = \left[\int_{\theta_1}^{\theta_1+\omega} \Big\{ 2\alpha_i s_i^+ s_i^- \left(k_{11}\cos^2\theta + k_{12}\sin 2\theta + k_{22}\sin^2\theta\right) + \left(s_i^-(s_i^+)' - s_i^+(s_i^-)'\right)\left[\frac{(k_{22}-k_{11})}{2}\sin 2\theta + k_{12}\cos 2\theta\right]\Big\}\, d\theta\right]^{-1}$$
$$\text{no summation on } i. \tag{4.6}$$

Before computing the inverse of the integral in the right-hand side of (4.6), one must check that the integral is not identically zero. For the Laplace equation it can be shown that this integral is nonzero, but a proof for a general heat conduction equation is not known to exist at this time.

Inserting (4.6) in (4.5), one obtains that

$$I_{\Gamma_R} = A_i. \tag{4.7}$$

Combining (4.7) with (4.3), one obtains the dual singular function method for the computation of any desired GFIF:

$$A_i = \frac{c_i}{R^{\alpha_i}} \int_{\theta_1}^{\theta_1+\omega} \Big\{ \left(r\frac{\partial \tau}{\partial r} + \alpha_i \tau\right)\left(k_{11}\cos^2\theta + k_{12}\sin 2\theta + k_{22}\sin^2\theta\right) s_i^- + \left(s_i^- \frac{\partial \tau}{\partial \theta} - \tau (s_i^-)'\right)\left[\frac{(k_{22}-k_{11})}{2}\sin 2\theta + k_{12}\cos 2\theta\right]\Big\}_{r=R} d\theta,$$
$$\text{no summation on } i \tag{4.8}$$

In practice, the solution τ is not known, but only its numerical approximation (by the FEM for example). It is known, however, that in the vicinity of the singularity the finite element approximation is not of high accuracy. Thus, instead of using the exact solution τ for evaluating the integral in (4.8), one uses its numerical approximation τ_{FE} along a circular curve *away from the singular point*, thus obtaining $(A_i)_{FE}$.

Remark 4.1. *Because the dual singular function method is based on the path-independent integral, the circular path used for the computation can be of any* *radius. It has been proven in [12] that using τ_{FE}, the error $A_i - (A_i)_{\mathrm{FE}}$ has a superconvergent property, i.e., $(A_i)_{\mathrm{FE}}$ approaches A_i at a rate that is twice as fast as the error in energy norm.*

Remark 4.2. *Computing $(A_i)_{\mathrm{FE}}$ by (4.8) involves the derivative $\frac{\partial \tau_{\mathrm{FE}}}{\partial r}$. This enlarges the approximation error in the GFIFs because numerical derivatives are associated with larger numerical errors.*

For the Laplace equation, $s_i^+ = s_i^- = s_i$ and the expression for the computation of A_i reduces to

$$A_i = \frac{1}{2\alpha_i R^{\alpha_i} \int_{\theta_1}^{\theta_1+\omega} s_i^2 d\theta} \int_{\theta_1}^{\theta_1+\omega} \left(r \frac{\partial \tau}{\partial r} + \alpha_i \tau \right)_{r=R} s_i \, d\theta, \quad \text{no summation on } i. \tag{4.9}$$

4.2 Computing GFIFs Using the Complementary Weak Form

Another efficient method for the computation of the GFIFs without the necessity of dual singular functions or numerical derivatives is based on the complementary weak formulation [180]. First, we derive the complementary weak formulation for the heat conduction equation, and thereafter use it for the computation of the GFIFs.

4.2.1 Derivation of the Complementary Weak Form

Instead of the solution τ as the primal function of interest, we consider now the flux vector $\boldsymbol{q}$, connected to τ by

$$\boldsymbol{q} = \begin{Bmatrix} q_1 \\ q_2 \end{Bmatrix} = [k]\nabla\tau \leftrightarrow q_\beta = k_{\beta\gamma}\partial_\gamma\tau. \tag{4.10}$$

With this notation, the heat conduction equation (2.1) can be stated as

$$-\nabla \cdot \boldsymbol{q} = Q \quad \text{in } \Omega, \tag{4.11}$$

with the boundary conditions (2.2) - (2.3):

$$\boldsymbol{q} \cdot \boldsymbol{n} = \hat{q}_n \quad \text{on } \Gamma_N, \tag{4.12}$$

$$\tau = \hat{\tau} \quad \text{on } \Gamma_D. \tag{4.13}$$

Multiply (4.11) by a function χ and integrate over Ω:

$$-\iint_{\Omega} \chi \nabla \cdot \boldsymbol{q} \, d\Omega = \iint_{\Omega} \chi Q \, d\Omega. \tag{4.14}$$

Using Green's theorem on the LHS of (4.14), we obtain

$$-\oint_{\partial\Omega} \chi \boldsymbol{q} \cdot \boldsymbol{n} \, d\Omega + \iint_{\Omega} \nabla\chi \cdot \boldsymbol{q} d\Omega = \iint_{\Omega} \chi Q \, d\Omega. \tag{4.15}$$

Let us concentrate our attention on the first term of (4.15). From Green's theorem,

$$\iint_{\Omega} \nabla\cdot([k]\nabla\tau)\chi \, d\Omega = \iint_{\Omega} \nabla\cdot([k]\nabla\chi)\tau \, d\Omega + \oint_{\partial\Omega} \chi([k]\nabla\tau)\cdot\boldsymbol{n} - \tau([k]\nabla\chi)\cdot\boldsymbol{n} \, d\Gamma, \tag{4.16}$$

we obtain that

$$\begin{aligned} -\oint_{\partial\Omega} \chi([k]\nabla\tau)\cdot\boldsymbol{n} \, d\Gamma = &-\iint_{\Omega} \nabla\cdot([k]\nabla\tau)\chi \, d\Omega + \iint_{\Omega} \nabla\cdot([k]\nabla\chi)\tau \, d\Omega \\ &-\oint_{\partial\Omega} \tau([k]\nabla\chi)\cdot\boldsymbol{n} \, d\Gamma. \end{aligned} \tag{4.17}$$

Introduce a new vector function $\boldsymbol{l}$ obtained from χ,

$$\boldsymbol{l} = [k]\nabla\chi, \tag{4.18}$$

and substitute it in (4.17) to obtain

$$-\oint_{\partial\Omega} \chi([k]\nabla\tau)\cdot\boldsymbol{n} \, d\Gamma = -\iint_{\Omega} \nabla\cdot\boldsymbol{q}\chi \, d\Omega + \iint_{\Omega} \nabla\cdot\boldsymbol{l}\tau \, d\Omega - \oint_{\partial\Omega} \tau\boldsymbol{l}\cdot\boldsymbol{n} \, d\Gamma. \tag{4.19}$$

Substitute (4.11) for the first term on the RHS of (4.19) (this is a strong argument that requires that $-\nabla\cdot\boldsymbol{q} = Q$ at each point in Ω), and require that $\nabla\cdot\boldsymbol{l} = 0$ in Ω, so that (4.19) becomes

$$-\oint_{\partial\Omega} \chi([k]\nabla\tau)\cdot\boldsymbol{n} \, d\Gamma = \iint_{\Omega} \chi Q \, d\Omega - \oint_{\partial\Omega} \tau\boldsymbol{l}\cdot\boldsymbol{n} \, d\Gamma. \tag{4.20}$$

Substitute (4.20) in (4.15) to obtain

$$-\oint_{\partial\Omega} \tau\boldsymbol{l}\cdot\boldsymbol{n} \, d\Gamma + \iint_{\Omega} \nabla\chi\cdot\boldsymbol{q} d\Omega = 0. \tag{4.21}$$

Observe that $\nabla \chi = [k]^{-1} \boldsymbol{l}$, and require that $\boldsymbol{l} \cdot \boldsymbol{n} = 0$ on Γ_N. Then (4.21) finally becomes

$$\iint_{\Omega} \left([k]^{-1}\boldsymbol{l}\right) \cdot \boldsymbol{q} \, d\Omega = \int_{\Gamma_D} \hat{\tau} \, \boldsymbol{l} \cdot \boldsymbol{n} \, d\Gamma. \tag{4.22}$$

Before summarizing the complementary weak formulation, we define the vector space called the "statically admissible space":

Definition 4.1. The vector space called "statically admissible space" is defined as

$$\mathcal{E}_c(\Omega) \stackrel{\text{def}}{=} \left\{ \boldsymbol{q} \mid \int_{\Omega} \boldsymbol{q} \cdot [k]^{-1} \boldsymbol{q} \, d\Omega < \infty, \ -\nabla \cdot \boldsymbol{q} = Q \ \text{ in } \Omega \right\}. \tag{4.23}$$

If $\Gamma_N \neq \varnothing$, then we introduce the following statically admissible subspaces:

$$\tilde{\mathcal{E}}_c(\Omega) \stackrel{\text{def}}{=} \{ \boldsymbol{q} \mid \boldsymbol{q} \in \mathcal{E}_c(\Omega), \ \boldsymbol{q} \cdot \boldsymbol{n} = \hat{q}_n \quad \text{on } \Gamma_N \}, \tag{4.24}$$

$$(\mathcal{E}_c)_0(\Omega) \stackrel{\text{def}}{=} \{ \boldsymbol{q} \mid \boldsymbol{q} \in \mathcal{E}_c(\Omega), \ \boldsymbol{q} \cdot \boldsymbol{n} = 0 \quad \text{on } \Gamma_N \}. \tag{4.25}$$

With this notation we are ready to introduce the complementary weak form:

$$\begin{gathered} \text{Seek } \boldsymbol{q} \in \mathcal{E}_c(\Omega) \text{ satisfying} \\ \mathcal{B}_c(\boldsymbol{q}, \boldsymbol{l}) = \mathcal{F}_c(\boldsymbol{l}) \quad \forall \boldsymbol{l} \in \mathcal{E}_c(\Omega), \end{gathered} \tag{4.26}$$

where $\mathcal{B}_c(\boldsymbol{q}, \boldsymbol{l}) \stackrel{\text{def}}{=} \iint_{\Omega} \left([k]^{-1}\boldsymbol{l}\right) \cdot \boldsymbol{q} \, d\Omega, \qquad \mathcal{F}_c(\boldsymbol{l}) \stackrel{\text{def}}{=} \int_{\Gamma_D} \hat{\tau} \, \boldsymbol{l} \cdot \boldsymbol{n} \, d\Gamma.$

In case $\Gamma_N \neq \varnothing$, then the complementary weak form requires the use of $\tilde{\mathcal{E}}_c(\Omega)$ and $(\mathcal{E}_c)_0(\Omega)$:

$$\begin{gathered} \text{Seek } \boldsymbol{q} \in \tilde{\mathcal{E}}_c(\Omega) \text{ satisfying} \\ \mathcal{B}_c(\boldsymbol{q}, \boldsymbol{l}) = \mathcal{F}_c(\boldsymbol{l}) \quad \forall \boldsymbol{l} \in (\mathcal{E}_c)_0(\Omega), \end{gathered}$$

and if homogeneous Neumann boundary conditions are prescribed, then the above should read:

$$\begin{gathered} \text{Seek } \boldsymbol{q} \in (\mathcal{E}_c)_0(\Omega) \text{ satisfying} \\ \mathcal{B}_c(\boldsymbol{q}, \boldsymbol{l}) = \mathcal{F}_c(\boldsymbol{l}) \quad \forall \boldsymbol{l} \in (\mathcal{E}_c)_0(\Omega). \end{gathered} \tag{4.27}$$

Detailed discussion on the complementary weak form and its relation to the primal weak form is given in [131], where it is shown that the exact energy can

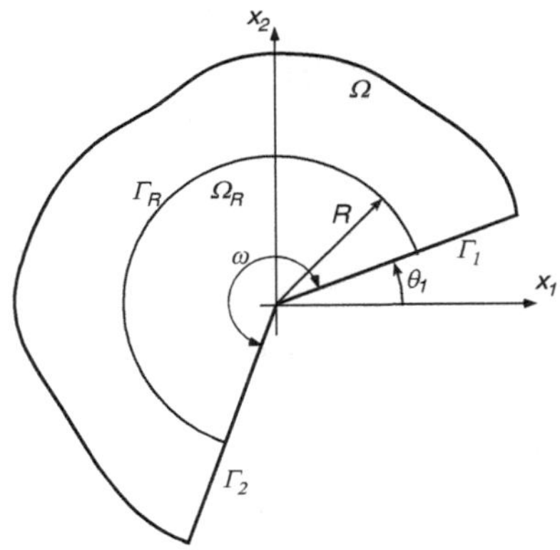

Fig. 4.1 The domain Ω_R.

be bounded from below as well as from above by the approximate energy computed by finite elements using the two weak forms. These bounds have been used for a-posteriori error estimations in which two finite element solutions were obtained over the same domain using both forms. These bounds, however, are global measures, which provide no information about the quality of the solution and its derivatives at specific points, and furthermore, each problem has to be solved twice, which is not practical in general.

4.2.2 *Using the Complementary Weak Formulation to Extract GFIFs*

For the extraction of the GFIFs, a circular subdomain Ω_R is considered centered at the singular point (see Figure 4.1):

$$\Omega_R = \{(r, \theta) \mid 0 \le r \le R, \quad \theta_1 \le \theta \le \theta_1 + \omega\}.$$

Assume that in Ω_R, which is in the vicinity of the singular point, $Q = 0$ and homogeneous boundary conditions are prescribed on Γ_1 and Γ_2. Furthermore, we assume that τ is known on Γ_R, which are Dirichlet type boundary conditions.

For applying the complementary weak form over Ω_R one needs first to construct a space of statically admissible vector functions. This can be constructed using the known eigenpairs.

By computing eigenfluxes by using the eigenpairs, these automatically satisfy the PDE and the boundary conditions on Γ_1 and Γ_2. Define the following statically admissible basis:

$$\boldsymbol{q}_i(r,\theta) = \begin{Bmatrix} (q_1)_i \\ (q_2)_i \end{Bmatrix} = [k]\nabla r^{\alpha_i} s_i^+(\theta) \tag{4.28}$$

$$= r^{\alpha_i - 1} \begin{Bmatrix} \left[\alpha_i(k_{11}\cos\theta + k_{12}\sin\theta)s_i^+(\theta) - (k_{11}\sin\theta - k_{12}\cos\theta)(s_i^+)'(\theta)\right] \\ \left[\alpha_i(k_{21}\cos\theta + k_{22}\sin\theta)s_i^+(\theta) - (k_{21}\sin\theta - k_{22}\cos\theta)(s_i^+)'(\theta)\right] \end{Bmatrix},$$

no summation on i,

which is linearly independent, and satisfies any homogeneous boundary conditions on Γ_1 and Γ_2. Any

$$\boldsymbol{q} = \sum_{i=1}^{N} A_i \boldsymbol{q}_i, \qquad \boldsymbol{l} = \sum_{i=1}^{N} b_i \boldsymbol{q}_i, \tag{4.29}$$

belongs to the statically admissible space (in $\mathcal{E}_c(\Omega_R)$ if homogeneous Dirichlet boundary conditions are given on $\Gamma_1 \cup \Gamma_2$, or in $(\mathcal{E}_c)_0(\Omega_R)$ if homogeneous Neumann boundary conditions are prescribed on Γ_1 or/and Γ_2). The coefficients A_i in (4.29) are the sought GFIFs. Defining $\boldsymbol{A}^T = (A_1, A_2, \ldots, A_N)$, and $\boldsymbol{b}^T = (b_1, b_2, \ldots, b_N)$, then after substituting (4.29), the complementary weak form can be stated as

$$\text{Seek } \boldsymbol{A} \text{ satisfying}$$
$$\boldsymbol{A}^T[B_c]\boldsymbol{b} = \{F_c\}^T\boldsymbol{b} \quad \forall \boldsymbol{b}, \tag{4.30}$$

or

$$\boldsymbol{A}^T = \{F_c\}^T[B_c]^{-1}, \tag{4.31}$$

where

$$(B_c)_{ij} = \int_0^R \int_{\theta_1}^{\theta_1+\omega} r^{\alpha_i+\alpha_j-2}$$
$$\times \begin{Bmatrix} \left[\alpha_i(k_{11}\cos\theta + k_{12}\sin\theta)s_i^+ - (k_{11}\sin\theta - k_{12}\cos\theta)(s_i^+)'\right] \\ \left[\alpha_i(k_{21}\cos\theta + k_{22}\sin\theta)s_i^+ - (k_{21}\sin\theta - k_{22}\cos\theta)(s_i^+)'\right] \end{Bmatrix}^T \begin{bmatrix} k_{11} & k_{12} \\ k_{21} & k_{22} \end{bmatrix}^{-1}$$
$$\times \begin{Bmatrix} \left[\alpha_j(k_{11}\cos\theta + k_{12}\sin\theta)s_j^+ - (k_{11}\sin\theta - k_{12}\cos\theta)(s_j^+)'\right] \\ \left[\alpha_j(k_{21}\cos\theta + k_{22}\sin\theta)s_j^+ - (k_{21}\sin\theta - k_{22}\cos\theta)(s_j^+)'\right] \end{Bmatrix} r dr\, d\theta$$
$$= \frac{R^{\alpha_i+\alpha_j}}{\alpha_i+\alpha_j} \int_{\theta_1}^{\theta_1+\omega} \begin{Bmatrix} \left[\alpha_i(k_{11}\cos\theta + k_{12}\sin\theta)s_i^+ - (k_{11}\sin\theta - k_{12}\cos\theta)(s_i^+)'\right] \\ \left[\alpha_i(k_{21}\cos\theta + k_{22}\sin\theta)s_i^+ - (k_{21}\sin\theta - k_{22}\cos\theta)(s_i^+)'\right] \end{Bmatrix}^T$$
$$\times \begin{bmatrix} k_{11} & k_{12} \\ k_{21} & k_{22} \end{bmatrix}^{-1} \begin{Bmatrix} \left[\alpha_j(k_{11}\cos\theta + k_{12}\sin\theta)s_j^+(-(k_{11}\sin\theta - k_{12}\cos\theta)(s_j^+)'\right] \\ \left[\alpha_j(k_{21}\cos\theta + k_{22}\sin\theta)s_j^+ - (k_{21}\sin\theta - k_{22}\cos\theta)(s_j^+)'\right] \end{Bmatrix} d\theta. \tag{4.32}$$

After some straightforward algebraic manipulations, (4.32) is simplified to

$$(B_c)_{ij} = \frac{R^{\alpha_i+\alpha_j}}{\alpha_i+\alpha_j}\int_{\theta_1}^{\theta_1+\omega}\Big\{\alpha_i\alpha_j\left(k_{11}\cos^2\theta + k_{12}\sin 2\theta + k_{22}\sin^2\theta\right)s_i^+ s_j^+$$
$$+ \left(k_{11}\sin^2\theta - k_{12}\sin 2\theta + k_{22}\cos^2\theta\right)(s_i^+)'(s_j^+)'$$
$$+ \left(\frac{1}{2}(k_{22}-k_{11})\sin 2\theta + k_{12}\cos 2\theta\right)\left(\alpha_i s_i^+(s_j^+)' + \alpha_j(s_i^+)'s_j^+\right)\Big\}\,d\theta,$$
$$\text{no summation on } i \text{ and } j,\ i,j = 1,2,\ldots,N \tag{4.33}$$

$$(F_c)_j = R^{\alpha_j}\int_{\theta_1}^{\theta_1+\omega}[\hat{\tau}]_R\Big[\alpha_j(k_{11}\cos^2\theta + k_{12}\sin 2\theta + k_{22}\sin^2\theta)s_j^+$$
$$+ \left(\frac{1}{2}(k_{22}-k_{11})\sin 2\theta + k_{12}\cos 2\theta\right)(s_j^+)'\Big]_R\,d\theta,$$
$$\text{no summation on } j = 1,2,\ldots,N \tag{4.34}$$

Remark 4.3. *The matrix* $[B_c]$ *is symmetric, so that* (4.31) *can be written also as* $A = [B_c]^{-1}\{F_c\}$.

Remark 4.4. *Computing the entries of the matrix* $[B_c]$ *involves only a path integral (1-D integration) and includes only terms associated with the eigenpairs and the heat conduction coefficients. Its dimension is determined by the number of GFIFs sought, thus is usually very small.*

Remark 4.5. *Computing the entries of the vector* $\{F_c\}$ *involves only a path integral (1-D integration) along an arc, and requires the knowledge of the exact solution along that arc in addition to the eigenpairs and the heat conduction coefficients. In practice, instead of the exact solution, an approximation* τ_{FE} *is used.*

Remark 4.6. *Homogeneous Neumann boundary conditions* $\boldsymbol{q}\cdot\boldsymbol{n} = 0$ *on* Γ_1 *and* Γ_2 *in the framework of the complementary weak formulation have to be treated by constraining the statically admissible space. However, using the eigenpairs in constructing the statically admissible space, the constraints are automatically satisfied, because any* $\boldsymbol{q}_i$ *in* (4.28) *satisfies the condition* $\boldsymbol{q}\cdot\boldsymbol{n} = 0$ *on* Γ_1 *and* Γ_2.

The Laplace equation as a special case: For the specific example of the *Laplace equation*, we have $k_{ij} = \delta_{ij}$, so expressions for $[B_c]$ and $\{F_c\}$ are simplified to

$$(B_c)_{ij} = \frac{R^{\alpha_i+\alpha_j}}{\alpha_i+\alpha_j}\int_{\theta_1}^{\theta_1+\omega}\left(\alpha_i\alpha_j s_i^+ s_j^+ + (s_i^+)'(s_j^+)'\right)d\theta,$$
$$\text{no summation on } i \text{ and } j, \tag{4.35}$$

$$(F_c)_j = \alpha_j R^{\alpha_j}\int_{\theta_1}^{\theta_1+\omega}[\hat{\tau}]_R s_j^+\,d\theta, \qquad \text{no summation on } j \tag{4.36}$$

We need the following lemma to prove that the matrix $[B_c]$ is a diagonal matrix in the case of the Laplace equation.

Lemma 4.1. *Let $s_i^+(\theta)$ and $s_j^+(\theta)$ be the ith and jth eigenpairs of the Laplace equation in Ω_R with homogeneous boundary conditions on Γ_1 and Γ_2. Then*

$$\int_{\theta_1}^{\theta_1+\omega} (s_i^+)'(s_j^+)' d\theta = \begin{cases} 0, & i \neq j, \\ \alpha_i^2 \int_{\theta_1}^{\theta_1+\omega} (s_i^+)^2 d\theta, & i = j. \end{cases} \tag{4.37}$$

Proof. Multiply the Laplace equation by a function χ, integrate over the domain Ω_R, then use Green's theorem to obtain

$$\oint_{\partial\Omega_R} (\nabla\tau \cdot n)\chi \, d\Gamma - \iint_{\Omega_R} (\nabla\tau)\cdot(\nabla\chi) \, d\Omega = 0. \tag{4.38}$$

Take $\tau = r^{\alpha_i} s_i^+(\theta)$ and $\chi = r^{\alpha_j} s_j^+(\theta)$. Then these satisfy the homogeneous boundary conditions on Γ_1 and Γ_2, so that (4.38) becomes

$$\int_{\theta_1}^{\theta_1+\omega} \left[\frac{\partial r^{\alpha_i} s_i^+(\theta)}{\partial r} r^{\alpha_j} s_j^+(\theta) \right]_{r=R} R d\theta \tag{4.39}$$

$$- \iint_{\Omega_R} \left[\frac{\partial r^{\alpha_i} s_i^+(\theta)}{\partial r} \frac{\partial r^{\alpha_j} s_j^+(\theta)}{\partial r} + \frac{1}{r^2} \frac{\partial r^{\alpha_i} s_i^+(\theta)}{\partial \theta} \frac{\partial r^{\alpha_j} s_j^+(\theta)}{\partial \theta} \right] r dr d\theta = 0,$$

no summation on i and j,

which after integrating over r, becomes

$$R^{\alpha_i+\alpha_j} \left\{ \alpha_i \left(1 - \tfrac{\alpha_j}{\alpha_i+\alpha_j}\right) \int_{\theta_1}^{\theta_1+\omega} s_i^+ s_j^+ d\theta - \tfrac{1}{\alpha_i+\alpha_j} \int_{\theta_1}^{\theta_1+\omega} (s_i^+)'(s_j^+)' \, d\theta \right\} = 0,$$

no summation on i and j (4.40)

Equation (4.40) has to hold for any R, so that,

$$\alpha_i \left(1 - \tfrac{\alpha_j}{\alpha_i+\alpha_j}\right) \int_{\theta_1}^{\theta_1+\omega} s_i^+ s_j^+ d\theta - \tfrac{1}{\alpha_i+\alpha_j} \int_{\theta_1}^{\theta_1+\omega} (s_i^+)'(s_j^+)' \, d\theta = 0, \tag{4.41}$$

no summation on i and j.

For the Laplace equation $s_j^+ = s_j^-$, and using the orthogonality of the eigenfunction (1.21), the first integral in (4.40) is zero for $i \neq j$, so that

$$\int_{\theta_1}^{\theta_1+\omega} (s_i^+)'(s_j^+)' \, d\theta = 0, \qquad i \neq j. \tag{4.42}$$

In case $i = j$, (4.41) becomes

$$\alpha_i/2 \int_{\theta_1}^{\theta_1+\omega} (s_i^+)^2 d\theta - \frac{1}{2\alpha_i} \int_{\theta_1}^{\theta_1+\omega} [(s_i^+)']^2 \, d\theta = 0, \qquad \text{no summation on } i. \quad \blacksquare \tag{4.43}$$

With the help of Lemma 4.1, the matrix $[B_c]$ in (4.35) (for the Laplace equation) is diagonal, and its entries are easily computed by

$$(B_c)_{ij} = \begin{cases} 0, & i \neq j, \\ \alpha_i R^{2\alpha_i} \int_{\theta_1}^{\theta_1+\omega} (s_i^+)^2 d\theta, & i = j, \end{cases} \qquad \text{no summation on } i. \tag{4.44}$$

It can further be shown that for Dirichlet, Neumann, or Newton homogeneous boundary conditions the diagonal terms in $[B_c]$ are given by

$$(B_c)_{ii} = (\alpha_i/2) R^{2\alpha_i} \omega, \qquad \text{no summation on } i. \tag{4.45}$$

Problem 4.1. Use (1.11) and (1.14) - (1.15) to obtain the result in equation (4.45).

Because $[B_c]$ is diagonal, for the Laplace equation one can explicitly compute each of the A_i's, using (4.45) and (4.36):

$$A_i = \frac{2}{\omega R^{\alpha_i}} \int_{\theta_1}^{\theta_1+\omega} [\hat{\tau}]_R s_i^+ \, d\theta, \qquad \text{no summation on } i. \tag{4.46}$$

Because the solution τ is unknown, we replace $[\hat{\tau}]_R$ by its finite element approximation. Denoting the error in the finite element solution by $e = \tau - \tau_{\text{FE}}$, we can show that the error in the A_i's due to the use of the finite element solution is bounded by the finite element error in the energy norm.

Theorem 4.1. *The error in A_i due to replacing τ with τ_{FE} is bounded by the error in energy norm, $|A_i - (A_i)_{\text{FE}}| \leq C(R)\|e\|_{\mathcal{E}}$.*

Proof. Consider the difference between A_i and its finite element approximation, and because $\mathcal{F}_c$ is a linear form, one gets

$$\begin{aligned} A_i - (A_i)_{\text{FE}} &= \frac{2}{\omega R^{\alpha_i}} \int_{\theta_1}^{\theta_1+\omega} [\tau - \tau_{\text{FE}}]_R s_i^+ \, d\theta \\ &= \frac{2}{\omega R^{\alpha_i}} \int_{\theta_1}^{\theta_1+\omega} e(\theta) s_i^+ \, d\theta, \qquad \text{no summation on } i \end{aligned} \tag{4.47}$$

The eigenfunctions $s_i^+(\theta)$ are analytic continuous functions on Γ_R. Therefore they are normalized so that $|s_i^+(\theta)| \leq 1$, and (4.47) becomes

$$|A_i - (A_i)_{\text{FE}}| \leq \frac{2}{\omega R^{\alpha_i}} \int_{\theta_1}^{\theta_1+\omega} |e(\theta)|\, d\theta \equiv \frac{2}{\omega R^{\alpha_i}} \|e\|_{L_1(\Gamma_R)}. \tag{4.48}$$

Recalling that $\| \bullet \|_{L_r} \leq C \| \bullet \|_{L_s}$, $s \geq r \geq 1$, one obtains

$$|A_i - (A_i)_{\text{FE}}| \leq \frac{2C}{\omega R^{\alpha_i}} \|e\|_{L_2(\Gamma_R)} \leq \frac{2C}{\omega R^{\alpha_i}} \|e\|_{L_2(\partial\Omega_R)},$$

where C is a generic constant. We use the trace theorem now to obtain

$$|A_i - (A_i)_{\text{FE}}| \leq \frac{C}{\omega R^{\alpha_i}} \|e\|_{H^1(\Omega_R)} \leq \frac{C}{\omega R^{\alpha_i}} \|e\|_{H^1(\Omega)}. \tag{4.49}$$

Using Friedrich's inequality, it is shown that the $H^1(\Omega)$ norm is equivalent to the $H^1(\Omega)$ seminorm for the Laplace problem, so we finally obtain

$$|A_i - (A_i)_{\text{FE}}| \leq \frac{C}{\omega R^{\alpha_i}} \|e\|_{H^1(\Omega)} \leq \frac{C}{\omega R^{\alpha_i}} \|e\|_{\mathcal{E}}. \quad \blacksquare \tag{4.50}$$

Remark 4.7. *We conclude that the convergence rate of the approximated GFIFs for the Laplace equation is at least as fast as the convergence rate of the finite-element error in energy norm. Numerical experiments show that in practice, the convergence rate is as fast as the convergence of the finite-element energy (twice as fast as the convergence rate in energy norm), namely the method is superconvergent.*

The following numerical examples indicate that errors in the computed GFIFs converge much faster than the error in energy norm.

4.2.3 Extracting GFIFs Using the Complementary Weak Formulation and Approximated Eigenpairs

In general, instead of the the exact eigenpairs, which are not known, the modified Steklov method is applied and their approximation is used. The approximated eigenfunctions, being computed by the p-version of the finite element method, using elements of polynomial order p, are given by

$$(s_j^+)_{\text{FE}}(\theta(\xi)) = c_{ij} \Phi_i(\xi), \qquad i = 1, 2, \ldots, p+1, \tag{4.51}$$

where $\Phi_i(\xi)$ are the edge shape functions, and c_{ij} is the eigenvector corresponding to the jth eigenvalue. In the following we formulate the matrix $[B_c]$ and the vector $\{F_c\}$ for the case that the eigenpairs are only an approximation of the exact values. Denoting by $(\alpha_i)_{\text{FE}}$ the approximated eigenvalues, (4.33) becomes

$$
\begin{aligned}
(B_c)_{ij} &= \frac{R^{(\alpha_i)_{\mathrm{FE}}+(\alpha_j)_{\mathrm{FE}}}}{(\alpha_i)_{\mathrm{FE}} + (\alpha_j)_{\mathrm{FE}}} \\
&\times \int_{\theta_1}^{\theta_1+\omega} \Big\{ (\alpha_i)_{\mathrm{FE}}(\alpha_j)_{\mathrm{FE}} \left(k_{11}\cos^2\theta + k_{12}\sin 2\theta + k_{22}\sin^2\theta\right) \\
&\qquad \times c_{ik}\Phi_k(\theta) c_{j\ell}\Phi_\ell(\theta) \\
&\qquad + \left(k_{11}\sin^2\theta - k_{12}\sin 2\theta + k_{22}\cos^2\theta\right) c_{ik}\Phi_k(\theta)' c_{j\ell}\Phi_\ell(\theta)' \\
&\qquad + \left(\frac{1}{2}(k_{22} - k_{11})\sin 2\theta + k_{12}\cos 2\theta\right) \\
&\qquad \times \left((\alpha_i)_{\mathrm{FE}} c_{ik}\Phi_k(\theta) c_{j\ell}\Phi_\ell(\theta)' + (\alpha_j)_{\mathrm{FE}}\Phi_k(\theta)' c_{j\ell}\Phi_\ell(\theta)\right)\Big\}\, d\theta, \\
&\quad \text{no summation on } i \text{ and } j,\ i, j = 1, 2, \ldots, N
\end{aligned}
\tag{4.52}
$$

Assume that the finite element mesh used for computing the eigenpairs has nG elements in the circumferential direction and a polynomial degree p. Using a Gauss quadrature of N_G points, the explicit expression for each term in $[B_c]$ is given by

$$
\begin{aligned}
(B_c)_{ij} &= \frac{R^{(\alpha_i)_{\mathrm{FE}}+(\alpha_j)_{\mathrm{FE}}}}{(\alpha_i)_{\mathrm{FE}} + (\alpha_j)_{\mathrm{FE}}} \sum_{n=1}^{nG} \sum_{k,\ell=1}^{p+1} \sum_{m=1}^{N_G} w_m c_{ik}^{(n)} c_{j\ell}^{(n)} \\
&\quad \times \Big\{ \frac{\omega^{(n)}}{2} (\alpha_i)_{\mathrm{FE}}(\alpha_j)_{\mathrm{FE}} \Big(k_{11}^{(n)}\cos^2[\theta(\xi_m)] + k_{12}^{(n)}\sin[2\theta(\xi_m)] \\
&\qquad\qquad + k_{22}^{(n)}\sin^2[\theta(\xi_m)]\Big) \times \Phi_k(\xi_m)\Phi_\ell(\xi_m) \\
&\quad + \frac{2}{\omega^{(n)}} \left(k_{11}^{(n)}\sin^2[\theta(\xi_m)] - k_{12}^{(n)}\sin[2\theta(\xi_m)] + k_{22}^{(n)}\cos^2[\theta(\xi_m)]\right) \\
&\qquad \times \Phi_k'(\xi_m)\Phi_\ell'(\xi_m) \\
&\quad + \left(\frac{1}{2}(k_{22}^{(n)} - k_{11}^{(n)})\sin[2\theta(\xi_m)] + k_{12}^{(n)}\cos[2\theta(\xi_m)]\right) \\
&\qquad \times \left((\alpha_i)_{\mathrm{FE}}\Phi_k(\xi_m)\Phi_\ell'(\xi_m) + (\alpha_j)_{\mathrm{FE}}\Phi_k'(\xi_m), \Phi_\ell(\xi_m)\right)\Big\} \\
&\qquad \text{no summation on } i \text{ and } j, \qquad i, j = 1, 2, \ldots, N,
\end{aligned}
\tag{4.53}
$$

where w_m and ξ_m are the weights and abscissas of the Gauss quadrature, and $\omega^{(n)}$ is the opening angle of element n used for the computation of the eigenpairs.

Similarly, the expression for the terms in the vector $\{F_c\}$ in (4.34) is

$$(F_c)_i = R^{(\alpha_i)_{\text{FE}}} \sum_{n=1}^{nG} \sum_{k=1}^{p+1} \sum_{m=1}^{N_G} \hat{\tau}(\theta(\xi_m)) w_m c_{ik}^{(n)}$$
$$\times \Bigg[(\alpha_i)_{\text{FE}} \frac{\omega^{(n)}}{2} \left(k_{11}^{(n)} \cos^2[\theta(\xi_m)] + k_{12}^{(n)} \sin[2\theta(\xi_m)] + k_{22}^{(n)} \sin^2[\theta(\xi_m)] \right) \Phi_k(\xi_m)$$
$$+ \left(\frac{1}{2}(k_{22}^{(n)} - k_{11}^{(n)}) \sin[2\theta(\xi_m)] + k_{12}^{(n)} \cos[2\theta(\xi_m)] \right) \Phi'_k(\xi_m) \Bigg], \tag{4.54}$$

no summation on i, $\quad i = 1, 2, \ldots, N.$

Considering the Laplace equation, the above expressions simplify to

$$(B_c)_{ij} = \frac{R^{(\alpha_i)_{\text{FE}} + (\alpha_j)_{\text{FE}}}}{(\alpha_i)_{\text{FE}} + (\alpha_j)_{\text{FE}}} \sum_{n=1}^{nG} (\alpha_i)_{\text{FE}} (\alpha_j)_{\text{FE}} \sum_{k,\ell=1}^{p+1} c_{ik}^{(n)} c_{j\ell}^{(n)}$$
$$\times \left[\left(\frac{\omega^{(n)}}{2} \sum_{m=1}^{N_G} w_m \Phi_k(\xi_m) \Phi_\ell(\xi_m) \right) + \left(\frac{2}{\omega^{(n)}} \sum_{m=1}^{N_G} w_m \Phi'_k(\xi_m) \Phi'_\ell(\xi_m) \right) \right],$$

no summation on i and j, $\quad i, j = 1, 2, \ldots, N.$ (4.55)

Remark 4.8. $(B_c)_{ij} = 0$ *for* $i \neq j$, *and therefore only the diagonal terms are to be computed. The values* $(B_c)_{ij}$, $i \neq j$, *are computed also to assess the accuracy of the approximate eigenpairs.*

Expression (4.54) becomes for the Laplace equation

$$(F_c)_i = (\alpha_i)_{\text{FE}} R^{(\alpha_i)_{\text{FE}}} \sum_{n=1}^{nG} \sum_{k=1}^{p+1} c_{ik}^{(n)} \frac{\omega^{(n)}}{2} \sum_{m=1}^{N_G} w_m \Phi_k(\xi_m) \hat{\tau}(\theta(\xi_m)),$$

no summation on i. (4.56)

Note that the finite element discretization over the domain Ω may be different from the one used for the modified Steklov problem.

4.3 Numerical Examples: Extracting GFIFs Using the Complementary Weak Form

Numerical examples are provided to demonstrate the accuracy and efficiency of GFIFs extraction by the complementary weak form. We provide examples using both the analytical eigenpairs and the approximated eigenpairs.

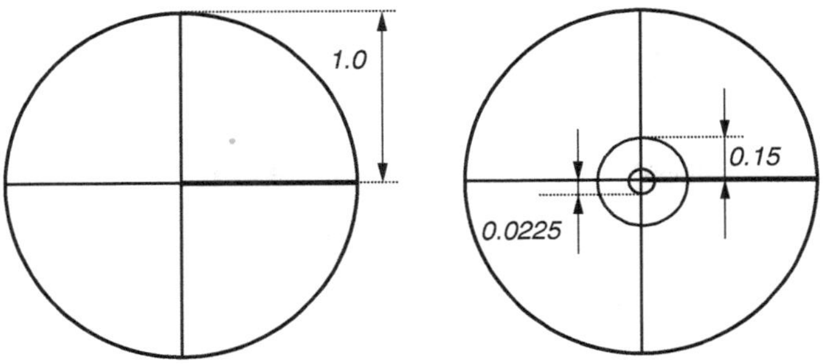

Fig. 4.2 The finite element mesh for Babuška's model problem.

Table 4.1 Computed values of the first three GFIFs, $R = 0.9$ and $N = 10$.

	$p=1$	$p=2$	$p=3$	$p=4$	$p=5$	$p=6$	$p=7$	$p=8$	$p=\infty$
DOF	12	36	64	104	156	220	296	384	∞
$\|e\|_{\mathcal{E}}$ (%)	34.5	16.7	12.8	11.3	10.3	9.5	8.9	8.4	0
$(A_1)_{\mathrm{FE}}$	−1.106022	−1.26095	−1.28694	−1.30458	−1.31351	−1.31990	−1.32479	−1.32849	−1.35812
e_{A_1} (%)	−18.56	−7.15	−5.24	−3.94	−3.28	−2.81	−2.45	−2.18	0
$(A_2)_{\mathrm{FE}}$	0.892975	0.970822	0.969563	0.970206	0.970089	0.970075	0.970091	0.970084	0.970087
e_{A_2} (%)	−7.9	0.075	−0.05	0.012	0.0002	−0.0012	0.0004	−0.0003	0
$(A_3)_{\mathrm{FE}}$	0.378148	0.445853	0.452560	0.452493	0.452697	0.452704	0.452706	0.452707	0.452707
e_{A_3} (%)	−16.4	−1.5	−0.03	−0.047	−0.002	−0.0007	−0.0002	0	0

4.3.1 Laplace equation with Newton BCs

Consider first the Laplace problem introduced in Section 3.4.2, solved by Babuška and Miller [12] using the dual singular function method. The domain is shown in Figure 3.5 over which the Laplace equation is to be solved with Newton boundary conditions on the upper and lower surfaces of the slit; see the problem statement in (3.51). The solution to the problem, accurate up to the sixth significant digit [12], is given by (3.52). This particular example problem was chosen to demonstrate that the proposed method has the same superconvergent properties as the extraction method proposed by Babuška and Miller [12]. We solved the problem using the p-version of the finite element method over the mesh shown in Figure 4.2, having two refinements toward the singular point. The trunk space was used as the trial function space in all computations. Using the shown mesh, we extract the GFIFs on the path $R = 0.9$ having 10 terms in the series. In Table 4.1 we summarize the approximated first three GFIFs, the corresponding number of degrees of freedom, and the relative error in energy norm.

The following conclusions may be drawn from the results shown in Table 4.1 and other numerical experiments performed:

1. Despite the presence of a strong ($r^{1/4}$-type) singularity, $(A_1)_{\mathrm{FE}}$ appears to be converging at a rate that is at least twice the convergence rate of the error in energy norm. This rate of convergence is approximately the same as that reported in [12].

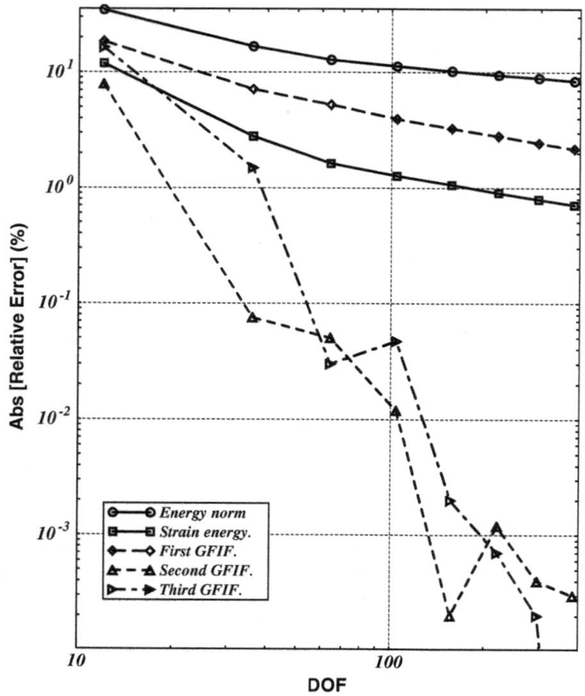

Fig. 4.3 Convergence of the relative error in energy norm, strain energy, and the GFIFs for Babuška's problem.

2. The GFIFs $(A_2)_{FE}$ and $(A_3)_{FE}$ are much more accurate than $(A_1)_{FE}$, and the observed convergence rate is considerably faster that the convergence of the error in energy norm.
3. For path radii taken far enough from the singular point, $R > 0.5$ in this example problem, the accuracy of the GFIFs is almost independent of R.
4. As expected, the number of terms considered in the series has no influence on the accuracy of the GFIFs (because the matrix $[B_c]$ is diagonal for the Laplace problem).

We present in Figure 4.3 the convergence of the GFIFs compared to the relative error in energy norm and the relative error in strain energy. Note that the rate of convergence in the first GFIF is faster than the rate of convergence of the energy norm, and at $p > 4$ is virtually the same as the rate of convergence of the strain energy. The second and third GFIFs converge much faster. It is seen that the first GFIF converges monotonically, which should not be expected in general.

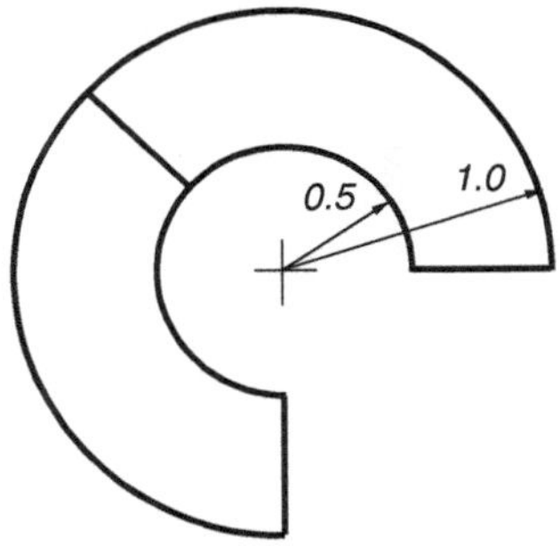

Fig. 4.4 Mesh for the computation of the approximate eigenpairs.

4.3.2 Laplace Equation with Homogeneous Neumann BCs: Approximate eigenpairs

The example discussed in this subsection is constructed to demonstrates the influence of the approximate eigenpairs on the accuracy of the extracted GFIFs.

Consider the $3\pi/2$ corner discussed in Section 2.3, shown in Figure 2.7 with homogeneous Neumann boundary conditions on the faces intersecting at $(x, y) = (0, 0)$ and Neumann BCs corresponding to the exact analytic solution, which is known, on the other boundaries (the derivatives in the$x$ and y directions are prescribed on the boundaries of the L-shaped domain).

First, an approximation to the eigenpairs has to be obtained. The modified Steklov method is used over a mesh containing two elements shown in Figure 4.4. As the p-level of the shape functions is increased over the mesh in Figure 4.4, a better approximation of the eigenpairs is obtained. We use the eigenpairs obtained when assigning p-levels 4, 5, 6, 7, and 8.

Once the approximate eigenpairs are available, a finite element solution is sought for the L-shaped domain. We construct a mesh containing the minimum possible number of elements over the L-shaped domain without any refinements in the vicinity of the singular point, as shown in Figure 2.7. The boundary conditions are imposed on the L-shaped boundaries, with the GFIFs chosen to be $A_1 = 1$, $A_2 = 1/2$, $A_3 = 1/3$, $A_4 = 1/4$, $A_5 = 1/5$, and $A_i = 0$, $(i = 6, 7, \ldots)$. The GFIFs were then extracted, taking R to be 0.9. The results of these computations are displayed in Table 4.2.

The following conclusions may be drawn from the results shown in Table 4.2:

1. The errors in the approximate ith eigenpair do not influence the accuracy of the jth GFIF. This is because the eigenfunctions are orthogonal.
2. The error in the GFIFs is always bounded by the error in energy norm when the error in the eigenpairs is less than 0.1%. Moreover, in this case the error in the GFIFs is virtually the same as if the exact eigenvalues had been used to extract the GFIFs.

Table 4.2 Relative error (%) in computed GFIFs for the L-shaped domain.

	$\|e\|_E$		$p=4$	$p=5$	$p=6$	$p=7$	$p=8$	
			6.02	4.65	3.74	3.10	2.62	extrapolated
p-level for e-val computation	GFIF #	Error in eigen-value						
	$\hat{A}_1$	0.0002	−1.63	−0.75	−0.38	−0.45	−0.31	−0.5
	$\hat{A}_2$	0.03	0.54	0.36	0.36	0.42	0.41	0.4
$p=4$	$\hat{A}_3$	0.39	6.79	6.84	6.87	6.87	6.84	6.87
	$\hat{A}_4$	0.73	7.12	7.68	7.64	7.52	7.56	7.56
	$\hat{A}_5$	17	28.5	25.55	26.15	34.4	26.8	26.85
	$\hat{A}_1$	$3 \cdot 10^{-6}$	−1.62	−0.73	−0.36	−0.43	−0.30	−0.48
	$\hat{A}_2$	$3 \cdot 10^{-5}$	0.12	−0.064	−0.064	0.001	−0.0096	−0.016
$p=5$	$\hat{A}_3$	0.035	0.54	0.63	0.61	0.61	0.61	0.62
	$\hat{A}_4$	0.71	7.2	7.76	7.72	7.6	7.64	7.6
	$\hat{A}_5$	2.4	29.5	26.65	27.5	28.7	28.1	28.35
	$\hat{A}_1$	$4.5 \cdot 10^{-8}$	−1.62	−0.73	−0.37	−0.43	−0.29	−0.48
	$\hat{A}_2$	$2.25 \cdot 10^{-5}$	0.122	−0.058	−0.058	0.0066	−0.0044	−0.01
$p=6$	$\hat{A}_3$	0.0018	−0.216	−0.135	−0.138	−0.160	−0.17	−0.147
	$\hat{A}_4$	0.0084	−0.52	0.009	−0.044	−0.136	−0.112	−0.116
	$\hat{A}_5$	0.49	6.2	3.6	4.19	5.05	4.69	4.77
	$\hat{A}_1$	$<10^{-9}$	−1.62	−0.73	−0.37	−0.43	−0.29	−0.48
	$\hat{A}_2$	$3 \cdot 10^{-8}$	0.13	−0.05	−0.05	−0.016	0.005	−0.001
$p=7$	$\hat{A}_3$	$7.8 \cdot 10^{-5}$	−0.08	−0.01	−0.0048	−0.027	−0.037	−0.037
	$\hat{A}_4$	0.0069	−0.71	−0.184	−0.235	−0.326	−0.302	−0.304
	$\hat{A}_5$	0.063	1.235	−1.235	−0.639	−0.212	−0.173	−0.0825

$p = 8$	$\hat{A}_1$	$<10^{-9}$	−1.62	−0.73	−0.37	−0.43	−0.30	−0.48
	$\hat{A}_2$	$<10^{-9}$	0.13	−0.05	−0.05	0.016	0.0046	−0.0015
	$\hat{A}_3$	$2.3 \cdot 10^{-6}$	−0.058	0.021	0.027	0.004	−0.0055	0.009
	$\hat{A}_4$	$3.7 \cdot 10^{-5}$	−0.427	0.099	0.047	−0.044	−0.021	−0.022
	$\hat{A}_5$	0.0064	2.07	−0.436	0.155	1.0	0.63	0.715
$p = \infty$	$\hat{A}_1$	0	−1.63	−0.73	−0.36	−0.43	−0.30	−0.48
	$\hat{A}_2$	0	−1.34	−0.05	−0.05	−0.014	−0.004	−0.016
	$\hat{A}_3$	0	0	0	0	0	0	0
	$\hat{A}_4$	0	−0.41	0.108	0.056	−0.03	−0.0096	−0.011
	$\hat{A}_5$	0	1.5	−0.92	−0.33	0.49	0.105	0.207

3. Using the coarsest mesh possible for the extraction of both the eigenpairs and the GFIFs, excellent results have been obtained. The relative error in the first five eigenpairs is less than 0.007%, and the relative error in the first five GFIFs is less than 0.7% when the relative error in energy norm is 2.6%.

4.3.3 Anisotropic Heat Conduction Equation with Newton BCs

Consider the anisotropic heat conduction equation

$$4\frac{\partial^2 \tau}{\partial x_1^2} + \frac{\partial^2 \tau}{\partial x_2^2} = 0$$

presented in Section 3.4.3, over the domain in Figure 3.8.

On the two edges Γ_1 and Γ_2, which meet at the origin of the coordinate system, homogeneous Neumann boundary conditions are applied and $\tau = 0$ is specified at $(0, 0)$.

Using the solution τ in (3.54), one may prescribe on the circular boundary the Neumann boundary condition

$$\begin{aligned} \nabla \boldsymbol{q} \cdot \boldsymbol{n} = \frac{\partial \tau}{\partial r} &= A_1 r^{-1/3} [2(1 + 3\sin^2\theta)]^{-2/3} \\ &\times \left\{ \left[\frac{2}{3}(1 + 3\cos^2\theta)(1 + 3\sin^2\theta) - \frac{3}{2}\sin^2 2\theta \right] \right. \\ &\times \left. \cos[\frac{2}{3}\arctan(2\tan\theta)] + 2\sin 2\theta \sin[\frac{2}{3}\arctan(2\tan\theta)] \right\}, \end{aligned} \quad (4.57)$$

and choose arbitrarily $A_1 = 1$, so the solution in the entire domain is given by (3.54) with $A_1 = 1$ and $A_i = 0$, for $i \geq 2$.

Let Ω be the unit circle sector shown in Figure 4.5, which is divided into six finite elements, such that the refined finite element layer around the singular point has radius 0.15.

To demonstrate the entire numerical procedure, we do not use the exact eigenpairs in our computations but their approximations obtained at $p = 8$ ($\alpha_1 = 0.666666675$, $\alpha_2 = 1.333333307$, and $\alpha_3 = 2.000000413$), see Section 3.4.3.

The first three GFIFs were extracted, taking R to be 0.5. The number of degrees of freedom, the error in energy norm, and the computed values of the three GFIFs are listed in Table 4.3. Of course, A_1 has to converge to 1, and A_2 and A_3 have to converge to 0.

We may see from Table 4.3 that the GFIFs converge strongly, although not monotonically. Our method yields solutions at p-level 2 or 3 that are within the range of precision normally needed in engineering computations.

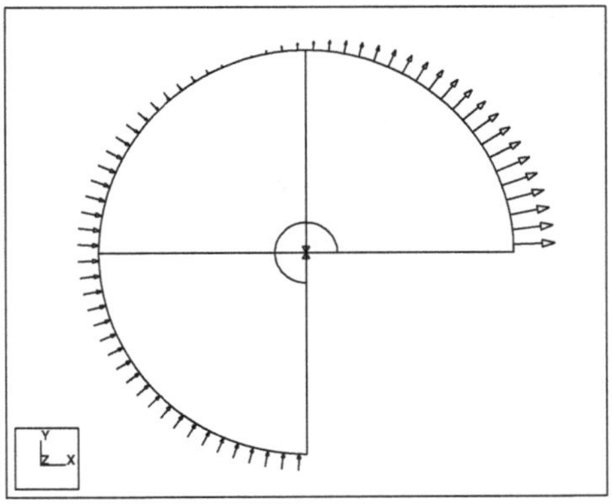

Fig. 4.5 Solution domain and mesh design (six elements) for model problem.

Table 4.3 First three GFIFs for the model problem with six-element mesh

	$p=1$	$p=2$	$p=3$	$p=4$	$p=5$	$p=6$	$p=7$	$p=8$
DOF	8	22	39	62	91	126	167	214
$\|e\|_{\mathcal{E}}(\%)$	33.95	7.02	3.74	2.23	1.65	1.32	1.10	0.93
A_1	0.8027116	0.9905142	0.9997004	0.9997054	0.9993824	0.9995902	0.9997769	0.9998464
A_2	0.1694275	0.0038147	0.0036318	$2.38e-4$	$1.5e-4$	$7.8e-5$	$4.6e-5$	$3.4e-5$
A_3	$1e-10$	$5e-10$	$4e-10$	$5e-10$	$5e-10$	$5e-10$	$5e-10$	$5e-10$

We have plotted the relative error in energy norm, the relative error in strain energy, and the absolute value of the relative error in A_1 on a log-log scale in Figure 4.6.

The convergence path of the GFIF A_1 follows closely that of the strain energy, which is a behavior referred to as "superconvergence."

Taking advantage of the strong convergence observed, we now use over the same domain only three finite elements, without the refined layer toward the singular point. The integration path was taken to be $R = 0.9$, and we plot the same data as in Figure 4.6 for the three-element mesh in Figure 4.7. The convergence curve of the GFIF A_1 is oscillating with mean approximately the strain energy convergence curve.

This anisotropic model problem clearly demonstrates the effectiveness and the superconvergent property of the proposed method for anisotropic materials.

4.3.4 *An Internal point at the Interface of Two Materials*

The problem presented in Section 3.4.4 is solved by a finite element mesh consisting of six elements, such that the inner elements have radius 0.15; see Figure 4.8. Using the eigenvalues obtained by the modified Steklov method at $p = 8$, we compute the first two GFIFs $(A_1)_{FE}$ and $(A_2)_{FE}$. These GFIFs, according to (3.61), have to converge to 1 as the number of degrees of freedom is increased. The number of degrees of freedom, the relative error in energy norm (%), the relative error in energy

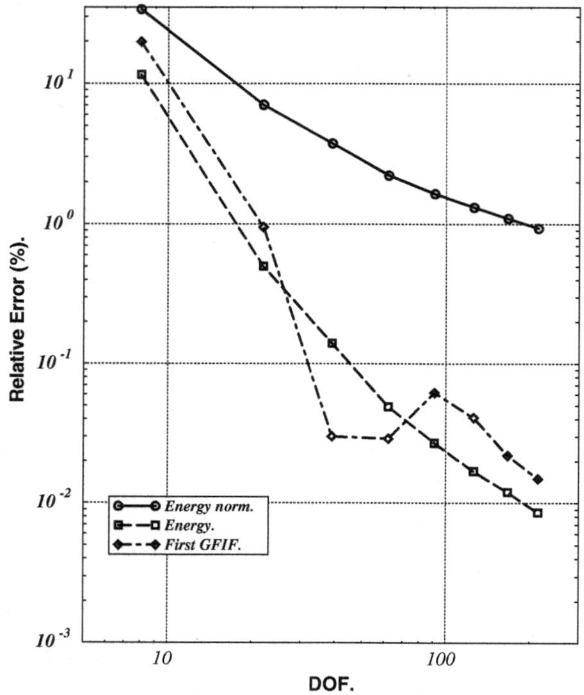

Fig. 4.6 Convergence of error in energy norm ($\|e\|_{\mathcal{E}}$), the strain energy ($\|e\|_{\mathcal{E}}^2$), and A_1 for the six-element mesh.

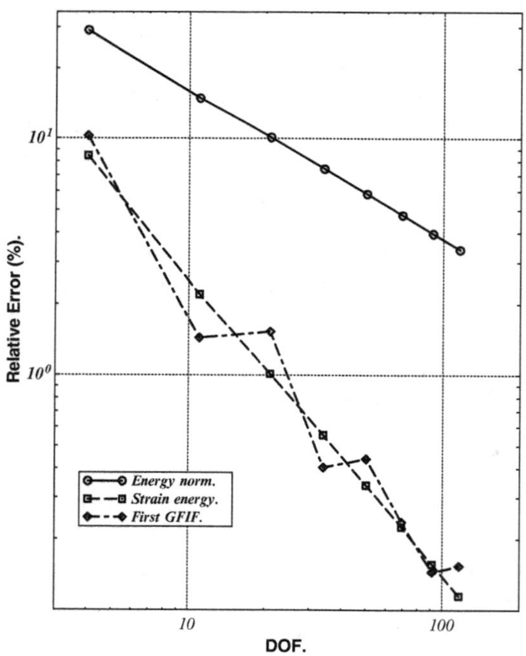

Fig. 4.7 Convergence of error in energy norm ($\|e\|_{\mathcal{E}}$), the strain energy ($\|e\|_{\mathcal{E}}^2$), and A_1 for the three-element mesh.

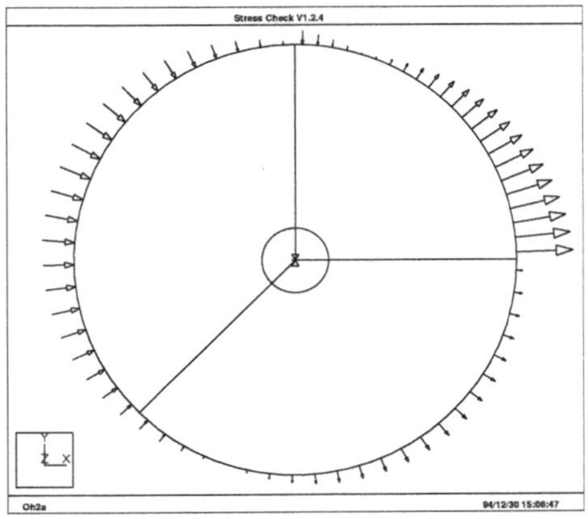

Fig. 4.8 FE mesh and boundary conditions for an internal point at the interface of two materials.

Table 4.4 First two GFIFs for scalar problem 3: Internal point at the interface of two materials.

	$p=1$	$p=2$	$p=3$	$p=4$	$p=5$	$p=6$	$p=7$	$p=8$
DOF	6	18	33	54	81	114	153	198
$\|e\|_{\mathcal{E}}$ (%)	73.68	10.52	8.47	1.28	0.75	0.51	0.41	0.34
$\|e\|^2_{\mathcal{E}}$ (%)	54.28	1.11	0.717	0.0164	0.0056	0.0026	0.00168	0.001156
$(A_1)_{FE}$	0.82729	0.98729	0.99934	0.99956	0.99973	0.99990	0.99993	0.99994
$(A_2)_{FE}$	0.19937	1.03126	1.03587	1.00040	0.99939	1.000006	1.00004	0.999997
$100((A_1)_{FE} - A_1)/A_1$	−17.27	−1.27	−0.065	−0.0434	−0.0273	−0.0103	−0.0067	−0.0059
$100((A_2)_{FE} - A_2)/A_2$	−80.06	3.12	3.587	0.0405	−0.0610	0.0005	0.0037	−0.00030

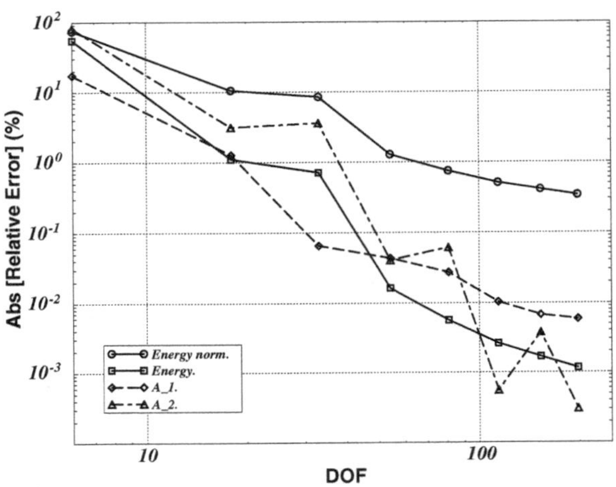

Fig. 4.9 Convergence of $\|e\|_{\mathcal{E}}$, the energy ($\|e\|^2_{\mathcal{E}}$), A_1 and A_2 for scalar problem 3.

(%), the computed value of the GFIFs, and the relative error in GFIFs (%) are listed in Table 4.4 for $R = 0.6$.

The data in Table 4.4 are plotted on a log-log scale in Figure 4.9. It is seen that the rate of convergence of the GFIF is faster than the rate of convergence in the energy norm, and although not monotonic, is similar to the rate of convergence of the energy.

Chapter 5
Eigenpairs for Two-Dimensional Elasticity

The two-dimensional elastic solution in the vicinity of a singular point has the same characteristics as presented for the heat conduction solution, namely, it can be expanded as a linear combination of eigenpairs and their coefficients:

$$\boldsymbol{u} = \sum_{i=1}^{I} \sum_{j=0}^{J} \sum_{\ell=0}^{L} A_{ij\ell} r^{\alpha_i + j} \ln^{\ell}(r) \boldsymbol{s}_{ij\ell}(\theta) + \boldsymbol{u}^{\text{reg}}, \tag{5.1}$$

r and θ being the polar coordinates of a system located in the singular point, and α_i and $\boldsymbol{s}(\theta)$ are the eigenpairs; M is zero except for cases in which the boundary near the singular point is curved (see Appendix C), and logarithmic terms may be present $J \neq 0$ only for special cases for which m multiple eigenvalues exist with fewer than m corresponding eigenvectors (the algebraic multiplicity is grater than the geometric multiplicity), or when inhomogeneous BCs are prescribed on the V-notch faces. This case is not rigorously discussed in this chapter, but several remarks are provided at the end of it and it is further addressed in the chapters that compute the eigenpairs numerically.

Near a singular point, in the case of an *isotropic* material, the completeness[1] of the eigenfunctions is ensured in the framework of a very general theory given by Kondratiev [95]. Kondratiev showed that the solution of any *even-order* elliptic boundary value problem near angular or conical points can be expressed as a series of eigenfunctions of the form $\boldsymbol{u}_i = r^{\alpha_i} (\ln r)^q \boldsymbol{s}_i(\theta)$.

In the 2-D isotropic elasticity case, Gregory [69] has provided a proof of completeness for a "free-free" wedge, and as proposed by Gregory, the proof could be extended to all other linear homogeneous boundary conditions. A detailed discussion on the behavior of the eigenvalues for isotropic, "free-free", 2-D corners was given by Vasilopoulos in [185].

[1]Completeness is defined in the sense that the analytic solution has an expansion as a sum of the eigenfunctions.

Z. Yosibash, *Singularities in Elliptic Boundary Value Problems and Elasticity and Their Connection with Failure Initiation*, Interdisciplinary Applied Mathematics 37, DOI 10.1007/978-1-4614-1508-4_5,

However, for anisotropic multimaterial corners a general proof of completeness of eigenfunctions does not exist. The only available proof (to the best of our knowledge) is the existence and uniqueness of the solution for the anisotropic bimaterial wedge presented by Zajaczkowski [212].

Throughout this book, it is implicitly assumed that even for the anisotropic multimaterial corner, the solution can be expanded in terms of a complete set of eigenfunctions as in the isotropic case. A mathematical proof is not available at present. However, some discussions and examples regarding the Steklov method are presented in [100].

To illustrate the solution characteristics and its derivation, we first use analytic methods, starting with an isotropic material in the vicinity of a reentrant corner.

5.1 Asymptotic Solution in the Vicinity of a Reentrant Corner in an Isotropic Material

Consider a reentrant corner in an isotropic material as shown in Figure 5.1. ***Notice that in this chapter the coordinate system is located so that the x_1 axis coincides with the bisector of the V-notch, so that the angle $-\omega/2 \leq \theta \leq \omega/2$ is measured from this axis. This notation is different from that in the previous chapters, but is common in the engineering literature.*** There are four different possible homogeneous boundary conditions on each of the boundaries Γ_1 and Γ_2

$$\begin{cases} u_n = 0 \\ u_t = 0 \end{cases} \qquad \text{hard clamped or fixed (HC)} \tag{5.2}$$

$$\begin{cases} u_n = 0 \\ T_t = 0 \end{cases} \qquad \text{soft clamped or symmetric (SC)} \tag{5.3}$$

$$\begin{cases} T_n = 0 \\ u_t = 0 \end{cases} \qquad \text{simply supported or antisymmetric (SS)} \tag{5.4}$$

$$\begin{cases} T_n = 0 \\ T_t = 0 \end{cases} \qquad \text{traction-free (TF)} \tag{5.5}$$

and ten possible combinations on the two boundaries, as follows: HC/HC, HC/SC, HC/SS, HC/TF, SC/SC, SC/SS, SC/TF, SS/SS, SS/TF, TF/TF.

There have been several investigations to analytically provide the eigenpairs for the 2-D elastic isotropic materials. To our knowledge, Williams [189] was the first to provide an analytical study on the stress singularities half a decade ago. Karp and Karal [93] studied the TF boundary conditions 40 years ago, providing the eigenpairs in a graphical form and the implicit equation to obtain them. Kalandiia [89] studied the HC, SC and TF boundary conditions with the use of the Airy stress function. Although many studies have been reported since then, we follow here Rössle's paper [153] to derive the implicit equations for the eigenvalues.

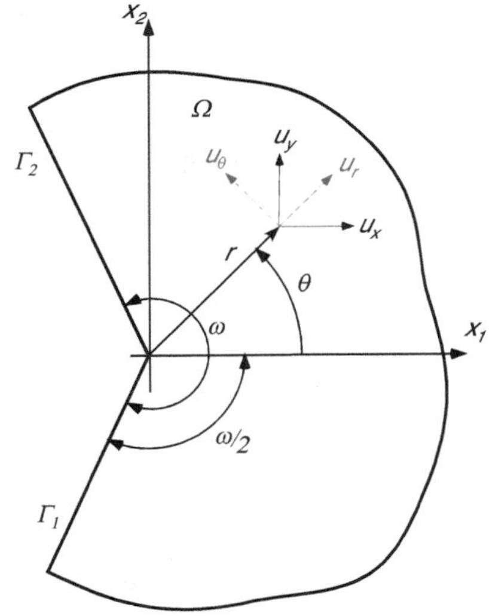

Fig. 5.1 Reentrant corner and notations for elasticity eigenpairs.

To prescribe the boundary conditions on Γ_1 and Γ_2 it is more natural to refer to a cylindrical coordinate system as well as to consider the displacements in the radial and circumferential directions. To distinguish these from the Cartesian quantities we denote them by

$$\tilde{\boldsymbol{u}} \stackrel{\text{def}}{=} \begin{Bmatrix} u_r \\ u_\theta \end{Bmatrix}, \qquad \tilde{\boldsymbol{\sigma}} \stackrel{\text{def}}{=} \begin{Bmatrix} \sigma_{rr} \\ \sigma_{\theta\theta} \\ \sigma_{r\theta} \end{Bmatrix}. \tag{5.6}$$

There is a simple connection between the displacements and stresses in cylindrical coordinates and their Cartesian counterparts

$$\boldsymbol{u} = \begin{bmatrix} \cos\theta & -\sin\theta \\ \sin\theta & \cos\theta \end{bmatrix} \tilde{\boldsymbol{u}}, \;\; \boldsymbol{\sigma} = [Q]\tilde{\boldsymbol{\sigma}}, \;\; \text{where } [Q] \stackrel{\text{def}}{=} \begin{bmatrix} \cos^2\theta & \sin^2\theta & -\sin 2\theta \\ \sin^2\theta & \cos^2\theta & \sin 2\theta \\ \frac{1}{2}\sin 2\theta & -\frac{1}{2}\sin 2\theta & \cos 2\theta \end{bmatrix}. \tag{5.7}$$

The Navier-Lamé system (1.45) is expressed in cylindrical base vectors (with displacements in the r and θ directions), under the assumption of plane-strain:

$$(\lambda + 2\mu)\left(\partial_{rr}u_r + \frac{1}{r}\partial_r u_r - \frac{1}{r^2}u_r\right) + \mu\frac{1}{r^2}\partial_{\theta\theta}u_r + (\lambda+\mu)\frac{1}{r}\partial_{r\theta}u_\theta$$
$$-(\lambda + 3\mu)\frac{1}{r^2}\partial_\theta u_\theta = 0, \tag{5.8}$$

$$\mu\left(\partial_{rr}u_\theta + \frac{1}{r}\partial_r u_\theta - \frac{1}{r^2}u_\theta\right) + (\lambda + 2\mu)\frac{1}{r^2}\partial_{\theta\theta}u_\theta + (\lambda+\mu)\frac{1}{r}\partial_{r\theta}u_r$$
$$+(\lambda + 3\mu)\frac{1}{r^2}\partial_\theta u_r = 0, \tag{5.9}$$

Equations (5.8) - (5.9) hold also for the plane-stress situation after one replaces the Lamé constant λ by $\tilde{\lambda} = 2\mu\lambda(\lambda + 2\mu)^{-1}$; see [125, Sec. 25].

Assume a solution of the form (this is formally obtained using the Mellin transform with respect to r on the system of equations (5.8) - (5.9)

$$\tilde{\boldsymbol{u}} = r^\alpha \begin{Bmatrix} s_r(\theta) \\ s_\theta(\theta) \end{Bmatrix} \stackrel{\text{def}}{=} r^\alpha \tilde{\boldsymbol{s}}(\theta). \tag{5.10}$$

After substituting (5.10) into (5.8) - (5.9), and dividing by $r^{\alpha-2}$, one reduces the problem to a set of two ODEs for the unknown functions $s_r(\theta), s_\theta(\theta)$:

$$\mu s_r'' + (\lambda + 2\mu)(\alpha^2 - 1)s_r + [\alpha(\lambda + \mu) - (\lambda + 3\mu)]s_\theta' = 0, \tag{5.11}$$

$$(\lambda + 2\mu)s_\theta'' + \mu(\alpha^2 - 1)s_\theta + [\alpha(\lambda + \mu) + (\lambda + 3\mu)]s_r' = 0. \tag{5.12}$$

For solving the system (5.11) - (5.12) we consider a solution vector of the form

$$\tilde{\boldsymbol{s}}(\theta) = e^{\imath\beta\theta} \begin{Bmatrix} C \\ \imath D \end{Bmatrix}, \tag{5.13}$$

where C, D, and β are unknowns. Substituting (5.13) into (5.11) - (5.12) a set of two homogeneous algebraic equations is obtained:

$$C\left[(\alpha^2 - 1)(\lambda + 2\mu) - \beta^2\mu\right] - D\beta\left[(\lambda + \mu)\alpha - (\lambda + 3\mu)\right] = 0, \tag{5.14}$$

$$C\beta\left[\alpha(\lambda + \mu) + (\lambda + 3\mu)\right] + D\left[-\beta^2(\lambda + 2\mu) + \mu(\alpha^2 - 1)\right] = 0. \tag{5.15}$$

Expressing C in terms of D and α in (5.14),

$$C = D\beta\frac{(\lambda + \mu)\alpha - (\lambda + 3\mu)}{(\alpha^2 - 1)(\lambda + 2\mu) - \beta^2\mu},$$

and inserting into (5.15) one obtains

$$D\left[\beta^4 - 2\beta^2(\alpha^2 + 1) + (\alpha^2 - 1)^2\right] = 0. \tag{5.16}$$

For a nontrivial solution the expression in parenthesis has to be zero, yielding

$$\beta_1 = (1 + \alpha), \quad \beta_2 = -(1 + \alpha), \quad \beta_3 = (1 - \alpha), \quad \beta_4 = -(1 - \alpha). \tag{5.17}$$

Once the β's are found, the relationships between C and D can be expressed also. For example, for $\beta_1 = (1 + \alpha)$ one obtains $C_1 = D_1$, for $\beta_2 = -(1 + \alpha)$ then $C_2 = -D_2$, for $\beta_3 = (1 - \alpha)$ one obtains $D_3 = -C_3\frac{(\lambda+\mu)\alpha+(\lambda+3\mu)}{(\lambda+\mu)\alpha-(\lambda+3\mu)}$, and for

$\beta_4 = -(1-\alpha)$ one obtains $D_4 = C_4 \frac{(\lambda+\mu)\alpha+(\lambda+3\mu)}{(\lambda+\mu)\alpha-(\lambda+3\mu)}$. We may conclude that for $\alpha \neq 0$,

$$\tilde{s}(\theta) = C_1 e^{\imath(1+\alpha)\theta} \begin{Bmatrix} 1 \\ \imath \end{Bmatrix} + C_2 e^{-\imath(1+\alpha)\theta} \begin{Bmatrix} 1 \\ -\imath \end{Bmatrix} + e^{\imath(1-\alpha)\theta} \begin{Bmatrix} C_3 \\ \imath D_3(C_3) \end{Bmatrix}$$
$$+ e^{-\imath(1-\alpha)\theta} \begin{Bmatrix} C_4 \\ \imath D_4(C_4) \end{Bmatrix}. \tag{5.18}$$

For $\alpha = 0$ we have two double identical roots $\beta_1 = \beta_3 = 1$ and $\beta_2 = \beta_4 = -1$, so that the solution for $\alpha = 0$ is

$$\tilde{s}(\theta) = C_1 e^{\imath\theta} \begin{Bmatrix} 1 \\ \imath \end{Bmatrix} + C_2 e^{-\imath\theta} \begin{Bmatrix} 1 \\ -\imath \end{Bmatrix} + \theta e^{\imath\theta} \begin{Bmatrix} C_3 \\ \imath D_3(C_3) \end{Bmatrix} + \theta e^{-\imath\theta} \begin{Bmatrix} C_4 \\ \imath D_4(C_4) \end{Bmatrix}. \tag{5.19}$$

The solutions (5.18) - (5.19) can be also expressed in terms of sin and cos functions after some algebraic manipulations, and in view of (5.10), we can finally express the elastic solution in the vicinity of a singular point with different generic constants C_i,

$$\tilde{\boldsymbol{u}} = C_1 r^\alpha \begin{Bmatrix} \cos(1+\alpha)\theta \\ -\sin(1+\alpha)\theta \end{Bmatrix} + C_2 r^\alpha \begin{Bmatrix} \sin(1+\alpha)\theta \\ \cos(1+\alpha)\theta \end{Bmatrix}$$
$$+ C_3 r^\alpha \begin{Bmatrix} [\lambda+3\mu-\alpha(\lambda+\mu)]\cos(1-\alpha)\theta \\ -[\lambda+3\mu+\alpha(\lambda+\mu)]\sin(1-\alpha)\theta \end{Bmatrix}$$
$$+ C_4 r^\alpha \begin{Bmatrix} [\lambda+3\mu-\alpha(\lambda+\mu)]\sin(1-\alpha)\theta \\ [\lambda+3\mu+\alpha(\lambda+\mu)]\cos(1-\alpha)\theta \end{Bmatrix}, \qquad \alpha \neq 0, \tag{5.20}$$

and for $\alpha = 0$ one obtains after manipulating (5.19),

$$\tilde{\boldsymbol{u}} = C_1 \begin{Bmatrix} \cos\theta \\ -\sin\theta \end{Bmatrix} + C_2 \begin{Bmatrix} \sin\theta \\ \cos\theta \end{Bmatrix} + C_3 \begin{Bmatrix} -(\lambda+3\mu)\theta\cos\theta + \mu\sin\theta \\ (\lambda+3\mu)\theta\sin\theta + (\lambda+2\mu)\cos\theta \end{Bmatrix}$$
$$+ C_4 \begin{Bmatrix} (\lambda+3\mu)\theta\sin\theta + \mu\cos\theta \\ (\lambda+3\mu)\theta\cos\theta - (\lambda+2\mu)\sin\theta \end{Bmatrix}, \qquad \alpha = 0. \tag{5.21}$$

Problem 5.1. Derive (5.20) from (5.18), and (5.21) from (5.19).

So far, α is undetermined, as well as the four constants C_1, C_2, C_3, C_4. This is the stage at which the boundary conditions on Γ_1 and Γ_2 are to be considered, enabling us to determine α and two of the four coefficients. In cylindrical coordinates, the normal and tangential directions on Γ_1 are simply the $-\theta$, and r directions, and on Γ_2 these are the θ and r directions. Thus, $u_n = u_r$, $u_t = u_\theta$, $T_n = T_\theta = \sigma_{\theta\theta}$, and $T_t = T_r = \sigma_{r\theta}$.

Because we wish to consider traction boundary conditions (three of the four possible homogeneous boundary conditions (5.3) - (5.5) involve tractions), we use Hooke's constitutive law for the plane-strain situation (relations between the stresses and strains, where the strains are expressed in terms of the displacements):

$$\sigma_{rr} = (\lambda + 2\mu)\partial_r u_r + \lambda \frac{1}{r}(\partial_\theta u_\theta + u_r), \tag{5.22}$$

$$\sigma_{\theta\theta} = (\lambda + 2\mu)\frac{1}{r}(\partial_\theta u_\theta + u_r) + \lambda \partial_r u_r, \tag{5.23}$$

$$\sigma_{r\theta} = \mu\left(\frac{1}{r}\partial_\theta u_r + \partial_r u_\theta - \frac{1}{r}u_\theta\right). \tag{5.24}$$

Notice that Hooke's law (5.22) - (5.24) holds for plane-stress situation also if one replaces the Lamé constant λ by $\tilde{\lambda} = 2\mu\lambda(\lambda + 2\mu)^{-1}$. Inserting (5.10) in (5.22) - (5.24) one obtains

$$\sigma_{rr} = r^{\alpha-1}\left\{(\lambda + 2\mu)\alpha s_r + \lambda(s'_\theta + s_r)\right\}, \tag{5.25}$$

$$\sigma_{\theta\theta} = r^{\alpha-1}\left\{(\lambda + 2\mu)(s'_\theta + s_r) + \lambda\alpha s_r\right\}, \tag{5.26}$$

$$\sigma_{r\theta} = r^{\alpha-1}\left\{\mu\left(s'_r + (\alpha - 1)s_\theta\right)\right\}. \tag{5.27}$$

In view of (5.20) (respectively (5.21)), the stresses are

$$\begin{aligned}\begin{Bmatrix}\sigma_{rr}\\ \sigma_{\theta\theta}\\ \sigma_{r\theta}\end{Bmatrix} &= C_1 2\alpha\mu r^{\alpha-1}\begin{Bmatrix}\cos(1+\alpha)\theta\\ -\cos(1+\alpha)\theta\\ -\sin(1+\alpha)\theta\end{Bmatrix} + C_2 2\alpha\mu r^{\alpha-1}\begin{Bmatrix}\sin(1+\alpha)\theta\\ -\sin(1+\alpha)\theta\\ \cos(1+\alpha)\theta\end{Bmatrix}\\ &+ C_3 2\alpha\mu r^{\alpha-1}\begin{Bmatrix}(\lambda+\mu)(3-\alpha)\cos(1-\alpha)\theta\\ (\lambda+\mu)(1+\alpha)\cos(1-\alpha)\theta\\ (\lambda+\mu)(1-\alpha)\sin(1-\alpha)\theta\end{Bmatrix}\\ &+ C_4 2\alpha\mu r^{\alpha-1}\begin{Bmatrix}(\lambda+\mu)(3-\alpha)\sin(1-\alpha)\theta\\ (\lambda+\mu)(1+\alpha)\sin(1-\alpha)\theta\\ -(\lambda+\mu)(1-\alpha)\cos(1-\alpha)\theta\end{Bmatrix}, \qquad \alpha \neq 0,\end{aligned} \tag{5.28}$$

and,

$$\begin{aligned}\begin{Bmatrix}\sigma_{rr}\\ \sigma_{\theta\theta}\\ \sigma_{r\theta}\end{Bmatrix} &= C_1\begin{Bmatrix}0\\0\\0\end{Bmatrix} + C_2\begin{Bmatrix}0\\0\\0\end{Bmatrix} + C_3\frac{2\mu}{r}\begin{Bmatrix}\lambda\sin\theta\\ (\lambda+2\mu)\sin\theta\\ -(\lambda+2\mu)\cos\theta\end{Bmatrix}\\ &+ C_4\frac{2\mu}{r}\begin{Bmatrix}\lambda\cos\theta\\ (\lambda+2\mu)\cos\theta\\ (\lambda+2\mu)\sin\theta\end{Bmatrix}, \qquad \alpha = 0.\end{aligned} \tag{5.29}$$

The last two terms in (5.29) have to vanish, because they produce infinite strain energy; thus $C_3 = C_4 = 0$. The first two terms are associated with a rigid-body motion in the x_1 and x_2 directions, producing of course zero stresses.

In [185], V-notches with TF/TF boundary conditions are studied by the Airy stress function method, where it is demonstrated that the $\alpha = 0$ case represents a state of zero stress, i.e., a rigid-body motion, so can be neglected. Therefore we will concentrate our attention on the case $\alpha \neq 0$.

As an example for deriving the explicit eigenequation, consider the TF/TF situation, for which the homogeneous boundary conditions have to be satisfied for all r, so in view of (5.28), the four conditions on the stresses for $\theta = \pm\omega/2$ are

$$\begin{Bmatrix} T_\theta \\ T_r \end{Bmatrix}_{\theta=\pm\frac{\omega}{2}} = \begin{Bmatrix} \sigma_{\theta\theta} \\ \sigma_{r\theta} \end{Bmatrix}_{\theta=\pm\frac{\omega}{2}} = \begin{Bmatrix} 0 \\ 0 \end{Bmatrix}, \tag{5.30}$$

and explicitly,

$$-2\alpha\mu\cos\frac{\omega(1+\alpha)}{2}C_1 - 2\alpha\mu\sin\frac{\omega(1+\alpha)}{2}C_2 + 2\alpha\mu(\lambda+\mu)(1+\alpha)\cos\frac{\omega(1-\alpha)}{2}C_3$$
$$+2\alpha\mu(\lambda+\mu)(1+\alpha)\sin\frac{\omega(1-\alpha)}{2}C_4 = 0,$$

$$-2\alpha\mu\sin\frac{\omega(1+\alpha)}{2}C_1 + 2\alpha\mu\cos\frac{\omega(1+\alpha)}{2}C_2 + 2\alpha\mu(\lambda+\mu)(1-\alpha)\sin\frac{\omega(1-\alpha)}{2}C_3$$
$$-2\alpha\mu(\lambda+\mu)(1-\alpha)\cos\frac{\omega(1-\alpha)}{2}C_4 = 0,$$

$$-2\alpha\mu\cos\frac{-\omega(1+\alpha)}{2}C_1 - 2\alpha\mu\sin\frac{-\omega(1+\alpha)}{2}C_2 + 2\alpha\mu(\lambda+\mu)(1+\alpha)\cos\frac{-\omega(1-\alpha)}{2}C_3$$
$$+2\alpha\mu(\lambda+\mu)(1+\alpha)\sin\frac{-\omega(1-\alpha)}{2}C_4 = 0,$$

$$-2\alpha\mu\sin\frac{-\omega(1+\alpha)}{2}C_1 + 2\alpha\mu\cos\frac{-\omega(1+\alpha)}{2}C_2 + 2\alpha\mu(\lambda+\mu)(1-\alpha)\sin\frac{-\omega(1-\alpha)}{2}C_3$$
$$-2\alpha\mu(\lambda+\mu)(1-\alpha)\cos\frac{-\omega(1-\alpha)}{2}C_4 = 0.$$

Define $\gamma \overset{\text{def}}{=} \frac{\omega(1+\alpha)}{2}$, so that $\frac{\omega(1-\alpha)}{2} = \omega - \gamma$, and using the odd/even properties of the trigonometric functions $\sin(-\gamma) = -\sin\gamma$, $\cos(-\gamma) = \cos\gamma$, the above equations can be written in matrix form:

$$2\alpha\mu\begin{bmatrix} -\cos\gamma & -\sin\gamma & (\lambda+\mu)(1+\alpha)\cos(\omega-\gamma) & (\lambda+\mu)(1+\alpha)\sin(\omega-\gamma) \\ -\sin\gamma & \cos\gamma & (\lambda+\mu)(1-\alpha)\sin(\omega-\gamma) & -(\lambda+\mu)(1-\alpha)\cos(\omega-\gamma) \\ -\cos\gamma & \sin\gamma & (\lambda+\mu)(1+\alpha)\cos(\omega-\gamma) & -(\lambda+\mu)(1+\alpha)\sin(\omega-\gamma) \\ \sin\gamma & \cos\gamma & -(\lambda+\mu)(1-\alpha)\sin(\omega-\gamma) & -(\lambda+\mu)(1-\alpha)\cos(\omega-\gamma) \end{bmatrix}\begin{Bmatrix} C_1 \\ C_2 \\ C_3 \\ C_4 \end{Bmatrix} = \mathbf{0}. \tag{5.31}$$

By adding and subtracting rows as follows—to row 1 add row 3, to row 2 add row 4, from row 3 subtract row 1, and from row 4 subtract row 2—one obtains:

$$4\alpha\mu \begin{bmatrix} -\cos\gamma & 0 & (\lambda+\mu)(1+\alpha)\cos(\omega-\gamma) & 0 \\ 0 & \cos\gamma & 0 & -(\lambda+\mu)(1-\alpha)\cos(\omega-\gamma) \\ 0 & \sin\gamma & 0 & -(\lambda+\mu)(1+\alpha)\sin(\omega-\gamma) \\ \sin\gamma & 0 & -(\lambda+\mu)(1-\alpha)\sin(\omega-\gamma) & 0 \end{bmatrix} \begin{Bmatrix} C_1 \\ C_2 \\ C_3 \\ C_4 \end{Bmatrix} = \mathbf{0}. \tag{5.32}$$

Then interchanging rows and columns, (5.32) can be brought to the following form:

$$4\alpha\mu \begin{bmatrix} -\cos\gamma & (\lambda+\mu)(1+\alpha)\cos(\omega-\gamma) & 0 & 0 \\ \sin\gamma & -(\lambda+\mu)(1-\alpha)\sin(\omega-\gamma) & 0 & 0 \\ 0 & 0 & \cos\gamma & -(\lambda+\mu)(1-\alpha)\cos(\omega-\gamma) \\ 0 & 0 & \sin\gamma & -(\lambda+\mu)(1+\alpha)\sin(\omega-\gamma) \end{bmatrix} \begin{Bmatrix} C_1 \\ C_3 \\ C_2 \\ C_4 \end{Bmatrix} = \mathbf{0}, \tag{5.33}$$

where C_1 and C_3 are connected and C_2 and C_4 are connected, and furthermore, C_1, C_3 are independent of C_2, C_4. For a nontrivial solution of the homogeneous equation (5.31), the determinant of the matrix has to vanish. The determinant is the product of two determinants of 2×2 submatrices, resulting in

$$\begin{aligned} &[(1-\alpha)\cos\gamma\sin(\omega-\gamma) - (1+\alpha)\sin\gamma\cos(\omega-\gamma)] \\ &\quad \times [(1+\alpha)\cos\gamma\sin(\omega-\gamma) - (1-\alpha)\sin\gamma\cos(\omega-\gamma)] = 0. \end{aligned} \tag{5.34}$$

Equation (5.33) can be simplified:

$$\sin^2(\alpha\omega) - \alpha^2\sin^2\omega = [\sin(\alpha\omega) - \alpha\sin\omega]\,[\sin(\alpha\omega) + \alpha\sin\omega] = 0 \quad \text{TF/TF.} \tag{5.35}$$

This equation has an infinite number of pairs, one of them determining the connection between C_1 and C_3 and the other the connection between C_2 and C_4.

Problem 5.2. Derive (5.35) from (5.33).

The completeness of the eigenfunctions is ensured in the framework of the general theory of Kondratiev [95], where it is shown that for an even-order elliptic boundary value problem the solution in the vicinity of a V-notch can be expressed as a series of eigenfunctions of the form $r^{\alpha_i}(\ln r)^j \boldsymbol{s}_i(\theta)$, with $\boldsymbol{s}_i(\theta)$ being smooth functions. The case of logarithmic singularities appears when the geometric multiplicity is larger than the algebraic multiplicity (multiple α_i having the same eigenfunctions), when the curves intersecting at the singular point are not straight lines, or in the case of inhomogeneous boundary conditions on Γ_1 and/or Γ_2, and will be addressed in the next subsection. For two-dimensional elasticity, Gregory [69] provided a proof of completeness of the cases of infinite strip and a wedge.

In a similar manner, the equations for all other nine homogeneous boundary conditions can be obtained (see [153] and the references therein):

$$\sin^2(\alpha\omega) - \alpha^2 \frac{\lambda+\mu}{\lambda+3\mu}^2 \sin^2\omega = 0 \qquad \text{HC/HC} \tag{5.36}$$

$$\sin 2(\alpha\omega) - \alpha \frac{\lambda+\mu}{\lambda+3\mu}^2 \sin 2\omega = 0 \qquad \text{HC/SC} \tag{5.37}$$

$$\sin 2(\alpha\omega) + \alpha \frac{\lambda+\mu}{\lambda+3\mu}^2 \sin 2\omega = 0 \qquad \text{HC/SS} \tag{5.38}$$

$$\sin^2(\alpha\omega) - \frac{(\lambda+\mu)^2}{(\lambda+\mu)(\lambda+3\mu)} + \alpha^2 \frac{\lambda+\mu}{\lambda+3\mu} \sin^2\omega = 0 \qquad \text{HC/TF} \tag{5.39}$$

$$\cos^2(\alpha\omega) - \cos^2\omega = [\sin(1+\alpha)\omega][\sin(1-\alpha)\omega] \qquad \text{SC/SC} \tag{5.40}$$

$$\cos 2(\alpha\omega) + \cos^2\omega = [\cos(1+\alpha)\omega][\cos(1-\alpha)\omega] \qquad \text{SC/SS} \tag{5.41}$$

$$\sin 2(\alpha\omega) + \alpha \sin 2\omega = 0 \qquad \text{SC/TF} \tag{5.42}$$

$$\cos^2(\alpha\omega) - \cos^2\omega = \sin^2\omega - \sin^2(\alpha\omega) \qquad \text{SS/SS} \tag{5.43}$$

$$\sin 2(\alpha\omega) - \alpha \sin 2\omega = 0 \qquad \text{SS/TF} \tag{5.44}$$

Some of the equations (5.35) - (5.44) may have solutions α that are complex (complex eigenvalues) that appear in conjugate pairs. In this case, the eigenfunctions as well as the coefficients of the asymptotic expansion are complex conjugates.

The distribution of the zeros of equations (5.35) - (5.44) as a function of the opening angle ω is important for determining the strength of singularities (regularity of the displacements), and to find the opening angle beyond which the stresses are singular. A detailed analysis of the dependence of the eigenvalues on the opening angle ω can be found in [185].

In Table 5.1 we summarize the critical opening angles (angles at which the stresses become singular), and the minimum eigenvalue obtained at $\omega = 2\pi$ (crack). The distribution of zeros of the ten different homogeneous boundary-condition combinations are provided in Figures 2-10 of [153].

Remark 5.1. *Cracks in homogeneous materials are of major importance in engineering applications, and it is important to know that for these cases all eigenvalues are real (no oscillatory terms [57] even if the material is anisotropic) and no logarithmic terms (no terms of the type $r^\alpha \ln(r)$) exist if same homogeneous boundary conditions are prescribed on both faces of the crack [43]. The absence of the logarithmic terms extends also to cracks in three-dimensional domains.*

Table 5.1 Summary of critical angles and minimal eigenvalues for a crack ($\omega = 2\pi$).

B.C.	Critical angle ω	$\alpha_{\min}$
TF/TF	180°	1/2
HC/HC	180°	1/2
HC/SC	90°	1/4
HC/SS	90°	1/4
HC/TF	61.696804...° (*)	1/4
SC/SC	90°	$0 + \varepsilon$ (**)
SC/SS	45°	$0 + \varepsilon$ (**)
SC/TF	90°	1/4
SS/SS	90°	$0 + \varepsilon$ (**)
SS/TF	128.726699...° (*)	1/4

(*) Angle depends on the Poisson ratio and plane stress/strain condition. These values were computed for plane stress and $\nu = 0.29$.

(**) The eigenvalue can be as close to zero as one wishes, and this value may be obtained at other values of ω besides 2π.

5.2 The Particular Case of TF/TF BCs

In the case of a TF/TF reentrant corner, the constants C_1 and C_3 are determined independently of C_2 and C_4, and (5.35) represents two different possible eigenpairs:

$$\sin(\alpha^{(\mathrm{I})}\omega) - \alpha^{(\mathrm{I})} \sin\omega = 0, \tag{5.45}$$

$$\sin(\alpha^{(\mathrm{II})}\omega) + \alpha^{(\mathrm{II})} \sin\omega = 0. \tag{5.46}$$

The superscripts I and II denote the eigenvalues α_i associated with the relationship between C_1 and C_3, and with the relationship between C_2 and C_4 respectively. For each positive eigenvalue α_i, the negative eigenvalue $-\alpha_i$ also satisfies the mathematical equations. However these are excluded because they represent displacement fields that are infinite at the crack tip and thus are "nonphysical." Because there is an infinite number of eigenvalues, the solution will consist also of an infinite number of terms, as shown in the sequel.

For each $\alpha_i^{(\mathrm{I})}$ the relationship between C_1 and C_3 is obtained from either the first or second equation in (5.33). For $i = 1, 3, 5$, the second equation of (5.33) is used, whereas for $i = 2, 4, 6$, the first equation is used (this is because of the specific case of a crack $\omega = 2\pi$):

$$\begin{aligned}(\lambda+\mu)(1-\alpha_i^{(\mathrm{I})}) \sin\left[\omega(1-\alpha_i^{(\mathrm{I})})/2\right] C_{3i} &= \sin\left[\omega(1+\alpha_i^{(\mathrm{I})})/2\right] C_{1i},\\ i &= 1,3,5,\dots,\\ (\lambda+\mu)(1+\alpha_i^{(\mathrm{I})}) \cos\left[\omega(1-\alpha_i^{(\mathrm{I})})/2\right] C_{3i} &= \cos\left[\omega(1+\alpha_i^{(\mathrm{I})})/2\right] C_{1i},\\ i &= 2,4,6,\dots,\end{aligned} \tag{5.47}$$

and for $\alpha_i^{(\mathrm{II})}$, the relation between C_2 and C_4 is obtained from the fourth or third equation in (5.33):

$$
\begin{aligned}
(\lambda+\mu)(1+\alpha_i^{(\mathrm{II})})\sin\left[\omega(1-\alpha_i^{(\mathrm{II})})/2\right]C_{4i} &= \sin\left[\omega(1+\alpha_i^{(\mathrm{II})})/2\right]C_{2i},\\
i &= 1,3,5,\ldots,\\
(\lambda+\mu)(1-\alpha_i^{(\mathrm{II})})\cos\left[\omega(1-\alpha_i^{(\mathrm{II})})/2\right]C_{4i} &= \cos\left[\omega(1+\alpha_i^{(\mathrm{II})})/2\right]C_{2i},\\
i &= 2,4,6,\ldots.
\end{aligned}
\tag{5.48}
$$

5.2.1 A TF/TF Reentrant Corner (V-Notch)

In the general case of a TF/TF V-notch one is usually interested in the first one or two terms that produce singular stresses. Thus, for the first two eigenvalues $\alpha_1^{(\mathrm{I})}$ and $\alpha_1^{(\mathrm{II})}$ (5.47) and (5.48) are inserted into the expression for the displacements (5.20) and stresses in (5.28):

$$
\tilde{\boldsymbol{\sigma}} = C_{11}2\alpha_1^{(\mathrm{I})}\mu r^{\alpha_1^{(\mathrm{I})}-1}\left\{\begin{array}{c}
\cos(1+\alpha_1^{(\mathrm{I})})\theta+\frac{(3-\alpha_1^{(\mathrm{I})})}{(1-\alpha_1^{(\mathrm{I})})}\frac{\sin[\omega(1+\alpha_1^{(\mathrm{I})})/2]}{\sin[\omega(1-\alpha_1^{(\mathrm{I})})/2]}\cos(1-\alpha_1^{(\mathrm{I})})\theta\\
-\left[\cos(1+\alpha_1^{(\mathrm{I})})\theta-\frac{(1+\alpha_1^{(\mathrm{I})})}{(1-\alpha_1^{(\mathrm{I})})}\frac{\sin[\omega(1+\alpha_1^{(\mathrm{I})})/2]}{\sin[\omega(1-\alpha_1^{(\mathrm{I})})/2]}\cos(1-\alpha_1^{(\mathrm{I})})\theta\right]\\
-\left[\sin(1+\alpha_1^{(\mathrm{I})})\theta-\frac{\sin[\omega(1+\alpha_1^{(\mathrm{I})})/2]}{\sin[\omega(1-\alpha_1^{(\mathrm{I})})/2]}\sin(1-\alpha_1^{(\mathrm{I})})\theta\right]
\end{array}\right\}
$$

$$
+C_{21}2\alpha_1^{(\mathrm{II})}\mu r^{\alpha_1^{(\mathrm{II})}-1}\left\{\begin{array}{c}
\sin(1+\alpha_1^{(\mathrm{II})})\theta+\frac{(3-\alpha_1^{(\mathrm{II})})}{(1+\alpha_1^{(\mathrm{II})})}\frac{\sin[\omega(1+\alpha_1^{(\mathrm{II})})/2]}{\sin[\omega(1-\alpha_1^{(\mathrm{II})})/2]}\sin(1-\alpha_1^{(\mathrm{II})})\theta\\
-\left[\sin(1+\alpha_1^{(\mathrm{II})})\theta-\frac{\sin[\omega(1+\alpha_1^{(\mathrm{II})})/2]}{\sin[\omega(1-\alpha_1^{(\mathrm{II})})/2]}\sin(1-\alpha_1^{(\mathrm{II})})\theta\right]\\
\cos(1+\alpha_1^{(\mathrm{II})})\theta-\frac{(1-\alpha_1^{(\mathrm{II})})}{(1+\alpha_1^{(\mathrm{II})})}\frac{\sin[\omega(1+\alpha_1^{(\mathrm{II})})/2]}{\sin[\omega(1-\alpha_1^{(\mathrm{II})})/2]}\cos(1-\alpha_1^{(\mathrm{II})})\theta
\end{array}\right\},
\tag{5.49}
$$

$$
\tilde{\boldsymbol{u}} = C_{11}r^{\alpha_1^{(\mathrm{I})}}\left\{\begin{array}{c}
\cos(1+\alpha_1^{(\mathrm{I})})\theta+\frac{[\lambda+3\mu-\alpha_1^{(\mathrm{I})}(\lambda+\mu)]}{(\lambda+\mu)(1-\alpha_1^{(\mathrm{I})})}\frac{\sin[\omega(1+\alpha_1^{(\mathrm{I})})/2]}{\sin[\omega(1-\alpha_1^{(\mathrm{I})})/2]}cos(1-\alpha_1^{(\mathrm{I})})\theta\\
-\sin(1+\alpha_1^{(\mathrm{I})})\theta-\frac{[\lambda+3\mu-\alpha_1^{(\mathrm{I})}(\lambda+\mu)]}{(\lambda+\mu)(1-\alpha_1^{(\mathrm{I})})}\frac{\sin[\omega(1+\alpha_1^{(\mathrm{I})})/2]}{\sin[\omega(1-\alpha_1^{(\mathrm{I})})/2]}\sin(1-\alpha_1^{(\mathrm{I})})\theta
\end{array}\right\}
$$

$$
+C_{21}r^{\alpha_1^{(\mathrm{II})}}\left\{\begin{array}{c}
\sin(1+\alpha_1^{(\mathrm{II})})\theta+\frac{[\lambda+3\mu-\alpha_1^{(\mathrm{II})}(\lambda+\mu)]}{(\lambda+\mu)(1+\alpha_1^{(\mathrm{II})})}\frac{\sin[\omega(1+\alpha_1^{(\mathrm{II})})/2]}{\sin[\omega(1-\alpha_1^{(\mathrm{II})})/2]}\sin(1-\alpha_1^{(\mathrm{II})})\theta\\
\cos(1+\alpha_1^{(\mathrm{II})})\theta+\frac{[\lambda+3\mu-\alpha_1^{(\mathrm{II})}(\lambda+\mu)]}{(\lambda+\mu)(1+\alpha_1^{(\mathrm{II})})}\frac{\sin[\omega(1+\alpha_1^{(\mathrm{II})})/2]}{\sin[\omega(1-\alpha_1^{(\mathrm{II})})/2]}\cos(1-\alpha_1^{(\mathrm{II})})\theta
\end{array}\right\}
$$

The eigenstresses are normalized so that for mode I $S^{\mathrm{I}}_{\theta\theta}(\theta = 0) = 1$, and for mode II: $S^{\mathrm{II}}_{r\theta}(\theta = 0) = 1$. Thus, $\sigma_{\theta\theta}(r, \theta = 0) = A_1 r^{\alpha_1^{(\mathrm{I})}-1}$ and $\sigma_{r\theta}(r, \theta = 0) = A_2 r^{\alpha_1^{(\mathrm{II})}-1}$. Let us define the normalization factor by

$$S^{I}_{\theta\theta}(\theta = 0) \stackrel{\text{def}}{=} \frac{(1+\alpha_1^{(\mathrm{I})})}{(1-\alpha_1^{(\mathrm{I})})} \frac{\sin[\omega(1+\alpha_1^{(\mathrm{I})})/2]}{\sin[\omega(1-\alpha_1^{(\mathrm{I})})/2]} - 1, \tag{5.50}$$

$$S^{\mathrm{II}}_{r\theta}(\theta = 0) \stackrel{\text{def}}{=} 1 - \frac{(1-\alpha_1^{(\mathrm{II})})}{(1+\alpha_1^{(\mathrm{II})})} \frac{\sin[\omega(1+\alpha_1^{(\mathrm{II})})/2]}{\sin[\omega(1-\alpha_1^{(\mathrm{II})})/2]}. \tag{5.51}$$

Then one finally obtains

$$\begin{Bmatrix} \sigma_{rr} \\ \sigma_{\theta\theta} \\ \sigma_{r\theta} \end{Bmatrix} =$$

$$A_1 r^{\alpha_1^{(\mathrm{I})}-1} \begin{Bmatrix} \left[\cos(1+\alpha_1^{(\mathrm{I})})\theta + \frac{(3-\alpha_1^{(\mathrm{I})})}{(1-\alpha_1^{(\mathrm{I})})} \frac{\sin[\omega(1+\alpha_1^{(\mathrm{I})})/2]}{\sin[\omega(1-\alpha_1^{(\mathrm{I})})/2]} \cos(1-\alpha_1^{(\mathrm{I})})\theta\right] / S^{I}_{\theta\theta}(\theta = 0) \\ \left[-\cos(1+\alpha_1^{(\mathrm{I})})\theta + \frac{(1+\alpha_1^{(\mathrm{I})})}{(1-\alpha_1^{(\mathrm{I})})} \frac{\sin[\omega(1+\alpha_1^{(\mathrm{I})})/2]}{\sin[\omega(1-\alpha_1^{(\mathrm{I})})/2]} \cos(1-\alpha_1^{(\mathrm{I})})\theta\right] / S^{I}_{\theta\theta}(\theta = 0) \\ \left[-\sin(1+\alpha_1^{(\mathrm{I})})\theta + \frac{\sin[\omega(1+\alpha_1^{(\mathrm{I})})/2]}{\sin[\omega(1-\alpha_1^{(\mathrm{I})})/2]} \sin(1-\alpha_1^{(\mathrm{I})})\theta\right] / S^{I}_{\theta\theta}(\theta = 0) \end{Bmatrix}$$

$$+A_2 r^{\alpha_1^{(\mathrm{II})}-1} \begin{Bmatrix} \left[\sin(1+\alpha_1^{(\mathrm{II})})\theta + \frac{(3-\alpha_1^{(\mathrm{II})})}{(1+\alpha_1^{(\mathrm{II})})} \frac{\sin[\omega(1+\alpha_1^{(\mathrm{II})})/2]}{\sin[\omega(1-\alpha_1^{(\mathrm{II})})/2]} \sin(1-\alpha_1^{(\mathrm{II})})\theta\right] / S^{\mathrm{II}}_{r\theta}(\theta = 0) \\ \left[-\sin(1+\alpha_1^{(\mathrm{II})})\theta + \frac{\sin[\omega(1+\alpha_1^{(\mathrm{II})})/2]}{\sin[\omega(1-\alpha_1^{(\mathrm{II})})/2]} \sin(1-\alpha_1^{(\mathrm{II})})\theta\right] / S^{\mathrm{II}}_{r\theta}(\theta = 0) \\ \left[\cos(1+\alpha_1^{(\mathrm{II})})\theta - \frac{(1-\alpha_1^{(\mathrm{II})})}{(1+\alpha_1^{(\mathrm{II})})} \frac{\sin[\omega(1+\alpha_1^{(\mathrm{II})})/2]}{\sin[\omega(1-\alpha_1^{(\mathrm{II})})/2]} \cos(1-\alpha_1^{(\mathrm{II})})\theta\right] / S^{\mathrm{II}}_{r\theta}(\theta = 0) \end{Bmatrix}, \tag{5.52}$$

and the corresponding displacements are

$$\begin{Bmatrix} u_r \\ u_\theta \end{Bmatrix} =$$

$$\frac{A_1 r^{\alpha_1^{(\mathrm{I})}}}{2\mu\alpha_1^{(\mathrm{I})}} \begin{Bmatrix} \left[\cos(1+\alpha_1^{(\mathrm{I})})\theta + \frac{[\lambda+3\mu-\alpha_1^{(\mathrm{I})}(\lambda+\mu)]}{(\lambda+\mu)(1-\alpha_1^{(\mathrm{I})})} \frac{\sin[\omega(1+\alpha_1^{(\mathrm{I})})/2]}{\sin[\omega(1-\alpha_1^{(\mathrm{I})})/2]} \cos(1-\alpha_1^{(\mathrm{I})})\theta\right] \\ / S^{I}_{\theta\theta}(\theta = 0) \\ \left[-\sin(1+\alpha_1^{(\mathrm{I})})\theta - \frac{[\lambda+3\mu+\alpha_1^{(\mathrm{I})}(\lambda+\mu)]}{(\lambda+\mu)(1-\alpha_1^{(\mathrm{I})})} \frac{\sin[\omega(1+\alpha_1^{(\mathrm{I})})/2]}{\sin[\omega(1-\alpha_1^{(\mathrm{I})})/2]} \sin(1-\alpha_1^{(\mathrm{I})})\theta\right] \\ / S^{I}_{\theta\theta}(\theta = 0) \end{Bmatrix}$$

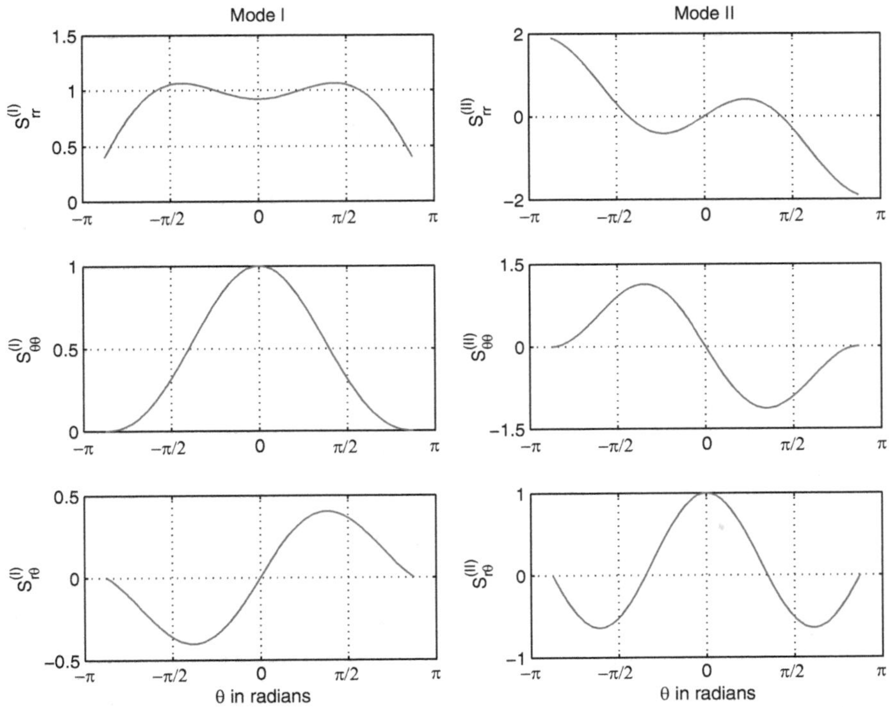

Fig. 5.2 Mode I and II polar eigenstresses for the TF/TF $\frac{7\pi}{4}$ V-notch.

$$+\frac{A_2 r^{\alpha_1^{(II)}}}{2\mu\alpha_1^{(II)}}\left\{\begin{array}{r}\left[\sin(1+\alpha_1^{(II)})\theta+\frac{[\lambda+3\mu-\alpha_1^{(II)}(\lambda+\mu)]}{(\lambda+\mu)(1+\alpha_1^{(II)})}\frac{\sin[\omega(1+\alpha_1^{(II)})/2]}{\sin[\omega(1-\alpha_1^{(II)})/2]}sin(1-\alpha_1^{(II)})\theta\right]\\ /S_{r\theta}^{II}(\theta=0)\\ \left[\cos(1+\alpha_1^{(II)})\theta+\frac{[\lambda+3\mu+\alpha_1^{(II)}(\lambda+\mu)]}{(\lambda+\mu)(1+\alpha_1^{(II)})}\frac{\sin[\omega(1+\alpha_1^{(II)})/2]}{\sin[\omega(1-\alpha_1^{(II)})/2]}cos(1-\alpha_1^{(II)})\theta\right]\\ /S_{r\theta}^{II}(\theta=0)\end{array}\right\}. \tag{5.53}$$

As an example we present in Figure 5.2 the eigenstresses and in Figure 5.3 the eigendisplacements for mode I and mode II for a V-notch with a solid angle of $\omega = \frac{7\pi}{4}$, and $E = 1$ and $\nu = 0.36$.

Remark 5.2. *The expressions in* (5.52) - (5.53) *are valid under the assumption of plane-strain. These hold also for the plane-stress situation if one replaces the Lamé constant* λ *by* $\tilde{\lambda} = 2\mu\lambda(\lambda + 2\mu)^{-1}$.

Remark 5.3. *The eigenstresses and eigendisplacements in a Cartesian coordinate system are derived in [177] from an Airy stress function in terms of a complex variable by the methods of Muskhelishvili [125].*

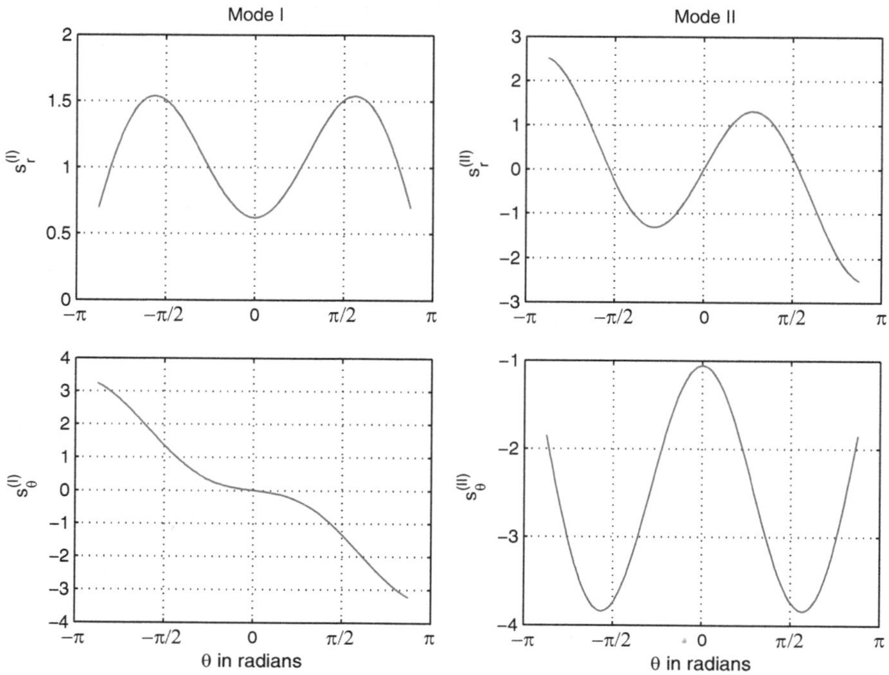

Fig. 5.3 Mode I and II polar eigendisplacements for the TF/TF $\frac{7\pi}{4}$ V-notch.

Table 5.2 First two eigenvalues for selected angles ω.

Solid Angle ω	2π (crack)	$\frac{11\pi}{6}$ (330°)	$\frac{7\pi}{4}$ (315°)	$\frac{5\pi}{3}$ (300°)	$\frac{3\pi}{2}$ (270°)	$\frac{4\pi}{3}$ (240°)
$\alpha_1^{(I)}$	1/2	0.5014530	0.5050097	0.5122214	0.5444837	0.6157311
$\alpha_1^{(II)}$	1/2	0.5981918	0.6597016	0.7309007	0.9085292	1.148913

Remark 5.4. *The eigenstresses (for the TF/TF BCs) in* (5.52) *are independent of the material properties and thus hold for both plane-strain and plane-stress, whereas the eigendisplacements in* (5.53) *depend on the material properties.*

When $\omega \neq 2\pi$, then not all roots are real, and multiple roots may exist. From the engineering viewpoint, V-notch solid angles up to $\frac{4\pi}{3}$ (240°) are of greatest importance, and in these cases the smallest roots are real; see a summary in Table 5.2.

For a V-notch solid angle smaller than 1.43028π (257.45°), then $\alpha_1^{(II)} > 1$ and the mode-II stress components are bounded, whereas mode-I stress components are bounded for $\omega < \pi$.

5.2.2 A TF/TF Crack

An important particular case of engineering importance is the case of a crack $\omega = 2\pi$ such that (5.45) - (5.46) are further simplified to

$$[\sin(2\alpha\pi)]^2 = 0 \rightarrow \begin{cases} \alpha_i^{(\mathrm{I})} = \frac{1}{2}i, & i = 0, 1, 2, 3, \ldots, \\ \alpha_i^{(\mathrm{II})} = \frac{1}{2}i, & i = 0, 1, 2, 3, \ldots \end{cases} \tag{5.54}$$

For the TF/TF crack, the first two zero eigenvalues are associated with rigid body motion, translation in the x_1 and x_2 directions. The first two nonzero eigenvalues are $\alpha_1^{(\mathrm{I})} = \alpha_1^{(\mathrm{II})} = 1/2$, and give rise to a singular stress field at the crack tip. The third eigenvalue $\alpha_2^{(\mathrm{I})} = 1$ is associated with the *T-stress*, a constant stress field parallel to the crack, and the fourth eigenvalue $\alpha_2^{(\mathrm{II})} = 1$ is associated with a rigid-body rotation, producing a zero state of stress.

For $i = 2, 4, 6$ the first and third equations in (5.33) are used for the relations between C_{3i} and C_{1i} and between C_{4i} and C_{2i}. For a TF/TF crack one obtains

$$C_{31} = \frac{-2}{\lambda + \mu} C_{11}, \quad C_{41} = \frac{-2}{3(\lambda + \mu)} C_{21}, \quad C_{32} = \frac{1}{2(\lambda + \mu)} C_{12}, \tag{5.55}$$

$$C_{22} = 0, \quad C_{42} = \text{const}, \quad C_{33} = \frac{2}{(\lambda + \mu)} C_{13}, \quad C_{43} = \frac{-2}{5(\lambda + \mu)} C_{23}.$$

Substituting $\alpha_i^{(\mathrm{I})} = i/2$, $\alpha_i^{(\mathrm{II})} = i/2$ and the various constants (5.47), (5.48), and (5.55) in (5.28), the first three terms in the series expansion of the stresses in the vicinity of a TF/TF crack tip are

$$\tilde{\boldsymbol{\sigma}} = \begin{Bmatrix} \sigma_{rr} \\ \sigma_{\theta\theta} \\ \sigma_{r\theta} \end{Bmatrix} = -C_{11}\mu r^{-1/2} \begin{Bmatrix} 5\cos\theta/2 - \cos 3\theta/2 \\ 3\cos\theta/2 + \cos 3\theta/2 \\ \sin\theta/2 + \sin 3\theta/2 \end{Bmatrix}$$

$$-\frac{1}{3} C_{21}\mu r^{-1/2} \begin{Bmatrix} 5\sin\theta/2 - 3\sin 3\theta/2 \\ 3\sin 3\theta/2 + 3\sin\theta/2 \\ -\cos\theta/2 - 3\cos 3\theta/2 \end{Bmatrix} + 2C_{12}\mu \begin{Bmatrix} \cos 2\theta + 1 \\ 1 - \cos 2\theta \\ -\sin 2\theta \end{Bmatrix} + \mathcal{O}(r^{1/2}). \tag{5.56}$$

For consistency with the classical fracture mechanics literature, the notation for the constants may be changed to obtain the expressions commonly used:

$$\frac{K_{\mathrm{I}}}{4\sqrt{2\pi}} \stackrel{\text{def}}{=} -C_{11}\mu, \qquad \frac{K_{\mathrm{II}}}{4\sqrt{2\pi}} \stackrel{\text{def}}{=} +\frac{1}{3} C_{21}\mu, \qquad T \stackrel{\text{def}}{=} -4C_{12}\mu. \tag{5.57}$$

Table 5.3 First two terms in the asymptotic expansion of a TF/TF crack used in LEFM.

Quant.	Mode I	Mode II
u_r	$\frac{K_I\sqrt{r}}{\mu\sqrt{2\pi}}\frac{1}{4}\left[(2\kappa-1)\cos\frac{\theta}{2}-\cos\frac{3\theta}{2}\right]$	$\frac{K_{II}\sqrt{r}}{2\mu\sqrt{2\pi}}\sin\frac{\theta}{2}\left[2-\kappa+3\cos\theta\right]$
u_θ	$\frac{K_I\sqrt{r}}{\mu\sqrt{2\pi}}\frac{1}{4}\left[-(2\kappa+1)\sin\frac{\theta}{2}+\sin\frac{3\theta}{2}\right]$	$\frac{-K_{II}\sqrt{r}}{2\mu\sqrt{2\pi}}\cos\frac{\theta}{2}\left[2+\kappa-3\cos\theta\right]$
σ_{rr}	$\frac{K_I}{\sqrt{2\pi r}}\frac{1}{4}\left[5\cos\frac{\theta}{2}-\cos\frac{3\theta}{2}\right]$	$\frac{-K_{II}}{\sqrt{2\pi r}}\frac{1}{4}\left[5\sin\frac{\theta}{2}-3\sin\frac{3\theta}{2}\right]$
$\sigma_{\theta\theta}$	$\frac{K_I}{\sqrt{2\pi r}}\frac{1}{4}\left[3\cos\frac{\theta}{2}+\cos\frac{3\theta}{2}\right]$	$\frac{-K_{II}}{\sqrt{2\pi r}}\frac{1}{4}\left[3\sin\frac{\theta}{2}+3\sin\frac{3\theta}{2}\right]$
$\sigma_{r\theta}$	$\frac{K_I}{\sqrt{2\pi r}}\frac{1}{4}\left[\sin\frac{\theta}{2}+\sin\frac{3\theta}{2}\right]$	$\frac{K_{II}}{\sqrt{2\pi r}}\frac{1}{4}\left[\cos\frac{\theta}{2}+3\cos\frac{3\theta}{2}\right]$
u_1	$\frac{K_I\sqrt{r}}{2\mu\sqrt{2\pi}}\cos\frac{\theta}{2}\left[\kappa-1+2\sin^2\frac{\theta}{2}\right]$	$\frac{K_{II}\sqrt{r}}{2\mu\sqrt{2\pi}}\sin\frac{\theta}{2}\left[\kappa+1+2\cos^2\frac{\theta}{2}\right]$
u_2	$\frac{K_I\sqrt{r}}{2\mu\sqrt{2\pi}}\sin\frac{\theta}{2}\left[\kappa+1-2\cos^2\frac{\theta}{2}\right]$	$\frac{-K_{II}\sqrt{r}}{2\mu\sqrt{2\pi}}\cos\frac{\theta}{2}\left[\kappa-1-2\sin^2\frac{\theta}{2}\right]$
σ_{11}	$\frac{K_I}{\sqrt{2\pi r}}\cos\frac{\theta}{2}\left[1-\sin\frac{\theta}{2}\sin\frac{3\theta}{2}\right]$	$\frac{-K_{II}}{\sqrt{2\pi r}}\sin\frac{\theta}{2}\left[2+\cos\frac{\theta}{2}\cos\frac{3\theta}{2}\right]$
σ_{22}	$\frac{K_I}{\sqrt{2\pi r}}\cos\frac{\theta}{2}\left[(1+\sin\frac{\theta}{2}\sin\frac{3\theta}{2}\right]$	$\frac{K_{II}}{\sqrt{2\pi r}}\sin\frac{\theta}{2}\cos\frac{\theta}{2}\cos\frac{3\theta}{2}$
σ_{12}	$\frac{K_I}{\sqrt{2\pi r}}\sin\frac{\theta}{2}\cos\frac{\theta}{2}\cos\frac{3\theta}{2}$	$\frac{K_{II}}{\sqrt{2\pi r}}\cos\frac{\theta}{2}\left[1-\sin\frac{\theta}{2}\sin\frac{3\theta}{2}\right]$

where $\kappa = 3 - 4\nu$ for plane-strain, and $\kappa = (3-\nu)/(1+\nu)$ for plane-stress.

The Cartesian stress tensor in the vicinity of a TF/TF crack is obtained by (5.7), using the notation in (5.57), the polar stress tensor (5.56), and trigonometrical relations

$$\boldsymbol{\sigma} = \begin{Bmatrix}\sigma_{11}\\ \sigma_{22}\\ \sigma_{12}\end{Bmatrix} = \frac{K_I}{\sqrt{2\pi r}}\begin{Bmatrix}\cos\frac{\theta}{2}\left(1-\sin\frac{\theta}{2}\sin\frac{3\theta}{2}\right)\\ \cos\frac{\theta}{2}\left(1+\sin\frac{\theta}{2}\sin\frac{3\theta}{2}\right)\\ \sin\frac{\theta}{2}\cos\frac{\theta}{2}\cos\frac{3\theta}{2}\end{Bmatrix}$$
$$+\frac{K_{II}}{\sqrt{2\pi r}}\begin{Bmatrix}-\sin\frac{\theta}{2}\left(2+\cos\frac{\theta}{2}\cos\frac{3\theta}{2}\right)\\ \sin\frac{\theta}{2}\cos\frac{\theta}{2}\cos\frac{3\theta}{2}\\ \cos\frac{\theta}{2}\left(1-\sin\frac{\theta}{2}\sin\frac{3\theta}{2}\right)\end{Bmatrix} + T\begin{Bmatrix}1\\0\\0\end{Bmatrix} + \mathcal{O}(r^{1/2}) \tag{5.58}$$

The Cartesian singular stress field was first derived in [87] and [190]. The third term is called the *T-stress* [148], and is a constant value independent of r, θ.

We summarize in Table 5.3 the first two eigendisplacements and eigenstresses (mode I and mode II) in the asymptotic expansion as commonly used in linear elastic fracture mechanics. The eigenstresses are normalized so that for mode I, $\sigma_{\theta\theta}(\theta = 0) = \sigma_{11}(\theta = 0) = 1$, and for mode II, $\sigma_{r\theta}(\theta = 0) = \sigma_{12}(\theta = 0) = 1$. A graphical representation of the mode I and mode II Cartesian eigenstresses is shown in Figure 5.4 and polar eigenstresses in Figure 5.5.

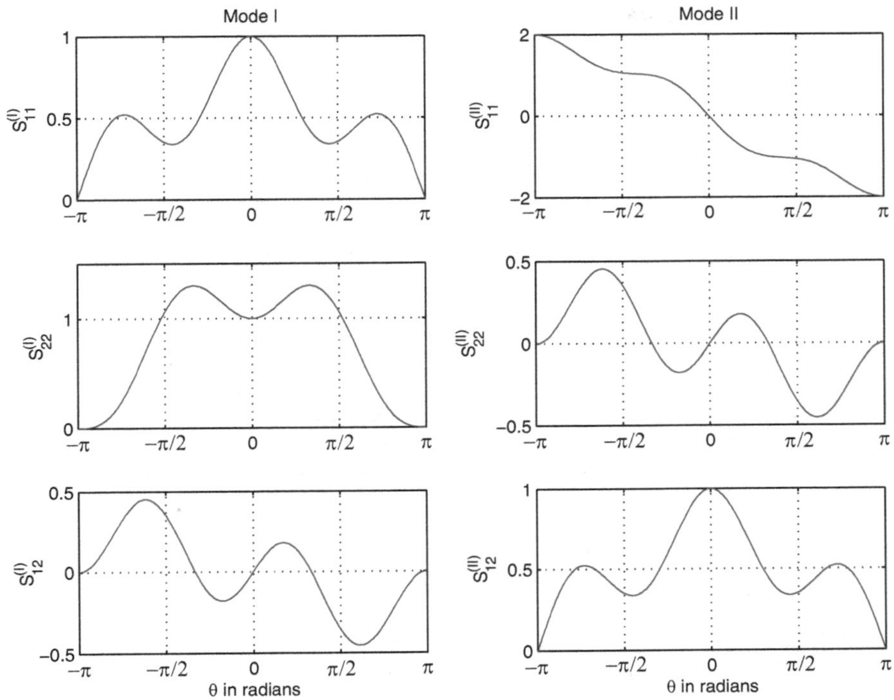

Fig. 5.4 Mode I and II Cartesian eigenstresses for the TF/TF crack.

Problem 5.3. For a TF/TF crack ($\omega = 2\pi$) derive the Cartesian stress tensor in (5.58) using (5.56).

Problem 5.4. A popular expression for the singular part of the stress tensor for a TF/TF crack is

$$\begin{Bmatrix} \sigma_{rr} \\ \sigma_{\theta\theta} \\ \sigma_{r\theta} \end{Bmatrix} = \frac{K_{\mathrm{I}}}{\sqrt{2\pi r}} \begin{Bmatrix} \cos\frac{\theta}{2}\left[1+\sin^2\frac{\theta}{2}\right] \\ \cos^3\frac{\theta}{2} \\ \frac{1}{2}\cos\frac{\theta}{2}\sin\theta \end{Bmatrix} + \frac{K_{\mathrm{II}}}{\sqrt{2\pi r}} \begin{Bmatrix} \cos\frac{\theta}{2}\left[\frac{3}{2}\sin\theta - 2\tan\frac{\theta}{2}\right] \\ -\frac{3}{2}\cos\frac{\theta}{2}\sin\theta \\ \frac{1}{2}\cos\frac{\theta}{2}(3\cos\theta - 1) \end{Bmatrix}.$$

Show that the above expressions are identical to the ones in Table 5.3.

Problem 5.5. For a TF/TF crack ($\omega = 2\pi$) derive the displacement vector $\tilde{\boldsymbol{u}} \stackrel{\mathrm{def}}{=} \{u_r, u_\theta\}^T$ up to (and including) the term of order $r^{3/2}$, expressing it in terms of K_{I}, K_{II} and other constants.

It can be shown that the displacements in a Cartesian coordinate system can be also presented by the following asymptotic series:

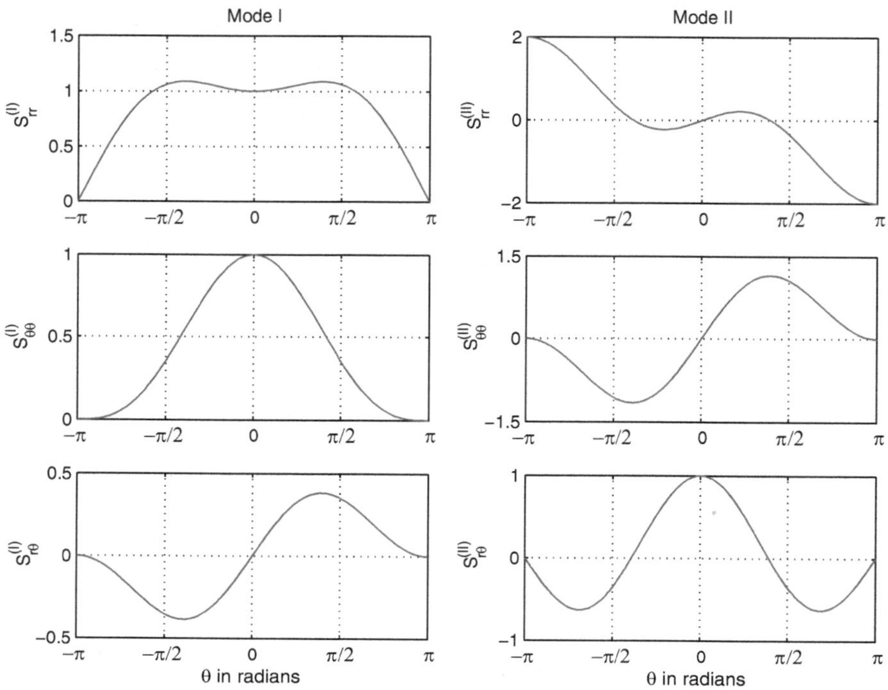

Fig. 5.5 Mode I and II polar eigenstresses for the TF/TF crack.

$$
\begin{aligned}
u_1 &= \sum_{i=1}^{\infty} a_{i1} \frac{r^{i/2}}{2\mu} \left[(\kappa + \frac{i}{2} + (-1)^i) \cos \frac{i\theta}{2} - \frac{i}{2} \cos \frac{(i-4)\theta}{2} \right] \\
&\quad - a_{i2} \frac{r^{i/2}}{2\mu} \left[(\kappa + \frac{i}{2} - (-1)^i) \sin \frac{i\theta}{2} - \frac{i}{2} \sin \frac{(i-4)\theta}{2} \right], \\
u_2 &= \sum_{i=1}^{\infty} a_{i1} \frac{r^{i/2}}{2\mu} \left[(\kappa - \frac{i}{2} - (-1)^i) \sin \frac{i\theta}{2} + \frac{i}{2} \sin \frac{(i-4)\theta}{2} \right] \\
&\quad + a_{i2} \frac{r^{i/2}}{2\mu} \left[(\kappa - \frac{i}{2} + (-1)^i) \cos \frac{i\theta}{2} + \frac{i}{2} \cos \frac{(i-4)\theta}{2} \right].
\end{aligned} \tag{5.59}
$$

Remark 5.5. *For a crack in a homogeneous isotropic or anisotropic domain having* the same boundary conditions on its two faces, *the asymptotic expansion contains neither oscillatory terms (complex eigenvalues) nor logarithmic terms [42, 57, 96]; i.e., the singular functions all behave as half-integer powers* $r^{\frac{1}{2}+n}\mathbf{s}(\theta)$, *and furthermore, without any logarithmic terms [43].*

The eigenvalues are half-integer exponents also for straight or curved crack fronts, in two dimensions as well as in three-dimensions (edge singularities). The fact that the eigenvalues are half-integer exponents in all these cases does not imply *that the eigenfunctions are the same, and in fact these are different functions for the different cases [43].*

In the case that the boundary conditions on the two sides of the crack are not the same, and in particular in mixed Dirichlet - Neumann (clamped - traction free) BCs, the real part of the exponents of singularity have the form $1/4 + \imath q + n/2$ *with real* q *and integer* n. *This is valid for general anisotropic elasticity too [42].*

Some remarks on the T-stress

The third term in the series expansion (5.58), giving rise to a constant stress parallel to the crack face, has an engineering relevance because it may affect the path stability of slightly curved or kinked cracks [48, 108], and plays an important role in determining the size and shape of a crack tip plastic zone. The displacements associated with the T-stress (both plane-strain and plane-stress) are

$$\begin{Bmatrix} u_1 \\ u_2 \end{Bmatrix} = \frac{Tr}{2(1+\nu)\mu} \begin{Bmatrix} (1-\nu^2)\cos\theta \\ -(\nu+\nu^2)\sin\theta \end{Bmatrix}. \tag{5.60}$$

In [48] a stability analysis of the path of a crack under mode I loading concludes that the straight crack path is stable only if $T < 0$. More recent studies such as [108] provide different conclusions, namely that the straight crack growth under mode I loading remains stable up to a strictly positive threshold $T_c > 0$, as shown in several experimental observations.

5.2.3 *A TF/TF Crack at a Bimaterial Interface*

A bimaterial interface is a composite of two homogeneous materials with continuity of tractions and displacements maintained across the interface. Solutions for interface cracks have been studied extensively in the literature and deserve special attention. Consider a crack at a bimaterial interface as shown in Figure 5.6, with the upper material denoted by the index 1. The singularity of the stress field at the tip of a crack at the interface of two isotropic materials was analytically determined by Williams in [191] and further investigated in [60–62, 150, 164].

The characteristic equation for the computation of the eigenvalues may be obtained by considering an Airy stress function, or applying Muskhelishvili's complex functions. In both cases, because continuity of tractions and *displacements* across the interface is required, the characteristic equation for the determination of the eigenpairs depends on the state of plane-stress or plane-strain (for a TF/TF crack in an isotropic material, the eigenpairs are independent of material properties and state of plane-stress/strain). For example, in the case of *plane-stress* and isotropic materials, the characteristic equation is [191]

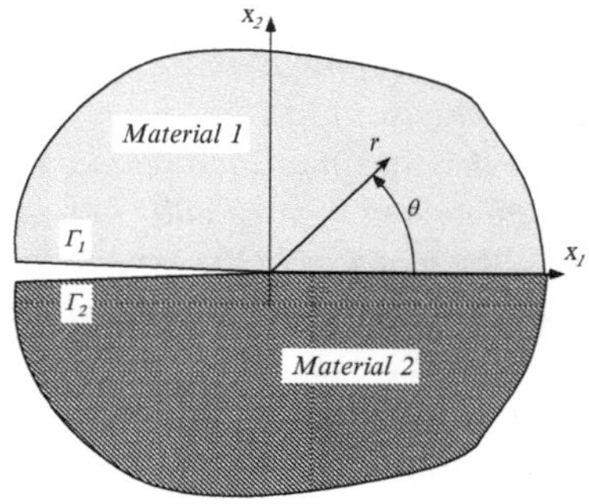

Fig. 5.6 Crack at a bimaterial interface and notation.

Table 5.4 ε for representative material combinations, plane-strain condition

Material 1/2	Al_2O_3/Cu		MgO/Au	Si/Cu	MgO/Ni	Al_2O_3/Ti		Al_2O_3/Nb
ε	−0.028		−0.0036	0.0105	0.0049	−0.039		−0.019
Material	Au	Ti	Ni	MgO	Cu	Al_2O_3	Nb	Si
μ [GPa]	29.3	43.4	80.8	128.3	47.8	179.2	37.7	68.8
ν	0.417	0.322	0.314	0.175	0.345	0.207	0.392	0.220

$$\cot^2(n\alpha) + \frac{1}{4}\left[\frac{\frac{E_1}{E_2}(1-\nu_2) - (1-\nu_1)}{1+\frac{E_1}{E_2}}\right]^2 = 0. \tag{5.61}$$

There are two families of eigenvalues for a crack at a bimaterial interface: integers $\alpha_n = (n-1)/2$, $n = 1, 3, 5, \ldots$, which do not contribute to the singular behavior of the stresses, and complex eigenvalues that come in conjugate pairs:

$$\alpha_n = \alpha_n^{\Re} + \imath\alpha^{\Im} = \frac{n-1}{2} + \imath\varepsilon, \quad n = 2, 4, 6, \ldots, \tag{5.62}$$

$$\alpha_{n+1} = \alpha_n^{\Re} - \imath\alpha^{\Im} = \frac{n-1}{2} - \imath\varepsilon, \quad n = 2, 4, 6, \ldots, \tag{5.63}$$

with $\imath \overset{\text{def}}{=} \sqrt{-1}$. The imaginary part $\alpha^{\Im}$ is determined by the material properties of the two materials, and is also denoted in the engineering literature by ε:

$$\alpha^{\Im} \equiv \varepsilon = \frac{1}{2\pi}\ln\left(\frac{\kappa_1\mu_2 + \mu_1}{\kappa_2\mu_1 + \mu_2}\right), \tag{5.64}$$

where $\kappa_i = (3-\nu_i)/(1+\nu_i)$ for plane-stress and $\kappa_i = (3-4\nu_i)$ for plane-strain.

For a *plane-strain* situation, typical values of ε are presented in Table 5.4 for six representative material combinations taken from [82]. Notice in (5.64) that ε reverses sign when materials 1 and 2 are interchanged (but this does not make any difference in the solution, since the complex eigenvalues appear in conjugate pairs).

For the nth and $(n+1)$st complex eigenvalues, the corresponding eigenfunctions are also complex conjugates $\tilde{\boldsymbol{s}}_n(\theta) = \tilde{\boldsymbol{s}}_n^{\Re}(\theta) \pm \imath \tilde{\boldsymbol{s}}_n^{\Im}(\theta)$, and so are the generalized stress intensity factors, denoted by $A_n \pm \imath A_{n+1}$. We may therefore address these two terms in the solution:

$$\begin{aligned}\tilde{\boldsymbol{u}}_{n,n+1} &= (A_n + \imath A_{n+1})\, r^{\alpha_n^{\Re} + \imath\alpha_n^{\Im}} (\tilde{\boldsymbol{s}}_n^{\Re}(\theta) + \imath \tilde{\boldsymbol{s}}_n^{\Im}(\theta)) \\ &\quad + (A_n - \imath A_{n+1})\, r^{\alpha_n^{\Re} - \imath\alpha_n^{\Im}} (\tilde{\boldsymbol{s}}_n^{\Re}(\theta) - \imath \tilde{\boldsymbol{s}}_n^{\Im}(\theta)). \qquad (5.65)\end{aligned}$$

Notice that

$$\begin{aligned}r^{\alpha_n^{\Re} \pm \imath\alpha_n^{\Im}} &= r^{\alpha_n^{\Re}}\, r^{\pm \imath\alpha_n^{\Im}} = r^{\alpha_n^{\Re}}\, e^{\ln(r^{\pm \imath\alpha_n^{\Im}})} = r^{\alpha_n^{\Re}}\, e^{\pm \imath\alpha_n^{\Im} \ln(r)} \\ &= r^{\alpha_n^{\Re}} \left[\cos(\alpha_n^{\Im} \ln(r)) \pm \imath \sin(\alpha_n^{\Im} \ln(r))\right]. \qquad (5.66)\end{aligned}$$

Substituting (5.66) in (5.65) and defining $\delta_n \stackrel{\text{def}}{=} \alpha_n^{\Im} \ln(r)$, we finally obtain:

$$\begin{aligned}\tilde{\boldsymbol{u}}_{n,n+1} &= r^{\alpha_n^{\Re}} \Big\{ 2A_n \left[\cos\delta_n \tilde{\boldsymbol{s}}_n^{\Re}(\theta) - \sin\delta_n \tilde{\boldsymbol{s}}_n^{\Im}(\theta)\right] \\ &\quad - 2A_{n+1} \left[\cos\delta_n \tilde{\boldsymbol{s}}_n^{\Im}(\theta) + \sin\delta_n \tilde{\boldsymbol{s}}_n^{\Re}(\theta)\right] \Big\}. \qquad (5.67)\end{aligned}$$

The stresses can be easily computed from (5.67):

$$\begin{aligned}\tilde{\boldsymbol{\sigma}}_{n,n+1} =& (A_n + \imath A_{n+1}) r^{\alpha_n^{\Re} - 1 + \imath\alpha_n^{\Im}} (\tilde{\boldsymbol{S}}_n^{\Re}(\theta) + \imath \tilde{\boldsymbol{S}}_n^{\Im}(\theta)) \\ &+ (A_n - \imath A_{n+1}) r^{\alpha_n^{\Re} - 1 - \imath\alpha_n^{\Im}} (\tilde{\boldsymbol{S}}_n^{\Re}(\theta) - \imath \tilde{\boldsymbol{S}}_n^{\Im}(\theta)) \\ =& A_n r^{\alpha_n^{\Re} - 1} \left[2(\tilde{\boldsymbol{S}}_n^{\Re}(\theta)\cos\delta_n - \tilde{\boldsymbol{S}}_n^{\Im}(\theta)\sin\delta_n)\right] \\ &+ A_{n+1} r^{\alpha_n^{\Re} - 1} \left[-2(\tilde{\boldsymbol{S}}_n^{(\Re}(\theta)\sin\delta_n + \tilde{\boldsymbol{S}}_n^{\Im}(\theta)\cos\delta_n)\right]. \qquad (5.68)\end{aligned}$$

Note the following consequences associated with (5.67):

- The strength of the singularity is determined by the real part $\alpha_n^{\Re}$ of the complex eigenvalue.
- In the close vicinity of the singular point $\ln(r) \stackrel{r\to 0}{\to} \infty$, therefore $\delta_n \stackrel{r\to 0}{\to} \infty$, so the expressions $\cos\delta_n, \sin\delta_n$ oscillate with increasingly higher frequency when approaching the singular point.
- The exponential-oscillatory singularity is unrealistic, since it implies that the lower face of the crack interpenetrates the upper face of the crack when $r \to 0$.
- The so-called mode I loading excites both generalized stress intensity factors A_n and A_{n+1} (similarly for mode II loading).
- In Figure 5.7 we illustrate the typical behavior of the stress and displacements as one approaches the singular point, i.e., as $r \to 0$.

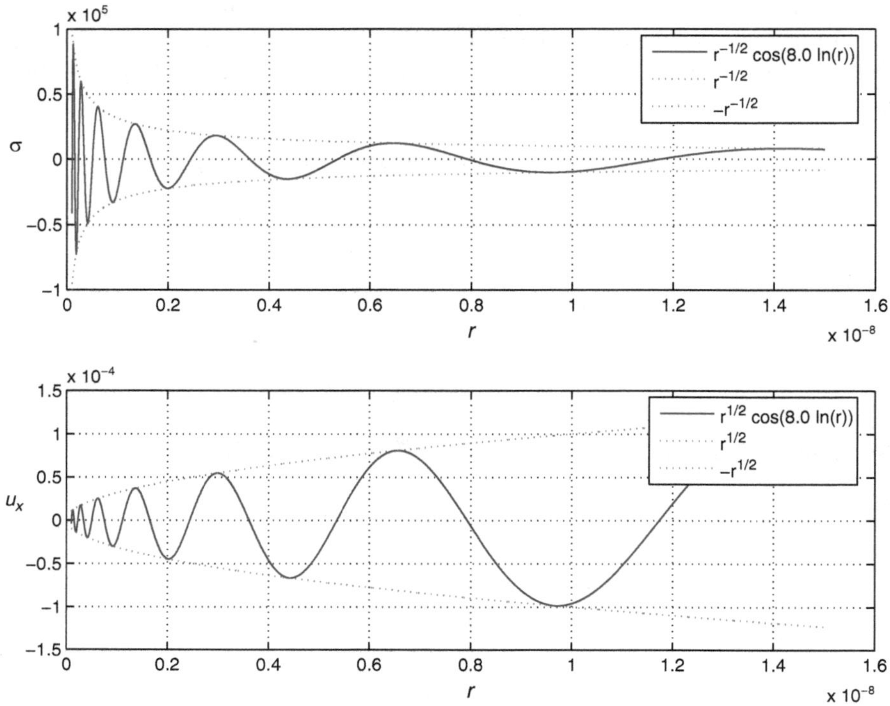

Fig. 5.7 Typical behavior of stresses and displacements in the vicinity of a crack tip at a bimaterial interface due to complex eigenpairs.

Following [55], the Cartesian displacements and stresses are given as

$$\boldsymbol{u} = \sum_{n=0} r^{\frac{n+1}{2}} \left\{\Re(A_n r^{\iota\varepsilon})\boldsymbol{s}_n^{\mathrm{I}}(\theta) + \Im(A_n r^{\iota\varepsilon})\boldsymbol{s}_n^{\mathrm{II}}(\theta)\right\}, \tag{5.69}$$

$$\boldsymbol{\sigma} = \sum_{n=0} r^{\frac{n-1}{2}} \left\{\Re(A_n r^{\iota\varepsilon})\boldsymbol{S}_n^{\mathrm{I}}(\theta) + \Im(A_n r^{\iota\varepsilon})\boldsymbol{S}_n^{\mathrm{II}}(\theta)\right\}, \tag{5.70}$$

where $A_n = A_n^{\mathrm{I}} + \iota A_n^{\mathrm{II}}$.
The eigenfunctions in *material 1* for $n = 0, 2, 4, \ldots$ are:

$$\boldsymbol{s}_n^{\mathrm{I}}(\theta) = \frac{1}{\mu_1\left[(n+1)^2 + 4\varepsilon^2\right]\cosh\pi\varepsilon} \tag{5.71}$$

$$\begin{cases} (n+1)\left[\sinh\varepsilon(\pi-\theta) - \frac{\kappa_1-1}{2}e^{-\varepsilon(\pi-\theta)}\right]\cos\frac{(n+1)\theta}{2} \\ (n+1)\left[\cosh\varepsilon(\pi-\theta) + \frac{\kappa_1-1}{2}e^{-\varepsilon(\pi-\theta)}\right]\sin\frac{(n+1)\theta}{2} \end{cases}$$

$$
\left.\begin{aligned}
&+\tfrac{1}{2}\left[(n+1)^2+4\varepsilon^2\right]e^{-\varepsilon(\pi-\theta)}\sin\theta\sin\tfrac{(n-1)\theta}{2}\\
&-\tfrac{1}{2}\left[(n+1)^2+4\varepsilon^2\right]e^{-\varepsilon(\pi-\theta)}\sin\theta\cos\tfrac{(n-1)\theta}{2}\\
&+2\varepsilon\left[\cosh\varepsilon(\pi-\theta)+\tfrac{\kappa_1-1}{2}e^{-\varepsilon(\pi-\theta)}\right]\sin\tfrac{(n+1)\theta}{2}\\
&-2\varepsilon\left[\sinh\varepsilon(\pi-\theta)-\tfrac{\kappa_1-1}{2}e^{-\varepsilon(\pi-\theta)}\right]\cos\tfrac{(n+1)\theta}{2}
\end{aligned}\right\},
$$

$$
\begin{aligned}
\boldsymbol{s}_n^{\mathrm{II}}(\theta) = {} & \frac{1}{\mu_1\left[(n+1)^2+4\varepsilon^2\right]\cosh\pi\varepsilon} \\
& \left\{\begin{aligned}
&(n+1)\left[\cosh\varepsilon(\pi-\theta)+\tfrac{\kappa_1-1}{2}e^{-\varepsilon(\pi-\theta)}\right]\sin\tfrac{(n+1)\theta}{2}\\
&(n+1)\left[\sinh\varepsilon(\pi-\theta)-\tfrac{\kappa_1-1}{2}e^{-\varepsilon(\pi-\theta)}\right]\cos\tfrac{(n+1)\theta}{2}
\end{aligned}\right.\\
& \begin{aligned}
&+\tfrac{1}{2}\left[(n+1)^2+4\varepsilon^2\right]e^{-\varepsilon(\pi-\theta)}\sin\theta\cos\tfrac{(n-1)\theta}{2}\\
&-\tfrac{1}{2}\left[(n+1)^2+4\varepsilon^2\right]e^{-\varepsilon(\pi-\theta)}\sin\theta\sin\tfrac{(n-1)\theta}{2}
\end{aligned}\\
& \left.\begin{aligned}
&-2\varepsilon\left[\sinh\varepsilon(\pi-\theta)-\tfrac{\kappa_1-1}{2}e^{-\varepsilon(\pi-\theta)}\right]\cos\tfrac{(n+1)\theta}{2}\\
&+2\varepsilon\left[\cosh\varepsilon(\pi-\theta)+\tfrac{\kappa_1-1}{2}e^{-\varepsilon(\pi-\theta)}\right]\sin\tfrac{(n+1)\theta}{2}
\end{aligned}\right\},
\end{aligned}
\tag{5.72}
$$

$$
\begin{aligned}
\boldsymbol{S}_n^{\mathrm{I}}(\theta) = {} & \frac{1}{\cosh\pi\varepsilon}\left\{\begin{aligned}
&-\left[\sinh\varepsilon(\pi-\theta)-e^{-\varepsilon(\pi-\theta)}\right]\cos\tfrac{(n-1)\theta}{2}\\
&\left[\sinh\varepsilon(\pi-\theta)+e^{-\varepsilon(\pi-\theta)}\right]\cos\tfrac{(n-1)\theta}{2}\\
&\sinh\varepsilon(\pi-\theta)\sin\tfrac{(n-1)\theta}{2}
\end{aligned}\right.\\
& \left.\begin{aligned}
&-\tfrac{1}{2}e^{-\varepsilon(\pi-\theta)}\sin\theta\left[(n-1)\sin\tfrac{(n-3)\theta}{2}-2\varepsilon\cos\tfrac{(n-3)\theta}{2}\right]\\
&+\tfrac{1}{2}e^{-\varepsilon(\pi-\theta)}\sin\theta\left[(n-1)\sin\tfrac{(n-3)\theta}{2}-2\varepsilon\cos\tfrac{(n-3)\theta}{2}\right]\\
&-\tfrac{1}{2}e^{-\varepsilon(\pi-\theta)}\sin\theta\left[(n-1)\cos\tfrac{(n-3)\theta}{2}+2\varepsilon\sin\tfrac{(n-3)\theta}{2}\right]
\end{aligned}\right\},
\end{aligned}
\tag{5.73}
$$

$$
\begin{aligned}
\boldsymbol{S}_n^{\mathrm{II}}(\theta) = {} & \frac{1}{\cosh\pi\varepsilon}\left\{\begin{aligned}
&\left[\cosh\varepsilon(\pi-\theta)+e^{-\varepsilon(\pi-\theta)}\right]\sin\tfrac{(n-1)\theta}{2}\\
&-\left[\cosh\varepsilon(\pi-\theta)-e^{-\varepsilon(\pi-\theta)}\right]\sin\tfrac{(n-1)\theta}{2}\\
&\cosh\varepsilon(\pi-\theta)\cos\tfrac{(n-1)\theta}{2}
\end{aligned}\right.\\
& \left\{\begin{aligned}
&+\tfrac{1}{2}e^{-\varepsilon(\pi-\theta)}\sin\theta\left[(n-1)\cos\tfrac{(n-3)\theta}{2}+2\varepsilon\sin\tfrac{(n-3)\theta}{2}\right]\\
&-\tfrac{1}{2}e^{-\varepsilon(\pi-\theta)}\sin\theta\left[(n-1)\cos\tfrac{(n-3)\theta}{2}+2\varepsilon\sin\tfrac{(n-3)\theta}{2}\right]\\
&-\tfrac{1}{2}e^{-\varepsilon(\pi-\theta)}\sin\theta\left[(n-1)\sin\tfrac{(n-3)\theta}{2}-2\varepsilon\cos\tfrac{(n-3)\theta}{2}\right]
\end{aligned}\right\}
\end{aligned}
\tag{5.74}
$$

and for $n = 1, 3, 5, \ldots$ the eigenfunctions in *material 1* are:

$$\mathbf{s}_n^{\mathrm{I}}(\theta) = \frac{1}{\mu_1(n+1)(1+\zeta)} \begin{Bmatrix} (\kappa_1+1)\cos\frac{(1+n)\theta}{2} - (n+1)\sin\theta\sin\frac{(n-1)\theta}{2} \\ (\kappa_1-1)\sin\frac{(1+n)\theta}{2} - (n+1)\sin\theta\cos\frac{(n-1)\theta}{2} \end{Bmatrix}, \tag{5.75}$$

$$\mathbf{s}_n^{\mathrm{II}}(\theta) = \frac{1}{\mu_1(n+1)(1+\zeta)} \begin{Bmatrix} (\kappa_1-1)\sin\frac{(1+n)\theta}{2} + (n+1)\sin\theta\cos\frac{(n-1)\theta}{2} \\ -(\kappa_1+1)\cos\frac{(1+n)\theta}{2} - (n+1)\sin\theta\sin\frac{(n-1)\theta}{2} \end{Bmatrix}, \tag{5.76}$$

$$\mathbf{S}_n^{\mathrm{I}}(\theta) = \frac{1}{1+\zeta} \begin{Bmatrix} 4\cos\frac{(n-1)\theta}{2} - (n-1)\sin\theta\sin\frac{(n-3)\theta}{2} \\ (n-1)\sin\theta\sin\frac{(n-3)\theta}{2} \\ -2\sin\frac{(n-1)\theta}{2} - (n-1)\sin\theta\cos\frac{(n-3)\theta}{2} \end{Bmatrix} \tag{5.77}$$

$$\mathbf{S}_n^{\mathrm{II}}(\theta) = \frac{1}{1+\zeta} \begin{Bmatrix} 2\sin\frac{(n-1)\theta}{2} + (n-1)\sin\theta\cos\frac{(n-3)\theta}{2} \\ 2\sin\frac{(n-1)\theta}{2} - (n-1)\sin\theta\cos\frac{(n-3)\theta}{2} \\ -(n-1)\sin\theta\sin\frac{(n-3)\theta}{2} \end{Bmatrix}, \tag{5.78}$$

where $\zeta = (\kappa_1+1)\mu_2/(\kappa_2+1)\mu_1$. Expressions for the bottom half-plane having index 2 are obtained simply by changing the index 1 to 2, π to $-\pi$, and ζ to ζ^{-1} in (5.71)-(5.78).

In "engineering notation," the first two GSIFs, $A_1 \pm \imath A_2$, associated with the eigenvalues $\frac{1}{2} \pm \varepsilon$ are denoted by $\mathbf{K} \stackrel{\text{def}}{=} \frac{K_1}{\sqrt{2\pi}} \pm \imath \frac{K_2}{\sqrt{2\pi}}$. From this definition one notes that the dimension of $\mathbf{K}$ is [stress][length]$^{1/2-\imath\varepsilon}$, which is "unnatural."

For this reason, a characteristic length ℓ is introduced and the following definition is adopted:

$$\mathbf{K}\ell^{\imath\varepsilon} = |\mathbf{K}|e^{\imath\psi}, \tag{5.79}$$

where the dimension of $\mathbf{K}\ell^{\imath\varepsilon}$ is the usual [stress][length]$^{1/2}$ dimension and so is the dimension of the amplitude $|\mathbf{K}| = \sqrt{K_1^2 + K_2^2}$. The phase angle ψ is a measure of the relative proportion of shear to normal stresses at the characteristic length ℓ ahead of the crack tip. It is defined through the relation [149]

$$\psi = \arctan\left(\frac{\Im\left(\mathbf{K}\ell^{\imath\varepsilon}\right)}{\Re\left(\mathbf{K}\ell^{\imath\varepsilon}\right)}\right). \tag{5.80}$$

The phase angle ψ is an important parameter in the characterization of interfacial fracture toughness, and the characteristic length ℓ associated with a factor-of-10 change affects little the phase angle for the small ε [149]. Therefore, in reporting the phase angle for a given loading configuration, the characteristic length ℓ can be taken as the crack length or a specimen dimension. For example, if ψ_2 is associated with one characteristic length ℓ_2 and ψ_1 is associated with another characteristic length ℓ_1, then

$$\psi_2 = \psi_1 + \varepsilon \ln\left(\frac{\ell_2}{\ell_1}\right). \tag{5.81}$$

Remark 5.6. *Unlike the treatment of cracks in isotropic materials, tension and shear effects are inseparable in the vicinity of an interface crack tip.*

5.3 Power-Logarithmic or Logarithmic Singularities with Homogeneous BCs

In addition to real and complex eigenvalues, in 2-D elasticity without body forces, there is a possibility of power-logarithmic singularities although homogeneous boundary conditions are prescribed on the reentrant corner faces. This topic is reviewed by Sinclair in [165]. Such power-logarithmic singularities can manifest themselves in solutions of the form

$$\boldsymbol{u} \propto r^\alpha \ln r \rightarrow \boldsymbol{\sigma} \propto r^{\alpha-1} \ln r + r^{\alpha-1}, \tag{5.82}$$

and logarithmic singularities give rise to solutions of the form

$$\boldsymbol{u} \propto r \ln r \rightarrow \boldsymbol{\sigma} \propto \ln r. \tag{5.83}$$

The power-logarithmic singularities may occur when repeated roots exist for the eigenvalue equations (5.36-5.44), at transition loci separating regions of real and complex eigenvalues, i.e., at the transition from two real roots to two roots that are complex conjugates, or vice versa. This "transition" occurs when the opening angle ω is changed, for example. The existence of power-logarithmic and logarithmic singularities is associated with the rank deficiency of the matrix resulting from satisfying the boundary conditions, shown for the TF/TF case in (5.31), i.e., when the geometric multiplicity is smaller than the algebraic multiplicity of an eigenvalue.

Logarithmic singularities $\boldsymbol{u} \propto r \ln r$ (with an eigenvalue $\alpha = 1$) are the weakest stress singularities possible in elasticity, and consequently the hardest to detect. Sinclair [165] summarizes the situation under which these occur, as provided in Table 5.5.

Mathematical analysis on stable asymptotics in the neighborhood of angles where power-logarithmic singularities occur can be found in [116] for scalar equations.

Power-logarithmic singularities have been extensively investigated analytically for isotropic bimaterial interfaces [28, 52–54, 136] where the methods proposed are mathematically cumbersome. There is also a vast literature on power-logarithmic singularities due to *inhomogeneous BCs*; interested readers are referred to the review

Table 5.5 Cases for which logarithmic singularities occur

Boundary conditions	Configuration specifications
HC/HC	$\omega = \omega^*,\ \kappa = 1$, with $\omega^* = \tan\omega^*$, see Note 1.
TF/TF	$\omega = \pi - \omega_\kappa, 2\pi - \omega_\kappa, \kappa = -\frac{\tan\omega}{\omega}$, with $\omega_\kappa = \arcsin\frac{\sqrt{\kappa+1}}{2}$
TF/TF	$\kappa = 1 + 2\cos(2\omega) - 2\frac{\sin(2\omega)}{\omega},\ \omega \neq \pi, 2\pi$

Note 1: For $\kappa = 1$ the TF/TF eigen-values coincide with those for HC/HC eigen-values. Thus $\omega^* = 257.5^\circ$ determines the angle for the termination of anti-symmetric power singularities with TF/TF BCs conditions.

by Sinclair [165] and the work by Chen [37], which discusses the power-logarithmic singularities *due to surface tractions* for isotropic bimaterial wedges.

Numerical algorithms that trigger the existence of power-logarithmic singularities in two-dimensional elasticity are provided in [44, 138, 154] and the modified Steklov method, as will be demonstrated by a numerical example. Most numerical methods that trigger the existence of power-logarithmic singularities are based on the mathematical observation that m multiple eigenvalues exist with fewer than m corresponding eigenvectors (the algebraic multiplicity is higher than the geometric multiplicity). For the modified Steklov method we use a routine that determines whether the rank of the matrices $[K_S]$ and $[M_R]$ is the same as their dimension. Following this step, one may look at the computed eigenvalues (assuming that the eigenproblem is not singular and may be solved using generalized eigenproblem routines) and associated eigenvectors and determine whether the algebraic multiplicity of the repeated eigenvalue is the same as the geometric multiplicity. Other methods, such as the matrix method presented in [109, chapter VI.5], can also be used.

5.4 Modified Steklov Eigenproblem for Elasticity

For multimaterial interfaces and anisotropic materials, the eigenpairs are not available analytically, and numerical methods should be used. One of the well-known methods for the numerical computation of eigenpairs was introduced by Leguillon and Sanchez-Palencia in [109], resulting in a quadratic eigenproblem. *This method is described in detail in Section 13.3.1.*

Herein we extend the modified Steklov method, already introduced in Section 3.2 for scalar elliptic problems, to the elasticity system. Consider again the artificial subdomain Ω_R shown in Figure 3.1 in the vicinity of the singular point P. Assuming that no body forces are present in Ω_R, the equilibrium equations (1.44) are

$$\left.\begin{aligned} \partial_1\sigma_{11} + \partial_2\sigma_{12} &= 0 \\ \partial_1\sigma_{12} + \partial_2\sigma_{22} &= 0 \end{aligned}\right\} \quad in\ \Omega_R. \tag{5.84}$$

On the boundaries Γ_1 and Γ_2 we consider homogeneous boundary conditions, that may be traction-free, clamped (zero displacements), or a combination of these:

$$\left.\begin{aligned} u_n &= 0, \quad u_t = 0 \\ T_n &= 0, \quad T_t = 0 \end{aligned}\right\} \quad on \ \Gamma_i, \quad i = 1, 2. \tag{5.85}$$

To obtain the boundary conditions on Γ_R and Γ_{R^*} we observe that in Ω_R, being in the vicinity of P, displacements are of the form[2]

$$\boldsymbol{u} \propto r^{\alpha} \begin{Bmatrix} s_1^+(\theta) \\ s_2^+(\theta) \end{Bmatrix}. \tag{5.86}$$

Using (5.86), on Γ_R and Γ_{R^*} one obtains

$$(\partial \boldsymbol{u}/\partial n) = (\alpha/R)\boldsymbol{u}, \quad (x_1, x_2) \in \Gamma_R, \tag{5.87}$$

$$(\partial \boldsymbol{u}/\partial n) = -(\alpha/R^*)\boldsymbol{u}, \quad (x_1, x_2) \in \Gamma_{R^*}. \tag{5.88}$$

The second-order system of PDEs (stresses are expressed in terms of displacement derivatives through Hooke's law) given by (5.84) together with the four boundary conditions in (5.85), (5.87), (5.88) is the strong (classical) Steklov problem, where α and the eigenfunctions $\boldsymbol{u}$ are sought.

The weak form for the elasticity problem, suited for FE implementation, is obtained by multiplying the two equations in (5.84) by two test functions v_1 and v_2, adding the equations, integrating over Ω_R, and using Green's theorem (see (1.50)-(1.57)):

$$\text{Seek} \quad \boldsymbol{u} \in \mathcal{E}(\Omega_R) \quad \text{such that}$$

$$\mathcal{B}(\boldsymbol{u}, \boldsymbol{v}) \stackrel{\text{def}}{=} \int_{\Omega} ([D]\boldsymbol{v})^T [E][D]\boldsymbol{u} \, d\boldsymbol{x} = \int_{\partial\Omega_N} \hat{T}^T \boldsymbol{v} d\Gamma \stackrel{\text{def}}{=} \mathcal{F}(\boldsymbol{v}) \qquad \forall \boldsymbol{v} \in \mathcal{E}(\Omega_R).$$

In case of homogeneous displacements boundary conditions, the space $\mathcal{E}$ has to be replaced by $\mathcal{E}_o$. We now analyze the linear form $\mathcal{F}(\boldsymbol{v})$. Because either traction-free or clamped boundary conditions are prescribed on $\Gamma_1 \cup \Gamma_2$, then

$$\int_{\Gamma_1 \cup \Gamma_2} \{v_n \, , \, v_t\} \begin{Bmatrix} T_n \\ T_t \end{Bmatrix} d\Gamma = 0. \tag{5.89}$$

[2]Under special circumstances, $\boldsymbol{u}$ may also have additional terms such as $r^{\alpha} ln(r) \begin{Bmatrix} s_1^+(\theta) \\ s_2^+(\theta) \end{Bmatrix}$, but this case will be discussed later.

We may proceed and evaluate $\mathcal{F}(\boldsymbol{v})$ on $\Gamma_R \cup \Gamma_{R^*}$. However, we first establish some relationships that will be useful later on. Cauchy's law provides the relations between the stress tensor and the traction vector on a boundary:

$$\begin{Bmatrix} T_1 \\ T_2 \end{Bmatrix} = [A_2]\boldsymbol{\sigma}, \quad [A_2] = \begin{bmatrix} \cos\theta & 0 & \sin\theta \\ 0 & \sin\theta & \cos\theta \end{bmatrix}, \quad \boldsymbol{\sigma} = \begin{Bmatrix} \sigma_{11} \\ \sigma_{22} \\ \sigma_{12} \end{Bmatrix}. \tag{5.90}$$

The traction in the normal and tangential directions can be expressed in terms of the stresses in the x_1 and x_2 directions:

$$\begin{Bmatrix} T_n \\ T_t \end{Bmatrix} = [A_1]\,[A_2]\,\boldsymbol{\sigma} = [A_3]\boldsymbol{\sigma}, \tag{5.91}$$

where

$$[A_1] = \begin{bmatrix} \cos\theta & \sin\theta \\ -\sin\theta & \cos\theta \end{bmatrix}, \quad [A_3] = \begin{bmatrix} \cos^2\theta & \sin^2\theta & 2\sin\theta\cos\theta \\ -\sin\theta\cos\theta & \sin\theta\cos\theta & \cos^2\theta - \sin^2\theta \end{bmatrix}.$$

Because on Γ_R and Γ_{R^*} the displacements are related to the derivatives of the displacements with respect to r and θ, we express the differential operator $[D]$ in terms of r and θ:

$$[D] \equiv \left[D^{(r,\theta)}\right] \stackrel{\Delta}{=} \begin{bmatrix} \cos\theta\frac{\partial}{\partial r} - \frac{\sin\theta}{r}\frac{\partial}{\partial\theta} & 0 \\ 0 & \sin\theta\frac{\partial}{\partial r} + \frac{\cos\theta}{r}\frac{\partial}{\partial\theta} \\ \sin\theta\frac{\partial}{\partial r} + \frac{\cos\theta}{r}\frac{\partial}{\partial\theta} & \cos\theta\frac{\partial}{\partial r} - \frac{\sin\theta}{r}\frac{\partial}{\partial\theta} \end{bmatrix}. \tag{5.92}$$

Combining (5.91) and (5.92), one obtains

$$\begin{Bmatrix} T_n \\ T_t \end{Bmatrix} = [A_3][E][D^{(r,\theta)}]\boldsymbol{u}. \tag{5.93}$$

The important step now is the use of (5.87-5.88) to replace $\frac{\partial \boldsymbol{u}}{\partial r}$ with $(\alpha/R)\boldsymbol{u}$ on Γ_R. Then we may explicitly compute $[D^{(r,\theta)}]\boldsymbol{u}$ on the boundary $r = R$:

$$\left[D^{(r,\theta)}\right]\boldsymbol{u}\Big|_{r=R} = \begin{Bmatrix} \frac{\alpha\cos\theta}{r}u_1 - \frac{\sin\theta}{r}\frac{\partial u_1}{\partial\theta} \\ \frac{\alpha\sin\theta}{r}u_2 + \frac{\cos\theta}{r}\frac{\partial u_2}{\partial\theta} \\ \frac{\alpha}{r}(\sin\theta u_1 + \cos\theta u_2) + \frac{1}{r}\left(\cos\theta\frac{\partial u_1}{\partial\theta} - \sin\theta\frac{\partial u_2}{\partial\theta}\right) \end{Bmatrix}_{r=R} \stackrel{\text{def}}{=} \partial\boldsymbol{u}\big|_{r=R}. \tag{5.94}$$

With this definition, (5.93) may be rewritten as

$$\begin{Bmatrix} T_n \\ T_t \end{Bmatrix} = [A_3][E]\partial\boldsymbol{u}. \tag{5.95}$$

We have now all the expressions necessary to evaluate $\mathcal{F}(\boldsymbol{v})$ on $\Gamma_R \cup \Gamma_{R^*}$:

$$\mathcal{F}(\boldsymbol{v}) = \int_{\Gamma_R \cup \Gamma_{R^*}} \{v_n \ , \ v_t\} \begin{Bmatrix} T_n \\ T_t \end{Bmatrix} d\Gamma = \int_{\Gamma_R \cup \Gamma_{R^*}} \boldsymbol{v}^T [A_1]^T [A_3][E] \partial \boldsymbol{u} \, d\Gamma. \quad (5.96)$$

We split $\partial \boldsymbol{u}$ into two expressions:

$$\partial \boldsymbol{u} = \left(\frac{\alpha}{r} [A_5] + \frac{1}{r} [D^{(\theta)}] \right) \boldsymbol{u}, \quad (5.97)$$

$$[A_5] = \begin{bmatrix} \cos\theta & 0 \\ 0 & \sin\theta \\ \sin\theta & \cos\theta \end{bmatrix}, \quad [D^{(\theta)}] = \begin{bmatrix} -\sin\theta \frac{\partial}{\partial\theta} & 0 \\ 0 & \cos\theta \frac{\partial}{\partial\theta} \\ \cos\theta \frac{\partial}{\partial\theta} & -\sin \frac{\partial}{\partial\theta} \end{bmatrix}.$$

Combining (5.96) and (5.97) for $r = R^*$, for example, yields

$$\mathcal{F}(\boldsymbol{v})|_{\Gamma_{R^*}} = -\int_{\Gamma_{R^*}} \boldsymbol{v}^T [A_1]^T [A_3][E] \partial \boldsymbol{u} \, d\Gamma = \int_{\theta_1}^{\theta_1+\omega} \boldsymbol{v}^T [A_1]^T [A_3][E] \partial \boldsymbol{u} \big]_{r=R^*} \, d\theta$$
$$\equiv \alpha \mathcal{M}_{R^*}(\boldsymbol{u}, \boldsymbol{v}) + \mathcal{N}_{R^*}(\boldsymbol{u}, \boldsymbol{v}),$$

so finally we obtain

$$\mathcal{F}(\boldsymbol{v}) = \alpha \left(\mathcal{M}_R(\boldsymbol{u}, \boldsymbol{v}) + \mathcal{M}_{R^*}(\boldsymbol{u}, \boldsymbol{v}) \right) + \mathcal{N}_R(\boldsymbol{u}, \boldsymbol{v}) + \mathcal{N}_{R^*}(\boldsymbol{u}, \boldsymbol{v}),$$

where

$$\mathcal{M}_R(\boldsymbol{u}, \boldsymbol{v}) = \int_{\theta_1}^{\theta_1+\omega} \left[\boldsymbol{v}^T [A_1]^T [A_3][E][A_5] \boldsymbol{u} \right]_{r=R} d\theta, \quad (5.98)$$

$$\mathcal{M}_{R^*}(\boldsymbol{u}, \boldsymbol{v}) = \int_{\theta_1}^{\theta_1+\omega} \left[\boldsymbol{v}^T [A_1]^T [A_3][E][A_5] \boldsymbol{u} \right]_{r=R^*} d\theta$$

$$\mathcal{N}_R(\boldsymbol{u}, \boldsymbol{v}) = \int_{\theta_1}^{\theta_1+\omega} \left[\boldsymbol{v}^T [A_1]^T [A_3][E][D^{(\theta)}] \boldsymbol{u} \right]_{r=R} d\theta, \quad (5.99)$$

$$\mathcal{N}_{R^*}(\boldsymbol{u}, \boldsymbol{v}) = \int_{\theta_1}^{\theta_1+\omega} \left[\boldsymbol{v}^T [A_1]^T [A_3][E][D^{(\theta)}] \boldsymbol{u} \right]_{r=R^*} d\theta,$$

Observe that $\mathcal{M}_{R^*}(\boldsymbol{u}, \boldsymbol{v})$ and $\mathcal{N}_{R^*}(\boldsymbol{u}, \boldsymbol{v})$ are similar to $\mathcal{M}_R(\boldsymbol{u}, \boldsymbol{v})$ and $\mathcal{N}_R(\boldsymbol{u}, \boldsymbol{v})$ except that the integrand is evaluated on $r = R^*$. We now can summarize and state the modified Steklov weak form as follows:

Seek $\quad \alpha \in \mathcal{C}, \ \boldsymbol{0} \neq \boldsymbol{u} \in \mathcal{E}(\Omega_R) \times \mathcal{E}(\Omega_R), \quad$ s. t. $\forall \boldsymbol{v} \in \mathcal{E}(\Omega_R) \times \mathcal{E}(\Omega_R)$,

$$\mathcal{B}(\boldsymbol{u}, \boldsymbol{v}) - \left(\mathcal{N}_R(\boldsymbol{u}, \boldsymbol{v}) + \mathcal{N}_{R^*}(\boldsymbol{u}, \boldsymbol{v}) \right) = \alpha \left(\mathcal{M}_R(\boldsymbol{u}, \boldsymbol{v}) + \mathcal{M}_{R^*}(\boldsymbol{u}, \boldsymbol{v}) \right). \quad (5.100)$$

Remark 5.7. *The bilinear form $\mathcal{N}(\boldsymbol{u}, \boldsymbol{v})$ is nonsymmetric with respect to $\boldsymbol{u}$ and $\boldsymbol{v}$. As a consequence, the symmetric properties of the weak form are destroyed. This means that in general, complex eigenvalues and eigenvectors exist.*

The weak form (5.100) is not a linear form, but a *sesquilinear* form, and the coefficient vectors that multiply the stiffness matrix can have complex entries. However, from the practical point of view, the formulation of the matrices corresponding to $\mathcal{B}$, $\mathcal{M}_R$, and $\mathcal{N}$ is not affected by this fact.

Also note that the formulation of the weak form (5.100) has not limited the domain Ω_R to be isotropic, and in fact, (5.100) can be applied to multimaterial anisotropic interface, as will be demonstrated by a numerical example.

Remark 5.8. *The weak form* (5.100) *does not exclude the existence of negative eigenvalues. This is because solutions of the form $r^{-\alpha} f(\theta)$ belong to the space $\mathcal{E}(\Omega_R)$. Therefore both the positive and negative eigenpairs will be obtained using formulation* (5.100).

Remark 5.9. *The domain Ω_R does not include singular points; hence no special refinements of the finite element mesh is required. Furthermore, Ω_R is much smaller in size than Ω.*

Remark 5.10. *By formulating the weak form over Ω_R, the singular point is excluded from the domain of interest such that the accuracy of the finite element solution does not deteriorate in its vicinity. A small subdomain is obtained, over which a finite element solution is smooth, and therefore the mesh does not have to be refined toward the singular point.*

5.4.1 Numerical Solution by p-FEMs

The numerical solution of the elasticity problem by means of FEMs is similar to that described briefly in Section 3.3 but more complicated owing to the existence of two fields $\boldsymbol{u} = (u_1, u_2)$. In the following, the expressions in (5.100) are reformulated in the framework of p-FEM.

The domain Ω_R is divided into finite elements through a meshing process. The polynomial basis and trial functions are defined on a standard element in the ξ, η plane such that $-1 < \xi < 1$, $-1 < \eta < 1$. These elements are then mapped by appropriate mapping functions onto the "real" elements. Let the displacement functions u_1, u_1 be expressed in terms of the basis functions in the standard element $\Phi_i(\xi, \eta)$:

$$\left.\begin{aligned} u_1(\xi, \eta) &= \textstyle\sum_{i=1}^{N} a_i \; \Phi_i(\xi, \eta) \\ u_2(\xi, \eta) &= \textstyle\sum_{i=1}^{N} a_{N+i} \; \Phi_i(\xi, \eta) \end{aligned}\right\} \tag{5.101}$$

or

$$\begin{Bmatrix} u_1 \\ u_2 \end{Bmatrix} = \begin{bmatrix} \Phi_1 \cdots \Phi_N & 0 & \cdots & 0 \\ 0 & \cdots & 0 & \Phi_1 \cdots \Phi_N \end{bmatrix} \begin{Bmatrix} a_1 \\ \vdots \\ a_N \\ a_{N+1} \\ \vdots \\ a_{2N} \end{Bmatrix} \stackrel{\text{def}}{=} [\Phi]\boldsymbol{a}, \tag{5.102}$$

where a_i are the amplitudes of the basis functions, and Φ_i are the shape functions.

Using (5.102), the unconstrained stiffness matrix corresponding to $\mathcal{B}(\boldsymbol{u}, \boldsymbol{v})$ is given by

$$[K] \stackrel{\text{def}}{=} \int\int_{\Omega_R} [D]^T [\Phi]^T [E][D][\Phi] d\Omega. \tag{5.103}$$

For simplicity, we assume traction-free boundary condition on Γ_1 and Γ_2, and concentrate our discussion first on $\mathcal{N}_R(\boldsymbol{u}, \boldsymbol{v})$. We start by evaluating the required expressions involved in the computation.

The mapping of ξ, $\eta = -1$ (side 1 of the standard element) onto Γ_R is given by

$$\theta = \frac{\theta_1 + \omega - \theta_1}{2}\xi + \frac{\theta_1 + \omega + \theta_1}{2} = \frac{\omega}{2}\xi + \frac{2\theta_1 + \omega}{2}, \tag{5.104}$$

so that $d\theta = \frac{\omega}{2} d\xi$, and the matrix $[D^{(\theta)}]d\theta$ becomes

$$[D^{(\theta)}]d\theta = \begin{bmatrix} -\sin\theta \frac{\partial}{\partial \xi} & 0 \\ 0 & \cos\theta \frac{\partial}{\partial \xi} \\ \cos\theta \frac{\partial}{\partial \xi} & -\sin\theta \frac{\partial}{\partial \xi} \end{bmatrix} d\xi. \tag{5.105}$$

On side 1, the basis and trial functions $\Phi_i(\xi, \eta)$ are nothing more than integrals of Legendre polynomials $P_i(\xi)$ for $i > 3$. The expression $\boldsymbol{v}^T [A_1]^T [A_3]$ in (5.99) is therefore given by

$$\{b_1 \cdots b_{2N}\} \begin{bmatrix} P_1 \cos\theta & 0 & P_1 \sin\theta \\ \vdots & \vdots & \vdots \\ P_N \cos\theta & 0 & P_N \sin\theta \\ 0 & P_1 \sin\theta & P_1 \cos\theta \\ \vdots & \vdots & \vdots \\ 0 & P_N \sin\theta & P_N \cos\theta \end{bmatrix} \stackrel{\text{def}}{=} \boldsymbol{b}^T [PC]. \tag{5.106}$$

The expression $[D^{(\theta)}]\boldsymbol{u}d\theta$ in (5.99), when using (5.105), becomes

$$\begin{bmatrix} -\sin\theta\frac{\partial P_1}{\partial\xi} & \cdots & -\sin\theta\frac{\partial P_N}{\partial\xi} & 0 & \cdots & 0 \\ 0 & \cdots & 0 & \cos\theta\frac{\partial P_1}{\partial\xi} & \cdots & \cos\theta\frac{\partial P_N}{\partial\xi} \\ \cos\theta\frac{\partial P_1}{\partial\xi} & \cdots & \cos\theta\frac{\partial P_N}{\partial\xi} & -\sin\theta\frac{\partial P_1}{\partial\xi} & \cdots & -\sin\theta\frac{\partial P_N}{\partial\xi} \end{bmatrix} \begin{Bmatrix} a_1 \\ \vdots \\ a_{2N} \end{Bmatrix} \stackrel{\text{def}}{=} [\partial P]\boldsymbol{a}. \tag{5.107}$$

Substituting (5.106) and (5.107) into (5.99), we finally have an expression for $\mathcal{N}_R(\boldsymbol{u},\boldsymbol{v})$:

$$\mathcal{N}_R(\boldsymbol{u},\boldsymbol{v}) = \boldsymbol{b}^T\left(\int_{-1}^{1}[PC][E][\partial P]d\xi\right)\boldsymbol{a} \stackrel{\text{def}}{=} \boldsymbol{b}^T[N_R]\boldsymbol{a}. \tag{5.108}$$

The entries of $[N_R]$ are computed using Gauss quadrature:

$$N_{R\,ij} = \sum_{m=1}^{M} W_m \sum_{\ell,k=1}^{3} PC_{i\ell}(\xi_m)\, E_{\ell k}\, \partial P_{kj}(\xi_m), \tag{5.109}$$

where W_m and ξ_m are the weights and abscissas of the Gauss quadrature points, respectively.

Using the same arguments as above, the expression $\mathcal{M}_R(\boldsymbol{u},\boldsymbol{v})$ in (5.98) is evaluated:

$$\mathcal{M}_R(\boldsymbol{u},\boldsymbol{v}) = \frac{\omega}{2}\boldsymbol{b}^T\left(\int_{-1}^{1}[PC][E][PC]^T d\xi\right)\boldsymbol{a} \stackrel{\text{def}}{=} \boldsymbol{b}^T[M_R]\boldsymbol{a} \tag{5.110}$$

and

$$M_{R\,ij} = \frac{\omega}{2}\sum_{m=1}^{M} W_m \sum_{\ell,k=1}^{3} PC_{i\ell}(\xi_m)\, E_{\ell k}\, PC_{jk}(\xi_m). \tag{5.111}$$

The matrices $[N_{R^*}]$ and $[M_{R^*}]$ are computed similarly having nonzero entries that correspond to DOFs on Γ_{R^*}.

Once $[K]$, $[N_R]$, $[N_{R^*}]$, $[M_R]$, and $[M_{R^*}]$ have been evaluated, the eigenpairs can be obtained using the same procedures described in Section 3.3.

Remark 5.11. *Although we derived our matrices as if only one finite element existed along the boundary Γ_R, the formulation for multiple finite elements is identical, and the matrices $[K]$, $[N_R]$, $[N_{R^*}]$, $[M_R]$, and $[M_{R^*}]$ are obtained by an assembly procedure.*

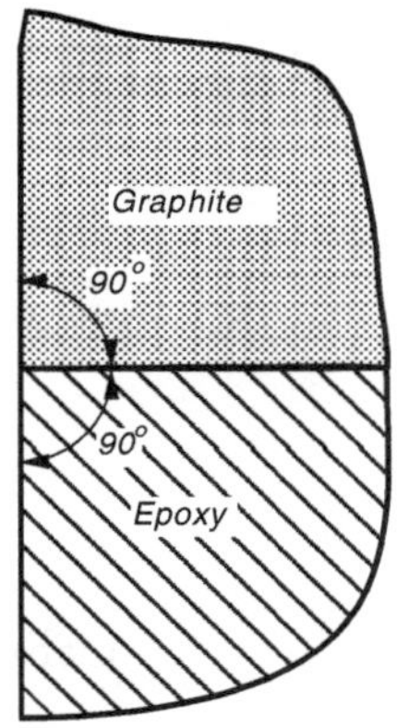

Fig. 5.8 Orthotropic bonded materials test problem.

5.4.2 Numerical Investigation: Two Bonded Orthotropic Materials

The test problem in this subsection consists of two orthotropic materials, graphite and adhesive (epoxy), bonded together, with plane-strain condition assumed. See Figure 5.8. This problem was chosen to demonstrate the modified Steklov method for anisotropic multimaterials with a singular point. The material properties are listed below.

	Graphite	Adhesive
$E_{11} \times 10^6$ psi	20.0	1.4
E_{22}	2.0	1.4
E_{33}	2.0	1.4
ν_{12}	0.450	0.3
ν_{23}	0.040	0.3
ν_{31}	0.045	0.3
$\mu_{12} \times 10^6$ psi	1.1	0.7

In [156] the Lekhnitskii stress potentials were used to solve the anisotropic problem for the two materials. The first four exact nonzero eigenvalues obtained are given in the following.

$$\begin{aligned}(\alpha_1)_{\text{EX}} &= 0.905 \pm 0.0000i,\\ (\alpha_2)_{\text{EX}} &= 1.000 \pm 0.0000i,\\ (\alpha_3)_{\text{EX}} &= 1.944 \pm 0.3051i,\\ (\alpha_4)_{\text{EX}} &= 2.475 \pm 0.9559i.\end{aligned}$$

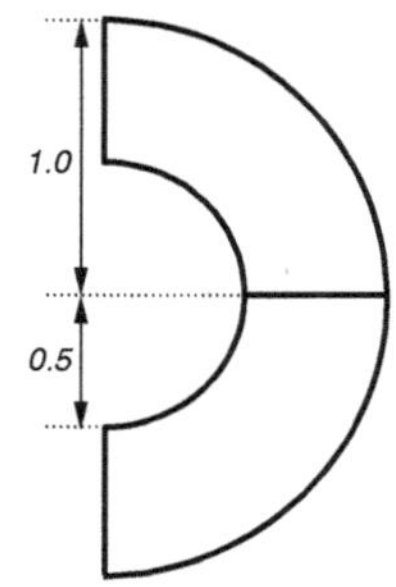

Fig. 5.9 Finite element mesh for the elasticity anisotropic problem.

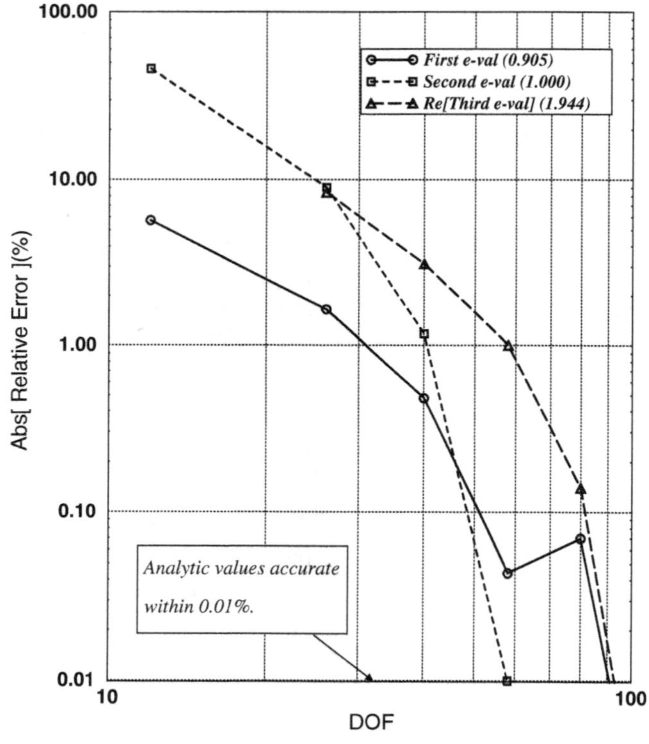

Fig. 5.10 Convergence of first three eigenvalue as the p-level increases.

The mesh used for this example problem has the minimum possible number of finite elements, i.e., one element in each subdomain. See Figure 5.9. The exact eigenvalues are given with an accuracy of up to the third digit, so that the accuracy of the numerical results may be assessed up to about 0.01% relative error.

Convergence curves for the first three eigenvalues are shown in Figure 5.10. The results demonstrate an excellent convergence rate for the coarsest mesh possible.

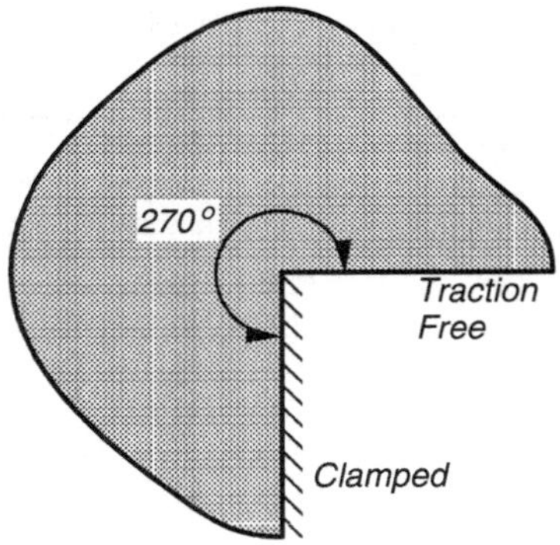

Fig. 5.11 Free-clamped wedge exciting power-logarithmic stress singularity.

Remarks on Robustness and Efficiency

We demonstrated that the proposed modified Steklov method performs very well when the eigenvalues are well separated. From the numerical point of view the treatment of eigenvalues that are close without being a double eigenvalue is more difficult. Usually these situations occur near bifurcation points, where the nature of the eigenvalues changes from complex to real or vice versa. As an example of the robustness of the modified Steklov method, we considered a bifurcation point for an isotropic material in a wedge with free-free boundary condition studied in [138], with $\theta_0 = 146.30854358°$ being such a bifurcation angle.

We performed two analyses, one with wedge angle 146.3085° and the other with wedge angle 146.3086°. The results obtained using the modified Steklov method and computational times indicated no degradation, and the bifurcation angle was determined with very high accuracy.

5.4.3 Numerical Investigation: Power-Logarithmic Singularity

The analytical conditions governing the occurrence of a power-logarithmic stress singularity, $\mathcal{O}(r^{\alpha-1}\ln r)$, are presented in [52], where it is shown that for a free-clamped wedge of a specific Poisson ratio, the power-logarithmic stress singularity is excited. For a $3\pi/4 = 270°$ free-clamped wedge (see Figure 5.11) in an isotropic material with Poisson ratio $\nu = 0.331046412$, the stress field contains a power-logarithmic singularity [52, Table 1]:

$$\boldsymbol{\sigma} = A_1 r^{0.34254974-1} \boldsymbol{S}_1(\theta) + A_2 r^{0.34254974-1} \ln r \boldsymbol{S}_2(\theta) + \text{higher-order terms.} \tag{5.112}$$

To demonstrate the robustness of the modified Steklov method with respect to power-logarithmic singularity types, we present the first two computed eigenpairs (these are computed on a four-element mesh). It is expected that the first two eigenvalues will collapse into a single one as the p-level (representing the number of

Table 5.6 First two FE eigenvalues which collapse to a single value $\alpha_1^{EX} = 0.34254974$.

	$p = 1$	$p = 2$	$p = 3$	$p = 4$	$p = 5$	$p = 6$	$p = 7$	$p = 8$
DOF	20	46	72	106	148	198	256	322
α_1^{FE}	0.29603441	0.31881456	0.33848389	0.34183181	0.34254967*	0.34251766	0.34253064	0.34254449
α_2^{FE}	0.38225179	0.36605285	0.34658798	0.34326833	0.34254967*	0.34258184	0.34256885	0.34255499

* Complex conjugates with imaginary part $\pm 0.000108i$.

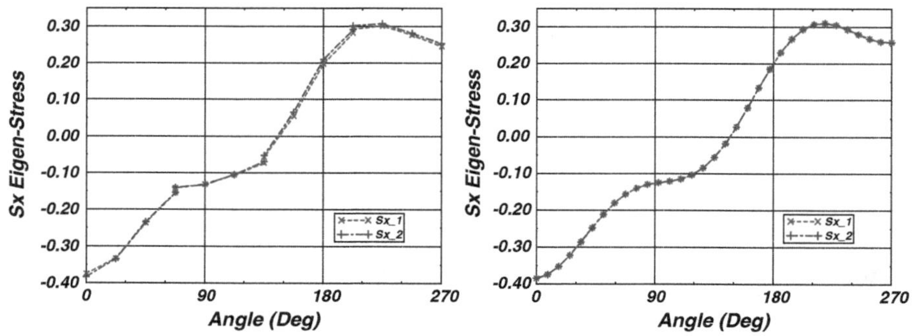

Fig. 5.12 First and second S_{11} eigenstresses ($S_{11}^{(1)}$, $S_{11}^{(2)}$) at $p = 3$ (left) and at $p = 8$ (right).

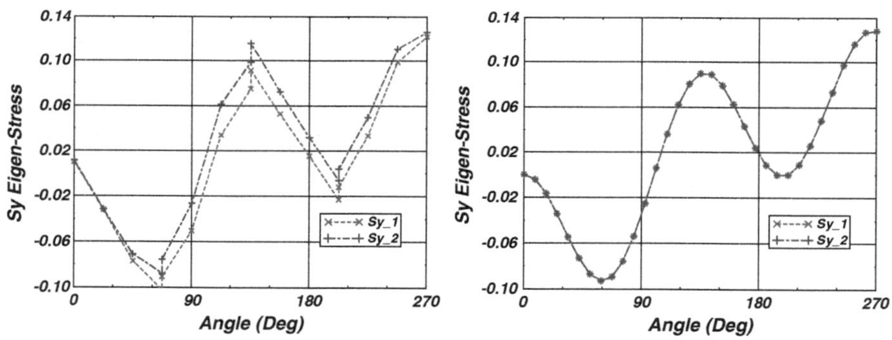

Fig. 5.13 First and second S_{22} eigenstresses ($S_{22}^{(1)}$, $S_{22}^{(2)}$) at $p = 3$ (left) and at $p = 8$ (right).

degrees of freedom) is increased, and the corresponding eigenstresses will become identical. In Table 5.6 we summarize the first two computed eigenvalues obtained as the p-level is increased from 1 to 8. The first and second eigenstresses S_{11} and S_{22} for p-levels 3 and 8 are shown in Figures 5.12–5.13. As observed, the first two eigenvalues and eigen-stresses collapse into one as the number of degrees of freedom is increased, indicating the presence of the power-logarithmic stress singularity.

Chapter 6
Computing Generalized Stress Intensity Factors (GSIFs)

Two efficient methods for extracting GSIFs from FE solutions are detailed in this chapter, and their performance is demonstrated by several numerical examples.

6.1 The Contour Integral Method, Also Known as the Dual-Singular Function Method or the Reciprocal Work Contour Method

Extraction of GSIFs by the contour integral method (CIM) [12,27,177] is one of the most accurate and efficient methods that provides independently any GSIF. Because it is based on a path-independent integral along a path that does not have to be in the close vicinity of the crack tip, the method may be applied to an FE solution that is not polluted by numerical errors and can be easily implemented as an FE post-solution operation. We first derive the path-independent integral; then we introduce the dual singular functions and prove their orthogonal properties with respect to the primal singular functions and finally use them to extract the GSIFs.

6.1.1 A Path-Independent Contour Integral

Consider an elastic *isotropic and homogeneous* 2D domain with a reentrant V-notch (or crack), with traction-free or clamped boundary conditions on V-notch faces and subjected to two different sets of tractions with corresponding displacements (system $*$ and $**$) away from the singular point. The traction of system $*$, $\boldsymbol{T}^{(*)}$, and the traction of system $**$, $\boldsymbol{T}^{(**)}$, act on the same boundary of the domain. The two systems result in two systems of displacement, $\boldsymbol{u}^{(*)}$ and $\boldsymbol{u}^{(**)}$.

Z. Yosibash, *Singularities in Elliptic Boundary Value Problems and Elasticity and Their Connection with Failure Initiation*, Interdisciplinary Applied Mathematics 37, DOI 10.1007/978-1-4614-1508-4_6, © Springer Science+Business Media, LLC 2012

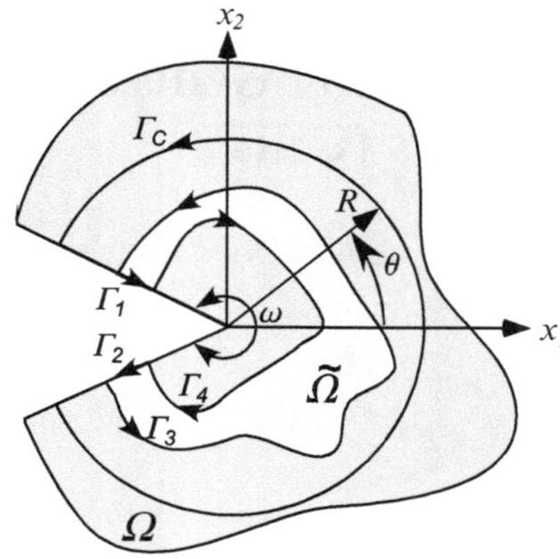

Fig. 6.1 The contour $\Gamma = \Gamma_1 \cup \Gamma_2 \cup \Gamma_3 \cup \Gamma_4$ in the vicinity of a V-notch.

Let us examine a closed contour $\Gamma = \Gamma_1 \cup \Gamma_2 \cup \Gamma_3 \cup \Gamma_4$, as illustrated in Figure 6.1. According to Betti's theorem

$$\int_{\Gamma} \left(\boldsymbol{T}^{(*)} \cdot \boldsymbol{u}^{(**)} - \boldsymbol{T}^{(**)} \cdot \boldsymbol{u}^{(*)} \right) d\Gamma = 0. \tag{6.1}$$

For either traction-free or homogeneous Dirichlet boundary conditions on the crack (or notch) surfaces,

$$\int_{\Gamma_1} \left(\boldsymbol{T}^{(*)} \cdot \boldsymbol{u}^{(**)} - \boldsymbol{T}^{(**)} \cdot \boldsymbol{u}^{(*)} \right) d\Gamma = \int_{\Gamma_2} \left(\boldsymbol{T}^{(*)} \cdot \boldsymbol{u}^{(**)} - \boldsymbol{T}^{(**)} \cdot \boldsymbol{u}^{(*)} \right) d\Gamma = 0, \tag{6.2}$$

so that equation (6.1) becomes

$$\oint_{\Gamma_3} \left(\boldsymbol{T}^{(*)} \cdot \boldsymbol{u}^{(**)} - \boldsymbol{T}^{(**)} \cdot \boldsymbol{u}^{(*)} \right) d\Gamma + \oint_{\Gamma_4} \left(\boldsymbol{T}^{(*)} \cdot \boldsymbol{u}^{(**)} - \boldsymbol{T}^{(**)} \cdot \boldsymbol{u}^{(*)} \right) d\Gamma = 0. \tag{6.3}$$

The integration along Γ_4 is in the clockwise direction, whereas the integration along Γ_3 is in the counter clockwise direction. By changing the direction of integration of Γ_4, one obtains

$$\oint_{\Gamma_3} \left(\boldsymbol{T}^{(*)} \cdot \boldsymbol{u}^{(**)} - \boldsymbol{T}^{(**)} \cdot \boldsymbol{u}^{(*)} \right) d\Gamma = \oint_{\Gamma_4} \left(\boldsymbol{T}^{(*)} \cdot \boldsymbol{u}^{(**)} - \boldsymbol{T}^{(**)} \cdot \boldsymbol{u}^{(*)} \right) d\Gamma. \tag{6.4}$$

Since Γ_3 and Γ_4 are randomly selected, the right-hand side as well as the left-hand side of (6.4) is constant along any path Γ_c that starts at one edge of the crack (or notch) and terminates at the other edge. Therefore the integral I_{Γ_c} defined by

$$I_{\Gamma_c}(\boldsymbol{u}^{(*)}, \boldsymbol{u}^{(**)}) \overset{\text{def}}{=} \oint_{\Gamma_c} \left(\boldsymbol{T}^{(*)} \cdot \boldsymbol{u}^{(**)} - \boldsymbol{T}^{(**)} \cdot \boldsymbol{u}^{(*)} \right) d\Gamma = \text{const}, \tag{6.5}$$

called the reciprocal work contour, is a path-independent integral. This path-independent integral for isotropic elasticity is analogous to (1.18) introduced for the Laplace problem.

6.1.2 Orthogonality of the Primal and Dual Eigenfunctions

For simplicity of presentation we consider real and simple eigenpairs (the CIM may be applied to complex eigenfunctions; see [196], where a crack at a bimaterial interface is addressed). Consider one term in the expansion of the displacements associated with the eigenvalue α_i, for example

$$\tilde{\boldsymbol{u}}_i^+ = \begin{Bmatrix} u_r^+ \\ u_\theta^+ \end{Bmatrix}_i = r^{\alpha_i} \tilde{\boldsymbol{s}}_i^+(\theta), \qquad \text{no summation on } i. \tag{6.6}$$

The eigenstress associated with (6.6) is denoted by

$$\tilde{\boldsymbol{\sigma}}_i^+ = \begin{Bmatrix} \sigma_{rr}^+ \\ \sigma_{\theta\theta}^+ \\ \sigma_{r\theta}^+ \end{Bmatrix}_i = r^{\alpha_i - 1} \tilde{\boldsymbol{S}}_i^+(\theta), \qquad \text{no summation on } i. \tag{6.7}$$

Similarly we may consider the displacements and stresses associated with a negative eigenvalue α_j (the dual solution):

$$\tilde{\boldsymbol{u}}_j^- = r^{-\alpha_j} \tilde{\boldsymbol{s}}_j^-(\theta), \qquad \text{no summation on } j. \tag{6.8}$$

The eigenstress associated with (6.8) is

$$\tilde{\boldsymbol{\sigma}}_j^- = -r^{-\alpha_j - 1} \tilde{\boldsymbol{S}}_j^-(\theta), \qquad \text{no summation on } j. \tag{6.9}$$

It is important to realize that $\tilde{\boldsymbol{u}}_i^+$ and $\tilde{\boldsymbol{\sigma}}_i^+$ as well as $\tilde{\boldsymbol{u}}_j^-$ and $\tilde{\boldsymbol{\sigma}}_j^-$ are displacements and stresses that satisfy the elasticity system and boundary conditions in the vicinity of the singular point. Although the dual solution is an inadmissible solution because it produces infinite displacements at the singular point and stresses that yield infinite strain energy, yet it can be used in mathematical manipulations, since it solves the elasticity equations and boundary conditions.

Consider now a circular arc of radius R centered on the V-notch tip, i.e., a path starting at Γ_2 and terminating at Γ_1 as shown in Figure 6.1. Along this path we have the following traction vectors corresponding to $\tilde{\boldsymbol{\sigma}}_i^+$ and $\tilde{\boldsymbol{\sigma}}_j^-$:

$$\boldsymbol{T}_i^+ = \begin{Bmatrix} T_r \\ T_\theta \end{Bmatrix}_i = \begin{Bmatrix} \sigma_{rr}^+ \\ \sigma_{r\theta}^+ \end{Bmatrix}_i = r^{\alpha_i - 1} \begin{Bmatrix} S_{rr}^+ \\ S_{r\theta}^+ \end{Bmatrix}_i, \qquad \boldsymbol{T}_j^- = -r^{-\alpha_j - 1} \begin{Bmatrix} S_{rr}^- \\ S_{r\theta}^- \end{Bmatrix}_j, \tag{6.10}$$

no summation on i, j.

Now, computing $I_{\Gamma_c}(\tilde{\boldsymbol{u}}_i^+, \tilde{\boldsymbol{u}}_j^-)$ along the circular path with $d\Gamma = Rd\theta$ yields (substitute (6.10), (6.6), and (6.8) in (6.5))

$$I_{\Gamma_c}(\tilde{\boldsymbol{u}}_i^+, \tilde{\boldsymbol{u}}_j^-) = R^{\alpha_i - \alpha_j} \int_{-\omega/2}^{\omega/2} \left[\tilde{\boldsymbol{S}}_i^+ \tilde{\boldsymbol{s}}_j^- - \tilde{\boldsymbol{S}}_j^- \tilde{\boldsymbol{s}}_i^+ \right] d\theta, \qquad \text{no summation on } i, j. \tag{6.11}$$

Because I_{Γ_c} is path-independent, it should not depend on R. Therefore, if $\alpha_i \neq \alpha_j$, the integral in (6.11) must vanish, i.e.,

$$\int_{-\omega/2}^{\omega/2} \left[\tilde{\boldsymbol{S}}_i^+ \tilde{\boldsymbol{s}}_j^- - \tilde{\boldsymbol{S}}_j^- \tilde{\boldsymbol{s}}_i^+ \right] d\theta = 0, \qquad \text{for } i \neq j, \qquad \text{no summation on } i, j. \tag{6.12}$$

Condition (6.12) is the orthogonality property of primal and dual eigenfunctions associated with different eigenvalues, and is a key property for the extraction of GSIF by the CIM. There may be cases of multiple identical eigenvalues with the same multiplicity of eigenfunctions, as in the important case of cracks. For example, let us consider an eigenvalue with multiplicity two, $\alpha_i^{(I)} = \alpha_i^{(II)}$, but in this case the eigenfunctions will be different, i.e., $\boldsymbol{s}_i^{(I)}(\theta) \neq \boldsymbol{s}_i^{(II)}(\theta)$. For the crack case we observed that $\boldsymbol{s}_i^{(I)}(\theta)$ is a symmetric function with respect with θ, whereas $\boldsymbol{s}_i^{(II)}(\theta)$ was an antisymmetric function. In this case, although $\alpha_i^{(I)} = \alpha_i^{(II)}$, the orthogonality condition in (6.12) still holds due to the multiplication of symmetric and antisymmetric functions, i.e., for cracks (6.12) reads

$$\int_{-\pi}^{\pi} \left[\tilde{\boldsymbol{S}}_i^{+(I)} \tilde{\boldsymbol{s}}_i^{-(II)} - \tilde{\boldsymbol{S}}_i^{-(II)} \tilde{\boldsymbol{s}}_i^{+(I)} \right] d\theta = 0, \qquad \text{no summation on } i. \tag{6.13}$$

The value of the path integral for the ith eigen-pair and its dual:

$$I_{\Gamma_c}(\tilde{\boldsymbol{u}}_i^+, \tilde{\boldsymbol{u}}_i^-) = \int_{-\omega/2}^{\omega/2} \left[\tilde{\boldsymbol{S}}_i^+ \tilde{\boldsymbol{s}}_i^- - \tilde{\boldsymbol{S}}_i^- \tilde{\boldsymbol{s}}_i^+ \right] d\theta, \qquad \text{no summation on } i. \tag{6.14}$$

is independent of the path Γ_c.

The orthogonality property (6.12) and the path-independent property (6.14) hold for the Cartesian representation of $\boldsymbol{s} = (s_1 \; s_2)^T$ and $\boldsymbol{S} = (S_{11} \; S_{22} \; S_{12})^T$, i.e.,

$$\int_{-\omega/2}^{\omega/2} \left[\boldsymbol{S}_i^+ \boldsymbol{s}_j^- - \boldsymbol{S}_j^- \boldsymbol{s}_i^+ \right] d\theta = 0, \quad \text{for } i \neq j, \quad \text{no summation on } i, j, \tag{6.15}$$

$$I_{\Gamma_c}(\boldsymbol{u}_i^+, \boldsymbol{u}_i^-) = \int_{-\omega/2}^{\omega/2} \left[\boldsymbol{S}_i^+ \boldsymbol{s}_i^- - \boldsymbol{S}_i^- \boldsymbol{s}_i^+ \right] d\theta, \quad \text{no summation on } i. \tag{6.16}$$

6.1.3 *Extracting GSIFs (A_i's) Using the CIM*

The method of extracting the first generalized stress intensity factor using the reciprocal work contour integral was first presented by Stern and Soni in [172] for an isotropic 2-D $\pi/2$ corner with homogeneous Dirichlet boundary conditions on one edge ($\boldsymbol{u}|_{\theta=0} = 0$) and traction-free boundary conditions on the other edge ($\boldsymbol{T}|_{\theta=\frac{\pi}{2}} = 0$). In their work, only *the first eigenpair was considered*, and therefore *only the first generalized stress intensity factor A_1 is nonzero*. Sinclair et al. [166] and Szabó and Babuška [177] extended the CIM method for the extraction of any GSIF for V-notches with a solid angle of ω, as shown in Figure 6.1. The algorithm for the extraction of any of the GSIFs' A_i's in (5.1) is developed using the path-independent integral I_{Γ_c} and the orthogonal property of the primal and dual eigenpairs. We assume that the edges intersecting at the singular point are straight, that homogeneous boundary conditions are applied on these edges, that no logarithmic terms are present in the asymptotic expansion, and that eigen-values are real (none of these assumptions are mandatory, but these allow a clearer and simplified presentation of the method), so that (5.1) is simplified to

$$\tilde{\boldsymbol{u}} = \sum_{i=1}^{\infty} A_i r^{\alpha_i} \tilde{\boldsymbol{s}}_i^+(\theta). \tag{6.17}$$

Consider again the path independent integral along the circular arc of radius R in Figure 6.1 centered at the singular point, so that the exact solution along it is given by (6.17), and the stresses given by

$$\tilde{\boldsymbol{\sigma}} = \sum_{i=1}^{\infty} A_i r^{\alpha_i - 1} \tilde{\boldsymbol{S}}_i^+(\theta) \tag{6.18}$$

If one is interested in extracting the jth GSIF A_j, then an auxiliary function and an associated auxiliary stress are defined based on the jth dual eigenpair, multiplied by a factor C_j, which will be determined later

$$\tilde{\boldsymbol{w}} = C_j r^{-\alpha_j} \tilde{\boldsymbol{s}}_j^-(\theta), \qquad \tilde{\boldsymbol{\sigma}}^w = C_j r^{-\alpha_j - 1} \tilde{\boldsymbol{S}}_j^-(\theta). \tag{6.19}$$

Evaluating now $I_{\Gamma_c}(\tilde{\boldsymbol{u}}, \tilde{\boldsymbol{w}}_j)$ (substitute (6.17), (6.18) and (6.19) in (6.5)), and using the orthogonal property of the eigenfunctions and their duals (6.12), one obtains

$$I_{\Gamma_c}(\tilde{\boldsymbol{u}}, \tilde{\boldsymbol{w}}) = A_j C_j \int_{-\omega/2}^{\omega/2} \left[\tilde{\boldsymbol{S}}_j^+ \tilde{\boldsymbol{s}}_j^- - \tilde{\boldsymbol{S}}_j^- \tilde{\boldsymbol{s}}_j^+ \right] d\theta. \tag{6.20}$$

Letting

$$C_j = \left[\int_{-\omega/2}^{\omega/2} \left[\tilde{\boldsymbol{S}}_j^+ \tilde{\boldsymbol{s}}_j^- - \tilde{\boldsymbol{S}}_j^- \tilde{\boldsymbol{s}}_j^+ \right] d\theta \right]^{-1},$$

one obtains

$$I_{\Gamma_c}(\tilde{\boldsymbol{u}}, \tilde{\boldsymbol{w}}) = A_j, \tag{6.21}$$

where Γ_c is arbitrarily chosen. The exact solution, i.e., $\tilde{\boldsymbol{u}}$ and $\tilde{\boldsymbol{\sigma}}$, is not known, but one can use finite element methods to obtain an approximation, $\tilde{\boldsymbol{u}}^{\text{FE}}$ and $\tilde{\boldsymbol{\sigma}}^{\text{FE}}$. Using these approximations, together with the known auxiliary functions $\tilde{\boldsymbol{w}}$ and $\tilde{\boldsymbol{\sigma}}^w$, one may easily compute numerically the value of

$$I_{\Gamma_c}(\tilde{\boldsymbol{u}}^{\text{FE}}, \tilde{\boldsymbol{w}}) = A_j^{\text{FE}}$$

and thus obtain a numerical approximation of A_j.

The CIM belongs to a class of methods that utilize a special functional defined on the finite element solution and an auxiliary extraction function (the dual eigenfunction) in the post-solution phase. As such, the accuracy of the post-processed value A_j^{FE} is related to how well the finite element space is able to simultaneously approximate both the solution of the basic problem $\tilde{\boldsymbol{u}}^{\text{FE}}$ and the solution $\tilde{\boldsymbol{w}}$ of an auxiliary problem having the same characteristics as the basic problem. Babuška and Miller proved in [12] that the convergence of A_j^{FE} to the exact value is at least as fast as the strain energy (twice as fast as the convergence measured in energy norm) of the basic problem and therefore is superconvergent.

Another very important aspect of the CIM extraction technique is its insensitivity to the size of the extraction path. This property is of special importance because the quality of the FE solution is poor in the elements touching the singular point, and therefore the CIM extraction path may be outside of these elements. The superconvergence and robustness properties of the CIM will be demonstrated by numerical examples in Section 6.3.

6.1.3.1 Extracting GSIFs for a TF/TF V-Notch Using the CIM

Consider the first two terms which may be singular in the series expansion of the stresses and displacements for a TF/TF V-notch as shown in Figure 6.1, given in (5.52) and (5.53). These can also be written as

$$\begin{aligned} u_r = {} & A_1 \frac{r^{\alpha_1}}{2\alpha_1 \mu} \Bigg((\kappa - \alpha_1)\cos((1-\alpha_1)\theta) \\ & \qquad - \frac{\alpha_1^2 - 1}{\cos(\alpha_1\omega) - \alpha_1\cos(\omega)} \cos((1+\alpha_1)\theta) \Bigg) \\ & + A_2 \frac{r^{\alpha_2}}{2\alpha_2 \mu} \Bigg(-(\kappa - \alpha_2)\sin((1-\alpha_2)\theta) \\ & \qquad - \frac{\alpha_2^2 - 1}{\cos(\alpha_2\omega) - \alpha_2\cos(\omega)} \sin((1+\alpha_2)\theta) \Bigg), \end{aligned}$$

$$u_\theta = A_1 \frac{r^{\alpha_1}}{2\alpha_1 \mu} \Bigg(-(\kappa + \alpha_1)\sin((1-\alpha_1)\theta) + \frac{\alpha_1^2 - 1}{\cos(\alpha_1\omega) - \alpha_1\cos(\omega)} \sin((1+\alpha_1)\theta) \Bigg) - A_2 \frac{r^{\alpha_2}}{2\alpha_2 \mu} \Bigg((\kappa + \alpha_2)\cos((1-\alpha_2)\theta) + \frac{\alpha_2^2 - 1}{\cos(\alpha_2\omega) - \alpha_2\cos(\omega)} \cos((1+\alpha_2)\theta) \Bigg), \quad (6.22)$$

$$\sigma_{rr} = -A_1 r^{\alpha_1 - 1} \Bigg((3-\alpha_1)\cos((1-\alpha_1)\theta) - \frac{\alpha_1^2 - 1}{\cos(\alpha_1\omega) - \alpha_1\cos(\omega)} \cos((1+\alpha_1)\theta) \Bigg) + A_2 r^{\alpha_2 - 1} \Bigg((1+\alpha_2)\sin((1-\alpha_2)\theta) - \frac{\alpha_2^2 - 1}{\cos(\alpha_2\omega) - \alpha_2\cos(\omega)} \sin((1+\alpha_2)\theta) \Bigg),$$

$$\sigma_{\theta\theta} = A_1 r^{\alpha_1 - 1} \Bigg(-(1+\alpha_1)\cos((1-\alpha_1)\theta) - \frac{\alpha_1^2 - 1}{\cos(\alpha_1\omega) - \alpha_1\cos(\omega)} \cos((1+\alpha_1)\theta) \Bigg) + A_2 r^{\alpha_2 - 1} \Bigg((1+\alpha_2)\sin((1-\alpha_2)\theta) - \frac{\alpha_2^2 - 1}{\cos(\alpha_2\omega) - \alpha_2\cos(\omega)} \sin((1+\alpha_2)\theta) \Bigg),$$

$$\sigma_{r\theta} = -A_1 r^{\alpha_1 - 1} \Bigg((1-\alpha_1)\sin((1-\alpha_1)\theta) + \frac{\alpha_1^2 - 1}{\cos(\alpha_1\omega) - \alpha_1\cos(\omega)} \sin((1+\alpha_1)\theta) \Bigg) - A_2 r^{\alpha_2 - 1} \Bigg((\alpha_2 - 1)\cos((1-\alpha_2)\theta) + \frac{\alpha_2^2 - 1}{\cos(\alpha_2\omega) - \alpha_2\cos(\omega)} \cos((1+\alpha_2)\theta) \Bigg). \quad (6.23)$$

The auxiliary displacement and stress fields for the extraction of A_1 are associated with the first negative eigenvalue $-\alpha_1$:

$$
\begin{aligned}
w_{(\mathrm{I})\,r} &= C_1 \frac{r^{-\alpha_1}}{2\alpha_1\mu}\left((\kappa+\alpha_1)\cos((1+\alpha_1)\theta)\right.\\
&\qquad\left.-\frac{\alpha_1^2-1}{\alpha_1\cos(\omega)+\cos(\alpha_1\omega)}\cos((\alpha_1-1)\theta)\right),\\
w_{(\mathrm{I})\,\theta} &= C_1 \frac{r^{-\alpha_1}}{2\alpha_1\mu}\left(-(\kappa-\alpha_1)\sin((1+\alpha_1)\theta)\right.\\
&\qquad\left.-\frac{\alpha_1^2-1}{\alpha_1\cos(\omega)+\cos(\alpha_1\omega)}\sin((\alpha_1-1)\theta)\right),\\
\sigma^w_{(\mathrm{I})\,rr} &= -C_1 r^{-\alpha_1-1}\left((3+\alpha_1)\cos((1+\alpha_1)\theta)\right.\\
&\qquad\left.-\frac{\alpha_1^2-1}{\alpha_1\cos(\omega)+\cos(\alpha_1\omega)}\cos((\alpha_1-1)\theta)\right),\\
\sigma^w_{(\mathrm{I})\,\theta\theta} &= C_1 r^{-\alpha_1-1}\left(-(1-\alpha_1)\cos((1+\alpha_1)\theta)\right.\\
&\qquad\left.-\frac{\alpha_1^2-1}{\alpha_1\cos(\omega)+\cos(\alpha_1\omega)}\cos((\alpha_1-1)\theta)\right),\\
\sigma^w_{(\mathrm{I})\,r\theta} &= -C_1 r^{-\alpha_1-1}\left((1+\alpha_1)\sin((1+\alpha_1)\theta)\right.\\
&\qquad\left.-\frac{\alpha_1^2-1}{\alpha_1\cos(\omega)+\cos(\alpha_1\omega)}\sin((\alpha_1-1)\theta)\right).
\end{aligned}
\tag{6.24}
$$

with $C_1 = \frac{\sqrt{8\pi}\mu\alpha_1\left[\alpha_1\sin^2(\omega/2)+\sin^2(\alpha_1\omega/2)\right]}{(1+\kappa)(\sin(\omega)+\omega\cos(\alpha_1\omega))}$, chosen to satisfy

$$
A_1 = \int_{\Gamma_c}\sum_{i,j=r,\theta}\left(\sigma_{ij}w_{(\mathrm{I})i} - \sigma^w_{(\mathrm{I})ij}u_i\right)n_j\,R d\theta. \tag{6.25}
$$

The auxiliary displacement and stress fields for extractioning A_2 are associated with the second negative eigen-value $-\alpha_2$:

$$
\begin{aligned}
w_{(\mathrm{II})\,r} &= C_2 \frac{r^{-\alpha_2}}{2\alpha_2\mu}\left(-(\kappa+\alpha_2)\sin((1+\alpha_2)\theta)\right.\\
&\qquad\left.+\frac{\alpha_2^2-1}{\alpha_2\cos(\omega)-\cos(\alpha_2\omega)}\sin((1-\alpha_2)\theta)\right),
\end{aligned}
$$

$$w_{(\mathrm{II})\,\theta} = -C_2 \frac{r^{-\alpha_2}}{2\alpha_2 \mu}\left((\kappa - \alpha_2)\cos((1+\alpha_2)\theta) + \frac{\alpha_2^2 - 1}{\alpha_2 \cos(\omega) - \cos(\alpha_2 \omega)} \cos((1-\alpha_2)\theta)\right),$$

$$\sigma^{w}_{(\mathrm{II})\,rr} = -C_2 r^{-\alpha_2 - 1}\left(-(3+\alpha_2)\sin((1+\alpha_2)\theta) + \frac{\alpha_2^2 - 1}{\alpha_2 \cos(\omega) - \cos(\alpha_2 \omega)} \sin((1-\alpha_2)\theta)\right),$$

$$\sigma^{w}_{(\mathrm{II})\,\theta\theta} = C_2 r^{-\alpha_2 - 1}\left((1-\alpha_2)\sin((1+\alpha_2)\theta) + \frac{\alpha_2^2 - 1}{\alpha_2 \cos(\omega) - \cos(\alpha_2 \omega)} \sin((1-\alpha_2)\theta)\right),$$

$$\sigma^{w}_{(\mathrm{II})\,r\theta} = C_2 r^{-\alpha_2 - 1}\left(-(1+\alpha_2)\cos((1+\alpha_2)\theta) + \frac{\alpha_2^2 - 1}{\alpha_2 \cos(\omega) - \cos(\alpha_2 \omega)} \cos((1-\alpha_2)\theta)\right), \qquad (6.26)$$

with $C_2 = \frac{\sqrt{8\pi}\mu\alpha_2\left[\alpha_2 \sin^2(\omega/2) - \sin^2(\alpha_2\omega/2)\right]}{(1+\kappa)(\sin(\omega) - \omega\cos(\alpha_2\omega))}$, chosen to satisfy

$$A_2 = \int_{\Gamma_c} \sum_{i,j=r,\theta} \left(\sigma_{ij} w_{(\mathrm{II})\,i} - \sigma^{w}_{(\mathrm{II})\,ij} u_i\right) n_j \, R d\theta. \qquad (6.27)$$

Because the integrals (6.25) and (6.27) are path-independent, one may calculate them along any path Γ_c using numerical methods. Once the first set of displacements and stresses is replaced by finite element solution, numerical integration gives either the GSIF A_1 or A_2 depending on the auxiliary displacements and stresses used.

The extraction method of the stress intensity factors using the reciprocal work contour integral was independently developed by Carpenter for 2-D V-notched and cracked domains in [33]. In a later work Carpenter and Byers [35] extended the extraction method to 2-D bimaterial domains. Extracting GSIFs associated with high-order eigenvalues by the CIM was addressed by Carpenter in [34].

6.1.3.2 Extracting SIFs for a TF/TF Crack Using the CIM

To extract the stress intensity factors K_{I} and K_{II} for a TF/TF crack (see Table 5.3), the following auxiliary displacement and stress fields associated with the first negative eigenvalue $-1/2$ (see, e.g., [171]) are used

$$w_{(\mathrm{I})\,r} = \frac{1}{2\sqrt{2\pi r}(1+\kappa)}\left[(2\kappa+1)\cos\frac{3\theta}{2} - 3\cos\frac{\theta}{2}\right],$$

$$w_{(\mathrm{I})\,\theta} = \frac{1}{2\sqrt{2\pi r}(1+\kappa)}\left[-(2\kappa-1)\sin\frac{3\theta}{2} + 3\sin\frac{\theta}{2}\right],$$

$$\sigma^{w}_{(\mathrm{I})\,rr} = -\frac{\mu}{2\sqrt{2\pi r^3}(1+\kappa)}\left[7\cos\frac{3\theta}{2} - 3\cos\frac{\theta}{2}\right],$$

$$\sigma^{w}_{(\mathrm{I})\,\theta\theta} = -\frac{\mu}{2\sqrt{2\pi r^3}(1+\kappa)}\left[\cos\frac{3\theta}{2} + 3\cos\frac{\theta}{2}\right],$$

$$\sigma^{w}_{(\mathrm{I})\,r\theta} = -\frac{3\mu}{2\sqrt{2\pi r^3}(1+\kappa)}\left[\sin\frac{3\theta}{2} + \sin\frac{\theta}{2}\right], \tag{6.28}$$

$$w_{(\mathrm{II})\,r} = \frac{1}{2\sqrt{2\pi r}(1+\kappa)}\left[(2\kappa+1)\sin\frac{3\theta}{2} - \sin\frac{\theta}{2}\right],$$

$$w_{(\mathrm{II})\,\theta} = \frac{1}{2\sqrt{2\pi r}(1+\kappa)}\left[(2\kappa-1)\cos\frac{3\theta}{2} - \cos\frac{\theta}{2}\right],$$

$$\sigma^{w}_{(\mathrm{II})\,rr} = -\frac{\mu}{2\sqrt{2\pi r^3}(1+\kappa)}\left[7\sin\frac{3\theta}{2} - \sin\frac{\theta}{2}\right],$$

$$\sigma^{w}_{(\mathrm{II})\,\theta\theta} = -\frac{\mu}{2\sqrt{2\pi r^3}(1+\kappa)}\left[\sin\frac{3\theta}{2} + \sin\frac{\theta}{2}\right],$$

$$\sigma^{w}_{(\mathrm{II})\,r\theta} = \frac{\mu}{2\sqrt{2\pi r^3}(1+\kappa)}\left[\cos\frac{3\theta}{2} + \cos\frac{\theta}{2}\right]. \tag{6.29}$$

6.2 Extracting GSIFs by the Complementary Energy Method (CEM)

For general singular points in elastostatics, neither the eigenpairs nor the GSIFs are known explicitly. Nevertheless, because the eigenpairs may be determined numerically with very high accuracy, one may use these to extract the GSIFs using the CEM. The CEM for extracting GSIFs in a two-dimensional elastic body is similar to the methods presented in Section 4.2. To this end, one must first generate a "statically admissible space," i.e., a set of stresses that automatically satisfy the equilibrium equations. Let $\mathcal{E}_c(\Omega)$ be the statically admissible space defined by

$$\mathcal{E}_c(\Omega) = \left\{ \underset{\approx}{\sigma} \;\middle|\; \|\underset{\approx}{\sigma}\|_{L^2,\Omega} < \infty;\ \mathbf{div}\,\underset{\approx}{\sigma} = \boldsymbol{f} \text{ in } \Omega \right\}.$$

Here $\|\underset{\approx}{\boldsymbol{\sigma}}\|_{\mathbf{L}^2,\Omega} \stackrel{\text{def}}{=} \sqrt{\left(\underset{\approx}{\boldsymbol{\sigma}}, \underset{\approx}{\boldsymbol{\sigma}}\right)_{\mathbf{L}^2,\Omega}}$, where

$$\left(\underset{\approx}{\boldsymbol{\sigma}}, \underset{\approx 1}{\boldsymbol{\sigma}}\right)_{\mathbf{L}^2,\Omega} \equiv \iint_{\Omega} \underset{\approx}{\boldsymbol{\sigma}} : \underset{\approx 1}{\boldsymbol{\sigma}} \, d\Omega \stackrel{\text{def}}{=} \iint_{\Omega} (\sigma)_{ij} (\sigma_1)_{ij} \, d\Omega.$$

For inhomogeneous traction boundary conditions with $[n]\underset{\approx}{\boldsymbol{\sigma}} = \boldsymbol{T}$, on the boundary Γ_1 and/or Γ_2, the statically admissible space is defined by

$$\tilde{\mathcal{E}}_c(\Omega) = \left\{ \underset{\approx}{\boldsymbol{\sigma}} \;\middle|\; \underset{\approx}{\boldsymbol{\sigma}} \in \mathcal{E}_c(\Omega); \; [n]\underset{\approx}{\boldsymbol{\sigma}} = \boldsymbol{T} \text{ on } \Gamma_1 \text{ and/or } \Gamma_2 \right\}.$$

One may notice that $\mathcal{E}_c$ and $\tilde{\mathcal{E}}_c$ are not linear spaces in the sense that the resulting stress tensor from an addition of two stress tensors in the space does not belong to the same space. We also define the space $\mathcal{E}_c^H$ by

$$\mathcal{E}_c^H(\Omega) = \left\{ \underset{\approx}{\boldsymbol{\sigma}} \;\middle|\; \|\underset{\approx}{\boldsymbol{\sigma}}\|_{\mathbf{L}^2,\Omega} < \infty; \; \mathbf{div}\, \underset{\approx}{\boldsymbol{\sigma}} = \mathbf{0} \text{ in } \Omega \right\}.$$

When $[n]\underset{\approx}{\boldsymbol{\sigma}} = \mathbf{0}$ on Γ_1 and/or Γ_2, we define the space $\overset{\circ}{\mathcal{E}}{}_c^H(\Omega)$ by

$$\overset{\circ}{\mathcal{E}}{}_c^H(\Omega) = \left\{ \underset{\approx}{\boldsymbol{\sigma}} \;\middle|\; \underset{\approx}{\boldsymbol{\sigma}} \in \mathcal{E}_c^H(\Omega); \; [n]\underset{\approx}{\boldsymbol{\sigma}} = \mathbf{0} \text{ on } \Gamma_1 \text{ and/or } \Gamma_2 \right\}.$$

Any finite-dimensional subspaces of the above will be denoted by a subscript N, for example, $\mathcal{E}_{cN}$ is a subspace of $\mathcal{E}_c$ with $\dim\mathcal{E}_{cN} = N < \infty$.

Once the necessary statically admissible spaces have been defined we may introduce the complementary energy principle, also know as the dual (complementary) weak form [128, pp. 103–108]). It states that the stress tensor $\boldsymbol{\sigma}_{\text{EX}}$ that is the solution of the elasticity problem is found by solving

$$\begin{gathered} \text{Seek } \boldsymbol{\sigma}_{\text{EX}} \in \mathcal{E}_c(\Omega) \text{ such that} \\ \mathcal{B}_c(\boldsymbol{\sigma}_{\text{EX}}, \boldsymbol{\sigma}_1) = \mathcal{F}_c(\boldsymbol{\sigma}_1) \qquad \forall \boldsymbol{\sigma}_1 \in \mathcal{E}_c(\Omega), \end{gathered} \tag{6.30}$$

where

$$\mathcal{B}_c(\boldsymbol{\sigma}, \boldsymbol{\sigma}_1) = \iint_{\Omega} \boldsymbol{\sigma}^T [E]^{-1} \boldsymbol{\sigma}_1 d\Omega, \tag{6.31}$$

$$\mathcal{F}_c(\boldsymbol{\sigma}_1) = \int_{(\partial\Omega)_u} \boldsymbol{u}^T \cdot ([n]\boldsymbol{\sigma}_1) d\Gamma. \tag{6.32}$$

Here $[E]$ is the material matrix, $(\partial\Omega)_u$ denotes the part of the boundary where displacement boundary conditions $\boldsymbol{u}$ are prescribed, and $[n] = \begin{bmatrix} \cos\theta & 0 & \sin\theta \\ 0 & \sin\theta & \cos\theta \end{bmatrix}$.

When traction boundary conditions are prescribed on a part of the domain's boundary, the statically admissible space has to be restricted to automatically satisfy these boundary conditions:

$$\text{Seek } \boldsymbol{\sigma}_{\text{EX}} \in \tilde{\mathcal{E}}_c(\Omega) \text{ such that}$$

$$\mathcal{B}_c(\boldsymbol{\sigma}_{\text{EX}}, \boldsymbol{\sigma}_1) = \mathcal{F}_c(\boldsymbol{\sigma}_1) \qquad \forall \boldsymbol{\sigma}_1 \in \overset{\circ}{\mathcal{E}}_c(\Omega). \tag{6.33}$$

Equipped with the eigenfunctions, the CEM can be utilized to extract the GSIFs as follows: First one solves the elastostatic problem over the entire domain Ω by means of the finite element method based on the displacement formulation thus obtaining $\boldsymbol{u}_{\text{FE}}$. Second, a subdomain around the singular point is considered. Define S_R as the set of interior points of a circle of radius R, centered on the singular point P. Then Ω_R is defined by $\Omega \cap S_R$, and Γ_R is the *circular* part of its boundary, see Figure 4.1. The eigenpairs can be computed numerically to obtain $\alpha_i, \boldsymbol{s}_i$, for each singular point in the domain Ω. The trial and test statically admissible spaces are chosen to be linear combinations of the eigenstresses ($\boldsymbol{S}_i$), which are computed from the eigenpairs, using the stress-strain relationship and Hooke's law. These eigenstresses automatically satisfy the equilibrium equations and boundary conditions on the V-notch faces. The unknowns are the series coefficients, i.e., the GSIFs, which are extracted in the postsolution phase by post-processing the FE solution on Γ_R.

We represent the stress tensor in the vicinity of the singular point by (observe that the eigenstresses $\boldsymbol{S}_i$ in this representation may be normalized to have the maximum value of 1)

$$\boldsymbol{\sigma} = \sum_i A_i r^{\alpha_i - 1} \boldsymbol{S}_i, \qquad \boldsymbol{\sigma}_1 = \sum_j C_j r^{\alpha_j - 1} \boldsymbol{S}_j. \tag{6.34}$$

We can insert these into (6.31) evaluated over Ω_R to obtain

$$\mathcal{B}_c = \boldsymbol{A}^T [B_c] \boldsymbol{C} \tag{6.35}$$

with $\boldsymbol{A}^T = (A_1, A_2, \ldots)$ (similarly $\boldsymbol{C}$), and the ijth element of the "compliance matrix" $[B_c]$ is provided by

$$(B_c)_{ij} = \frac{R^{\alpha_i + \alpha_j}}{\alpha_i + \alpha_j} \int_{\theta_1}^{\theta_1 + \omega} \boldsymbol{S}_i^T [E]^{-1} \boldsymbol{S}_j \, d\theta. \tag{6.36}$$

Remark 6.1 *For an isotropic material, the compliance matrix* $[B_c]$ *is diagonal, e.g.,* $(B_c)_{ij} = 0,\ i \neq j$. *See also* (8.15).

Remark 6.2 *The elements of the compliance matrix* $[B_c]$ *do not depend on the FE solution over* Ω, *and may be precomputed using the eigenpairs and material properties in the vicinity of the singular point only.*

The eigenstress tensor, being derived from the eigenpairs, automatically satisfies the boundary conditions on all boundaries except Γ_R, so that the linear form (6.32) degenerates to an integral over the circular boundary Γ_R alone.

Replacing the vector $\boldsymbol{u}$ in (6.32) with the approximated finite element solution on Γ_R, $\boldsymbol{u}_{\text{FE}}$, then defining the vector $\boldsymbol{u}_0 \stackrel{\text{def}}{=} ((u_1)_{\text{FE}} \cos\theta,\ (u_2)_{\text{FE}} \sin\theta,\ (u_1)_{\text{FE}} \sin\theta + (u_2)_{\text{FE}} \cos\theta)^T$, the jth term of the load vector corresponding to the linear form (6.32) becomes

$$(F_c)_j = R \int_{\theta_1}^{\theta_1+\omega} \boldsymbol{u}_0^T [E] \boldsymbol{S}_j|_{(r=R)}\, d\theta. \tag{6.37}$$

In view of (6.36) and (6.37), the complementary energy principle (6.30) is represented by the finite dimensional statically admissible subspace $\mathcal{E}_{cN}$ in matrix form

$$\boldsymbol{A}^T [B_c] = \boldsymbol{F}_c^T. \tag{6.38}$$

Solving (6.38), one obtains an approximation for the the GSIFs, $(A_1, A_2, \ldots, A_N)$.

6.2.0.3 FE Implementation of the CEM for Extracting GSIFs

When the eigenpairs are computed numerically, and the possibility of complex eigenpairs is not excluded, then the ith eigenpair representing the displacement field in the x_1 direction is given by

$$\begin{aligned} u_1 &= \sum_i r^{\alpha_i^{(\mathcal{R})} + \imath\alpha_i^{(\mathcal{I})}} \left(s_i^{(\mathcal{R})}(\theta) + \imath s_i^{(\mathcal{I})}(\theta) \right) \\ &= \sum_i r^{\alpha_i^{(\mathcal{R})} + \imath\alpha_i^{(\mathcal{I})}} \sum_{m=1}^{p+1} \left(a_{im}^{(\mathcal{R})} + \imath a_{im}^{(\mathcal{I})} \right) \Phi_m(\theta(\xi)). \end{aligned} \tag{6.39}$$

The displacement u_2 is represented identically to (6.39), with a_{im} replaced by b_{im}. With this notation in hand, the ijth term of the compliance matrix associated with (6.36) is given by

$$(B_c)_{ji} = \int_0^R \int_{\theta_1}^{\theta_1+\omega} r^{(\alpha_i^{(\mathcal{R})} + \alpha_j^{(\mathcal{R})} - 1)} \sum_{\ell,k=1}^{3} (Du)_k^j E_{k\ell} (Du)_\ell^i \, dr d\theta, \tag{6.40}$$

$\alpha_i^{(\mathcal{R})}$ being the real part of the eigenpair, and $\mathbf{Du}^i$ a 3×1 vector corresponding to the ith eigenfunction and is given by

$$(Du)_1^i = \sum_{m=1}^{p+1} \Big[\cos\theta\cos\delta_i \left(\alpha_i^{(\mathcal{R})} a_{im}^{(\mathcal{R})} - \alpha_i^{(\mathcal{I})} a_{im}^{(\mathcal{I})}\right) - \cos\theta\sin\delta_i \left(\alpha_i^{(\mathcal{R})} a_{im}^{(\mathcal{I})} + \alpha_i^{(\mathcal{I})} a_{im}^{(\mathcal{R})}\right)\Big] \Phi_m(\theta(\xi)) + \Big[-a_{im}^{(\mathcal{R})} \sin\theta\cos\delta_i + a_{im}^{(\mathcal{I})} \sin\theta\sin\delta_i\Big] \Phi'_m(\xi) \frac{2}{\omega - \theta_1}, \tag{6.41}$$

$$(Du)_2^i = \sum_{m=1}^{p+1} \Big[\sin\theta\cos\delta_i \left(\alpha_i^{(\mathcal{R})} b_{im}^{(\mathcal{R})} - \alpha_i^{(\mathcal{I})} b_{im}^{(\mathcal{I})}\right) - \sin\theta\sin\delta_i \left(\alpha_i^{(\mathcal{R})} b_{im}^{(\mathcal{I})} + \alpha_i^{(\mathcal{I})} b_{im}^{(\mathcal{R})}\right)\Big] \Phi_m(\theta(\xi)) + \Big[b_{im}^{(\mathcal{R})} \cos\theta\cos\delta_i - b_{im}^{(\mathcal{I})} \cos\theta\sin\delta_i\Big] \Phi'_m(\xi) \frac{2}{\omega - \theta_1}, \tag{6.42}$$

$$\begin{aligned}(Du)_3^i = \sum_{m=1}^{p+1} \Big[&\sin\theta\cos\delta_i \left(\alpha_i^{(\mathcal{R})} a_{im}^{(\mathcal{R})} - \alpha_i^{(\mathcal{I})} a_{im}^{(\mathcal{I})}\right) - \sin\theta\sin\delta_i \left(\alpha_i^{(\mathcal{R})} a_{im}^{(\mathcal{I})} + \alpha_i^{(\mathcal{I})} a_{im}^{(\mathcal{R})}\right) \\ &+ \cos\theta\cos\delta_i \left(\alpha_i^{(\mathcal{R})} b_{im}^{(\mathcal{R})} - \alpha_i^{(\mathcal{I})} b_{im}^{(\mathcal{I})}\right) - \cos\theta\sin\delta_i \left(\alpha_i^{(\mathcal{R})} b_{im}^{(\mathcal{I})} + \alpha_i^{(\mathcal{I})} b_{im}^{(\mathcal{R})}\right)\Big] \Phi_m(\theta(\xi)) \\ &+ \Big[a_{im}^{(\mathcal{R})} \cos\theta\cos\delta_i - a_{im}^{(\mathcal{I})} \cos\theta\sin\delta_i - b_{im}^{(\mathcal{R})} \sin\theta\cos\delta_i + b_{im}^{(\mathcal{I})} \sin\theta\sin\delta_i\Big] \Phi'_m(\xi) \frac{2}{\omega - \theta_1},\end{aligned} \tag{6.43}$$

where, $\delta_i \stackrel{\text{def}}{=} \alpha_i^{(\mathcal{I})} \ln r$, and Φ_m are the shape functions on an edge. If α_i is complex, then the elements of the $(i+1)$th vector $\mathbf{Du}^{i+1}$ are

$$(Du)_1^{i+1} = \sum_{m=1}^{p+1} \Big[\cos\theta\cos\delta_i \left(\alpha_i^{(\mathcal{R})} a_{im}^{(\mathcal{I})} + \alpha_i^{(\mathcal{I})} a_{im}^{(\mathcal{R})}\right) + \cos\theta\sin\delta_i \left(\alpha_i^{(\mathcal{R})} a_{im}^{(\mathcal{R})} - \alpha_i^{(\mathcal{I})} a_{im}^{(\mathcal{I})}\right)\Big] \Phi_m(\theta(\xi)) - \Big[a_{im}^{(\mathcal{R})} \sin\theta\sin\delta_i + a_{im}^{(\mathcal{I})} \sin\theta\cos\delta_i\Big] \Phi'_m(\xi) \frac{2}{\omega - \theta_1}, \tag{6.44}$$

$$(Du)_2^{i+1} = \sum_{m=1}^{p+1} \Big[\sin\theta\cos\delta_i \left(\alpha_i^{(\mathcal{R})} b_{im}^{(\mathcal{I})} + \alpha_i^{(\mathcal{I})} b_{im}^{(\mathcal{R})}\right) + \sin\theta\sin\delta_i \left(\alpha_i^{(\mathcal{R})} b_{im}^{(\mathcal{R})} - \alpha_i^{(\mathcal{I})} b_{im}^{(\mathcal{I})}\right)\Big] \Phi_m(\theta(\xi)) + \Big[b_{im}^{(\mathcal{R})} \cos\theta\sin\delta_i + b_{im}^{(\mathcal{I})} \cos\theta\cos\delta_i\Big] \Phi'_m(\xi) \frac{2}{\omega - \theta_1}, \tag{6.45}$$

$$
\begin{aligned}
(Du)_3^{i+1} = \textstyle\sum_{m=1}^{p+1} \Big[& \sin\theta\cos\delta_i \left(\alpha_i^{(\mathcal{R})} a_{im}^{(\mathcal{I})} + \alpha_i^{(\mathcal{I})} a_{im}^{(\mathcal{R})}\right) \\
& + \sin\theta\sin\delta_i \left(\alpha_i^{(\mathcal{R})} a_{im}^{(\mathcal{R})} - \alpha_i^{(\mathcal{I})} a_{im}^{(\mathcal{I})}\right) \\
& + \cos\theta\cos\delta_i \left(\alpha_i^{(\mathcal{R})} b_{im}^{(\mathcal{I})} + \alpha_i^{(\mathcal{I})} b_{im}^{(\mathcal{R})}\right) \\
& + \cos\theta\sin\delta_i \left(\alpha_i^{(\mathcal{R})} b_{im}^{(\mathcal{R})} - \alpha_i^{(\mathcal{I})} b_{im}^{(\mathcal{I})}\right)\Big] \Phi_m(\theta(\xi)) \\
+ \Big[& a_{im}^{(\mathcal{R})} \cos\theta\sin\delta_i + a_{im}^{(\mathcal{I})} \cos\theta\cos\delta_i \\
& - b_{im}^{(\mathcal{R})} \sin\theta\sin\delta_i - b_{im}^{(\mathcal{I})} \sin\theta\cos\delta_i \Big] \Phi'_m(\xi) \frac{2}{\omega - \theta_1}.
\end{aligned}
\tag{6.46}
$$

The ith term of the load vector associated with (6.38) is given by

$$
(\mathcal{F}_c)_i = R \int_{\theta_1}^{\theta_1+\omega} \boldsymbol{u}_0^T [E] (\mathbf{Du})^i |_{(r=R)} \, d\theta. \tag{6.47}
$$

6.3 Numerical Examples: Extracting GSIFs by CIM and CEM

Numerical examples provided in the following demonstrate that the rate of convergence of the GSIFs is as fast as the convergence of the strain energy, therefore the CIM and CEM are "superconvergent."

6.3.1 *A Crack in an Isotropic Material: Extracting SIFs by the CIM and CEM*

Let us consider the edge-cracked panel in an isotropic material studied by Szabó and Babuška in [177], shown in Figure 6.2. Plane strain and Poisson's ratio of 0.3 are assumed. The tractions that exactly correspond to the stresses of mode I and mode II stress fields were applied on the sides of the solution domain, and the first and second GSIFs were computed by the CIM and the CEM. For the CIM the exact eigenpairs were used, whereas in the CEM, the approximated eigenpairs obtained by the modified Steklov method were used. These approximated eigenpairs are computed using a four-element mesh (not shown) at $p = 6$. The two eigenvalues obtained are $\alpha_1 = 0.49999967, \alpha_2 = 0.50000051$ (the exact values are $1/2$).

Selecting the first two GSIFs to be A (A is arbitrary), we define the normalized stress intensity factors $\tilde{A}_1$ and $\tilde{A}_2$ as follows:

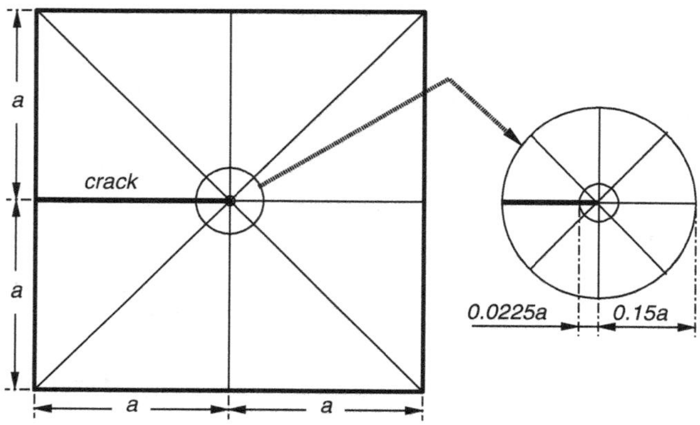

Fig. 6.2 Solution domain and mesh design for a crack in an isotropic material.

Table 6.1 First two GSIFs computed by the CEM for a crack in an isotropic material

	$p=1$	$p=2$	$p=3$	$p=4$	$p=5$	$p=6$	$p=7$	$p=8$
DOF	53	155	273	439	653	915	1225	1583
$\|e\|_E(\%)$	29.92	11.07	5.52	3.15	2.24	1.78	1.48	1.26
$\tilde{A}_1$	0.8144	0.9548	0.9912	0.99783	0.99795	0.99825	0.99862	0.99882
$\tilde{A}_2$	0.8317	0.9641	0.9946	0.99942	0.99888	0.99898	0.99926	0.99943

$$\tilde{A}_i \overset{\text{def}}{=} (A_i)_{\text{FE}}/A, \qquad i = 1, 2. \tag{6.48}$$

In this way, both normalized GSIFs have to converge to 1 as the number of degrees of freedom is increased.

The first two normalized GSIFs are computed with $R = 0.5a$, where $2a$ is the length of the side of the square. The number of degrees of freedom, the error in energy norm, and the computed values of the normalized GSIFs using the CEM are listed in Table 6.1. The relative error in energy norm, the relative error in strain energy, and the absolute value of the relative error in the first GSIF, computed by the CEM and by the CIM, are plotted against the number of degrees of freedom on a log-log scale in Figure 6.3. The same data for the second GSIF are shown in Figure 6.4.

It is seen that the rate of convergence of the GSIFs is faster than the rate of convergence in the energy norm and both the CIM and CEM have similar convergence patterns. Note that as the error in energy norm decreases, the CIM (based on the exact eigenpairs) performs better than CEM (based on the approximate eigenpairs). However, up to the relative error of approximately 0.1 percent, the performance of both methods is virtually the same. Therefore, the proposed extraction methods work well for levels of accuracy normally expected in engineering practice.

This example problem demonstrates the efficiency of the proposed extraction methods applied to isotropic materials.

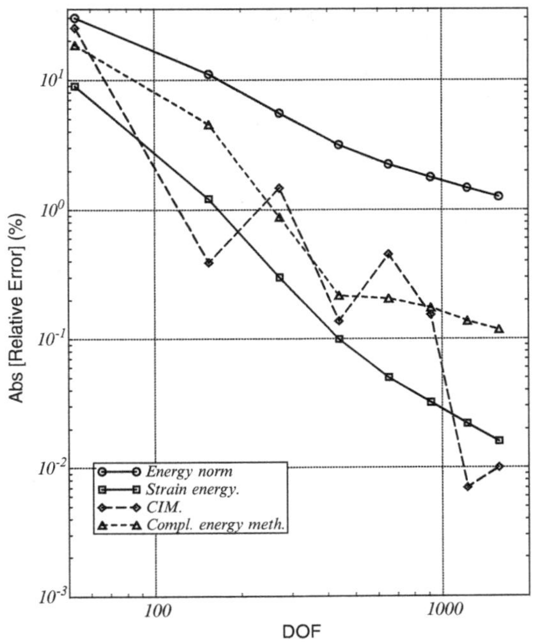

Fig. 6.3 Convergence of $\|e\|_{\mathcal{E}}$, the strain energy ($\|e\|^2_{\mathcal{E}}$), and A_1 for a crack in an isotropic material.

6.3.2 Crack at a Bimaterial Interface: Extracting SIFs by the CEM

In this example a crack between two homogeneous isotropic materials is considered. The displacements and the normal and shearing tractions are continuous along the ligament. A closed-form solution for the stress tensor is given in [174].[1] The stress intensity factors are computed for the example problem shown in Figure 6.5. Plane strain is assumed. The exact values for this problem, for $a/b \to 0$, $b/h = 1$ are $K_{\rm I} = 1.784122961$, $K_{\rm II} = -0.175277466$, including the $\sqrt{2\pi}$ factor [150]. The value of σ_0 in Figure 6.5 is 0.485714118 when other tractions are unity. Referring to Figure 6.5, we define

$$\varepsilon = -0.075811777690. \tag{6.49}$$

Ignoring terms that remain bounded at the right crack tip, the asymptotic stress fields in material 1 can be put into the form

$$\tilde{\sigma}_{ij} = \frac{1}{\sqrt{2\pi r}} \Big\{ K_{\rm I}[\cos(\varepsilon \ln r)\tilde{S}^{\Re}_{ij} + \sin(\varepsilon \ln r)\tilde{S}^{\Im}_{ij}] + K_{\rm II}[-\sin(\varepsilon \ln r)\tilde{S}^{\Re}_{ij} + \cos(\varepsilon \ln r)\tilde{S}^{\Im}_{ij}] \Big\}, \qquad i,j = r,\theta, \tag{6.50}$$

[1]The expressions for the displacements in [174] are not continuous across the interface at $\theta = 0$. Therefore they could not possibly be valid.

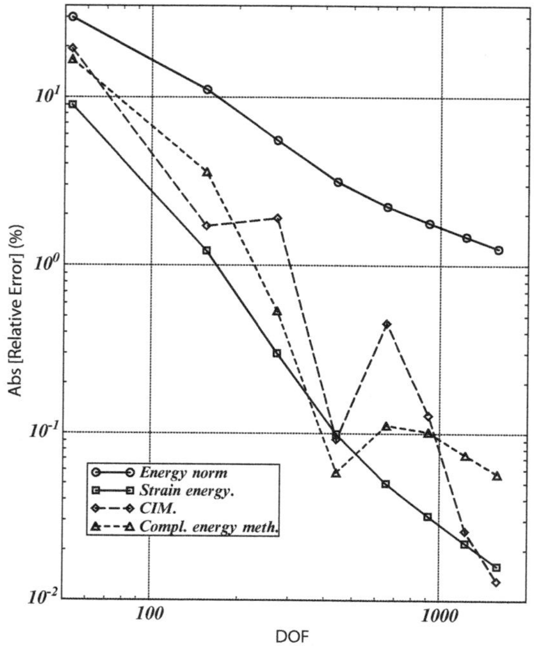

Fig. 6.4 Convergence of $\|e\|_{\mathcal{E}}$, the strain energy ($\|e\|^2_{\mathcal{E}}$), and A_2 for a crack in an isotropic material.

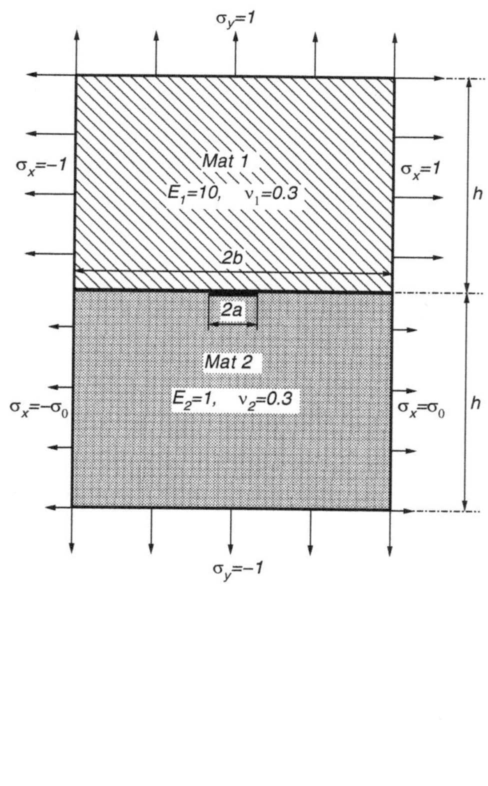

Fig. 6.5 Crack at a bimaterial interface, example problem.

where

$$\begin{aligned}
\tilde{S}_{rr}^{\Re} &= [-\sinh\varepsilon(\pi-\theta)\cos(3\theta/2) \\
&\quad + e^{-\varepsilon(\pi-\theta)}\cos(\theta/2)(1+\sin^2(\theta/2)+\varepsilon\sin(\theta))]/\cosh\varepsilon\pi, \\
\tilde{S}_{\theta\theta}^{\Re} &= [\sinh\varepsilon(\pi-\theta)\cos(3\theta/2) \\
&\quad + e^{-\varepsilon(\pi-\theta)}\cos(\theta/2)(\cos^2(\theta/2)-\varepsilon\sin(\theta))]/\cosh\varepsilon\pi, \\
\tilde{S}_{r\theta}^{\Re} &= [\sinh\varepsilon(\pi-\theta)\sin(3\theta/2) \\
&\quad + e^{-\varepsilon(\pi-\theta)}\sin(\theta/2)(\cos^2(\theta/2)-\varepsilon\sin(\theta))]/\cosh\varepsilon\pi, \\
\tilde{S}_{rr}^{\Im} &= [\cosh\varepsilon(\pi-\theta)\sin(3\theta/2) \\
&\quad - e^{-\varepsilon(\pi-\theta)}\sin(\theta/2)(1+\cos^2(\theta/2)-\varepsilon\sin(\theta))]/\cosh\varepsilon\pi, \\
\tilde{S}_{\theta\theta}^{\Im} &= [-\cosh\varepsilon(\pi-\theta)\sin(3\theta/2) \\
&\quad - e^{-\varepsilon(\pi-\theta)}\sin(\theta/2)(\sin^2(\theta/2)+\varepsilon\sin(\theta))]/\cosh\varepsilon\pi, \\
\tilde{S}_{r\theta}^{\Im} &= [\cosh\varepsilon(\pi-\theta)\cos(3\theta/2) \\
&\quad + e^{-\varepsilon(\pi-\theta)}\cos(\theta/2)(\sin^2(\theta/2)+\varepsilon\sin(\theta))]/\cosh\varepsilon\pi. \qquad (6.51)
\end{aligned}$$

The stress fields in material 2 can be obtained by replacing the combination $\varepsilon\pi$ to $-\varepsilon\pi$ and $-\varepsilon(\pi-\theta)$ to $\varepsilon(\pi+\theta)$ everywhere in (6.51). The first eigenvalues for this crack problem is a complex number and its conjugate given by $1/2 \pm i\varepsilon$. The accuracy and convergence behavior of the CEM are demonstrated on a bimaterial fracture mechanics problem.

Problem 6.1. The expression of the Cartesian eigenstresses for a bimaterial interface is given in a general setting in (5.70) and (5.73)–(5.74). Using these with $n = 1$, obtain expressions for the polar eigenstresses given by (6.50)–(6.51).

The first eigenvalues obtained by the modified Steklov method using four elements at $p = 8$ (the mesh is not shown) are $0.4999999990 \pm i0.07581177721$. These eigenvalues with their associated eigenvectors were used for computing the stress intensity factors. Due to symmetry, only half of the domain was discretized. The dimensions are taken to be $a/b = 1/20$ and $a/b = 1/40$, and $h/b = 1$. The finite element mesh for this example problem is shown in Figure 6.6. The polynomial level of the trial and test functions is increased over the shown mesh from 1 to 8. In Table 6.2 we summarize the relative error in the energy norm, and the stress intensity factors for the two values of a/b as the number of degrees of freedom is increased. The computations of the stress intensity factors were performed using an integration radius of $0.01a$. As we consider more terms in the asymptotic expansion, the influence of the integration radius is insignificant and one can use a larger radius, as employed in the first and third example problems. For example, using seven terms in the asymptotic expansion with an integration radius of $0.5a$, we obtain stress intensity factors that differ by less than 0.2% from those reported in Table 6.2. *If only two terms are considered, it is necessary to perform*

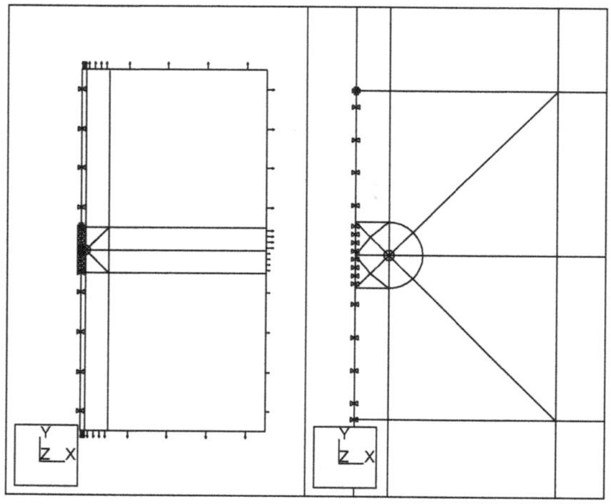

Fig. 6.6 Solution domain and mesh design for a crack at a bimaterial interface.

the integration along a radius that is close to the singular point. This phenomenon does not exist for isotropic materials where the $[B_c]$ *matrix is diagonal.* It is seen that the existence of complex eigenpairs has no influence on the performance of the CEM, and we are able to compute the SIFs with high accuracy. Of course, as $a/b \to 0$, the results become closer and closer to those presented in [150]. This example demonstrates that an accurate and efficient numerical solution of fracture mechanics problems even for complicated situations such as a crack at a bimaterial interface is possible.

6.3.3 Nearly Incompressible L-Shaped Domain: Extracting SIFs by the CEM

We demonstrate in the following that the accuracy of the CEM is insensitive to Poisson's ratio, and can be used for nearly incompressible materials with the same efficiency.

Let us consider the L-shaped plane elastic body presented in [179], having reentrant edges of length 1. See Figure 6.7. On the boundaries of the domain, tractions that correspond to the following exact stress field

$$\begin{aligned}\sigma_x = {} & A_1\alpha_1 r^{\alpha_1-1}\left\{[2 - Q_1(\alpha_1+1)]\cos(\alpha_1-1)\theta - (\alpha_1-1)\cos(\alpha_1-3)\theta\right\}\\ & + A_2\alpha_2 r^{\alpha_2-1}\left\{[2 - Q_2(\alpha_2+1)]\sin(\alpha_2-1)\theta - (\alpha_2-1)\sin(\alpha_2-3)\theta\right\},\end{aligned} \tag{6.52}$$

$$\begin{aligned}\sigma_y = {} & A_1\alpha_1 r^{\alpha_1-1}\left\{[2 + Q_1(\alpha_1+1)]\cos(\alpha_1-1)\theta + (\alpha_1-1)\cos(\alpha_1-3)\theta\right\}\\ & + A_2\alpha_2 r^{\alpha_2-1}\left\{[2 + Q_2(\alpha_2+1)]\sin(\alpha_2-1)\theta + (\alpha_2-1)\sin(\alpha_2-3)\theta\right\},\end{aligned} \tag{6.53}$$

Table 6.2 First two GSIFs for crack between dissimilar materials.

		$p=1$	$p=2$	$p=3$	$p=4$	$p=5$	$p=6$	$p=7$	$p=8$	Exact
DOF		105	387	845	1479	2289	3275	4437	5775	∞
$a/b=$	$\|e\|_{\mathcal{E}}(\%)$	1.5	0.52	0.21	0.10	0.06	0.04	0.02	0.02	0.0
1/40	K_{I}	1.724295	1.768182	1.777236	1.778712	1.779417	1.781130	1.782759	1.783864	1.784123
	K_{II}	−0.440176	−0.209461	−0.181435	−0.177027	−0.176215	−0.176050	−0.175979	−0.175942	−0.175277
$a/b=$	$\|e\|_{\mathcal{E}}(\%)$	2.95	0.94	0.32	0.14	0.08	0.06	0.05	0.04	0.0
1/20	K_{I}	1.741420	1.779560	1.784096	1.783564	1.783591	1.785092	1.786663	1.787754	1.784123
	K_{II}	−0.444001	−0.210455	−0.181885	−0.177378	−0.176543	−0.176377	−0.176307	−0.176269	−0.175277

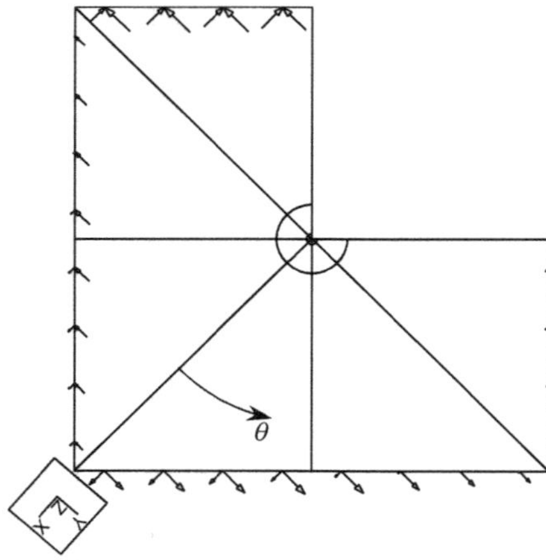

Fig. 6.7 Solution domain and mesh design for the L-shaped domain.

$$\tau_{xy} = A_1\alpha_1 r^{\alpha_1-1}\left\{(\alpha_1-1)\sin(\alpha_1-3)\theta + Q_1(\alpha_1+1)\sin(\alpha_1-1)\theta\right\}$$
$$+A_2\alpha_2 r^{\alpha_2-1}\left\{(\alpha_2-1)\cos(\alpha_2-3)\theta + Q_2(\alpha_2+1)\cos(\alpha_2-1)\theta\right\}, \quad (6.54)$$

are applied, where A_1 and A_2 are constants analogous to the mode I and model II stress intensity factors in linear elastic fracture mechanics; $\alpha_1 = 0.5444837368$, $Q_1 = 0.543075597$, $\alpha_2 = 0.9085291898$, $Q_2 = -0.218923236$ are constants determined so that the solution satisfies the equilibrium equations and the traction-free boundary conditions on the reentrant edges. In this example problem, plane-strain conditions are assumed with Young's modulus $E = 1$, and Poisson's ratio $\nu = 0.499999$. Defining $\epsilon = 0.5 - \nu$, the material becomes progressively more incompressible as $\epsilon \to 0$.

The eigenpairs were computed using a three-element mesh. The sensitivity of the modified Steklov method to the near-incompressibility condition was tested, with Poisson ratio ranging from 0.49 to 0.499999, i.e., ϵ ranging from 0.01 to 10^{-6}. The absolute value of the relative error (percent), $100 \times |\alpha_i - \alpha_i^{\mathrm{EX}}|/\alpha_i^{\mathrm{EX}}$, is plotted for the first and second eigenvalues as a function of ϵ on a log-log scale in Figure 6.8. The computed eigenpairs are seen to be insensitive with respect to the incompressibility condition up to $\nu = 0.499999$. As $\varepsilon \to 0$, the condition number of the matrix $([K] - [N_R] - [N_{R^*}])$ given in (3.31) increases rapidly, and scaling of the matrix is necessary. From an engineering point of view, values of $\nu = 0.499999$ are virtually the same as $\nu = 0.5$, so no scaling has been performed.

The eighteen-element mesh shown in Figure 6.7 and ν=0.499999, in conjunction with the approximate eigenvalues were used for extracting A_1 and A_2. A radius of $R = 0.9$ was used for the integration path. The absolute value of the relative error of A_1, A_2, the energy norm, and the strain energy are plotted against the number of degrees of freedom on a log-log scale in Figure 6.9.

These results show that the relative error in strain energy and that of A_1 and A_2 are of comparable magnitude, and they converge at approximately the same rate,

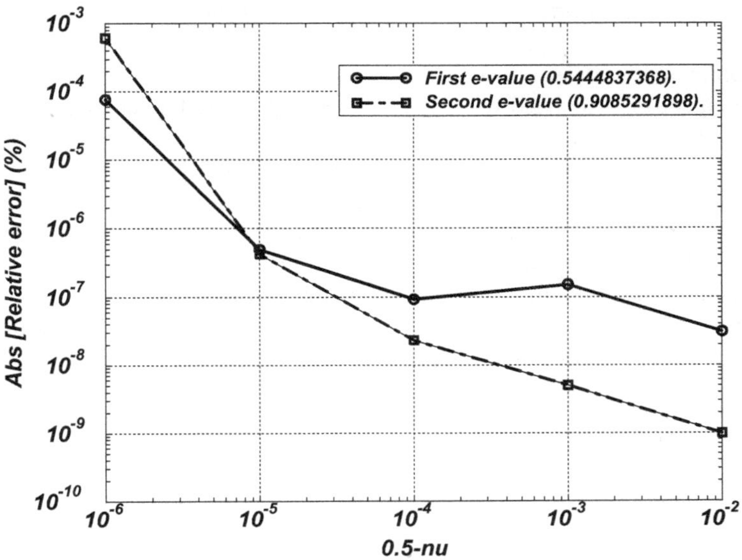

Fig. 6.8 Influence of Poisson's ratio on the accuracy of the eigenvalues computed by the modified Steklov method at $p = 8$.

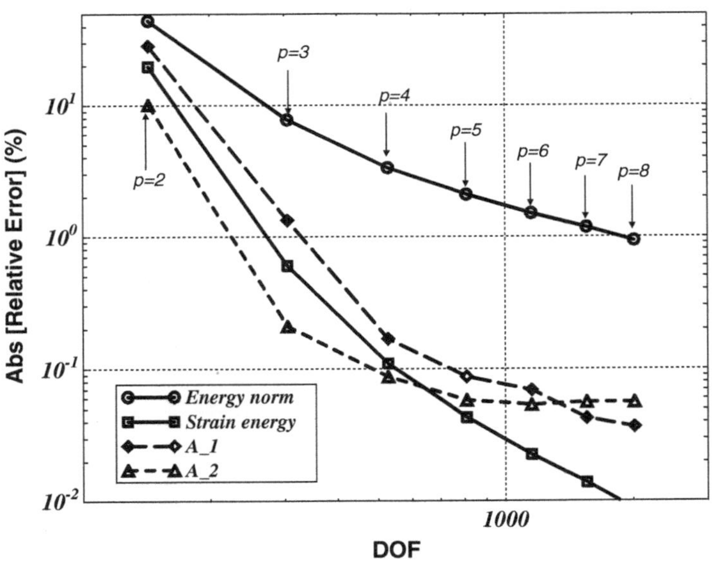

Fig. 6.9 Convergence of $\|e\|_{\mathcal{E}}$, the strain energy ($\|e\|^2_{\mathcal{E}}$), and A_1, A_2 for a nearly incompressible ($\nu = 0.499999$) L-shaped domain.

until the relative error drops below 0.1%. Below 0.1%, the contribution of the errors of the approximate eigenpairs to the computation of the GSIFs become significant. The coefficient A_2 fails to converge before A_1 fails, which is consistent with the fact that the eigenpair associated with A_2 is less accurate than that associated with A_1. Nevertheless, an excellent accuracy below 0.1% relative error can be achieved using the CEM, even when the material is nearly incompressible.

We may conclude that the CIM and CEM have four major advantages: (a) The error in the GSIFs exhibits superconvergence. (b) The methods are general in the sense that they are applicable to anisotropic materials and many types of singularities, including those associated with multimaterial interfaces. (c) The methods can be used in conjunction with any finite element analysis program. (d) The methods are efficient and robust.

Chapter 7
Thermal Generalized Stress Intensity Factors in 2-D Domains

Lately, methods that are capable of predicting failure initiation and propagation in structural components subjected to thermal loads have been sought. It is postulated, as in the theory of linear elastic fracture mechanics, that the methods should correlate experimental observed failures to parameters characterizing the thermoelastic stress field in the vicinity of failure initiation points. Failures due to thermal loading occur, for example, in integrated circuits, which are assemblages of dissimilar materials with different thermal and mechanical properties (addressed in Chapter 9). The mismatch of elastic constants and thermal expansion coefficients causes stress intensification at corners of interfaces and may lead to mechanical failure.

New approaches to predicting the initiation and extension of de-laminations in plastic-encapsulated LSI (large scale integrated circuit) devices, for example, are based on the computation of the thermal generalized stress intensity factors (TGSIFs) and the strength of the stress singularity [81, 85].

Although many studies have been reported in the past 30 years on thermoelastic crack problems in isotropic two-dimensional domains (see e.g [163, 173, 183] and the references therein), very little has been done on multimaterial corner interfaces. Especially, scant attention has been given to singularities affecting both the temperature flux field and the stress tensor. Recent publications on TGSIFs for a bonded interface between dissimilar materials, ignoring the singular behavior of the temperature fluxes, can be found, for example, in [63, 81, 114, 119].

In this chapter we address the uncoupled two-dimensional thermoelastic problem in the vicinity of singular points, and develop numerical methods for the computation of TGSIFs. We first compute the steady-state temperature distribution in the vicinity of the singularity, which is imposed in the elastic analysis as thermal loading, exciting the so-called thermal generalized stress intensity factors (TGSIFs). Both the temperature field and the stress tensor in the vicinity of the singular point may exhibit singular behavior.

The temperature and displacement fields are computed by p-FEMs, and the CEM is utilized in the post-processing stage in conjunction with Richardson's

Z. Yosibash, *Singularities in Elliptic Boundary Value Problems and Elasticity and Their Connection with Failure Initiation*, Interdisciplinary Applied Mathematics 37,
DOI 10.1007/978-1-4614-1508-4_7, © Springer Science+Business Media, LLC 2012

extrapolation to extract the TGSIFs. Importantly, the proposed method is applicable not only to singularities associated with crack tips, but also to multimaterial interfaces and inhomogeneous materials. Numerical results of crack-tip singularities (mode I, mode II, and mixed modes) and singular points associated with a two-material inclusion and a 90° dissimilar materials wedge, are presented. The strength of the flux and stress singularities are computed using the modified Steklov method. The complementary weak form is then applied in the post-processing phase over a series of subdomains with decreasing radii for computing the TGSIFs, and using Richardson's extrapolation, excellent results are obtained.

In Section 7.1, the classical and the corresponding complementary weak formulations of thermoelastic problems are presented. The results of the mathematical analysis are demonstrated in Section 7.2 by extracting the TGSIFs for several problems involving cracks in rectangular plates and cracks emanating from a circular hole, subjected to different temperature boundary conditions. In Section 7.2.4, TGSIFs for dissimilar isotropic elastic wedges perfectly bonded along their common interface, representing inclusion problems and a 90° two wedge interface problem are summarized.

7.1 Classical (Strong) and Weak Formulations of the Linear Thermoelastic Problem

Consider again the simply connected two-dimensional domain Ω with boundaries $\partial\Omega = \cup_i \Gamma_i$ that are analytic simple arc curves called edges, shown in Figure 4.1. The outward normal vector to the boundary is denoted by $\boldsymbol{n} \stackrel{\text{def}}{=} (n_1, n_2)$, and in matrix form, $[n] = \begin{bmatrix} n_1 & 0 & n_2 \\ 0 & n_1 & n_2 \end{bmatrix}$.

7.1.1 The Linear Thermoelastic Problem

The temperature field $\tau(x, y)$ in the vicinity of a singular point P is the solution of the linear heat conduction equation (1.34) (or in index notation (1.35)) discussed in Chapters 3-4. The heat conduction problem is solved in practice by means of the p-version of the finite element method over Ω. However, for purposes of mathematical analysis it is assumed that the exact solution can be found. In the vicinity of the singular point P, the special functional representation of τ is of special interest in the sequel.

There are two different approaches to considering the temperature field in the vicinity of P. The simplified approach assumes that for crack tips, for example, the presence of a singular point does not influence the heat flow continuum, because the two faces of the crack remain in contact, or are almost in contact (for small-deformation elasticity theory). In this case it is assumed that τ = constant in the vicinity of P.

The other approach, i.e., when the faces $\Gamma_1 \cup \Gamma_2$ are assumed to be either isolated or the temperature is kept constant, the temperature field admits in the vicinity of the singular point the well-known expansion

$$\tau(r,\theta) = \tau_0 + \sum_{i=1}^{\infty}\sum_{k=0}^{K} b_{iK} r^{\beta_i} \ln^s(k) s_{ik}(\theta), \tag{7.1}$$

where τ_0 is a constant, which may be zero, depending on temperature boundary conditions (except for inclusion problems, where it is almost always not zero). To distinguish between the GFIFs and GSIFs, we denote in this chapter the GFIFs by b_{is} and the eigenvalues associated with the singular temperature field by β_i. We also do not consider the special cases for which $K \neq 0$ (see [72, p. 264]). If grad τ is considered, the terms containing the $\ln r$ functions are less singular than terms containing r^{β_i} for $\beta_i < 1$.

Once the linear heat conduction problem is solved over Ω, one may proceed to the uncoupled linear thermoelasticity problem, and prescribe the exact temperature field as a thermal loading. Traction-free boundary conditions are assumed on Γ_1 and Γ_2. Thus the boundary conditions (due to thermal loading alone) are (see [167, Chapter 99]):

$$T_n = \mu\beta\tau \qquad \text{on } \Gamma_1 \cup \Gamma_2, \tag{7.2}$$

T_n being the normal traction on the boundary, τ the elevated temperature field with respect to the stress-free state,

$$\beta = \begin{cases} \frac{2(1+\nu)}{1-2\nu}\alpha, & \text{plane-strain} \\ \frac{2(1+\nu)}{1-\nu}\alpha, & \text{plane-stress} \end{cases}$$

and α the coefficient of linear thermal expansion. Again, it is assumed that the thermoelasticity problem can be solved analytically over Ω for purposes of mathematical analysis, but the p-version of the finite-element method IS used for performing the actual computations. On the boundary $\Gamma_o = \partial\Omega - \Gamma_1 - \Gamma_2$, displacement, traction or spring boundary conditions may be applied. However, on Γ_1 and Γ_2 we assume homogeneous mechanical boundary conditions.

Since our focus is the solution of the thermoelasticity problem in the vicinity of the singular point, we concentrate our attention in the subdomain Ω_R shown in Figure 4.1, assuming that the displacement field on Γ_R is available. The classical linear thermoelastic problem in an isotropic domain Ω_R described in terms of stresses is

$$\mathbf{div}\, \underset{\approx}{\sigma} = \beta \operatorname{grad}\tau \text{ in } \Omega_R, \tag{7.3}$$

$$T_n = \sigma_{\theta\theta} = \mu\beta\tau \text{ on } \Gamma_1 \cup \Gamma_2, \tag{7.4}$$

$$T_t = \sigma_{r\theta} = 0 \text{ on } \Gamma_1 \cup \Gamma_2, \tag{7.5}$$

$$\boldsymbol{u} \text{ prescribed on } \Gamma_R.$$

The stress tensor that is the solution to the thermoelastic problem in the vicinity of the singular point consists of three parts:

$\underset{\approx P}{\boldsymbol{\sigma}}$: a particular stress tensor solution satisfying the partial differential equation (7.3) and homogeneous boundary conditions on $\Gamma_1 \cup \Gamma_2$.
$\underset{\approx P}{\bar{\boldsymbol{\sigma}}}$: another particular stress tensor solution that satisfies the *homogeneous* partial differential equation (7.3), and the boundary conditions (7.4) and (7.5).
$\underset{\approx H}{\boldsymbol{\sigma}}$: a *homogeneous* stress tensor defined by

$$\underset{\approx H}{\boldsymbol{\sigma}} = \underset{\approx}{\boldsymbol{\sigma}} - \underset{\approx P}{\boldsymbol{\sigma}} - \underset{\approx P}{\bar{\boldsymbol{\sigma}}},$$

Thus $\underset{\approx H}{\boldsymbol{\sigma}}$ involves only the ordinary stress singularities, as shown in the following.

$\underset{\approx H}{\boldsymbol{\sigma}}$: The stress tensor that is the solution to the homogeneous equilibrium equations with homogeneous traction boundary conditions in the neighborhood of a singular point is known to be expanded in an asymptotic series (here we use the vector representation of the stress tensor)

$$\boldsymbol{\sigma}_H(r,\theta) = \sum_{i=1}^{\infty}\sum_{m=0}^{M} A_{im} r^{\alpha_i - 1} \ln^m r\, \mathbf{S}_{im}(\theta), \tag{7.6}$$

where A_{im} are called thermal generalized stress intensity factors (TGSIFs). We restrict our discussion to cases in which $M = 0$; thus (7.6) to be addressed is

$$\boldsymbol{\sigma}_H(r,\theta) = \sum_{i=1}^{\infty} A_i r^{\alpha_i - 1} \mathbf{S}_i(\theta). \tag{7.7}$$

$\underset{\approx P}{\boldsymbol{\sigma}}$ and $\underset{\approx P}{\bar{\boldsymbol{\sigma}}}$: Both particular stress tensors $\underset{\approx P}{\boldsymbol{\sigma}}$ and $\underset{\approx P}{\bar{\boldsymbol{\sigma}}}$ depend on the temperature field τ in the vicinity of the singular point but do not have to be actually computed. Their main role is the linkage to the theoretical framework providing the justification for their removal in a systematic way in the numerical algorithm.

As mentioned in Section 7.1.1, the simplified approach is to assume that the singular point has negligible influence on the distribution of the temperature in its vicinity. Therefore, the temperature distribution in a *closed vicinity of the crack tip* may be considered to be constant. In this case, the thermoelasticity problem to be solved becomes

$$\mathbf{div}\, \underset{\approx}{\boldsymbol{\sigma}} = \beta\, \mathrm{grad}\tau = 0 \qquad \text{in } \Omega_R, \tag{7.8}$$

$$T_n = \sigma_{\theta\theta} = \mu\beta\tau = C \qquad \text{on } \Gamma_1 \cup \Gamma_2, \tag{7.9}$$

$$T_t = \sigma_{r\theta} = 0 \qquad \text{on } \Gamma_1 \cup \Gamma_2. \tag{7.10}$$

The solution to the problem (7.8), (7.9), (7.10) consists of two parts only:

$$\underset{\approx}{\sigma} = \underset{\approx H}{\sigma} + \underset{\approx P}{\bar{\sigma}},$$

with the constant particular stress tensor $\underset{\approx P}{\bar{\sigma}}$ that satisfies the equilibrium equations (7.8), and the boundary conditions (7.9) and (7.10) given as

$$(\bar{\sigma}_{rr}\ \bar{\sigma}_{\theta\theta}\ \bar{\sigma}_{r\theta})_P^T = \begin{Bmatrix} C \\ C \\ 0 \end{Bmatrix} \stackrel{\text{def}}{=} \boldsymbol{\sigma}_o. \tag{7.11}$$

The second approach to considering the temperature field assumes that the temperature field is disturbed by the singular point. In this case, the temperature flux possesses singular behavior in the vicinity of the singular point, and the temperature field is given by (7.1). It is assumed that $\beta_1 \neq \alpha_i - 1$ for any i and $\beta_1 < 1$. By separation of variables, using the shift theorem for the equation in r, the stress vector $\boldsymbol{\sigma}_P$ in the vicinity of the singular point can be represented as follows:

$$\boldsymbol{\sigma}_P(r,\theta) = \boldsymbol{\sigma}_o + Br^{\beta_1}\mathbf{H}(\theta) + \mathcal{O}(r^{\beta_1+\epsilon}), \tag{7.12}$$

where B is a constant and $\epsilon > 0$. If $\beta_1 = \alpha_i - 1$, then (7.13) will also contain a $\ln(r)$ term

$$\boldsymbol{\sigma}_P(r,\theta) = \boldsymbol{\sigma}_o + Br^{\beta_1}\ln(r)\mathbf{F}_i(\theta) + r^{\beta_1}\mathbf{H}(\theta) + \mathcal{O}(r^{\beta_1+\epsilon}), \tag{7.13}$$

where B is chosen so as to satisfy Fredholm's alternative (see [65, pp. 78-80]). This case is not addressed in the following.

Since $\underset{\approx P}{\sigma}$ vanishes on the boundaries $\Gamma_1 \cup \Gamma_2$, the stress tensor $\underset{\approx P}{\bar{\sigma}}$ can be any tensor that produces only nonzero $\sigma_{\theta\theta}$, and this could be a linear function of θ multiplied by a function of r of the form $r^{\beta_1} + \mathcal{O}(r^{\beta_1+\epsilon})$.

The displacement vector corresponding to $\underset{\approx P}{\sigma} + \underset{\approx P}{\bar{\sigma}}$ is denoted by $\mathbf{u}_P$ and can be expressed in the vicinity of the singular point (as $r \to 0$) by

$$\boldsymbol{u}_P(r,\theta) = \boldsymbol{u}_o + r\mathbf{g}(\theta) + \mathcal{O}(r^{\beta_1+1+\epsilon}), \tag{7.14}$$

where $\boldsymbol{u}_o$ is a vector containing constants (describing the displacements of the singular point P, and $\mathbf{g}(\theta)$ is an analytic vector function of θ).

7.1.2 The Complementary Energy Formulation of the Thermoelastic Problem

The definition of the statically admissible space

$$\mathcal{E}_c(\Omega_R) = \left\{ \underset{\approx}{\sigma} \;\middle|\; \|\underset{\approx}{\sigma}\|_{\mathbf{L}^2,\Omega_R} < \infty;\ \mathbf{div}\,\underset{\approx}{\sigma} = \beta\ \mathrm{grad}\tau \text{ in } \Omega_R \right\}$$

and its variants are presented in Section 6.2, together with the dual (complementary) weak formulation for the (thermo-) elasticity problem over the subdomain Ω_R (see [128, pp. 103–108]):

$$\text{Seek } \boldsymbol{\sigma} \in \tilde{\mathcal{E}}_c(\Omega_R) \text{ such that}$$

$$\mathcal{B}_c(\boldsymbol{\sigma}, \boldsymbol{\sigma}_1) = \mathcal{F}_c(\boldsymbol{\sigma}_1) \qquad \forall \boldsymbol{\sigma}_1 \in \overset{\circ}{\mathcal{E}}_c(\Omega_R). \tag{7.15}$$

Any statically admissible stress vector $\boldsymbol{\sigma} \in \tilde{\mathcal{E}}_c(\Omega_R)$ can be written as an *arbitrary known function* from $\tilde{\mathcal{E}}_c(\Omega_R)$ and a suitably chosen function from $\overset{\circ}{\mathcal{E}}{}_c^H(\Omega_R)$. Therefore, we can write

$$\boldsymbol{\sigma} = \boldsymbol{\sigma}_H + \boldsymbol{\sigma}_P + \bar{\boldsymbol{\sigma}}_P, \tag{7.16}$$

$$\boldsymbol{\sigma}_H \in \overset{\circ}{\mathcal{E}}{}_c^H(\Omega_R), \qquad (\boldsymbol{\sigma}_P + \bar{\boldsymbol{\sigma}}_P) \in \tilde{\mathcal{E}}_c(\Omega_R).$$

With this notation, the dual weak form (7.15) can be restated in a more convenient manner

$$\text{Seek } \boldsymbol{\sigma}_H \in \overset{\circ}{\mathcal{E}}{}_c^H(\Omega_R) \text{ such that} \tag{7.17}$$

$$\mathcal{B}_c(\boldsymbol{\sigma}_H, \boldsymbol{\sigma}_1) = \mathcal{F}_c(\boldsymbol{\sigma}_1) - \mathcal{B}_c(\boldsymbol{\sigma}_P, \boldsymbol{\sigma}_1) - \mathcal{B}_c(\bar{\boldsymbol{\sigma}}_P, \boldsymbol{\sigma}_1) \qquad \forall \boldsymbol{\sigma}_1 \in \overset{\circ}{\mathcal{E}}{}_c^H(\Omega_R).$$

7.1.3 The Extraction Post-solution Scheme

For elastic problems, without thermal loading or body forces and with homogeneous boundary conditions on $\Gamma_1 \cup \Gamma_2$, the weak form (7.17) without the last two terms on the right-hand side is obtained. Computation of GSIFs for this case has been addressed in Chapter 6. We first attempted to use the methods in Chapter 6 for thermoelastic problems, using the CEM, without considering the particular stress tensors [201]. Mathematical analysis proved that the TGSIFs can be obtained at the limit when the radius of the subdomain Ω_R approaches zero. However, the error introduced in the extracted TGSIFs at a given finite R, due to neglecting $\boldsymbol{\sigma}_P$ and $\bar{\boldsymbol{\sigma}}_P$, was not properly investigated. Numerical experiments for crack-tip singularities and a singular point associated with an inclusion problem involving two dissimilar materials were presented to demonstrate that indeed at very small radii R one usually obtains good approximations of the TGSIFs. From the numerical point of view, this required a considerably refined mesh (with elements of an order of magnitude up to $\mathcal{O}(10^{-5})$) in the vicinity of the singular point. It has also been shown that for weak stress singularities even a very refined mesh did not provide satisfactory results.

Explicit computation of $\boldsymbol{\sigma}_P$ and $\bar{\boldsymbol{\sigma}}_P$ is a complicated and tedious task. Thus we wish to extract TGSIFs using only the knowledge of their functional representation in the r direction and a detailed analysis of the error introduced because of neglecting them (see [204]). An analysis of the error induced by not considering the particular solution provides the means to extrapolate to the limit $R = 0$ and obtain excellent results without the need for considerable mesh refinements. The model problem for the mathematical analysis will be a domain containing a reentrant corner in isotropic materials. Although this does not fully represent anisotropic materials and multimaterial interfaces, it still provides the necessary steps in the proof, which can be reconstructed in a more general case.

7.1.4 The Compliance Matrix, Load Vector and Extraction of TGSIFs

The stress vectors $\boldsymbol{\sigma}_H, \boldsymbol{\sigma}_1$ in (7.17), being in the space $\overset{\circ}{\mathcal{E}}{}_c^H(\Omega)$, can be represented by (7.7), and the elements of the compliance matrix corresponding to $\mathcal{B}_c(\boldsymbol{\sigma}_H, \boldsymbol{\sigma}_1)$ are given in (6.31). For isotropic materials, the compliance matrix is diagonal (see Remark 6.1):

$$(B_c)_{ij} = \begin{cases} \frac{R^{2\alpha_i}}{2\alpha_i} D_i, & i = j, \\ 0, & i \neq j, \end{cases} \qquad i, j = 1, 2, \ldots, N, \tag{7.18}$$

and D_i are constants that depend on the angle ω, the inverse of the material matrix $[E]^{-1}$, and the eigenstresses, but independent of R.

The load vector corresponding to the linear form $\mathcal{F}_c$ is to be evaluated only along the circular boundary Γ_R. This is because $\boldsymbol{\sigma}_1 \in \overset{\circ}{\mathcal{E}}{}_c^H$. Therefore $\boldsymbol{\sigma}_1 = \mathbf{0}$ at $\theta = \theta_1, \theta_1 + \omega$ if traction-free boundary conditions are considered. The displacement vector $\boldsymbol{u}$ as $r \to 0$ is given by the homogeneous part together with (7.14):

$$\boldsymbol{u} = \sum_{i=1}^{\infty} A_i r^{\alpha_i} \boldsymbol{s}_i(\theta) + \boldsymbol{u}_o + r\mathbf{g}(\theta) + \mathcal{O}(r^{\beta_1+1+\epsilon}). \tag{7.19}$$

To obtain the jth component of the load vector, one has to substitute (7.19) and the jth eigenstress in the series (7.7) in the expression for the linear form (6.32). Because the homogeneous eigen-stresses are orthogonal with respect to the integral along Γ_R for isotropic materials, one obtains

$$(F_c)_j = C_{1j} R^{\alpha_1+\alpha_j} + C_{3j} R^{\alpha_j+1} + C_{4j} R^{\alpha_j+\beta_1+1} + \text{h.o.t.} \qquad j = 1, 2, \ldots, N. \tag{7.20}$$

Here C_{ij} are constants independent of R that may be zero and depend on material properties, geometry, far field loading, and temperature field.

Evaluation of the other two terms in the load vector, corresponding to $\mathcal{B}_c(\boldsymbol{\sigma}_P, \boldsymbol{\sigma}_1)$ and $\mathcal{B}_c(\bar{\boldsymbol{\sigma}}_P, \boldsymbol{\sigma}_1)$, denoted by $\mathcal{B}_{c1}$ and $\mathcal{B}_{c2}$, is now considered. Substituting (7.12) in (6.31), and having $\boldsymbol{\sigma}_1 \in \overset{\circ}{\mathcal{E}}{}_c^H$, one obtains the jth element of the load vector corresponding to $\mathcal{B}_{c1}$:

$$
\begin{aligned}
(\mathcal{B}_{c1})_j &= R^{\alpha_j+1} \int_0^{\omega} \boldsymbol{\sigma}_o^T [E]^{-1} \boldsymbol{S}_j(\theta) d\theta \\
&\quad + R^{\alpha_j+\beta_1+1} \int_0^{\omega} \mathbf{H}^T(\theta)[E]^{-1} \boldsymbol{S}_j(\theta) d\theta + \text{h.o.t.} \\
&= C_{5j} R^{\alpha_j+1} + C_{6j} R^{\alpha_j+\beta_1+1} + \text{h.o.t.} \qquad j = 1, 2, \ldots, N. \qquad (7.21)
\end{aligned}
$$

The load vector corresponding to $\mathcal{B}_{c2}$ is very similar to that given in (7.21), without the first term, and we may add them together to obtain

$$
(\mathcal{B}_{c1})_j + (\mathcal{B}_{c2})_j = C_{5j} R^{\alpha_j+1} + C_{7j} R^{\alpha_j+\beta_1+1} + \text{h.o.t.} \qquad j = 1, 2, \ldots, N. \qquad (7.22)
$$

In view of (7.18), (7.20), and (7.22), the TGSIF A_1, for example, can be computed:

$$
\begin{aligned}
A_1 &= \frac{2\alpha_1}{D_1} R^{-2\alpha_1} \left[C_{11} R^{2\alpha_1} + C_{31} R^{\alpha_1+1} + C_{41} R^{\alpha_1+\beta_1+1} \right. \\
&\qquad \left. + C_{51} R^{\alpha_1+1} + C_{71} R^{\alpha_1+\beta_1+1} + \text{h.o.t.} \right] \\
&= \frac{2\alpha_1}{D_1} \left[C_{11} + C_{81} R^{1-\alpha_1} + C_{41} R^{1+\beta_1-\alpha_1} + \text{h.o.t.} \right]. \qquad (7.23)
\end{aligned}
$$

Examining (7.23) as $R \to 0$, one notices that the influence of the particular stress vector on A_1 is of the order of magnitude of $R^{1-\alpha_1}$ or $R^{1+\beta_1-\alpha_1}$ (depending on whether C_{81} is zero or not). The stress field in the neighborhood of a singular point is singular only if $\alpha_1 < 1$, and because $\beta_1 \geq 0$, the influence approaches zero as $R \to 0$. This suggests that the terms associated with the particular stress vectors could be neglected, contributing a relative error of order of magnitude $\mathcal{O}(R^{1-\alpha_1})$ or $\mathcal{O}(R^{1+\beta_1-\alpha_1})$ when computing A_1, for example. Using a finite subdomain of radius R_1, $(A_1)_1$ can be extracted by (7.23) neglecting the particular stress vector (only the first term on the right-hand side is considered). By repeating this extraction procedure over a sequence of decreasing subdomains of radii R_j, $R_j < R_{j-1} < \cdots < R_1$, one obtains a sequence of approximations denoted by $(A_1)_j$. Then, employing Richardson's extrapolation [145, pp. 94-95] with the error behaving as $R^{1-\alpha_1}$ or $R^{1+\beta_1-\alpha_1}$, A_1 can be extrapolated at the limit $R \to 0$. We can generate a table of A_1's, for example, by the recurrence relation

$$
(A_1)_j^{(m)} = (A_1)_{j-1}^{(m+1)} + \frac{(A_1)_{j-1}^{(m+1)} - (A_1)_{j-1}^{(m)}}{(R_j/R_{j+m})^{\gamma} - 1}, \qquad (7.24)
$$

where γ is either $1 - \alpha_1$ or $1 + \beta_1 - \alpha_1$, and the accuracy of A_1 improves as j and m increase (j corresponds to the radius R_j of the subdomains Ω_R, which is the row number in the generated table, and m corresponds to the column number; see Table 7.2, for example).

Remark 7.1 *If A_2 is to be computed, for example, the same procedure holds with γ either $1 - \alpha_2$ or $1 + \beta_1 - \alpha_2$ in* (7.24).

Remark 7.2 *The situation described in* (7.13) *will affect the third term in* (7.23), *and it seems that the leading term may sometimes be of the order of $R^{1+\beta_1-\alpha_i}[1 + \ln(R)]$. However, this is not the case due to Fredholm's alternative. This will be demonstrated by numerical examples on cracked domains, where the temperature field is proportional to $r^{1/2}$ and the homogeneous stress field is proportional to $r^{-1/2}$.*

Remark 7.3 *For anisotropic materials or mult-material interfaces, the matrix $[B_c]$ is fully populated, and an explicit equation like* (7.23) *is not obtainable. However, computation of the TGSIFs neglecting the particular stress tensor, for several R_j's and extrapolating to the limit, is still valid due to similar arguments (this will be shown by a numerical example). The mathematical analysis of this case is more cumbersome and is not provided here.*

7.1.5 Discretization and the Numerical Algorithm

For solving (7.17), the displacements along $(\partial\Omega_R)_u$ are to be substituted in (6.32). These are assumed for the mathematical analysis to be known, but in the numerical realization we substitute in (7.17) the displacements extracted from a finite element solution. Instead of $\boldsymbol{u}_{\mathrm{EX}}$ we obtain an approximation $\boldsymbol{u}_{\mathrm{FE}}$ by the "classical" finite element method (FEM) based on the principle of virtual work. Of course, $\boldsymbol{u}_{\mathrm{FE}}$ approximates the thermoelastic displacements field, with thermal loading being imposed on the domain of interest, and its accuracy can be controlled by p- or hp-extensions. The proposed procedure is a post-solution operation performed after the thermoelastic problem over the entire domain (Ω) has been solved by the FEM, and $\boldsymbol{u}_{\mathrm{FE}}$ having been obtained.

The weak form (7.17) is further discretized by choosing a statically admissible subspace $\overset{0}{\mathcal{E}}{}^{H}_{c\;N}(\Omega_R)$ constructed as a linear combination of the first N eigen-pairs according to (7.7), with A_i^{FE}, $i = 1, 2, \ldots, N$ being sought. In general, also the eigenpairs are not known explicitly and are computed by the modified Steklov method. Thus, the algorithm for computing TGSIFs is the following:

(a) Compute the smallest eigenvalue β_1 associated with the flux singularity (by the modified Steklov method) and the corresponding generalized flux intensity factor $(\mathrm{GFIF})_1$. If $(\mathrm{GFIF})_1 = 0$, use as β_1 the next smallest eigenvalue, i.e., β_2.

(b) Obtain a finite element solution of the thermal problem, then impose it as a thermal loading and solve the elasticity problem to obtain the displacements finite element solution $\boldsymbol{u}_{FE}$. Both the thermal and the elasticity solutions should have a small relative error measured in the energy norm.
(c) Extract TGSIFs by the CEM for *several values of integration radii* R_i. The radius of integration R_i should always be outside the first layer of elements that have a vertex in the singular point. At each integration radius R_i, the TGSIFs should be extracted using the finite element solutions corresponding to $p = 1, 2, \ldots, 8$, and estimated for $p \to \infty$. This ensures that discretization errors associated with replacing $\boldsymbol{u}_{EX}$ by $\boldsymbol{u}_{FE}$ are small.
(d) Use Richardson's extrapolation with the error behaving as either $R^{\beta_1+1-\alpha_1}$ or $R^{1-\alpha_1}$ to determine the first stress intensity factor and $R^{\beta_1+1-\alpha_2}$ or $R^{1-\alpha_2}$ to determine the second, as $R \to 0$.
(e) Examine the columns in the table generated by Richardson's extrapolation and ensure that the elements in the columns have similar values. Also examine the Richardson's extrapolation table for $p \to \infty$ in comparison with the table for $p = 8$ and ensure that the values are close (see the following example problems).
(f) Redo step (c), with a larger statically admissible space, i.e., increased number of homogeneous eigenstresses N, then redo steps (d)-(e). Check that the obtained TGSIFs are virtually independent of N. This ensures the reliability of the results.

In all examples, the integration is performed along a circle of radius R greater than the radius of the elements having a vertex at the singular point. This is because the finite element solution in the first group of elements at the singular point is not of high accuracy. Numerical examples are presented in the following to demonstrate that the proposed method performs well, resulting in accurate TGSIFs.

7.2 Numerical Examples

The temperature distribution is computed by solving the steady-state heat conduction problem, which is thereafter imposed as a thermal load in the elastostatic analysis. The trial space used in the p-FEM is the trunk space.

7.2.1 Central Crack in a Rectangular Plate

A rectangular plate with a central crack subjected to two different thermal loadings, for which numerical results are reported in previous publications, is considered. Analytical (exact) solutions to practical problems are very difficult, if not impossible to obtain, and to the best of our knowledge no analytic solutions are available to finite geometric models. The rectangular plate of width $2W$ and length $2L$ and a

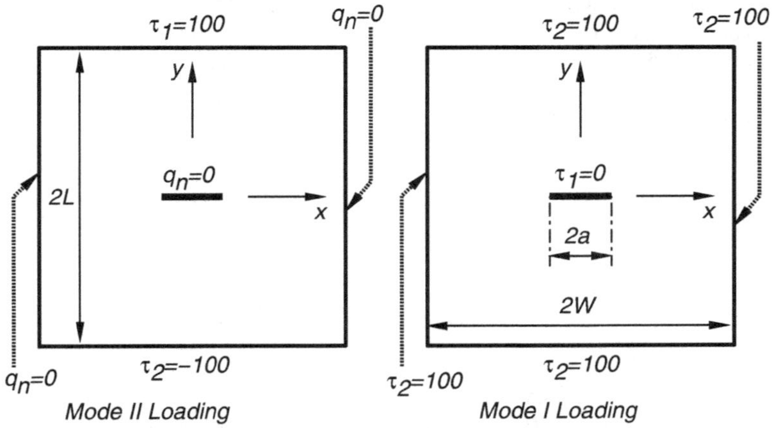

Fig. 7.1 Geometry and boundary conditions of a rectangular plate with a central crack.

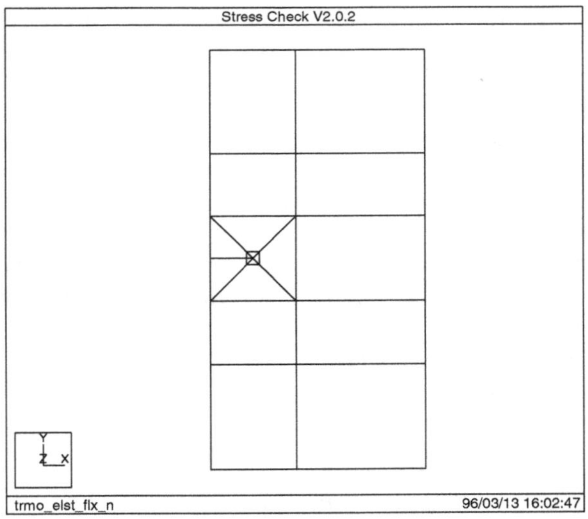

Fig. 7.2 The finite element mesh for the rectangular plate with a central crack.

central crack of length $2a = 2$ with $L/W = 1.0$ and $a/W = 0.2$ is solved for two different sets of thermal loadings representing pure mode I and pure mode II (see Figure 7.1).

The heat conduction coefficients are taken to be $k_{11} = k_{22} = 1$, $k_{12} = 0$ (results are independent of the heat conductivity for isotropic materials), and the mechanical material properties are Young's modulus $E = 1$; Poisson ratio $\nu = 0.3$; and the coefficient of linear thermal expansion $\alpha = 0.01$. A plane-strain condition is assumed. Taking advantage of the symmetry of the problem, only half of the model has been solved, imposing the following symmetry boundary conditions at $|y| \le L, x = 0$: $q_n = \frac{\partial \tau}{\partial x} = 0$, $u_x = 0$. The finite element mesh surrounding the crack tip contains *only one layer* graded in a geometric progression in the vicinity of the singular point with the grading factor 0.15, as shown in Figure 7.2. The results

Table 7.1 Convergence of the FE solution for mode I loading; center crack.

	Thermal Analysis		Thermo-Elastic Analysis	
	Estimated		Estimated	
p-level	DOF	$\|e\|_{\mathcal{E}(\Omega)}$ (%)	DOF	$\|e\|_{\mathcal{E}(\Omega)}$ (%)
1	12	30.55	45	73.23
2	44	12.31	129	18.81
3	81	8.15	223	10.28
4	137	5.61	355	6.42
5	212	4.43	525	4.50
6	306	3.69	733	3.53
7	419	3.18	979	2.89
8	551	2.80	1263	2.42

are compared with these previously published in [104, 112, 142, 173, 183, 201]. Sumi et al. [173] extracted the TSIFs using the modified-collocation and complex variable methods. Prasad et al. [142] reported the TSIFs obtained by employing the boundary element method (BEM), simultaneously solving the thermal and elasticity problems, then extracting the TSIFs using the path-independent J-integral. Lee et al. [104] used the BEM, first solving the thermal problem, then the elastic problem, and finally computing the TSIFs by the displacement extrapolation method. Tsai et al. [183] solved the mode I problem using the thermal weight function and the h-version finite element method. Liu et al. [112] used the BEM, coupling the direct boundary integral equations to the crack integral equation to extract the mode II TSIF. Yosibash showed in [201] that the TSIFs can be obtained by the CEM in a limit process as $R \to 0$. The method presented in [201] requires a very fine mesh in the neighborhood of singular points, resulting in many layers with very small radii $\mathcal{O}(R) = 0.0006$. Thus it is inefficient. Here these TSIFs are extracted at large Rs using the CEM, then using Richardson's extrapolation.

Mode I Thermal Loading

The quality of the finite element solution is summarized in Table 7.1. Examining the first eigenvalue of the thermal singularity and the first two eigenvalues of the elasticity problem, one observes that

$$\beta_1 + 1 - \alpha_1 = 1/2 + 1 - 1/2 = 1, \tag{7.25}$$

$$\beta_1 + 1 - \alpha_2 = 1/2 + 1 - 1/2 = 1. \tag{7.26}$$

Extracted values of $(K_{\mathrm{I}})_{\mathrm{FE}}$ at $p = 8$ and as $p \to \infty$ for different values of R are listed in Tables 7.2 and 7.3 respectively, together with the Richardson's extrapolated values as $R \to 0$. $(K_{\mathrm{II}})_{\mathrm{FE}}$ is 0.000000000 for all cases. By the mathematical analysis, the error in $(K_{\mathrm{I}})_{\mathrm{FE}}$ and $(K_{\mathrm{II}})_{\mathrm{FE}}$ behaves like R^1 as $R \to 0$. This power is used for Richardson's extrapolation. Extraction of $(K_{\mathrm{I}})_{\mathrm{FE}}$ at the smallest integration radius $R = 0.3$ is the least accurate because the integration path is the closest to the first layer of elements surrounding the singular point, and therefore contains the largest *numerical error*. One may gain further accuracy by adding an additional layer of elements. Numerical experiments support this last statement.

Table 7.2 $(K_{\mathrm{I}})_{\mathrm{FE}}$ at various values of R, $p = 8$, for mode I center crack problem and the extrapolated values as $R \to 0$.

R	$(K_{\mathrm{I}})_{\mathrm{FE}} \stackrel{\mathrm{def}}{=} K_{\mathrm{I}}^{(0)}$	$K_{\mathrm{I}}^{(1)}$	$K_{\mathrm{I}}^{(2)}$	$K_{\mathrm{I}}^{(3)}$
0.9	1.7506058183			
		0.8192715659		
0.7	1.5436426511		0.8001684752	
		0.8107813034		0.7882428331
0.5	1.3342536946		0.7922180471	
		0.8001737284		
0.3	1.1206217081			

Table 7.3 $(K_{\mathrm{I}})_{\mathrm{FE}}$ at various values of R, $p \to \infty^*$, for mode I center crack problem and the extrapolated values as $R \to 0$.

R	$(K_{\mathrm{I}})_{\mathrm{FE}} \stackrel{\mathrm{def}}{=} K_{\mathrm{I}}^{(0)}$	$K_{\mathrm{I}}^{(1)}$	$K_{\mathrm{I}}^{(2)}$
0.9	1.7522014811		
		0.8243398522	
0.7	1.5460100080		0.7997843595
		0.8134262999	
0.5	1.3367003771		

The value at $R = 0.3$ was not considered because it was away from the value obtained at $p = 8$. Therefore the extrapolated value is not accurate enough.

Tables 7.2 and 7.3 clearly demonstrate that the thermal stress intensity factor is extrapolated with high accuracy, even though the relative errors at finite values of R are very large. The significant reduction in the error already at the first step of the Richardson's algorithm, and the similarity of the results in each column, strongly support the mathematical analysis.

The extrapolated value is in excellent agreement with [142,173,201]. A summary of results obtained by other numerical methods compared with the extrapolated K_{I} is given in Table 7.7.

Mode II Thermal Loading

This subsection summarizes the TSIFs obtained when the domain is loaded by mode II thermal loading. The quality of the finite element solution is summarized in Table 7.4. Extracted values of $(K_{\mathrm{II}})_{\mathrm{FE}}$ at $p = 8$ and as $p \to \infty$ for different values of R are listed in Tables 7.5 and 7.6 respectively, together with the Richardson's extrapolated values as $R \to 0$. $(K_{\mathrm{I}})_{\mathrm{FE}}$ is 0.000000000 for all cases. Again, the error in $(K_{\mathrm{II}})_{\mathrm{FE}}$ behaves like R^1 as $R \to 0$. This power is used for Richardson's extrapolation. Tables 7.5 and 7.6 are very similar to Tables 7.2 and 7.3 of the previous mode I problem. Again one may notice the convergence achieved by the extrapolation algorithm, although the TSIFs at the various R's are of very low accuracy. The very similar values in Tables 7.5 and 7.6 demonstrate that the

Table 7.4 Convergence of the FE solution for mode II loading; center crack.

	Thermal Analysis		Thermoelastic Analysis	
	Estimated		Estimated	
p-level	DOF	$\|e\|_{\mathcal{E}(\Omega)}$ (%)	DOF	$\|e\|_{\mathcal{E}(\Omega)}$ (%)
1	21	8.98	45	669.68
2	62	4.44	129	35.11
3	108	2.72	223	30.03
4	173	2.02	355	7.72
5	257	1.58	525	5.65
6	360	1.32	733	4.39
7	482	1.13	979	3.66
8	623	0.99	1263	3.13

Table 7.5 $(K_{\mathrm{II}})_{\mathrm{FE}}$ at various values of R, $p = 8$, for mode II center crack problem and the extrapolated values as $R \to 0$.

R	$(K_{\mathrm{II}})_{\mathrm{FE}} \stackrel{\mathrm{def}}{=} K_{\mathrm{II}}^{(0)}$	$K_{\mathrm{II}}^{(1)}$	$K_{\mathrm{II}}^{(2)}$	$K_{\mathrm{II}}^{(3)}$
0.9	−0.0693129872			
		0.1246962094		
0.7	−0.0261998324		0.1219032328	
		0.1234548865		0.1212235415
0.5	0.0165586587		0.1214501053	
		0.1223092972		
0.3	0.0588589141			

Table 7.6 $(K_{\mathrm{II}})_{\mathrm{FE}}$ at various values of R, $p \to \infty$, for mode II center crack problem and the extrapolated values as $R \to 0$.

R	$(K_{\mathrm{II}})_{\mathrm{FE}} \stackrel{\mathrm{def}}{=} K_{\mathrm{II}}^{(0)}$	$K_{\mathrm{II}}^{(1)}$	$K_{\mathrm{II}}^{(2)}$	$K_{\mathrm{II}}^{(3)}$
0.9	−0.0693143367			
		0.1248325230		
0.7	−0.0261705901		0.1229332211	
		0.1239883888		0.1210350940
0.5	0.0167319753		0.1216678030	
		0.1226623398		
0.3	0.0591041211			

numerical error in $(K_{\mathrm{II}})_{\mathrm{FE}}$ is small at the given radii, and thus high confidence in the extrapolated results is achieved.

In Table 7.7 we summarize the results obtained, compared to these reported previously for the mode I and mode II loadings.

The results obtained here show good accuracy compared with these reported previously. The number of degrees of freedom in the FE model is half what was needed in [201] for obtaining similar accuracy in the TSIFs.

Table 7.7 Summary of mode I and mode II TSIFs.

Ref. Meth.	Tsai et al. [183] Weight fncn & FEM	Lee et al. [104] BEM	Prasad et al. [142] BEM	Sumi et al. [173] Complex var. & collocation	Liu et al. [112] BEM	Yosibash [201] Compl. Enrg. w/o Rich. Ext.	Present Method
K_{I}	0.8036	0.8593	0.7759	0.7759		0.7784	0.7998
K_{II}		0.1317	0.1207	0.1185	0.1324	0.1214	0.1210

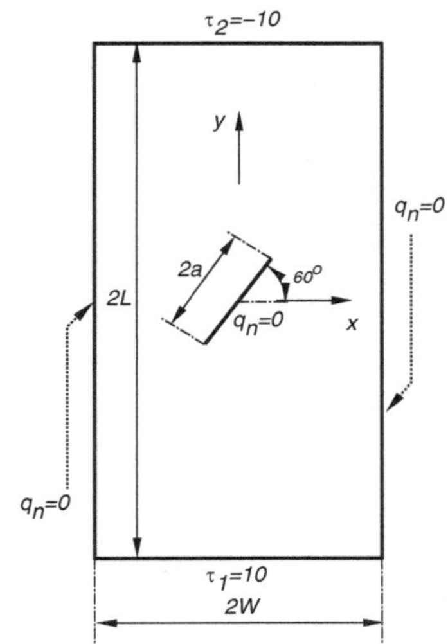

Fig. 7.3 Geometry and boundary conditions of a rectangular plate with a slanted crack.

7.2.2 A Slanted Crack in a Rectangular Plate

A rectangular plate with a central crack slanted at an angle of 60° to the x-axis is considered. The geometry and temperature boundary conditions are shown in Figure 7.3. The rectangular plate of width $2W = 2$ with $L/W = 2.0$ and $a/W = 0.3$ is solved for a temperature loading that gives rise to a mixed mode.

The heat conduction coefficients are taken to be $k_{11} = k_{22} = 1$, $k_{12} = 0$, and the mechanical material properties are Young's modulus $E = 2.184 \times 10^5$; Poisson ratio $\nu = 0.3$; and the coefficient of linear thermal expansion $\alpha = 1.67 \times 10^{-5}$. A plane-stress condition is assumed.

The results are compared with those obtained by Nakanishi et. al [126] by the complex variable method and reported in [124, pp. 1063–1067]. The finite element mesh surrounding the crack tip contains several layers graded in a geometric progression in the vicinity of the singular point with the grading factor 0.15. The finite element mesh is presented in Figure 7.4. The finite element discretization error in energy norm at $p = 8$ (1834 DOFs) for the thermal analysis is 0.18%, and for the thermoelastic analysis (3797 DOFs) is 1.15%. This model problem is used to demonstrate that the number of terms used to represent $\overset{0}{\mathcal{E}}{}_c^{H}$ has minor influence on the extracted TSIFs. Also, we show that Richardson's extrapolation for K_{I} assumes

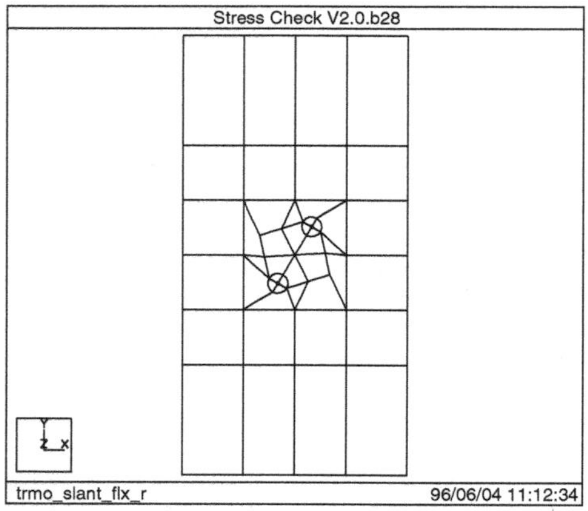

Fig. 7.4 The finite element mesh for the rectangular plate with a slanted crack.

that the error is $\mathcal{O}(R^{1-\alpha_1})$, whereas for K_{II} it is $\mathcal{O}(R^{1+\beta_1-\alpha_2})$. For the slant crack configuration we have $\alpha_1 = \alpha_2 = \beta_1 = 1/2$, thus γ in (7.24) is either 1/2 or 1. Extracted values of $(K_{\text{I}})_{\text{FE}}$ at $p \to \infty$ for different values of R are listed in Table 7.8 when $N = 2, 4, 6$ with $\gamma = 1/2$, and in Table 7.9 with $\gamma = 1$, together with the Richardson's extrapolated values as $R \to 0$. This is to demonstrate that the error in $(K_{\text{I}})_{\text{FE}}$ behaves like $R^{1/2}$ as $R \to 0$. Table 7.9 demonstrates that the coefficient C_8 is nonzero in (7.23) for K_{I}, whereas, as will be shown in Table 7.10, $C_8 = 0$ for K_{II}. This is possible because C_8 represents an integral of the corresponding eigenstress multiplied by the particular stress vector, and for some cases these are orthogonal. The similarity of the results along the columns in Table 7.8, starting at $K_{\text{I}}^{(3)}$, and the convergence of Richardson's extrapolation to the same value for $N = 2, 4$, and 6 ensures that accurate and reliable results are obtained. Although at finite radii of integration R_i the extracted TSIFs are extremely inaccurate, the extrapolated approximation $K_{\text{I}} = 0.0238, 0.0234$ is in excellent agreement with the value $K_{\text{I}} \approx 0.023$ given in [126] in a graph.

The value K_{II} is extrapolated at the limit $R \to 0$ with γ in (7.24) being 1. Extracted values of $(K_{\text{II}})_{\text{FE}}$ at $p \to \infty$ for different values of R are listed in Table 7.10 when $N = 2, 4, 6$ together with the Richardson's extrapolated values as $R \to 0$. The values in Table 7.10 clearly demonstrate the convergence of K_{II} to $-.642$ which is in excellent agreement with the value -0.64 reported in [126].

7.2.3 *A Rectangular Plate with Cracks at an Internal Hole*

A rectangular plate of width $2W$, length $2L$ with a circular hole of radius ρ and two cracks emanating from it is considered, and shown in Figure 7.5.

Table 7.8 $(K_\mathrm{I})_\mathrm{FE}$ at various values of R, at $p \to \infty$ taking $\gamma = 1/2$ for a plate with a slanted crack with $N = 2, 4, 6$, and extrapolated values as $R \to 0$.

	$(K_\mathrm{I})_\mathrm{FE} \overset{\mathrm{def}}{=} K_\mathrm{I}^{(0)}$			$K_\mathrm{I}^{(1)}$			$K_\mathrm{I}^{(2)}$			$K_\mathrm{I}^{(3)}$			$K_\mathrm{I}^{(4)}$		
R	$N=2$	$N=4$	$N=6$	$N=2$	$N=4$	$N=6$	$N=2$	$N=4$	$N=6$	$N=2$	$N=4$	$N=6$	$N=2$	$N=4$	$N=6$
0.085	−2.995	−2.536	−1.962												
				0.2071	0.2222	0.0261									
0.04	−1.990	−1.670	−1.338				−0.0305	−0.0345	0.0230						
				0.0847	0.0900	0.0246				0.0236	0.0236	0.0236			
0.02	−1.382	−1.154	−0.939				0.0051	0.0037	0.0234				0.02381	0.02388	0.02342
				0.0449	0.0468	0.0240				0.0237	0.0238	0.0235			
0.01	−0.964	−0.802	−0.657				0.0171	0.0167	0.0235						
				0.0310	0.0318	0.0237									
0.005	−0.673	−0.624	−0.457												

Table 7.9 $(K_\mathrm{I})_\mathrm{FE}$ at various values of R, at $p \to \infty$ taking $\gamma = 1$ for a plate with a slanted crack with $N = 2, 4, 6$, and extrapolated values as $R \to 0$.

	$(K_\mathrm{I})_\mathrm{FE} \stackrel{\mathrm{def}}{=} K_\mathrm{I}^{(0)}$			$K_\mathrm{I}^{(1)}$			$K_\mathrm{I}^{(2)}$			$K_\mathrm{I}^{(3)}$			$K_\mathrm{I}^{(4)}$		
R	$N=2$	$N=4$	$N=6$	$N=2$	$N=4$	$N=6$	$N=2$	$N=4$	$N=6$	$N=2$	$N=4$	$N=6$	$N=2$	$N=4$	$N=6$
0.085	−2.995	−2.536	−1.962												
				−1.09	−0.90	−0.78									
0.04	−1.990	−1.670	−1.338				−0.676	−0.558	−0.465						
				−0.77	−0.64	−0.54				−0.443	−0.365	−0.300			
0.02	−1.382	−1.154	−0.939				−0.470	−0.388	−0.320				−0.297	−0.243	−0.199
				−0.55	−0.45	−0.37				−0.310	−0.251	−0.205			
0.01	−0.964	−0.802	−0.657				−0.326	−0.268	−0.258						
				−0.38	−0.32	−0.26									
0.005	−0.673	−0.624	−0.457												

Table 7.10 $(K_{\mathrm{II}})_{\mathrm{FE}}$ at various values of R, at $p \to \infty$ for a plate with a slanted crack with $N = 2, 4, 6$, and extrapolated values as $R \to 0$.

	$(K_{\mathrm{II}})_{\mathrm{FE}} \overset{\mathrm{def}}{=} K_{\mathrm{II}}^{(0)}$			$K_{\mathrm{II}}^{(1)}$			$K_{\mathrm{II}}^{(2)}$			$K_{\mathrm{II}}^{(3)}$			$K_{\mathrm{II}}^{(4)}$		
R	$N = 2$	$N = 4$	$N = 6$	$N = 2$	$N = 4$	$N = 6$	$N = 2$	$N = 4$	$N = 6$	$N = 2$	$N = 4$	$N = 6$	$N = 2$	$N = 4$	$N = 6$
0.085	−0.318	−0.318	−0.345												
				−0.644	−0.644	−0.644									
0.04	−0.491	−0.491	−0.503				−0.642	−0.642	−0.642						
				−0.642	−0.642	−0.642				−0.646	−0.646	−0.646			
0.02	−0.567	−0.567	−0.573				−0.645	−0.645	−0.645				−0.642	−0.642	−0.642
				−0.645	−0.645	−0.644				−0.642	−0.642	−0.642			
0.01	−0.606	−0.606	−0.609				−0.643	−0.643	−0.643						
				−0.643	−0.643	−0.643									
0.005	−0.624	−0.624	−0.626												

Fig. 7.5 Geometry and boundary conditions of a rectangular plate with a crack at an internal hole.

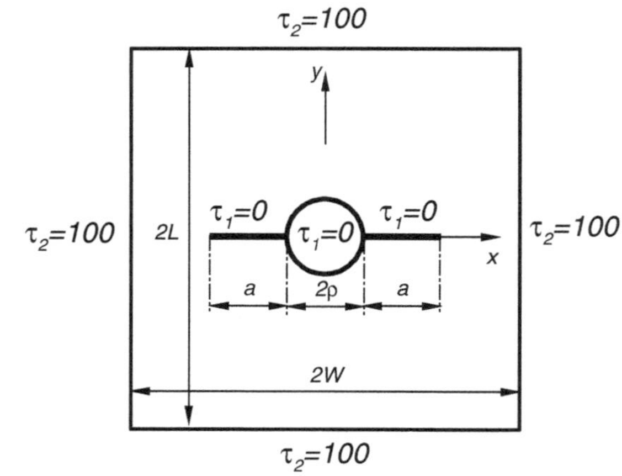

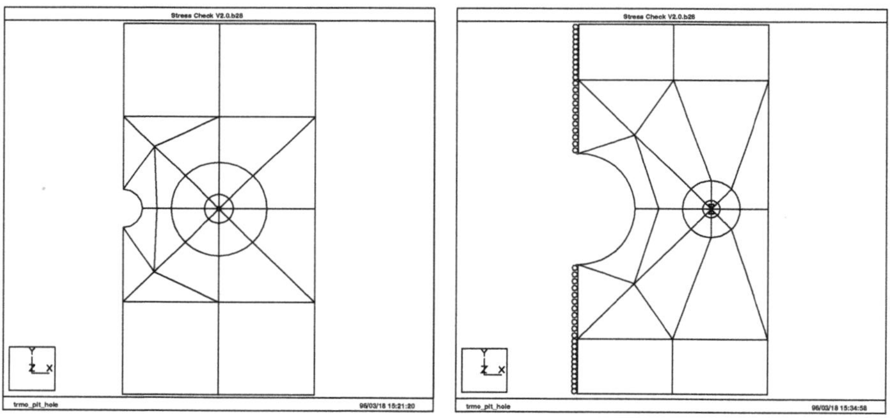

Mesh for $\rho/a = 0.25$ Mesh for $\rho/a = 0.75$

Fig. 7.6 Finite element meshes for the rectangular plate with a crack at an internal hole.

Dimensions of the analyzed problem are defined by $L/W = 1$, $a = 0.4W = 1$, and two different radii $\rho/a = 0.25$ and $\rho/a = 0.75$. The material properties are identical to those described in Section 7.2.1. This problem is analyzed for one set of thermal boundary conditions, representing mode I loading, namely $\tau_1 = 0$ on cracks and circular hole and $\tau_2 = 100$ on the outer boundary. Due to the symmetry, only half of the model has been considered, and the finite element meshes for the two different radii ρ are shown in Figure 7.6. The finite element mesh surrounding the crack tip contains two layers graded in a geometric progression in the vicinity of the singular point with the grading factor 0.15. The quality of the finite element solution is summarized in Table 7.11. Extracted values of $(K_I)_{FE}$ at $p = 8$ and as $p \to \infty$ for different values of R are listed in Table 7.12 for $\rho/a = 0.25$ and in

Table 7.11 Convergence of the FE solution for a plate with a crack at an internal hole.

	$\rho/a = 0.25$				$\rho/a = 0.75$			
	Thermal Analysis		Thermoelastic Analysis		Thermal Analysis		Thermoelastic Analysis	
p-level	Estimated DOF	$\|e\|_{\mathcal{E}(\Omega)}$ (%)	Estimated DOF	$\|e\|_{\mathcal{E}(\Omega)}$ (%)	Estimated DOF	$\|e\|_{\mathcal{E}(\Omega)}$ (%)	Estimated DOF	$\|e\|_{\mathcal{E}(\Omega)}$ (%)
1	27	18.34	91	73.09	27	19.83	91	102.20
2	97	5.58	267	15.18	97	6.56	267	33.31
3	179	3.66	467	11.31	179	5.15	467	21.71
4	303	2.32	751	4.70	303	2.11	751	9.94
5	469	1.65	1119	3.00	469	1.28	1119	4.60
6	677	1.30	1571	2.15	677	0.95	1571	2.38
7	927	1.07	2107	1.62	927	0.76	2107	1.47
8	1219	0.90	2727	1.26	1219	0.63	2727	0.96

Table 7.12 $(K_I)_{FE}$ at various values of R, $p = 8$ and $p \to \infty$ (in parentheses), for a plate with a crack at an internal hole, $\rho/a = 0.25$, and extrapolated values as $R \to 0$.

R	$(K_I)_{FE} \stackrel{\text{def}}{=} K_I^{(0)}$	$K_I^{(1)}$	$K_I^{(2)}$	$K_I^{(3)}$
0.5	1.4822 (1.4823)			
		0.8184 (0.8255)		
0.3	1.2167 (1.2196)		0.8018 (0.7997)	
		0.8051 (0.8049)		0.7950 (0.8011)
0.1	0.9423 (0.9431)		0.7957 (0.8010)	
		0.7973 (0.8016)		
0.05	0.8698 (0.8724)			

Table 7.13 $(K_I)_{FE}$ at various values of R, $p = 8$ and $p \to \infty$ (in parentheses), for a plate with a crack at an internal hole, $\rho/a = 0.75$, and extrapolated values as $R \to 0$.

R	$(K_I)_{FE} \stackrel{\text{def}}{=} K_I^{(0)}$	$K_I^{(1)}$	$K_I^{(2)}$	$K_I^{(3)}$
0.35	1.4083 (1.4084)			
		0.9523 (0.9532)		
0.15	1.1477 (1.1483)		0.9381 (0.9428)	
		0.9414 (0.9452)		0.9317 (0.9342)
0.08	1.0514 (1.0535)		0.9322 (0.9349)	
		0.9341 (0.9370)		
0.03	0.9781 (0.9807)			

Table 7.13 for $\rho/a = 0.75$ together with the Richardson's extrapolated values as $R \to 0$. Again, the error in $(K_I)_{FE}$ behaves like R^1 as $R \to 0$, and this power is used for Richardson's extrapolation. $(K_{II})_{FE}$ is 0.000000000 for all cases. The approximated K_I for the case $\rho/a = 0.25$ obtained by the BEM and reported in [142] is 0.806, which is in good agreement with our extrapolated value of 0.8011. For the case $\rho/a = 0.75$ we obtain $K_I = 0.9342$ which is again in good agreement with the value 0.941 reported in [142].

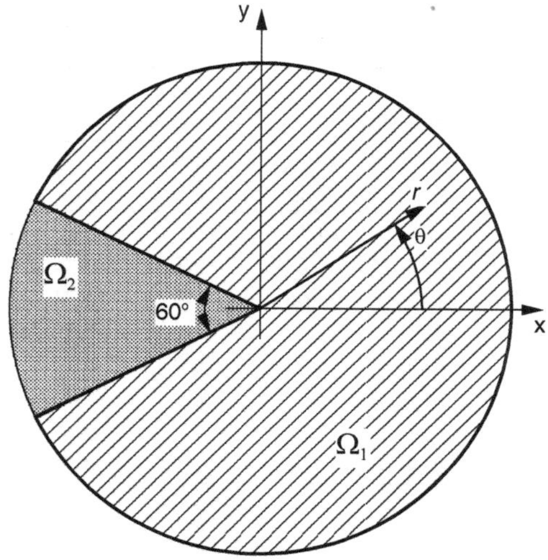

Fig. 7.7 Domain configuration of the inclusion problem.

7.2.4 Singular Points Associated with Multimaterial Interfaces

Composite bodies consisting of two dissimilar isotropic, homogeneous, and elastic wedges, perfectly bonded along all their interfaces (or some), are studied. Two examples are provided: an inclusion problem subjected to a temperature field that exhibits singular behavior of the temperature flux, and a body consisting of two dissimilar materials subjected to a uniform elevated temperature field.

7.2.4.1 An Inclusion Problem

Consider the unit circle domain Ω, divided into two sectors: Ω_1 occupying the sector $-5\pi/6 \leq \theta \leq 5\pi/6$ and Ω_2 occupying the sector $5\pi/6 \leq \theta \leq 7\pi/6$; see Figure 7.7. The heat conduction coefficients in Ω_1 are $k_{11} = k_{22} = 10$, $k_{12} = 0$, and in Ω_2 are $k_{11} = k_{22} = 1$, $k_{12} = 0$. A plane-strain condition is assumed with $\nu_1 = \nu_2 = 0.3$, $E_1 = 10$, $E_2 = 1$, and the coefficient of linear thermal expansion is $\alpha = 0.1$ in Ω_1 and $\alpha = 0.01$ in Ω_2.

The stress tensor in the domain is singular at $r = 0$ and can be written in the form

$$(\tilde{\sigma}_{ij}) = \frac{A_1}{\sqrt{2\pi}} r^{\alpha_1 - 1} \tilde{S}_{ij}^{(\mathrm{I})}(\theta) + \frac{A_2}{\sqrt{2\pi}} r^{\alpha_2 - 1} \tilde{S}_{ij}^{(\mathrm{II})}(\theta) + \mathcal{O}(r^{1+\epsilon}), \qquad ij = r, \theta, \tag{7.27}$$

where $\tilde{S}_{ij}^{(\mathrm{I})}(\theta)$ and $\tilde{S}_{ij}^{(\mathrm{II})}(\theta)$ are eigenstresses given by (see [36]):

$-5\pi/6 \leq \theta \leq 5\pi/6$:

$$\begin{aligned}
\tilde{S}_{rr}^{(\mathrm{I})}(\theta) &= -0.717604531\,\{0.401735588\cos[(1+\alpha_1)\theta] \\
&\qquad -1.561125474\cos[(\alpha_1-1)\theta]\}\,, \\
\tilde{S}_{rr}^{(\mathrm{II})}(\theta) &= -1.023570729\,\{-0.813197463\sin[(1+\alpha_2)\theta] \\
&\qquad -1.619121416\sin[(\alpha_2-1)\theta]\}\,, \\
\tilde{S}_{\theta\theta}^{(\mathrm{I})}(\theta) &= 0.717604531\,\{0.401735588\cos[(1+\alpha_1)\theta] \\
&\qquad +0.949198951\cos[(\alpha_1-1)\theta]\}\,, \\
\tilde{S}_{\theta\theta}^{(\mathrm{II})}(\theta) &= 1.023570729\,\{-0.813197463\sin[(1+\alpha_2)\theta] \\
&\qquad +1.235182867\sin[(\alpha_2-1)\theta]\}\,, \\
\tilde{S}_{r\theta}^{(\mathrm{I})}(\theta) &= 0.717604531\,\{0.401735588\sin[(1+\alpha_1)\theta] \\
&\qquad -0.305963261\sin[(\alpha_1-1)\theta]\}\,, \\
\tilde{S}_{r\theta}^{(\mathrm{II})}(\theta) &= -1.023570729\,\{-0.813197463\cos[(1+\alpha_2)\theta] \\
&\qquad -0.81397463\cos[(\alpha_2-1)\theta]\}\,,
\end{aligned}$$

$5\pi/6 \leq \theta \leq 7\pi/6$:

$$\begin{aligned}
\tilde{S}_{rr}^{(\mathrm{I})}(\theta) &= -\,0.000370516\,\{0.972611382\cos[(1+\alpha_1)(\pi-\theta)] \\
&\qquad -1.561125474\cos[(\alpha_1-1)(\pi-\theta)]\}\,, \\
\tilde{S}_{rr}^{(\mathrm{II})}(\theta) &= 0.000152306\,\{-1.240586478\sin[(1+\alpha_2)(\pi-\theta)] \\
&\qquad -1.619121416\sin[(\alpha_2-1)(\pi-\theta)]\}\,, \\
\tilde{S}_{\theta\theta}^{(\mathrm{I})}(\theta) &= 0.000370516\,\{0.972611382\cos[(1+\alpha_1)(\pi-\theta)] \\
&\qquad +0.949198951\cos[(\alpha_1-1)(\pi-\theta)]\}\,, \\
\tilde{S}_{\theta\theta}^{(\mathrm{II})}(\theta) &= -\,0.000152306\,\{-1.240586478\sin[(1+\alpha_2)(\pi-\theta)] \\
&\qquad +1.235182867\sin[(\alpha_2-1)(\pi-\theta)]\}\,, \\
\tilde{S}_{r\theta}^{(\mathrm{I})}(\theta) &= -\,0.000370516\,\{0.972611382\sin[(1+\alpha_1)(\pi-\theta)] \\
&\qquad -0.305963261\sin[(\alpha_1-1)(\pi-\theta)]\}\,, \\
\tilde{S}_{r\theta}^{(\mathrm{II})}(\theta) &= -\,0.000152306\,\{-1.240586478\cos[(1+\alpha_2)(\pi-\theta)] \\
&\qquad -0.191969274\cos[(\alpha_2-1)(\pi-\theta)]\}\,.
\end{aligned}$$

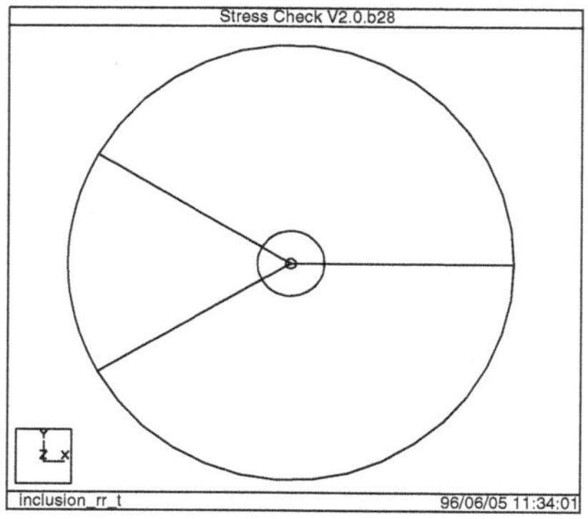

Fig. 7.8 Finite element mesh for inclusion problem.

The domain Ω is discretized by employing the finite element mesh shown in Figure 7.8, having several radial layers graded geometrically toward the singular point with a grading factor 0.15. The relative error in energy norm for the thermal analysis at $p = 8$ (1244 dof) is less than 0.4% and in the thermoelastic analysis (2474 dof) is less than 1.6%. Using two terms for spanning $\overset{0}{\mathcal{E}}{}^{H}_{c}$, $N = 2$, the error in the first two TGSIFs for any given radius R converges to zero as p is increased, with a relative error at $p = 8$ that is less than 0.01%.

The temperature distribution is first computed with the temperature boundary condition applied to the boundary of Ω_1:

$$\tau(\theta) = 100\frac{(5\pi/6 - |\theta|)}{5\pi/6}, \qquad -5\pi/6 \leq \theta \leq 5\pi/6,$$

and flux boundary condition on the boundary of Ω_2,

$$\frac{\partial\tau}{\partial r}(\theta) = \frac{(5\pi/6 - \theta)(5\pi/6 + \theta)}{(5\pi/6)^2}, \qquad -5\pi/6 \geq \theta \geq -\pi,\ \ 5\pi/6 \leq \theta \leq \pi.$$

The temperature distribution is then applied as a thermal load in the thermoelastic analysis, imposing clamped boundary conditions at $r = 1$:

$$\boldsymbol{u}(r = 1, \theta) = \boldsymbol{0}, \qquad -\pi \leq \theta \leq \pi.$$

For this problem we obtain $\alpha_1 = 0.6900333$ and $\alpha_2 = 0.7940938$, and since the temperature field in the vicinity of the singular point is of the form (7.1), with $\tau_0 \neq 0$, we use for Richardson's extrapolation the power $1-\alpha_1 = 0.3099667$ for A_1. Here A_2 is 0.00000000, and we report in Table 7.14 the normalized values of A_1 defined by: $A_1^* \overset{\text{def}}{=} -0.1040854 A_1$ obtained from the FE analysis corresponding to $p \to \infty$ and the Richardson's extrapolation. Unlike in [201], where this example problem demonstrated that no convergence was visible (without extrapolation) even

Table 7.14 A_1^* at various values of R, $p \to \infty$, for inclusion problem, and the extrapolated values as $R \to 0$.

R	$(A_1^*)_{FE} \overset{\text{def}}{=} (A_1^*)^{(0)}$	$(A_1^*)^{(1)}$	$(A_1^*)^{(2)}$	$(A_1^*)^{(3)}$	$(A_1^*)^{(4)}$	$(A_1^*)^{(5)}$
0.9	0.001209					
		0.055176				
0.5	0.010198		0.101188			
		0.077903		0.089350		
0.1	0.036791		0.094183		0.091774	
		0.086209		0.091173		0.091785
0.05	0.046346		0.092068		0.091782	
		0.089198		0.091636		
0.01	0.063177		0.091807			
		0.090529				
0.005	0.068466					

Table 7.15 Extrapolated A_1^* for different N's.

$N = 2$	$N = 4$	$N = 6$
0.091785	0.091510	0.091792

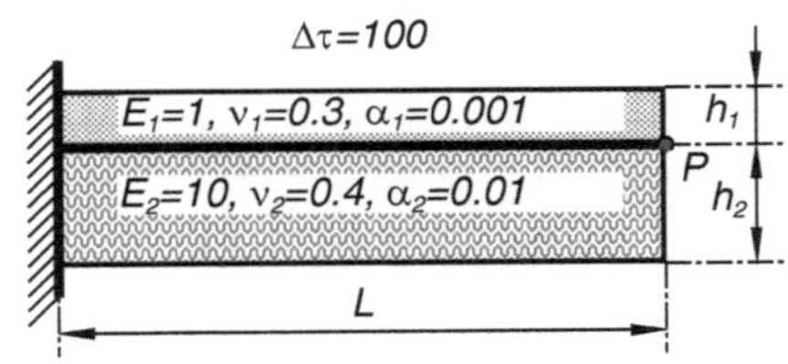

Fig. 7.9 Domain configuration of 90° dissimilar bonded wedges.

at a radius of $R = \mathcal{O}(10^{-5})$, here, the proposed extrapolation methodology provides good results with integration radii that are much larger.

To examine the influence of N on the extrapolated value of A_1^*, we summarize in Table 7.15 the extrapolated values obtained with $N = 2, 4, 6$.

This again demonstrates that A_1^* is virtually independent of N.

7.2.4.2 Two 90° Dissimilar Bonded Wedges

Two isotropic, homogeneous rectangular blocks of length $L = 10$, $L/h_1 = 10$, $L/h_2 = 5$, having different material properties are bonded together and clamped at their left boundary as shown in Figure 7.9. Starting from a uniform reference temperature, the body is heated uniformly by $\Delta\tau = 100$. Due to mismatch of the coefficients of thermal expansion of the two materials, the thermoelastic stress field is singular at point P (there are other singular points at the left boundary that are not of primary interest). Under the assumption of plane stress, a thermoelastic finite element analysis was performed, imposing a uniform temperature field over the domain. The domain is discretized by employing the finite element mesh shown in

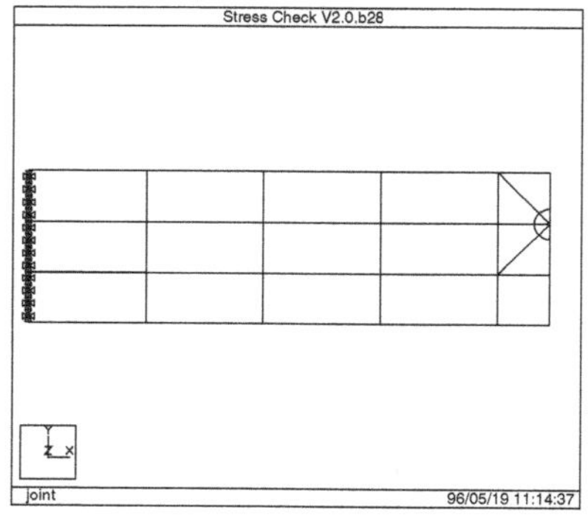

Fig. 7.10 Finite element mesh for 90° dissimilar bonded wedges.

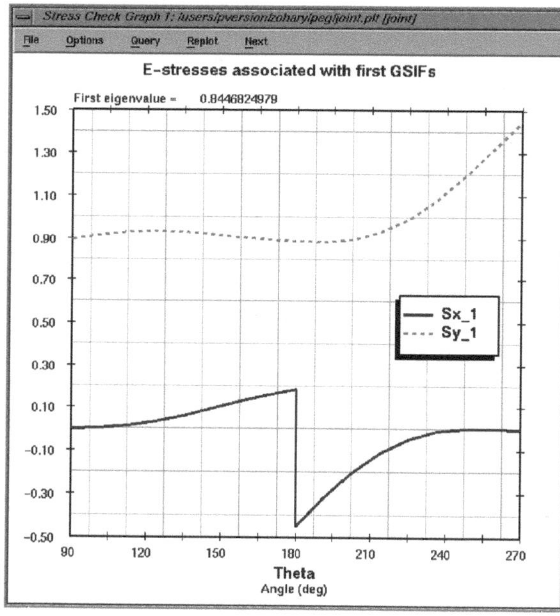

Fig. 7.11 $(S_{11})_1(\theta)$ and $(S_{22})_1(\theta)$ for 90° dissimilar bonded wedges.

Figure 7.10, having several radial layers graded geometrically toward the singular point P with a grading factor 0.15.

The eigenvalues and eigenstresses are extracted by the modified Steklov method. Only the first eigenvalue associated with point P, $\alpha_1 = 0.8446825$, imposes a weak singularity ($\alpha_2 \geq 1$), and the x and y components of the homogeneous first eigenstress vector (i.e., $(S_{11})_1(\theta)$ and $(S_{22})_1(\theta)$) are given in Figure 7.11. A uniform temperature distribution over the whole domain, not being influenced by the presence of the singular point P, results in a particular stress field as presented in (7.11). Thus, we use for Richardson's extrapolation the power $1 - \alpha_1 = 0.1553175$ for finding A_1.

Table 7.16 A_1 at various values of R, $p \to \infty$, for 90° dissimilar bonded wedges, and the extrapolated values as $R \to 0$.

R	$(A_1)_{FE} \stackrel{\text{def}}{=} (A_1)^{(0)}$	$(A_1)^{(1)}$	$(A_1)^{(2)}$	$(A_1)^{(3)}$
0.01	1.90078442			
		3.00341515		
0.005	2.01332472		3.04267731	
		3.01209911		3.00248408
0.002	2.14581500		3.03059243	
		3.01618944		
0.001	2.23465000			

In Table 7.16 the values of A_1 obtained from the FE analysis with $N = 2$, corresponding to $p \to \infty$ and the Richardson's extrapolation, are summarized. Comparing the extrapolated value of A_1 with these obtained at any finite R, we observe that the value obtained even at $R = 0.01$ is off by more than 27%.

The same problem was also considered by Bank-Sills and Ishbir [17] by a conservative M-integral. It may be noted that the eigenfunctions in [17] and here are normalized differently, so that a factor of 0.89 should be applied to our results to be compared to those in [17]. After applying this factor, the difference between the results is about 0.2%.

Chapter 8
Failure Criteria for Brittle Elastic Materials

The successful use of linear elastic fracture mechanics theory in predicting brittle fracture in isotropic domains with *cracks* is attributed to the successful correlation of a single parameter, namely the stress intensity factor, with experimental observations for the determination of failure initiation or crack propagation. For cracks in two-dimensional domains made of isotropic materials, catastrophic fracture is associated with the first coefficient in the expansion of the displacements/stress field in the vicinity of a crack tip. Usually "mode I" SIF, $K_{\rm I} = A_1 \times \sqrt{2\pi}$, determines the onset of fracture, i.e., when it equals the *fracture toughness* ($K_{\rm I\it c}$ a material-dependent parameter), fracture occurs. This criterion was first suggested by Irwin [87]. Typical values of K_{Ic} for "metals" and other brittle materials are given in Table 8.1. The feasibility of using the single parameter, i.e., check whether $K_{\rm I} > K_{\rm I\it c}$, to determine the onset of failure is a result of the universal nature of the stress tensor in the vicinity of the crack tip. For crack tips under mode I loading, a duality exists between the Irwin criterion and the Griffith criterion. The later is based on a critical value $\mathcal{G}_c$ of the energy release rate $\mathcal{G}$ defined as the derivative of the potential energy with respect to the crack length [70] (see also Appendix F).

Recently, failure laws for two-dimensional domains containing V-notches, multimaterial interfaces, or orthotropic materials have attracted major interest because of the relation to composite materials and electronic devices. A sufficient simple and reliable condition for predicting failure initiation (crack formation) in the above mentioned-cases, involving singular points, is still a topic of active research and interest. In such points the stress tensor is infinite under the assumption of linear elasticity. A typical example of a singular point is the reentrant V-notch tip, for which a crack tip is a particular case when the V-notch opening angle is 2π.

For V-notches and multimaterial interfaces, a considerable amount of research activity has been recently conducted for establishing a failure criterion applicable to brittle materials. The works of Dunn et al. [58,59] provide experimental correlation of A_1 (and possibly A_2), see (5.53)-(5.52), to fracture initiation in the case that the first two eigenvalues $\alpha_1^{(\rm I)}$ and $\alpha_1^{(\rm II)}$ are real and the singular points are V-notches in isotropic materials. Hattori et al. [80,81] propose a two-parameter failure law based

Z. Yosibash, *Singularities in Elliptic Boundary Value Problems and Elasticity and Their Connection with Failure Initiation*, Interdisciplinary Applied Mathematics 37,
DOI 10.1007/978-1-4614-1508-4_8,

Table 8.1 Typical K_{Ic} data at room temperature.

Material	K_{Ic} MPa$\sqrt{\text{mm}}$	K_{Ic} Ksi$\sqrt{\text{in}}$
PMMA	32.5	9.3
Alumina-7%Zirconia	129.6	37
4340 steel	150	43
6Al-4V titanium	122	35
7075-T651 aluminum	94	27

on A_1 and the exponent $\alpha_1^{(I)}$ for V-notch configurations, and demonstrated that a good correlation is achieved with experimental results. Reedy et al. in [146] and the references therein correlated failures of adhesive-bonded butt tensile joints with A_1. In all these cases a V-notch configuration is addressed, and no special attention is devoted to the connection of A_i, α_i, and $s^{(i)}(\theta)$. Novozhilov [130] proposed a simple failure criterion based on the average normal stress along the anticipated path of the crack formation. The validity of Novozhilov's criterion has been examined by Seweryn [158] by experiments performed on V-notch samples. Leguillon [106, 107] proposed a criterion for failure initiation at a sharp V-notch based on a combination of the Griffith energy criterion for a crack, and the strength criterion for a straight edge. This approach provides a criterion similar to Novozhilov's criterion, and shows good agreement with experimental observations in [59]. A recent work by Seweryn and Lukaszewicz [159] addresses some of the failure criteria in the vicinity of a V-notch tip under mixed mode loading. Here, we examine two of the most promising failure criteria (and discuss them in more detail in the sequel), the one proposed by Leguillon based on the strain energy release rate and strength, and the one based on averaged stress by Novoshilov-Seweryn, and suggest a simplified failure criterion based on the strain-energy density. This criterion was presented in [3], called the SED criterion. It proposes as the failure criterion the critical value of the average strain energy in a sector in the vicinity of the singular point over the volume of this sector. The same criterion has also been studied by Lazzarin and Zambardi in [101], there called the "finite-volume-energy" criterion, and good correlation to experimental data of other publications is demonstrated. The SED is rigorously treated here from the mathematical point of view, bringing its formulation to a contour integral using the expansion in (5.53), and validated by our own experiments. Although we call this failure criterion the SED criterion, it is a different criterion from the SED criterion of Sih (see, e.g., [162]), as will be explained in Section 8.1.4. Other failure criteria have been proposed, among them the "theory of critical distances" by Taylor [182] (which is not discussed here) and the "cohesive process zone model" by Gomez and Elices [66, 67] which allows extension of classical methods, based on linear elastic fracture mechanics, to rounded V-notches and contained plasticity. The cohesive process zone model's predictive capability has been demonstrated by many experimental results on brittle materials containing V- and U-notches.

Cases when α_1 is complex are not addressed in this chapter. The complex representation of the displacement field in a neighborhood of a general two-dimensional

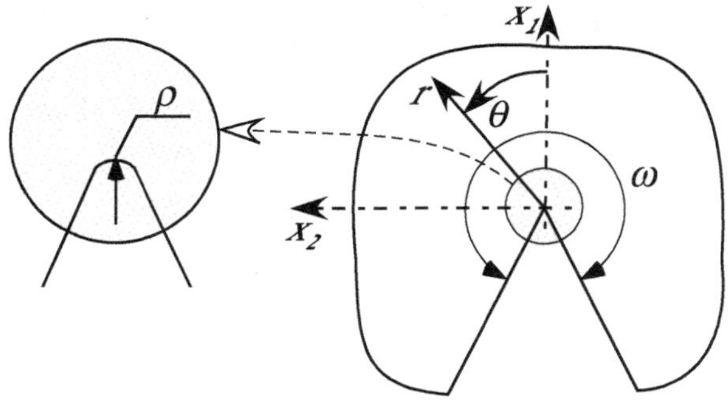

Fig. 8.1 The singular point and notation.

singular point poses several difficulties in establishing a failure law. Therefore, it is desirable that any proposed failure criterion have the following properties:

(a) Independence of the units used. That is, given the criterion for two different opening angles (ω_1 and ω_2) in a unit system, let us say F_1 and F_2, then a change of units must not change the ratio F_1/F_2.
(b) Unique applicability to real and complex eigenvalues (as in the case of cracks in a bimaterial interface).
(c) Unique applicability to single and mixed-mode loading.
(d) Degeneration to known failure criteria for cracks when the reentrant V-notch angle is 2π, and to strength criteria for a straight edge (V-notch angle of π).

We review four failure initiation criteria and compare them, including their advantages and drawbacks. Specifically, we formulate and discuss in detail the "strain-energy density" (SED) failure criterion. The validity of the various criteria is investigated by comparison to experimental observations. Sets of experiments performed on composite ceramic Alumina-7%Zirconia, and Poly-Methyl-Methacrylate (PMMA known as Plexiglas) V-notched specimens are summarized in Section 8.2. These mimic 2-D domains under plane-strain conditions, made of brittle materials, and loaded so as to produce a "mode I" stress field in a neighborhood of the singular point. The necessary information for the various failure criteria is extracted from p-finite element analysis simulating the experimental data, and documented in Section 8.3. Using the extracted information, the validity of the various failure laws in predicting the experimental observations is investigated.

Since in reality, no V-notch tip is sharp, but has a rounded (finite) tip, we examined quantitatively the influence of the V-notch tip radius on failure initiation in [110, 144]. We also extended the failure criteria to mixed mode loadings in [143, 207]. In Section 8.4 we present a practical application of the failure criteria for an easy estimation of the fracture toughness, but first let us consider the 2-D domain having a V-notch reentrant corner as described in Figure 8.1, assuming that the V-notch tip is sharp, i.e., $\rho = 0$ (this assumption is removed in [110, 144] and in Section 8.4).

8.1 On Failure Criteria Under Mode I Loading

Four failure criteria applicable to *sharp* V-notches subject to mode I loading in isotropic materials are described in this section.

8.1.1 Novozhilov-Seweryn Criterion

The failure criterion proposed by Novozhilov [130] and expanded by Seweryn [158] suggests that one consider the average normal stress along the anticipated path of the failure. Failure occurs when the average stress equals a material-dependent value, denoted by σ_c, which is the stress at failure without the presence of a notch. A characteristic length scale is introduced, denoted by d_0 (independent of ω), along which the average stress is considered. Let us assume that the failure will occur along axis x_1 in Figure 8.1. The average normal stress to axis x_1 is σ_{22} given in (1.57). Integrating along a distance d_0, the average stress is defined by

$$\begin{aligned}\sigma_c &= \frac{1}{d_0}\int_0^{d_0}\sum_{i=1}^{\infty} A_i\, r^{\alpha_i-1} S_{22}^{(i)}(\theta=0^\circ)dx_1 \\ &= \sum_{1=1}^{\infty}\frac{A_1}{\alpha_i}d_0^{\alpha_i-1}\; S_{22}^{(i)}(\theta=0^\circ)\quad (x_1 \equiv r \text{ along } x_2) \qquad (8.1)\end{aligned}$$

Assume that $d_0 \ll 1$. Then all terms for $i \geq 2$ are negligible in comparison with the first term in the series, so that (8.1) simplifies to:

$$\sigma_c = \frac{A_1}{\alpha_1}d_0^{\alpha_1-1}\; S_{22}^{(1)}(\theta=0^\circ). \qquad (8.2)$$

For the well-known particular case of a crack, failure occurs when

$$A_1 S_{22}^{(1)}(\theta=0^\circ) = \frac{K_{Ic}}{\sqrt{2\pi}}, \qquad (8.3)$$

where K_{Ic} is the fracture toughness. Eliminating A_1 from (8.3), and substituting $\alpha_1 = 1/2$ in (8.2) for a crack, one obtains

$$d_0 = \frac{2}{\pi}\frac{K_{Ic}^2}{\sigma_c^2}. \qquad (8.4)$$

Returning to (8.2) and substituting d_0 from (8.4), we finally obtain the Novoshilov-Seweryn failure criterion stating that failure occurs at the instant when

Table 8.2 Values of $(2\pi)^{1-\alpha} A_1 S_{22}^{(1)}(\theta = 0°)$ [MPa m$^{1-\alpha}$] and K_{I_c} [MPa $\sqrt{m}$] computed by (8.6) from [158].

ω	360°	320°	300°	280°	260°	240°	220°	200°
Plexiglas								
Computed K_{Ic}	1.859	1.789	1.960	1.892	1.752	1.561	1.594	2.493
$(2\pi)^{1-\alpha} A_1 S_{22}^{(1)}(\theta = 0°)$	1.866	1.851	2.167	2.436	3.059	4.347	8.861	28.60
Duralumin								
Computed K_{Ic}	53.39	55.85	56.67	56.24	56.10	53.40	49.62	53.18
$(2\pi)^{1-\alpha} A_1 S_{22}^{(1)}(\theta = 0°)$	53.51	57.10	60.53	66.34	80.15	102.00	150.44	291.81

$$A_1 S_{22}^{(1)}(\theta = 0°) = \alpha_1 \sigma_c \left(\frac{2}{\sqrt{2\pi}} \frac{K_{Ic}}{\sigma_c} \right)^{2-2\alpha_1}. \tag{8.5}$$

We call the value $k_c \stackrel{\text{def}}{=} A_1 S_{22}^{(1)}(\theta = 0°)$ the critical generalized stress intensity factor (GSIF$_c$).

The criterion proposed by Novozhilov [130] has the advantage of reducing to the well-known classical failure criterion for cracks, when $\alpha_1 = 1/2$

$$A_1 S_{22}^{(1)}(\theta = 0°) = \frac{K_{Ic}}{\sqrt{2\pi}},$$

and the usual strength criterion for a straight edge (when $\omega = \pi$, so that $\alpha_1 = 1$),

$$A_1 S_{22}^{(1)}(\theta = 0°) = \sigma_c.$$

Seweryn [158] examined the validity of Novozhilov's failure criterion by performing experiments on V-notch samples made of Plexiglas and Duralumin having V-notch angles of $\omega = 2\pi$ to π, with a tip radius $\rho = 0.01$ mm. Using the critical load at failure, the generalized stress intensity factor at failure can be computed, and by using (8.5), one can predict the fracture toughness

$$K_{Ic} = \frac{\sqrt{2\pi}\sigma_c}{2} \left(\frac{A_1 S_{22}^{(1)}(\theta = 0°)}{\alpha_1 \sigma_c} \right)^{\frac{1}{2-2\alpha_1}}. \tag{8.6}$$

Because the K_{Ic} value is not known, it has been computed from the results obtained for the various opening angles. Table 8.2 summarizes the results reported in [158]. As can be observed in Table 8.2, as the opening angle ω decreases, the criterion's validity deteriorates. Recently, the Novoshilov-Seweryn failure criterion has been extended to mixed mode failure by Seweryn and Lukaszewicz in [159], where its validity compared to other failure criteria in predicting the failure load and direction is demonstrated by comparison to experimental observations.

Table 8.3 Values of $K(\omega)$ as reported in [107].

ω	360°	330°	315°	300°	270°	240°	210°	195°	180°
$K(\omega)$	0.00248	0.00243	0.00242	0.00237	0.00212	0.00176	0.00128	0.00098	0.00069

8.1.2 Leguillon's Criterion

Leguillon [106, 107] proposed a criterion for failure initiation at a sharp V-notch based on a combination of the Griffith energy criterion for a crack and the strength criterion for a straight edge. This approach is based on the change of the potential energy in a notched specimen due to a virtual creation of a small crack in the direction that generates maximum change in potential energy. Here as well, a characteristic length scale is introduced, which is the length of the created crack ℓ_0:

$$\ell_0 = \frac{\left(S^{(1)}_{\theta\theta}(\theta_0)\right)^2}{K(\omega)} \frac{K^2_{I_c}}{\sigma^2_c}. \tag{8.7}$$

For a V-notch in an isotropic material under symmetric loading, the *critical material-dependent parameter* k_c is given by

$$k_c \stackrel{\text{def}}{=} A_1 S^{(1)}_{\theta\theta}(\theta = 0°) = \left(\frac{\mathcal{G}_c}{K(\omega)}\right)^{1-\alpha_1} \sigma_c^{2\alpha_1 - 1}, \tag{8.8}$$

where $\mathcal{G}_c$ is the fracture energy per unit surface and σ_c is the 1-D stress at brittle failure (*strength*), both being material properties. The parameter $K(\omega)$ depends on the local geometry and boundary conditions in a neighborhood of the V-notch tip, the eigenvalue α and its corresponding eigenfunction, and the material properties (E and ν in isotropic materials). It is important to realize that $K(\omega)$ is not the generalized stress intensity factor for the V-notch, but is computed by an integration procedure as detailed in [144] and the appendix of [107]. For example, in Table 2 in [107], the following values of $K(\omega)$ for a V-notch in PMMA, which is an isotropic homogeneous material with $E = 2.3$ GPa and $\nu = 0.36$, are given; see Table 8.3.

Based on Table 8.3 and the expression for evaluating $K(\omega)$, for any traction-free reentrant V-notch configuration in an isotropic homogeneous material with new material properties E^{new} and ν^{new} the new values of $K(\omega)$ may be easily obtained:

$$K^{\text{new}}(\omega) = K(\omega)\frac{2.3}{1 - 0.36^2}\frac{1 - (\nu^{\text{new}})^2}{E^{\text{new}}}, \qquad E^{\text{new}} \text{ in GPa}. \tag{8.9}$$

For example, for Alumina-7%Zirconia with $E \approx 360$ GPa and $\nu = 0.23$, the new values for $K(\omega)$ are given in Table 8.4.

Correlation of the current criterion with experimental observations in PMMA V-notched specimens as well as in bimaterial wedges shows good agreement.

Table 8.4 Values of $K(\omega)$ for Alumina-7%Zirconia.

ω	330°	300°	270°	240°
$K(\omega)$	1.68683E-05	1.64518E-05	1.47164E-05	1.22174E-05

8.1.3 Dunn et al. Criterion

Dunn et al. [59] proposed use of the GSIF at the instance of failure (named k_c) as the single parameter to be correlated to failures. Values of k_c were obtained at failure using experiments done on PMMA specimens [59] and single-crystal silicon [175]. This method requires the evaluation of k_c for each V-notch opening angle. Furthermore, its applicability for large opening angle is questionable (as $\omega \rightarrow \pi$, approaching a straight edge, the eigenstresses tend to be constant; thus the GSIF is meaningless).

Although the addressed criteria provide good correlation to experimental observation, there are some difficulties in applying them because:

1. The units of the critical "stress intensity factor" are somewhat entangled.
2. It is difficult to generalize the methods to a mixed-mode loading.
3. The critical stress intensity factor (K_{Ic}) for the material of interest has to be known.
4. A fracture stress σ_c has to be assumed. This value may be taken as σ_Y (yield stress), and for brittle materials it is *supposed* to be also the stress at fracture. However, even for brittle materials, the stress at fracture is higher than the conventional definition of σ_Y.

To overcome these difficulties, a simpler failure criterion is described in the next subsection.

8.1.4 The Strain Energy Density (SED) Criterion

It is conceivable to assume that failure initiates when the average elastic strain energy contained in a sector having the singular point as its center, over the volume of this sector, reaches a critical value. This averaged elastic strain energy density, which we call the strain energy density (SED) [206] and Lazzarin and Zambardi [101] call the by finite-volume energy, reminds the well-known SED criterion of Sih and Macdonald [162]. However these are considerably different in several respects. The SED of Sih is a *pointwise value* evaluated at any point on an arc located at a radius R away from the crack tip and is usually applied to crack tip singularities. Because it is a function of θ, a minimum value of Sih's SED can be found at a given angle θ_c. Thus, Sih's SED may be used as a criterion for predicting the crack propagation direction, as well as a failure criterion.

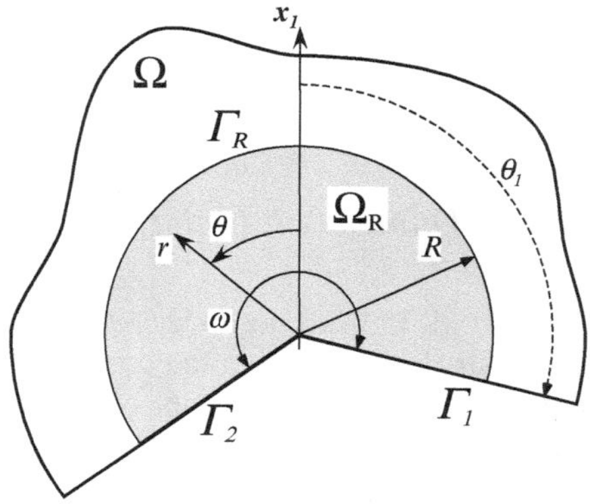

Fig. 8.2 The SED domain of interest and notation.

This pointwise minimum, correlated to a critical material-dependent parameter, is the failure criterion. The SED failure criterion proposed herein is an *avaraged value*, it is not aimed at predicting directions of crack propagation, but at predicting failure initiation at a specific critical value independent of the opening angle of the V-notch tip.

Consider the "usual" circular sector Ω_R of radius R centered at the singular point,

$$\Omega_R \stackrel{\text{def}}{=} \{(r,\theta)|0 \le r \le R,\ \theta_1 \le \theta \le \theta_1 + \omega\},$$

with traction free boundary conditions on the faces intersecting at the singular point. See Figure 8.2.

The elastic strain energy $\mathcal{U}(\boldsymbol{u})[R]$ in a 2-D domain of constant thickness b under the assumption of plane-strain is defined as

$$\mathcal{U}(\boldsymbol{u})[R] \stackrel{\text{def}}{=} \tfrac{1}{2} b \iint_{\Omega_R} \sigma_{\beta\gamma}\varepsilon_{\beta\gamma}\, d\Omega, \tag{8.10}$$

(one should keep in mind that summation notation is implied unless otherwise specified). For an isotropic material under the plane-strain assumption, Hooke's law is

$$\sigma_{\beta\gamma} = \lambda\delta_{\beta\gamma}\varepsilon_{\xi\xi} + 2\mu\varepsilon_{\beta\gamma}. \tag{8.11}$$

Substituting (8.11) in (8.10), we use the kinematic connections between strains and displacements $\varepsilon_{\beta\gamma} = \frac{1}{2}(\partial_\beta u_\gamma + \partial_\gamma u_\beta)$. Then using Green's theorem, we transform the area integral into a boundary integral on $\partial\Omega_R$

$$\mathcal{U}(\boldsymbol{u})[R] = \tfrac{1}{2} b \int_{\partial\Omega_R} \left(2\mu\varepsilon_{\beta\gamma} + \lambda\varepsilon_{\xi\xi}\delta_{\beta\gamma}\right) n_\gamma u_\beta d\Gamma. \tag{8.12}$$

Here n_γ is the γth component of the outward normal vector to the boundary $\partial\Omega_R$. Along the two straight lines Γ_1 and Γ_2 this integral is zero because of the traction-free boundary conditions. Reusing the strain-stress connection, we finally express the strain energy in Ω_R by a 1-D integral

$$\mathcal{U}(\boldsymbol{u})[R] = \frac{1}{2} b \int_{\theta_1}^{\theta_1+\omega} \left[\sigma_{\beta\gamma} n_\gamma u_\beta\right]_{r=R} R\, d\theta. \tag{8.13}$$

On Γ_R, the outward normal vector is $(\cos\theta,\ \sin\theta)$, so (8.13) becomes

$$\mathcal{U}(\boldsymbol{u})[R] = \frac{1}{2} b \sum_{k,\ell} A_k A_\ell R^{\alpha_k+\alpha_\ell} \int_{\theta_1}^{\theta_1+\omega} \Big[S_{11}^{(k)}(\theta) s_1^{(\ell)}(\theta) \cos\theta$$
$$+ S_{12}^{(k)}(\theta) \left(s_1^{(\ell)}(\theta) \sin\theta + s_2^{(\ell)}(\theta) \cos\theta \right) + S_{22}^{(k)}(\theta) s_2^{(\ell)}(\theta) \sin\theta \Big] d\theta. \tag{8.14}$$

For isotropic materials with traction-free boundary conditions in a neighborhood of the singular point, the following orthogonality holds:

$$\int_{\theta_1}^{\theta_1+\omega} \left[S_{11}^{(k)} s_1^{(\ell)} \cos\theta + S_{12}^{(k)} \left(s_1^{(\ell)} \sin\theta + s_2^{(\ell)} \cos\theta \right) + S_{22}^{(k)} s_2^{(\ell)} \sin\theta \right] d\theta = 0$$
$$\text{for} \quad k \neq \ell, \tag{8.15}$$

which simplifies (8.14) to

$$\mathcal{U}(\boldsymbol{u})[R] = \tfrac{1}{2} b \sum_k A_k^2 R^{2\alpha_k} \int_{\theta_1}^{\theta_1+\omega} \Big[S_{11}^{(k)} s_1^{(k)} \cos\theta + S_{12}^{(k)} \left(s_1^{(k)} \sin\theta + s_2^{(k)} \cos\theta \right)$$
$$+ S_{22}^{(k)} s_2^{(k)} \sin\theta \Big] d\theta. \tag{8.16}$$

Problem 8.1. Demonstrate that (8.15) is valid for the following example: consider a two-dimensional isotropic domain in a state of plane-strain containing a V-notch with a solid angle $\omega = 3\pi/2$ such that $\theta_1 = -\omega/2$ and $\theta_1 + \omega = \omega/2$. For this case the stresses and displacements are explicitly given by (5.52)-(5.53) with the first two eigenvalues $\alpha_1^{(\mathrm{I})} = 0.5444837$ and $\alpha_1^{(\mathrm{II})} = 0.9085292$. Taking $\boldsymbol{S}^{(\mathrm{I})}(\theta)$ corresponding to $\alpha_1^{(\mathrm{I})} = 0.5444837$ and $\boldsymbol{s}^{(\mathrm{II})}(\theta)$ corresponding to $\alpha_1^{(\mathrm{II})} = 0.9085292$, show that the integral in (8.15) is zero.

We define the strain energy density (SED) as

$$\mathrm{SED}[R] \stackrel{\text{def}}{=} \frac{\mathcal{U}(\boldsymbol{u})[R]}{b \times \Omega_R}.$$

Using (8.16), we finally obtain:

$$\mathrm{SED}[R] = \frac{2}{\omega}\sum_k A_k^2 R^{2\alpha_k-2}\int_{\theta_1}^{\theta_1+\omega}\left[S_{11}^{(k)}s_1^{(k)}\cos\theta + S_{12}^{(k)}\left(s_1^{(k)}\sin\theta + s_2^{(k)}\cos\theta\right)\right.$$
$$\left. + S_{22}^{(k)}s_2^{(k)}\sin\theta\right]d\theta. \tag{8.17}$$

SED[R] depends of course on a typical length size R, and it should be small enough that Ω_R is within the K-dominance region, ensuring that the singular terms represent the exact solution. To ensure that this holds, $\mathcal{U}(\boldsymbol{u})[R]$ is computed first using (8.16), followed by a second computation using the stress and strain tensors according to (8.10). If the two computations provide different results, the domain Ω_R is too large, and only one term in the asymptotic expansion does not represent well the stress field within the sector of radius R. On the other hand, R should be large enough so that it is large compared to the plastic radius r_p and the V-notch tip radius ρ.

Of course, the value of SED[R] has to be in the range of the two extremes obtained for $\omega = 2\pi$ (a crack) and $\omega = \pi$ (a straight edge). The expressions for SED$[R]_{\mathrm{straight}}$ and SED$[\mathrm{R}]_{\mathrm{crack}}$ in terms of $K_{\mathrm{I}c}$ and σ_c are derived next.

8.1.4.1 Computation of the Critical SED$[R]_{\mathrm{crack}}$ for a Crack and SED$[R]_{\mathrm{straight}}$ for a Straight Edge, and the Material Characteristic Integration Radius R_{mat}

The two extreme values of the critical SED[R] are obtained for the case of a crack ($\omega = 360°$) and a straight edge ($\omega = 180°$). We first derive these values for an isotropic material, under mode I loading, and then use these equations to determine a material characteristic integration radius denoted by R_{mat}.

For a specimen with a straight edge, failure occurs at the instant when the remote uniaxial stress is equal to σ_c. In this case, the state of stresses at any point will be $\sigma_{22} = \sigma_c$, and other stress components are zero. The strain energy can be expressed as

$$\mathcal{U}(\boldsymbol{u})[R] = \tfrac{1}{2}b\int_{-\pi/2}^{\pi/2}\int_0^R \sigma_{\beta\gamma}\varepsilon_{\beta\gamma}\, r d\theta dr = \tfrac{1}{2}b\int_{-\pi/2}^{\pi/2}\int_0^R \sigma_c\frac{\sigma_c}{E} r d\theta dr = \frac{b}{4E}\pi R^2\sigma_c^2, \tag{8.18}$$

so that we finally obtain the upper limit to the SED

$$\mathrm{SED}[R]_{\mathrm{straight}} = \frac{\sigma_c^2}{2E}. \tag{8.19}$$

Under mode I loading, for a specimen containing a crack, the stress tensor at the instance of fracture in the vicinity of a crack tip is given by

$$\sigma_{11} = \frac{K_{\mathrm{Ic}}}{\sqrt{2\pi r}} \cos\frac{\theta}{2}\left[1 - \sin\frac{\theta}{2}\sin\frac{3\theta}{2}\right],$$
$$\sigma_{22} = \frac{K_{\mathrm{Ic}}}{\sqrt{2\pi r}} \cos\frac{\theta}{2}\left[1 + \sin\frac{\theta}{2}\sin\frac{3\theta}{2}\right],$$
$$\sigma_{12} = \frac{K_{\mathrm{Ic}}}{\sqrt{2\pi r}} \sin\frac{\theta}{2}\cos\frac{\theta}{2}\cos\frac{3\theta}{2}. \tag{8.20}$$

Expressing the strains in terms of stresses using the plane-strain constitutive law, (8.21) becomes

$$\mathcal{U}(\boldsymbol{u})[R] = \tfrac{1}{2}\frac{(1-\nu)b}{E}\int_{-\pi}^{\pi}\int_{0}^{R}\left\{(1+\nu)\left[\sigma_{11}^2 + \sigma_{22}^2\right] - 2\nu\sigma_{11}\sigma_{22} + 2\sigma_{12}^2\right\} r dr d\theta. \tag{8.21}$$

Substituting (8.20) in the expression of the strain energy (8.21) yields

$$\mathcal{U}(\boldsymbol{u})[R] = \frac{b(1+\nu)(5-8\nu)}{8E} K_{\mathrm{Ic}}^2 R, \tag{8.22}$$

and one finally obtains the lower limit to the SED

$$\mathrm{SED}[R]_{\mathrm{crack}} = \frac{(1+\nu)(5-8\nu)}{8\pi R E} K_{\mathrm{Ic}}^2. \tag{8.23}$$

The SED at failure is postulated to be a material property, therefore independent of ω, i.e., for $\omega = 2\pi$ and for $\omega = \pi$ one should obtain the same critical SED. Thus, by equating (8.19) with (8.23), one obtains

$$\frac{\sigma_c^2}{2E} = \frac{(1+\nu)(5-8\nu)}{8\pi R E} K_{\mathrm{Ic}}^2. \tag{8.24}$$

Equation (8.24) holds for a specific integration radius R_{mat}, which is given after trivial algebraic manipulation:

$$R_{\mathrm{mat}} = \frac{(1+\nu)(5-8\nu)}{4\pi}\left(\frac{K_{\mathrm{Ic}}}{\sigma_c}\right)^2. \tag{8.25}$$

This integration radius R_{mat} must of course be larger than the plastic radius and the V-notch tip radius ρ, and smaller than the K-dominance zone. In any case, given the value of $\mathrm{SED}[R_1]$, one can easily determine the value of SED for a domain having a different radius R_2 by the following simple equation derived from (8.17)

$$\mathrm{SED}[R_2] = \mathrm{SED}[R_1]\left(\frac{R_1}{R_2}\right)^{2-2\alpha_1}. \tag{8.26}$$

Finally, substituting (8.25) into (8.17), and in view of (8.19) we may state the SED failure law:

$$\frac{2}{\omega}\sum_k A_k^2 \left(\frac{(1+\nu)(5-8\nu)}{4\pi}\right)^{2\alpha_k-2} \left(\frac{K_{\mathrm{Ic}}}{\sigma_c}\right)^{4\alpha_k-4}$$
$$\times \int_{\theta_1}^{\theta_1+\omega} \left[S_{11}^{(k)} s_1^{(k)} \cos\theta + S_{12}^{(k)} \left(s_1^{(k)} \sin\theta + s_2^{(k)} \cos\theta\right)\right.$$
$$\left. + S_{22}^{(k)} s_2^{(k)} \sin\theta\right] d\theta \le \frac{\sigma_c^2}{2E}. \qquad (8.27)$$

8.2 Materials and Experimental Procedures

The validity of the various failure criteria has to be assessed by a set of experiments. This section presents a series of experiments that were performed on two kinds of brittle materials: the composite ceramic Alumina-7%Zirconia and the polymer PMMA, both having a linear elastic constitutive law. The tests were carried out on V-notched specimens loaded by three-point and four-point bending.

8.2.1 Experiments with Alumina-7%Zirconia

A set of experiments was performed on V-notched specimens under a tight control of the geometric dimensions (including the V-notch tip radius ρ). Over 70 specimens were considered with four V-notch opening angles $\omega = 330°, 300°, 270°$ and $240°$, each having three different tip radii $\rho = 0.03, 0.06$, and 0.1 mm. The specimens were loaded so to produce a pure mode I stress field in the vicinity of the V-notch tip. The geometry and the loading of the various Alumina-7%Zirconia specimens are presented in Figure 8.3. TPB (three-point bending) loading was also applied to some of the specimens where a single load was applied opposite to the V-Notch tip.

The notch length a was approximately 5 mm, and varies slightly from specimen to specimen, and $a_o \approx 2.5$ mm (see Figure 8.3) for the specimens with the double opening angles $\omega = 330°, 300°$. The precise dimensions for each specimen were measured and used later on for the computations. Some representative specimens with various V-notch opening angles are shown in Figure 8.4.

Physical properties, Young modulus E, and Poisson ratio ν were measured using ultrasonic techniques, while density was determined by conventional methods. Table 8.5 summarizes the values measured on a sample of the specimens, with $\nu = 0.236$ being obtained with minor changes in the third digit.

We used the values 357 or 350 GPa as the Young modulus (the value of 350 GPa has been assigned to the specimens for which we did not measure their material properties according to the value reported in the literature) with $\nu = 0.236$ in our analysis in Section 8.3.

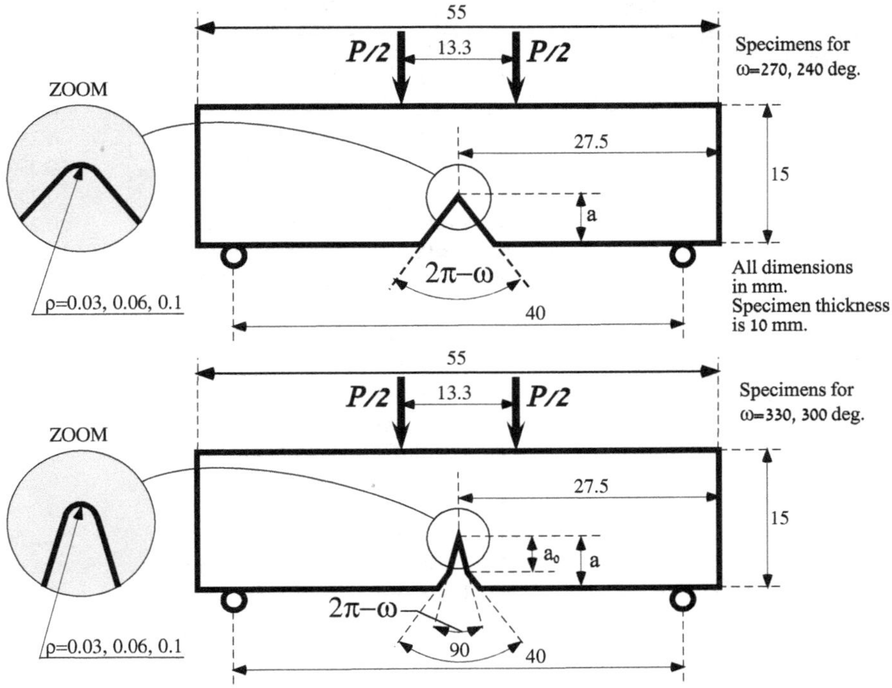

Fig. 8.3 Specimen geometry and loading configuration (FPB type) for the Alumina-7%Zirconia case.

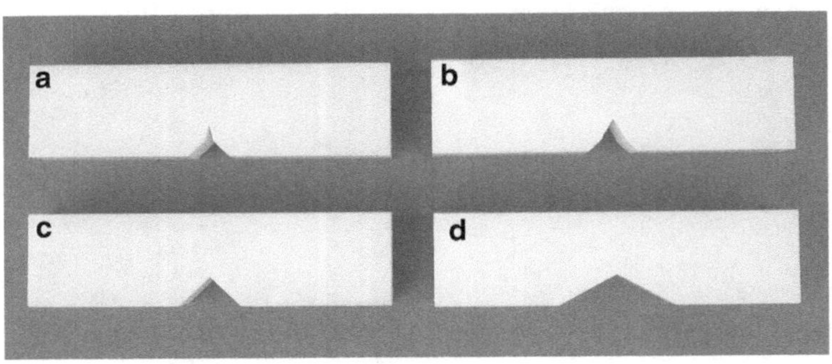

Fig. 8.4 Alumina-7%Zirconia specimens with various radii and notch angles; (a) 0.06mm, 330° (b) 0.1 mm, 300° (c) 0.1 mm, 270° (d) 0.1 mm, 240°.

The specimens were subject to a quasistatic loading (crosshead velocity was 0.5 mm/min) using a computerized MTS servo-hydraulic machine, with sensitive load cell of 10 kiloNewton full scale. V-notch opening displacement was measured by a crack opening displacement (COD) gauge (full scale of 0.25 mm), which was mounted at the V-notch intersection with the free edge. At microscopic scales, acoustic emission (AE) techniques were used to monitor events during loading from

Table 8.5 Measured E, density and ν for selected specimens

Specimen	E [GPa]	density [g/cm^3]	ν
30-FPB-0.1-1	356.52	3.963	0.235
30-FPB-0.1-2	356.89	3.965	0.236
30-FPB-0.1-3	355.30	3.962	0.233
30-TPB-0.1-4	357.90	3.970	0.235
30-FPB-0.03-4	360.24	3.975	0.239
30-FPB-0.03-5	359.85	3.975	0.236
30-FPB-0.03-6	357.31	3.976	0.233
60-FPB-0.1-1	354.00	3.963	0.235
60-FPB-0.1-2	367.31	3.965	0.238
60-FPB-0.1-3	369.53	3.962	0.238
60-FPB-0.1-4	355.20	3.970	0.235
60-FPB-0.03-2	357.71	3.964	0.238
60-FPB-0.03-3	356.50	3.968	0.234
90-FPB-0.1-5	355.50	3.956	0.235
90-FPB-0.1-6	358.70	3.958	0.235
90-FPB-0.1-7	356.06	3.951	0.235
90-FPB-0.03-3	354.50	3.959	0.239
90-FPB-0.03-4	355.40	3.959	0.239
90-FPB-0.03-5	338.50	3.898	0.235
90-FPB-0.03-6	349.70	3.940	0.238
Average	356.63	3.960	0.236

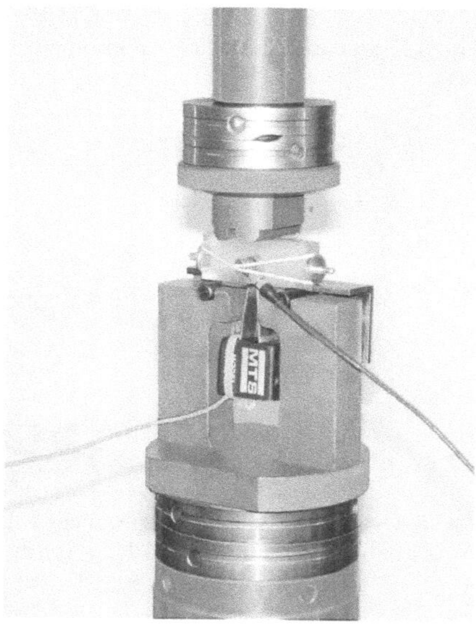

Fig. 8.5 Four-point-bending test fixture with acoustic emission transducer and crack opening displacement gauge (PMMA specimen).

undesired sources. It enables us to point out inhomogeneities in the microstructure such as high porosity/microcracks/impurities, and thus provide the means to exclude specimens with abnormal behavior (intense AE activity). The experimental setup including the AE sensor is illustrated in Figure 8.5.

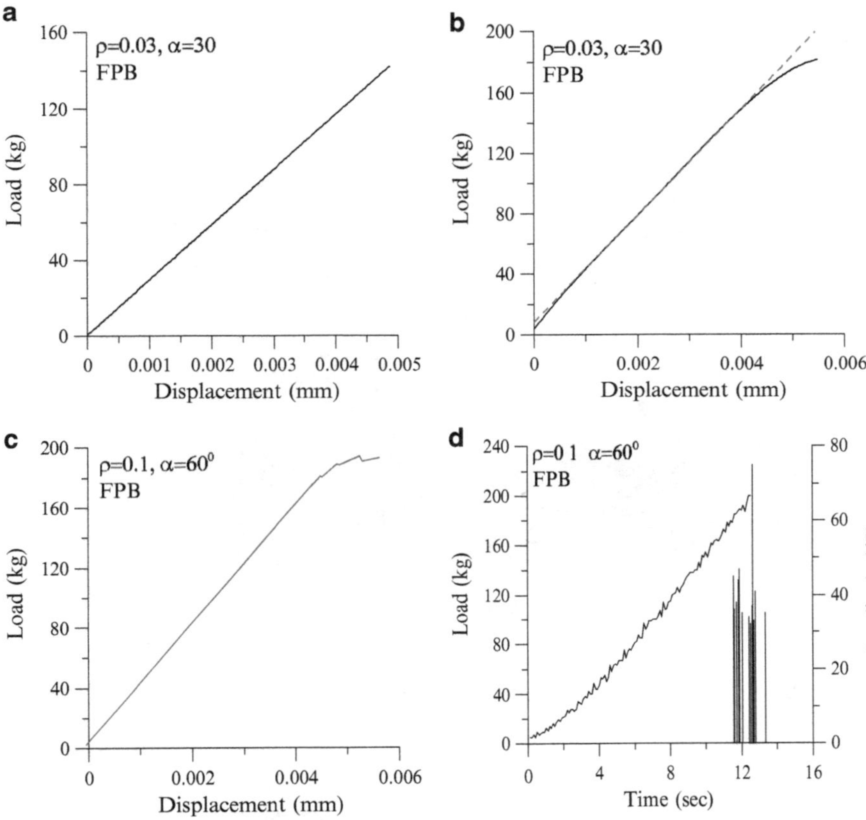

Fig. 8.6 Load displacement curves for linear-elastic behavior (a) and nonlinear response (b-c). (d) shows load and acoustic emission amplitude vs. time for the (c) case.

Examples of load versus displacement behavior obtained for typical specimens are shown in Figure 8.6. Specimens that exhibited a linear elastic response are illustrated in Figure 8.6(a), while specimens with nonlinear load displacement characterized by slow crack growth, both smooth and noncontinuous one are depicted in Figures 8.6(b-c) respectively. A case in which intense AE activity is pronounced in the later stage of loading is shown in Figure 8.6(d). Specimens that exhibited nonlinear load-displacement behavior also had large acoustic emission counts prior to failure, evidently due to impurities and microcracks. These specimens were excluded.

Details of V-notch angle and tip radius were documented optically for each specimen before and after fracture, as shown in Figure 8.7. This systematic procedure was done in order to eliminate specimens with macroscopic defects due to manufacturing problems (irregularities and non-symmetrical in the notch radius). In addition, cracks, that originated far from the notch root were also discarded from our calculations; see such an example in Figure 8.8 (a). In comparison, a crack that initiated close to the notch root is shown in Figure 8.8 (b).

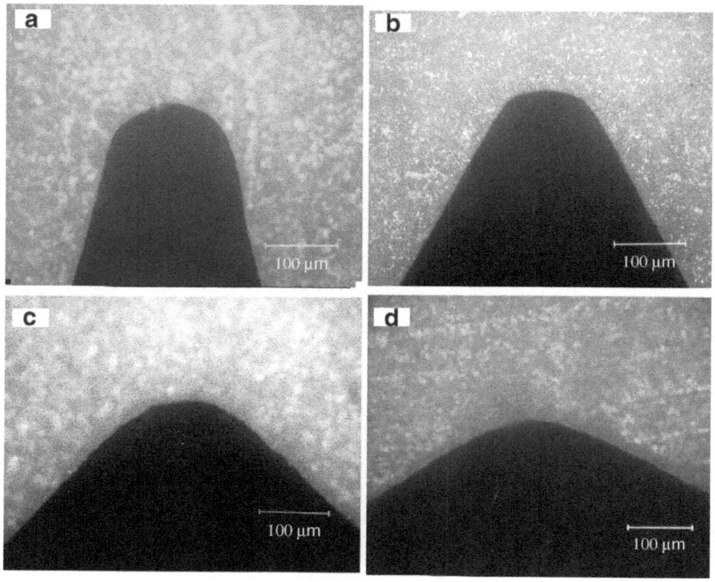

Fig. 8.7 The 0.1 mm notch tip radius profile for different notch angles; (a) 330°, (b) 300°, (c) 270°, (d) 240°.

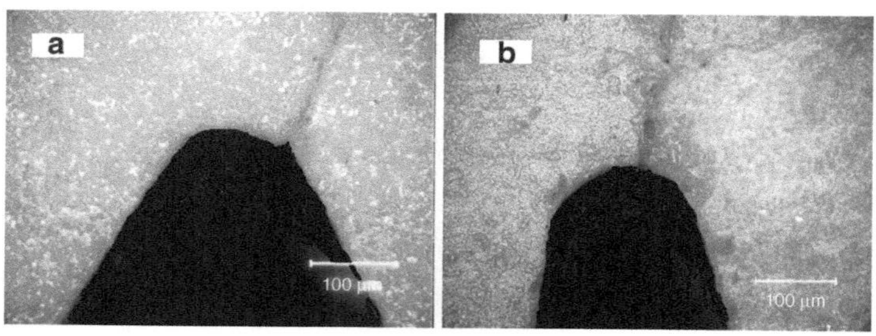

Fig. 8.8 Crack initiation at the edge (a), or at the notch root (b).

For the remaining specimens ("good results"), the values of the notch tip radius ρ, the fracture load P, and Young modulus E are listed in Table 8.6. The last four columns in the table are computed values addressed in the next section. Specimens denoted by "TPB" were loaded in three-point-bending mode, at the middle of the span.

8.2.2 Experiments with PMMA

Dunn et al. [59] carried out a set of experiments on 3PB notched PMMA specimens, with notch angles of 300°, 270° and 240° for various V-notched depths (a/h from

Table 8.6 Summary of the experiments for the "good Alumina-7%Zirconia specimens", and GSIFs at failure (k_c).

Specimen	ρ mm	P N	E GPa	SED[0.062 mm] $\frac{N}{m^2}$	$A_1 S_{22}^{(1)}(0°)$ Computed by (8.8)	$A_1 S_{22}^{(1)}(0°)$ Computed by (8.5) MPa $\times m^{1-\alpha_1}$	$(A_1)_{cr} S_{22}^{(1)}(0°)$ Experiments
$\omega = 330°, \alpha_1 = 0.501453$							
AO30003-16	0.031	1815	350	173547	1.643	1.662	1.942
30-TPB-0.03-3	0.040	1436	350	164253	1.643	1.662	1.863
AO30003-11	0.041	1628	350	137811	1.643	1.662	1.732
AO30006-13	0.060	1628	350	140944	1.643	1.662	1.751
30-TPB-0.1-4	0.060	1439	356.7	127912	1.627	1.662	1.690
30-TPB-0.1-5	0.060	1413	350	103800	1.643	1.662	1.509
AO3001-12	0.100	1844	350	179506	1.643	1.662	1.976
$\omega = 300°, \alpha_1 = 0.512221$							
15.4.02 - 11	0.060	1753	350	155625	1.859	1.870	2.010
15.4.02 - 12	0.060	1701	350	148996	1.859	1.870	1.967
15.4.02 - 15	0.060	1603	350	128253	1.859	1.870	1.826
15.4.02 - 18	0.060	1680	350	137102	1.859	1.870	1.888
A06001-12	0.100	1785	350	168763	1.859	1.870	2.095
A06001-13	0.100	1903	350	196644	1.859	1.870	2.262

(continued)

Table 8.6 (continued)

Specimen	ρ mm	P N	E GPa	SED[0.062 mm] $\frac{N}{m^2}$	$A_1 S_{22}^{(1)}(0°)$ Computed by (8.8)	$A_1 S_{22}^{(1)}(0°)$ Computed by (8.5) MPa $\times m^{1-\alpha_1}$	$(A_1)_{cr} S_{22}^{(1)}(0°)$ Experiments
90-FPB-0.03-5	0.035	1853	338.5	164962	2.774	2.655	2.844
90-TPB-0.03-7	0.050	1292	350	132235	2.732	2.655	2.596
90-TPB-0.06-1	0.060	1523	350	158707	2.732	2.655	2.842
90-TPB-0.06-2	0.060	1642	350	184986	2.732	2.655	3.071
90-TPB-0.06-3	0.067	1461	350	145375	2.732	2.655	2.722
90-FPB-0.1-5	0.100	2167	356.7	183066	2.708	2.655	3.077
90-FPB-0.1-6	0.100	2244	356.7	198267	2.708	2.655	3.201
90-TPB-0.1-3	0.100	1724	350	197814	2.732	2.655	3.176
$\omega = 240°, \alpha_1 = 0.6157311$							
120-TPB-0.03-5	0.062	1962	350	171724	6.087	5.688	6.329
120-TPB-0.06-1	0.080	1927	350	175779	6.087	5.688	6.403
120-TPB-0.06-2	0.080	1805	350	157698	6.087	5.688	6.065
120-TPB-0.06-3	0.080	1958	350	178687	6.087	5.688	6.456
120-FPB-0.1-3	0.100	2928	350	255128	6.087	5.688	7.705
120-FPB-0.1-5	0.100	2892	350	265188	6.087	5.688	7.853
120-TPB-0.1-6	0.100	2053	350	236806	6.087	5.688	7.435

Table 8.7 Summary of the experimental results for PMMA. Results for $\omega = 315°$ are from [206], whereas all others are from [59].

P N	SED[0.0158 mm] $\frac{N}{m^2}$	$A_1 S_{22}^{(1)}(0°)$ Computed by (8.8)	$A_1 S_{22}^{(1)}(0°)$ Computed by (8.5) MPa $\times m^{1-\alpha_1}$	$(A_1)_{cr} S_{22}^{(1)}(0°)$ Experiments
$\omega = 315°, \alpha_1 = 0.5050$				
376	5286756	0.427	0.432	0.586
379	5382800	0.427	0.432	0.591
$\omega = 300°, \alpha_1 = 0.5122$				
597	5390215	0.469	0.471	0.563
613	5683008	0.469	0.471	0.578
613	5683008	0.469	0.471	0.578
611	5645982	0.469	0.471	0.576
457	6164536	0.469	0.471	0.600
475	6659708	0.469	0.471	0.624
450	5977146	0.469	0.471	0.591
441	5740440	0.469	0.471	0.579
296	4311136	0.469	0.471	0.502
316	4913610	0.469	0.471	0.534
316	4913610	0.469	0.471	0.534
308	4667772	0.469	0.471	0.521
268	5919545	0.469	0.471	0.586
256	5401490	0.469	0.471	0.560
250	5151263	0.469	0.471	0.547
261	5614547	0.469	0.471	0.571

0.1 to 0.4) and V-notch tip radius less than 0.0254 mm. The material properties of the PMMA reported in [59] are $E = 2.3$ GPa and $\nu = 0.36$, with the failure stress being $\sigma_c = 124$ MPa. The failure load values are summarized in Tables 8.7 and 8.8 in the first column. These loads are for different V-notch depths, but using the load at failure and specific geometric dimensions, the GSIF at failure is computed and reported in the last column of the tables. No results for the angle $\omega = 315°$ are reported in [59]. We also tested similar PMMA specimens with $a/h = 0.235$ and V-notch tip radius 0.03 mm, which were loaded up to fracture in three-point bending. The results are reported in the first two rows of Table 8.7.

8.3 Verification and Validation of the Failure Criteria

To validate the various failure criteria, we constructed FE models of the various specimens tested, loaded by the load that caused the fracture. An example of the FE mesh for $\omega = 3\pi/4$ and the zoomed portion in a neighborhood of the notch tip is shown in Figure 8.9.

Table 8.8 Summary of the experimental results for PMMA from Dunn et. al [59] (continued).

P N	SED[0.0158 mm] $\frac{\text{N}}{\text{m}^2}$	$A_1 S_{22}^{(1)}(0°)$ Computed by (8.8)	$A_1 S_{22}^{(1)}(0°)$ Computed by (8.5) MPa $\times \text{m}^{1-\alpha_1}$	$(A_1)_{cr} S_{22}^{(1)}(0°)$ Experiments
$\omega = 270°, \alpha_1 = 0.5445$				
691	6369229	0.713	0.693	0.877
691	6369229	0.713	0.693	0.877
680	6168060	0.713	0.693	0.863
495	6072337	0.713	0.693	0.853
490	5950282	0.713	0.693	0.845
488	5901807	0.713	0.693	0.841
397	6355569	0.713	0.693	0.873
406	6646931	0.713	0.693	0.893
410	6778549	0.713	0.693	0.901
310	6431300	0.713	0.693	0.878
319	6810633	0.713	0.693	0.904
319	6810633	0.713	0.693	0.904
$\omega = 240°, \alpha_1 = 0.6157$				
875	6326762	1.717	1.604	2.011
886	6486834	1.717	1.604	2.036
886	6486834	1.717	1.604	2.036
880	6400874	1.717	1.604	2.022
652	5971698	1.717	1.604	1.951
660	6119142	1.717	1.604	1.975
646	5862296	1.717	1.604	1.933
671	6324814	1.717	1.604	2.008
468	4881352	1.717	1.604	1.764
521	6049561	1.717	1.604	1.964
521	6049561	1.717	1.604	1.964
519	6003203	1.717	1.604	1.956
321	3830550	1.717	1.604	1.564
340	4297431	1.717	1.604	1.656
339	4272189	1.717	1.604	1.651
345	4424756	1.717	1.604	1.681

A geometric progression of the elements with a factor of 0.17 toward the singular point ensures optimal convergence rates. The polynomial degree was increased over each element from 1 to 8, and the numerical error measured in energy norm is monitored. As a post-solution operation, the eigenpairs $(\alpha_\ell, \boldsymbol{u}^{(\ell)}(\theta))$ and the GSIFs A_ℓ's were extracted.

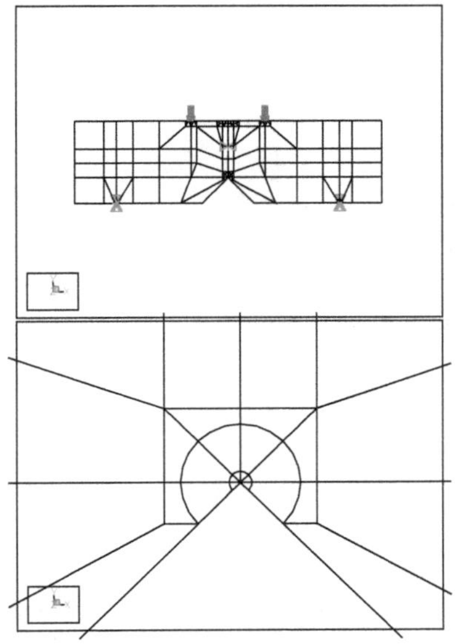

Fig. 8.9 FE mesh for an $\omega = 3\pi/4$ specimen. On the left the whole model, on the right the zoomed region around the singular point.

8.3.1 *Analysis of the Alumina-7%Zirconia Test Results*

To check the validity of Novoshilov-Seweryn's and Leguillon's criteria, the GSIF was computed at the failure point and compared to the k_c computed by (8.5) and (8.8) using K_{Ic} and σ_c. The values for $K(\omega)$ used in Leguillon's criterion are taken from Table 8.4. For the Alumina-7% Zirconia, $K_{Ic} = 4.1$ MPa$\sqrt{\text{m}}$ (see also [151]) and $\sigma_c = 290$ MPa. The predicted values for the "sharp" V-notch and the GSIF at failure (computed as $k_c = A_1 \times S_{11}^{(1)}(\theta = 90°)$) are summarized in the last three columns of Table 8.6. These results are also plotted in Figure 8.10, and show a good correlation between the predicted values (by the Leguillon and Novoshilov-Seweryn failure criteria) and experimental observations.

As noticed, the validity of both tested criteria is very good at large solid angles, and deteriorates as the solid angle decreases. Also, both criteria assume a sharp V-notch tip, and therefore, as the V-notch radius ρ increases the prediction is less accurate. This trend is best illustrated in Figure 8.11, where the GSIF at failure for the specimens with $\omega = 3\pi/2$ is plotted as a function of ρ.

Using the eigenpairs, A_1, and the integration radius computed by (8.25) $R_{\text{mat}} = 0.062$ mm, the SED in the vicinity of the singular points was computed and issummarized in the fourth column of Table 8.6. The chosen R_{mat} is four times the size of the approximate plastic radius, which is $r_p = \frac{1}{\pi}\left(\frac{K_{Ic}}{\sigma_c}\right)^2 = 0.0155$ mm. The SED was computed using A_1 and the first eigenpair, and once again using the stress and strain tensors in the circular region surrounding the V-notch tip. The differences in the two results were less than 3% in all cases, thus ensuring that the first term in the asymptotic expansion suffices to describe the quantity of interest with good

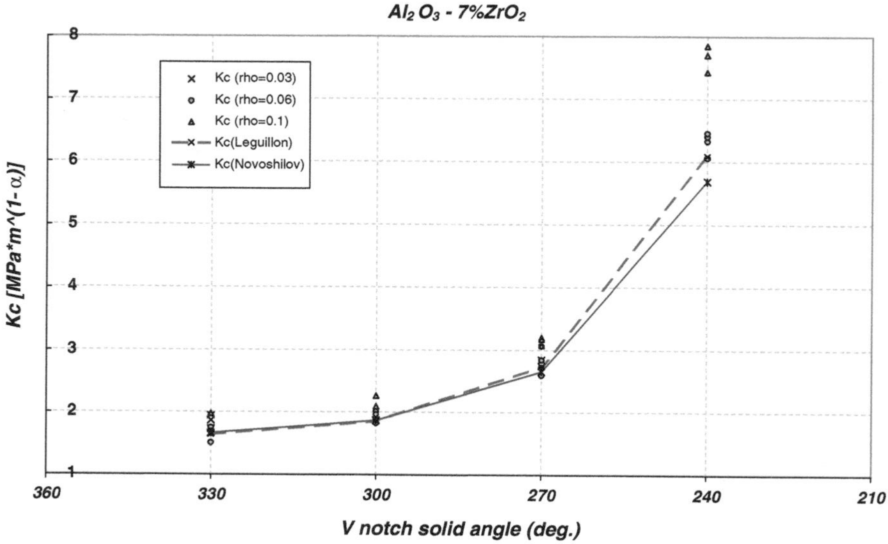

Fig. 8.10 Predicted GSIFs (k_c) at failure using Novoshilov's and Leguilon's criteria, and GSIFs in tested Alumina-7%Zirconia specimens.

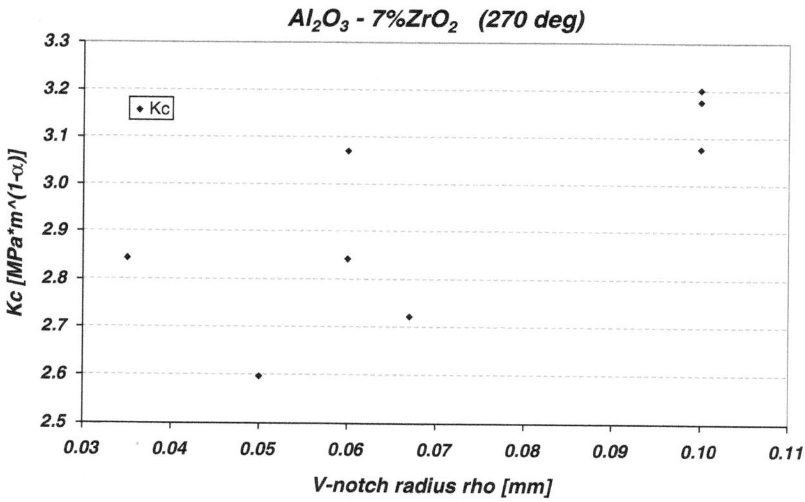

Fig. 8.11 GSIFs as a function of ρ for the $\omega = 3\pi/2$ at failure for Alumina-7%Zirconia.

accuracy. SED_{cr}[0.062 mm] as a function of the V-notch opening angle ω is shown in Figure 8.12.

Because the SED is proportional to the square of A_1, the sensitivity of the results to changes in this parameter is more pronounced. Computing SED[0.062 mm]$_{crack}$=SED[0.062 mm]$_{straight}$ $\approx$ 1.201E5 [N/m^2], it is clear that values of SED[0.062 mm]$_{cr}$ obtained for all angles are within the anticipated range. Of

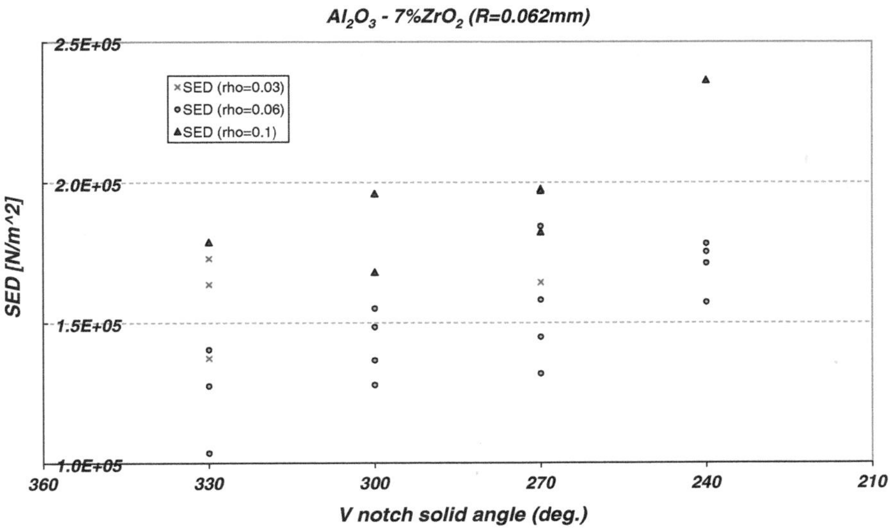

Fig. 8.12 SED_{cr}[0.062 mm] in tested Alumina-7%Zirconia specimens.

course, the V-notch radius causes a higher SED_{cr} than the calculated SED (assuming $\rho = 0$). $\mathrm{SED}[R]_{cr}$ at any R can be easily computed from the data presented herein by

$$\mathrm{SED}[R]_{cr} = \left(\frac{R}{0.062\ \mathrm{mm}}\right)^{2\alpha_1 - 2} \mathrm{SED}[0.062\ \mathrm{mm}]_{cr}.$$

The influence of ρ on the critical SED was also examined and depicted for the case $\omega = 3\pi/2$ in Figure 8.13. Because the values of ρ are close to these of R_{mat}, its influence is pronounced.

8.3.2 *Analysis of the PMMA Tests*

Similarly to the analysis described in previous subsection, the validity of Novoshilov-Seweryn's and Leguillon's criteria was also evaluated for the PMMA specimens. The values for $K(\omega)$ used in Leguillon's law are taken from Table 8.3. For the PMMA material, $K_{\mathrm{Ic}} = 1.028\ \mathrm{MPa}\sqrt{\mathrm{m}}$, $\sigma_c = 124$ MPa, $E = 2.3$ GPa, and $\nu = 0.36$ (see [59]). The predicted values and the GSIF at failure are summarized in the last three columns of Tables 8.7 and 8.8. These results are also plotted in Figure 8.14, and also show a good correlation of the predicted values and experimental observations.

Again, the validity of both criteria tested is good at small opening angles, and deteriorates as the opening angle increases.

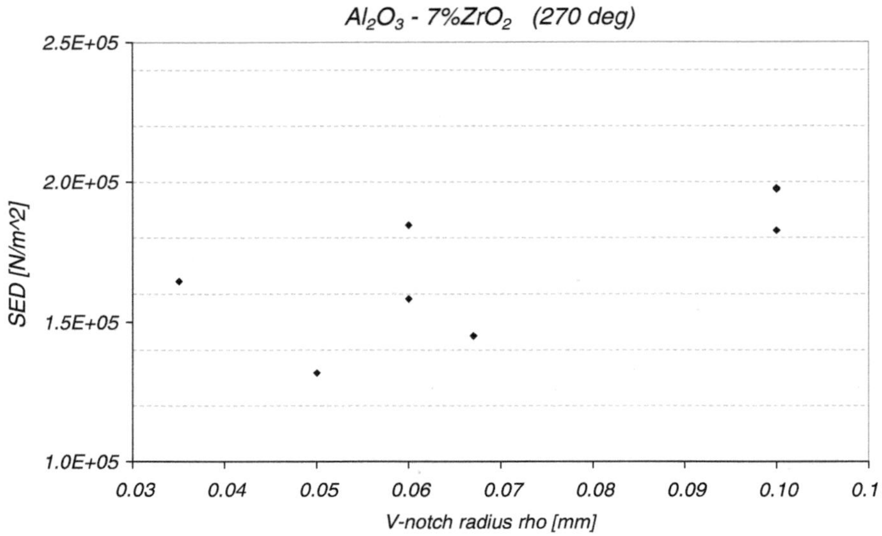

Fig. 8.13 SED_{cr}[0.062 mm] for the $\omega = 3\pi/2$ Alumina-7%Zirconia specimens as a function of ρ.

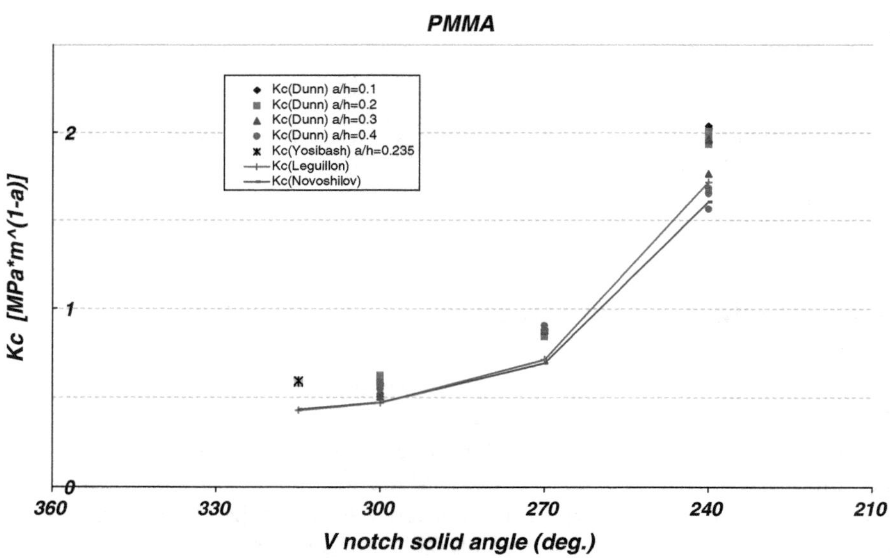

Fig. 8.14 Predicted GSIFs at failure using Novoshilov's and Leguilon's criteria, and GSIFs in PMMA specimens.

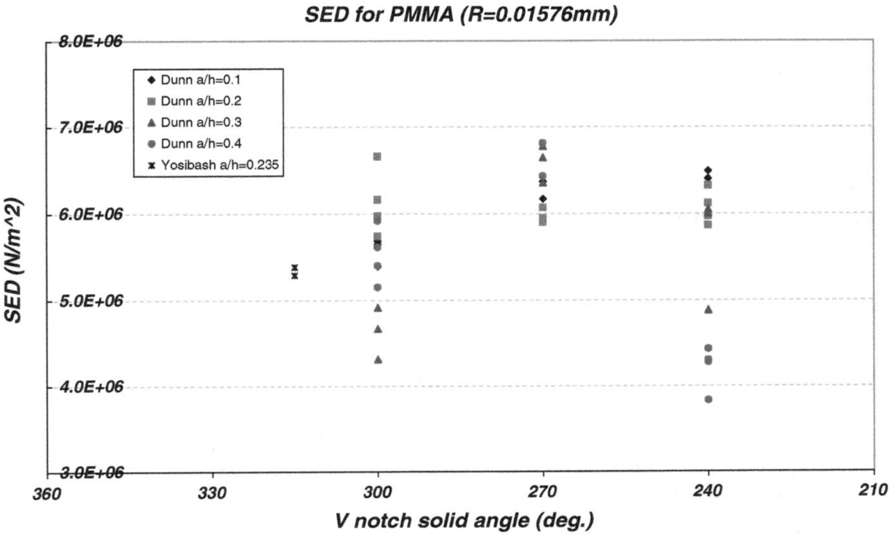

Fig. 8.15 SED_{cr}[0.0158 mm] in PMMA specimens.

SED in the vicinity of the singular points was also computed and is summarized in the second column of Tables 8.7 and 8.8. For the PMMA, the integration radius computed by (8.25) and used in our computations is $R_{\mathrm{mat}} = 0.0158$ mm. The plastic radius for the PMMA is approximately $r_p = 0.0215$ mm which is the same order of magnitude as $R_{\mathrm{mat}} = 0.0158$ mm. SED_{cr}[0.0158 mm] as a function of the V-notch opening angle ω is shown in Figure 8.15. The values of SED[0.0158 mm]$_{\mathrm{crack}}$= SED[0.0158 mm]$_{\mathrm{straight}}$ $\approx$ 3.34E6 [N/m^2], which is a lower bound to the SED obtained in the experiments (again probably due to the V-Notch radius).

In summary, the validity of four failure criteria for predicting failure initiation at V-notch sharp tips was examined and compared with experimental observations. All assume a mathematical sharp tip, namely a small blunt tip, and thus a higher k_c is obtained in the experiments compared to the predicted values. Nevertheless, both the Novoshilov-Seweryn and Leguillon criteria seem to predict well the observed failures, but as the opening angle increases, their validity deteriorates. This may be attributed to the inexact measurement of σ_c and the blunt tip radius. Leguillon's criterion outperforms the Novoshilov-Seweryn criterion, and it has been refined to include ρ dependency so to match better the experimental observations - see [110]. Table 8.9 summarizes the assumed crack length increment ℓ_0 of Leguillon criterion, the path over which the stress is averaged d_0 in case of the Novoshilov-Seweryn for the two elastic brittle materials considered, and R_{mat} used in the SED computations. The values show that indeed ℓ_0, d_0, and R_{mat} are small and of comparable orders of magnitude.

The SED criterion is more similar to the Dunn criteria in terms of the needed values of the critical SED for a large range of ω's. However, it is not unit-dependent,

Table 8.9 Values of ℓ_0, d_0 and R_{mat} for PMMA and Al_2O_3-7%ZrO_2.

	ℓ_0 from (8.7) [mm]		d_0 from (8.4) [mm]		R_{mat} from (8.25) [mm]	
ω	PMMA	Al_2O_3-7%ZrO_2	PMMA	Al_2O_3-7%ZrO_2	PMMA	Al_2O_3-7%ZrO_2
330°	0.0105	0.031	0.0431	0.127	0.0158	0.062
315°	0.0106	0.031	0.0431	0.127	0.0158	0.062
300°	0.0108	0.032	0.0431	0.127	0.0158	0.062
270°	0.121	0.036	0.0431	0.127	0.0158	0.062
240°	0.146	0.043	0.0431	0.127	0.0158	0.062

and does not require the knowledge of K_{Ic} or σ_c for the material of interest. A practical application of the SED criterion for predicting failure initiation in electronic devices under thermoelastic loading [205] is provided in the next chapter.

Using it, one can compute R_{mat} and obtain a SED that is independent of the reentrant opening angle. We have seen that the predicted SED_{cr} is a lower estimate of the experimental observations, and the scatter in SED_{cr} is wider.

The failure criteria were extended to mixed mode loading in [143]. The predicted failure load and failure initiation angle by three mixed mode failure criteria for brittle elastic V-notched structures were computed. The validity of three failure criteria for predicting failure initiation at *sharp* V-notch tips under mixed mode loading was examined and compared to experimental results. The experiments included loading of specimens made of two different elastic materials (PMMA and MACOR) under three- and four-point bending conditions that induces a state of mixed mode at the V-notch tip. Mixed mode experimental results reported in [159] for PMMA specimens with a large range of V-notch solid angles ω and mode mixity values were also examined. All criteria seem to predict well both the failure load and crack initiation angle. For failure load prediction there is no definite "best criterion" but the SED is easier to apply requiring only the calculation of the material notch integration radius and the calculation of SED or the resulting A_{1c}. The crack initiation angle is best predicted using Leguillon's criterion.

8.4 Determining Fracture Toughness of Brittle Materials Using Rounded V-Notched Specimens

The cornerstone of fracture mechanics in brittle materials is the plane-strain fracture toughness K_{Ic}. Conventional methods for its determination may require complicated procedures due to the need to introduce a sharp crack into specimens. One important possible engineering application of V-notch failure criteria (which usually have a rounded V-notch tip), is the determination of the fracture toughness of "hard materials" such as ceramics. Ceramics are highly brittle materials and therefore susceptive to crack formation so that introducing the pre-cracking is extremely difficult [129]. The standardized test specimens for K_{Ic} determination include the compact test specimen, disk-shaped compact test specimen, single edge-notched

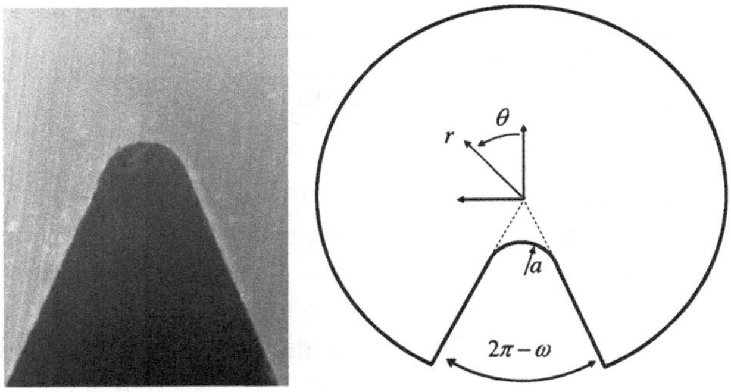

Fig. 8.16 Rounded V-notched PMMA specimen - Zoom on notch tip area and coordinate system

bend-specimen, middle tension specimen and arc-shaped specimen [5]. In many cases the crack is obtained by a simple saw cut, which may or may not be followed by fatigue loading but if the notch-root radius is too large, it leads to over estimation of the actual fracture toughness. As a result of this difficulty other experimental methods for fracture toughness determination such as the indentation strength method and the Chevron-Notched specimen (CNS) have been developed [5]. The indentation method typically results in an overestimation of the fracture toughness value and is generally not as accurate as traditional standardized mechanical test specimens. The CNS requires no pre-cracking but has a complicated design that leads to high machining cost.

In this section we present an alternative and simpler method to determine the fracture toughness accurately using mode I loading experiments conducted on specimens with rounded V-notches rather than by introducing a sharp crack. Using Leguillon's failure criterion developed for rounded V-notched uncracked components [110] and a process of "reverse engineering" it is possible to determine with reasonable accuracy the fracture toughness. Such an approach for estimating the fracture toughness using V-notched specimens also appears in [158] for pure mode I loading and a small V-notch opening angle ($\omega \leq 300°$) (see also [206]). Furthermore, the failure criterion in [158] assumes a sharp notch tip therefore the experimental specimens used to determine the fracture toughness had to contain a very small notch tip; radius $a \leq 0.01$ mm. In [141] a method to estimate the influence of the U-notch tip radius in standard test specimens has been proposed, introducing a correction to fracture toughness values obtained in single edge-notched bend specimens (SENB), showing that the notch tip radius may influence greatly the results.

8.4.1 The Failure Criterion for a Rounded V-Notch Tip

Consider specimens having a rounded V-notch as shown in Figure 8.16 with a rounded V-notched tip with radius ρ. The coordinate system is placed at the

Table 8.10 $\Delta H_{11}(\mu)$ for $\omega = 315°$, $E = 1, \nu = 0.3$.

μ	0	0.4	0.6	0.8	1.0	1.2	2.0	2.5	3.0	3.5	4.0
$\Delta H_{11}(\mu, \theta)$	0	1.27	2.26	3.32	4.40	5.51	10.00	12.82	15.66	18.49	21.31

intersection of the V-notch faces. In reality V-notches are never truly sharp but contain a small notch tip radius. We presented in Section 8.1.2 the Leguillon's criterion for the prediction of fracture initiation at a *sharp* V-notch tip. A correction factor to the sharp notch prediction accounting for a small radius at the V-notch tip has been presented in [110, 144], bringing prediction results closer to the observed experimental values. Because of this small radius at the V-notch tip, a correction factor $\Delta H_{11}(\mu_0)$ can be computed (see [144]) that accounts for the change in the energy release rate, and is a function of the ratio $\mu_0 = \ell_0/\rho$, where ℓ_0 is given in (8.7). Under mode I loading, the critical GSIF can be computed [144] as

$$A_{1c}^{\text{blunt}} = \sqrt{\frac{G_c \ell_0}{\rho^{2\alpha_1} \Delta H_{11}(\mu_0)}}, \tag{8.28}$$

where G_c is the critical ERR and α_1 is the V-notch mode I singularity exponent (eigenvalue). Here $\Delta H_{11}(\mu_0)$ depends on the local geometry (ω) and boundary conditions in a neighborhood of the rounded V-notch tip and is computed by an integration procedure as shown in [144]. Values of $\Delta H_{11}(\mu_0)$ for $\omega = 315°$ are tabulated in Table 8.10 and for other ω's are provided in Figure 8.17, taken from [144]. Figure 8.17 contains plots of $\Delta H_{11}(\mu)$ for different notch opening angles.

Remark 8.1 *Given ΔH_{11} for E and ν, one may easily obtain ΔH_{11} for any E_{new} and ν_{new} by the following relationship*

$$\Delta H_{11}^{new}(\omega) = \Delta H_{11}(\omega) \frac{E}{1-\nu^2} \frac{1-\nu_{new}^2}{E_{new}}. \tag{8.29}$$

8.4.2 Estimating the Fracture Toughness From Rounded V-Notched Specimens

The blunt V-notch failure criterion is exploited to estimate the fracture toughness using experimental data from mode I loading. From an experiment on the V-notched specimen one attains the failure load. Mode I GSIF at failure is obtained by an FE analysis of the specimen having a *sharp* V-notch, to which we apply the observed failure load. For the given critical mode I GSIF, one can compute the critical incremental length ℓ_0 at failure by first determining $\sigma_{\theta\theta}^{(1)}$ [144]:

$$\sigma_c = A_{1c}^{\text{blunt}} \rho^{\alpha_1 - 1} \sigma_{\theta\theta}^{(1)} \rightarrow \sigma_{\theta\theta}^{(1)} = \frac{\sigma_c}{A_{1c}^{\text{blunt}} \rho^{\alpha_1 - 1}}. \tag{8.30}$$

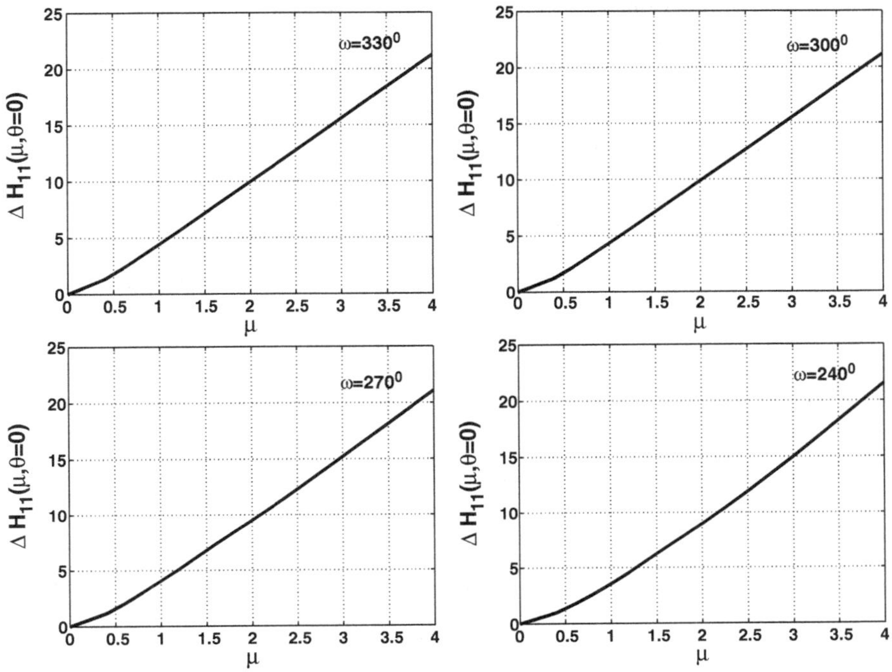

Fig. 8.17 $\Delta H_{11}(\mu)$ for different values of ω (From [144]).

The eigenstress function $\sigma_{\theta\theta}^{(1)}$ is a decreasing function of ℓ (see [144]) independent of elastic material parameters but depending on the V-notch opening angle. In Figure 8.18 as an example we plot $\sigma_{\theta\theta}^{(1)}$ as a function of $\mu = \frac{\ell}{\rho}$ for $\omega = 315°$ ($\sigma_{\theta\theta}^{(1)}$ in Figure 8.18 is independent of the elastic properties of the material and can therefore be used for any pure mode I loading case, provided $\omega = 315°$). From (8.30) after computing $\sigma_{\theta\theta}^{(1)}$ one may extract μ_0 from Figure 8.18 and then compute $\ell_0 = \rho\mu_0$.

Once ℓ_0 is known, one can reformulate (8.28) in order to obtain

$$G_c = \frac{(A_{1c}^{\text{blunt}})^2(\Delta H_{11}(\mu_0))\rho^{2\alpha_1}}{\ell_0}$$

$$\Downarrow$$

$$K_{\text{Ic}} = \sqrt{\left(\frac{(A_{1c}^{\text{blunt}})^2(\Delta H_{11}(\mu_0))\rho^{2\alpha_1}}{\ell_0}\right)\frac{E}{(1-\nu^2)}}. \tag{8.31}$$

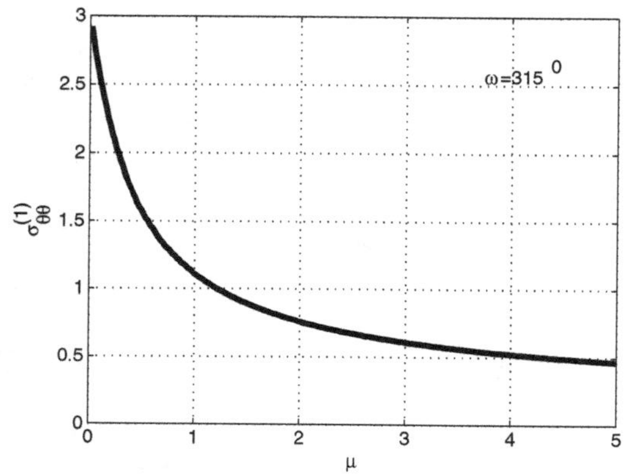

Fig. 8.18 $\sigma_{\theta\theta}^{(1)}$ for $\omega = 315°$.

Using (8.31) one can compute the estimated value of the fracture toughness. In summary, the fracture toughness determination process is conducted using the following five steps:

(1) Obtain failure load from experiments on rounded V-notched specimens.
(2) Generate an FE-model of the experimental specimen with a *sharp* V-notch and extract the generalized stress intensity factor A_{1c}^{blunt} at failure.
(3) From (8.30) compute the value of $\sigma_{\theta\theta}^{(1)}$.
(4) Extract the value of μ_0 from graphs of $\sigma_{\theta\theta}^{(1)}$ versus μ for the relevant ω.
(5) Compute the fracture toughness using (8.31).

In the following we demonstrate that the estimated values computed by (8.31) are very close to those obtained from standard experimental procedures on pre-cracked specimens.

8.4.3 Experiments on Rounded V-Notched Specimens in the Literature

To demonstrate the validity of the proposed method, we consider experimental data from several sources. Pure mode I experiments on PMMA V-notched specimens at different test temperatures (which effect the fracture toughness value) are reported in [59, 67]. In [206], 4PB specimens made of Alumina-7% Zirconia for different V-notch opening angles are reported. Mixed mode experiments on PMMA specimens are given in [59, 143] and on MACOR (glass ceramics) in [143]. The PMMA experiments in [159] and the MACOR experiments in [143] as well as pure mode I experiments in [158] are not used here because the fracture toughness values reported in both sources are estimated from experiments on V-notched specimens and therefore do not provide K_{Ic} of pre-cracked specimens in order to validate our proposed method.

Table 8.11 Failure loads reported in [206].

$2\pi - \omega$	Average Failure Load N	
deg	$\rho = 0.06$ mm	$\rho = 0.1$ mm
30	1493 ± 117	1844 ± 0
60	1684 ± 62	1844 ± 83
90	1582 ± 84	2045 ± 280
120	1962 ± 0	2624 ± 490

8.4.3.1 Experiments on Alumina-7% Zirconia from [206]

Experiments on 4PB Alumina-7% Zirconia bar specimens containing rounded V-notches with material properties $E = 360$ GPa , $\sigma_c = 290$ MPa and $\nu = 0.236$ are reported in [206]. The fracture toughness is $K_{Ic} = 4.1$ MPa$\sqrt{\text{m}}$ which is the standard value reported for Alumina-7% Zirconia [51, 151]. For our analysis we consider the specimens with notch tip radii $\rho = 0.06, 0.1$ mm. Table 8.11 summarizes the results.

Remark 8.2 *In [206], values of notch radii ranging from $\rho = 0.03$ to $\rho = 0.1$ were reported. Here we consider only the experiments for $\rho = 0.06, 0.1$ mm because they make up the majority of the reported experiments for each ω.*

One can observe that the average failure load obtained for $2\pi - \omega = 90°$ is lower then the average load obtained for $2\pi - \omega = 60°$, which is in contradiction to the expected rise in force needed to break the specimen when the opening angle increases. As a consequence, it is expected that an underestimation of the fracture toughness will be obtained for $2\pi - \omega = 90°$.

8.4.3.2 Experiments on PMMA [59]

Experiments on three point bending (3PB) V-notched PMMA bar specimens with a wide range of V-notch opening angles and notch depths are reported in [59]. The following material properties were reported $E = 2300$ MPa, $\sigma_c = 124$ MPa, and $\nu = 0.36$. K_{Ic} was determined using four SENB pre-cracked specimens, and the average value is $K_{Ic} = 1.02$ MPa$\sqrt{\text{m}}$ with a standard deviation of 0.12 MPa$\sqrt{\text{m}}$. The notch tip radius is $\rho = 0.0254$ mm] for $\omega = 300°$ and smaller for $\omega = 270°, 240°$. Since the notch tip radius is not specified exactly for all experimental specimens a constant value of $\rho = 0.0254[mm]$ was used (so in our computations for $\omega = 270°, 240°$ we expect an overestimation of the fracture toughness). Table 8.12 summarizes the experimental values obtained.

Table 8.12 Notch depth and average A_{1c} [59].

notch depth	Average A_{1c} MPa $(\text{mm})^{1-\alpha_1}$		
mm	$\omega = 300°$	$\omega = 270°$	$\omega = 240°$
1.78	16.85 ± 0.22	20.51 ± 0.25	29.10 ± 0.37
3.56	17.61 ± 0.6	19.93 ± 0.19	28.26 ± 1.04
5.33	15.43 ± 0.53	20.93 ± 0.47	27.50 ± 1.42
7.11	16.74 ± 0.53	21.09 ± 0.47	23.60 ± 1.64

Table 8.13 PMMA experimental results for $\omega = 270°$ [67]

Notch depth mm	Notch Radius mm	Average Failure Load N
5	0.05	1190 ± 10
10	0.06	770 ± 20
14	0.04	510 ± 20
20	0.06	190 ± 10

8.4.3.3 Experiments on PMMA Reported in [67]

Experiments on 3PB V-notched PMMA bar specimens are reported in [67]. The experiments were conducted at a temperature of $T = -60°\text{C}$ for different V-notch opening angles, notch depths, and notch tip radii. The material properties are $E = 5005$ MPa , $\sigma_c = 128.4$ MPa, and $\nu = 0.4$. The average fracture toughness from nine separate experiments using CT and SENB specimens was $K_{\text{I}c} = 1.7$ $\text{MPa}\sqrt{\text{m}}$ with the lowest value obtained as $K_{\text{I}c} = 1.5 MPa\sqrt{m}$ and the highest $K_{\text{I}c} = 1.77 MPa\sqrt{m}$. Table 8.13 summarizes experimental results for $\omega = 270°$.

8.4.4 Estimating the Fracture Toughness

To estimate the fracture toughness, high-order finite element models representing the experimental specimens were generated, and A_{1c}^{blunt} was computed when the average experimental failure load was applied. Next $\sigma_{\theta\theta}^{(1)}$ was determined for the pure mode I loading case from (8.30). We follow by using the computed value to extract μ_0 from Figures such as Figure 8.19, which were obtained using an asymptotic expansion method [144]. We can then calculate $\ell_0 = \rho\mu_0$. The final step is using (8.31) to compute the value of the fracture toughness.

8.4.4.1 Estimated Fracture Toughness for Alumina-7%Zirconia [206]

Table 8.14 summarizes computed and estimated fracture toughness values. In Figure 8.20 the estimated $K_{\text{I}c}$ is shown in comparison to the values obtained using the standard SENB specimen. The upper and lower bounds of the experimental values exhibit the standard scatter of 10%, which is usually reported for the

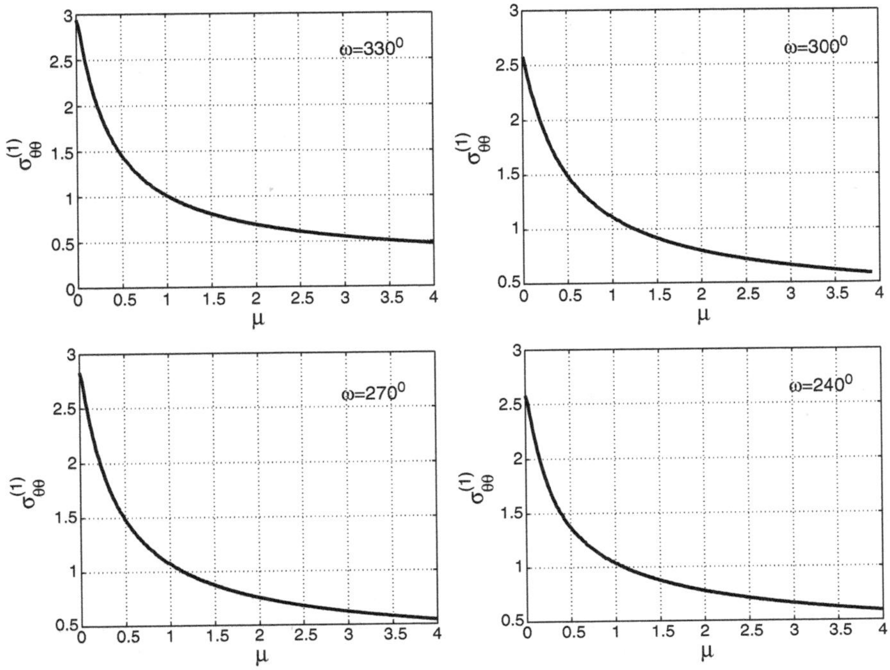

Fig. 8.19 Rounded notch mode I stress curve for different opening angles [144].

Table 8.14 Analysis results for 4PB Alumina-7%Zirconia specimens reported in [206].

	$\sigma_{\theta\theta}^{(1)}$		μ_0		$K_{Ic}^{\text{Estimated}}$ MPa $\sqrt{m}$		
$2\pi - \omega$ deg	$\rho = 0.06$ mm	$\rho = 0.1$ mm	$\rho = 0.06$ mm	$\rho = 0.1$ mm	$\rho = 0.06$ mm	$\rho = 0.1$ mm	K_{Ic} MPa $\sqrt{m}$ Experimental
30	1.4489	1.5134	0.6	0.6	3.713	3.784	4.1 ± 0.41
60	1.2879	1.5090	0.71	0.51	4.002	4.032	4.1 ± 0.41
90	1.8097	1.7118	0.32	0.38	3.303	3.040	4.1 ± 0.41
120	1.3674	1.2442	0.5	0.61	3.877	4.098	4.1 ± 0.41

method [188]. As can be seen the estimated K_{Ic}'s are well within the normal scatter of experimental results. As expected, for reasons stated in the previous section, there is an underestimation of K_{Ic} for $\omega = 270°$.

8.4.4.2 Estimated Fracture Toughness for PMMA [59]

Table 8.15 summarizes computed values and estimated fracture toughness values. In Figure 8.21, the estimated value of the fracture toughness is shown in comparison to the values obtained using a standard SENB specimen as reported in [59]. The fracture toughness estimated for $\omega = 300°$ is very close to the average experimental

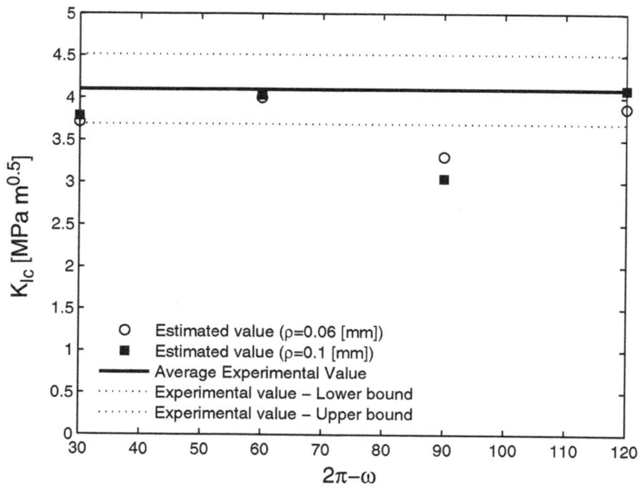

Fig. 8.20 Experimental and estimated fracture toughness values for Alumina-7%Zirconia specimens [206]

results. A slight overestimation of the fracture toughness for some of the specimens with $\omega = 270°, 240°$ can be seen, but one must remember that the exact value of the notch tip radius for that experimental batch were unknown and an overestimation of the fracture toughness was expected.

8.4.4.3 Estimated Values for PMMA Reported in [67]

Table 8.16 summarizes computed values and estimated fracture toughness values. In Figure 8.22, the estimated value of the fracture toughness is shown in comparison to the values obtained using standard SENB and CT specimens as reported in [67]. The fracture toughness estimated by the proposed method is well within the normal scattering of standard experimentally determined values and gives an overestimation of 5% at most with respect to the average value of experimental results.

Using the algorithm we have outlined herein we demonstrate that the estimated values obtained for a wide range of experimental data from several sources lie well within the conventional experimental scatter usually observed in standard fracture toughness experiments. Furthermore, if we compare estimated values for the same material type such as PMMA, it can be seen that when one reduces small plastic effects at the notch tip radius by lowering the test temperature as reported in [67], making the material more brittle, the estimated fracture toughness is closer to the average value obtained in standard pre-cracked specimens.

Table 8.15 Analysis results for PMMA specimens [59].

Notch Depth	$\sigma_{\theta\theta}^{(1)}$			μ_0			$K_{Ic}^{\text{Estimated}}$ MPa $\sqrt{\text{m}}$			K_{Ic} MPa $\sqrt{\text{m}}$
mm	$\omega = 300°$	$\omega = 270°$	$\omega = 240°$	$\omega = 300°$	$\omega = 270°$	$\omega = 240°$	$\omega = 300°$	$\omega = 270°$	$\omega = 240°$	Experimental
1.78	1.2267	1.1346	1.0389	0.8	0.88	1.05	1.08	1.15	1.18	1.02 ± 0.1
3.56	1.1737	1.1676	1.0697	0.85	0.85	1.05	1.10	1.10	1.14	1.02 ± 0.1
5.33	1.3396	1.1118	1.0993	0.7	0.9	1.05	1.05	1.16	1.12	1.02 ± 0.1
7.11	1.2347	1.1034	1.2810	0.8	0.93	0.6	1.07	1.15	1.27	1.02 ± 0.1

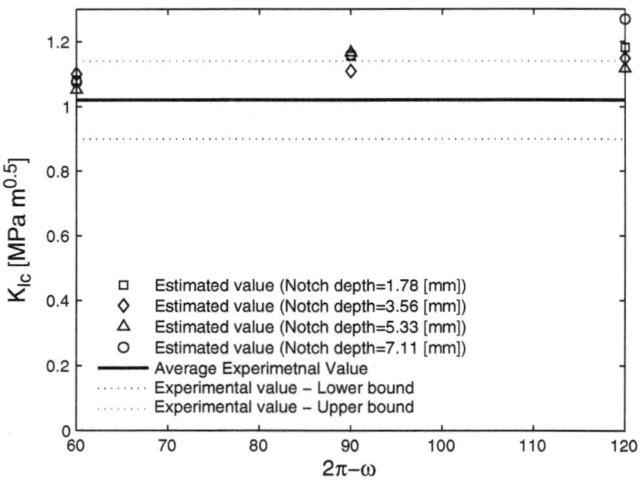

Fig. 8.21 Experimental and estimated fracture toughness values for PMMA specimens [59]

Table 8.16 Analysis results for PMMA specimens [67] for $\omega = 270°$.

Notch depth mm	$\sigma_{\theta\theta}^{(1)}$	μ_0	$K_{\mathrm{Ic}}^{\mathrm{Estimated}}$ MPa $\sqrt{\mathrm{m}}$	$K_{\mathrm{Ic}}^{\mathrm{Experimental}}$ MPa $\sqrt{\mathrm{m}}$
5	1.0938	0.94	1.77	1.7 ± 0.1
10	1.1819	0.8	1.78	1.7 ± 0.1
14	0.9994	1.11	1.68	1.7 ± 0.1
20	1.1949	0.78	1.78	1.7 ± 0.1

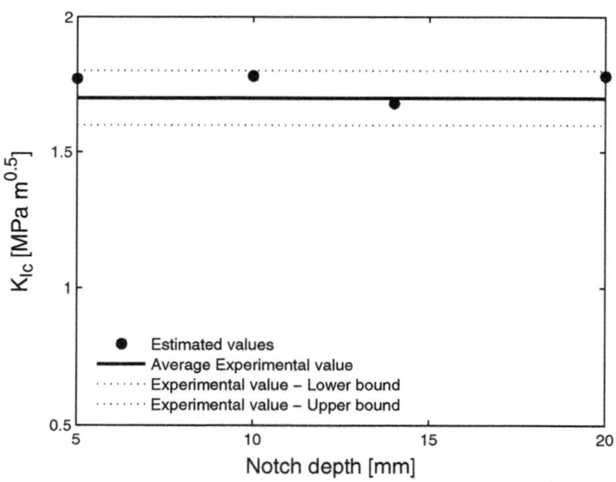

Fig. 8.22 Experimental and estimated fracture toughness values for PMMA specimens [67].

Chapter 9
A Thermoelastic Failure Criterion at the Micron Scale in Electronic Devices

Here we demonstrate the application of the SED failure criterion to a "real-life" engineering problem involving thermoelasticity effects in a microscale electronic device discussed in [205]. The fabrication of microelectronic devices (chips) is a multistep process aimed at creating a layered structure made of semiconductors, metals, and insulators. Thin aluminum interconnect lines are fabricated by sputtering technology on top of which the passivation is deposited by PECVD (plasma enhanced chemical vapor deposition). At this last step of the fabrication process, the wafer is heated to approximately 400°C, and the passivation Si_3N_4 layer is deposited to cover the metallic lines. Then, the wafer is cooled to room temperature, at which stage mechanical failures in the form of cracks emanating at V-notch tips are sometimes encountered. A typical layered structure before and after the passivation layer is deposited is shown in Figure 9.1. The cracks are often detected on "test chips" placed on the silicon wafer (typically of diameter 6, 8, or 12 inches) among the many chips fabricated on the same wafer. These "test chips" are manufactured to represent the worst possible configurations, which increase their affinity to failure. That is, if failure does not begin in them (mechanical, functional, etc.), all other chips on the wafer are fail-safe (see Figure 9.2). These cracks, emanating in the passivation layer at reentrant corners, are due to the thermal loading caused when the wafer is cooled in the last step of fabrication. Thermal stresses in confined metal lines during thermal cycling have been experimentally investigated by Moske et al. [122], where it was demonstrated that these can lead to damage formation in the passivation. The cause for the cracks is identified as a mismatch of the elastic constants and thermal expansion coefficients between the metal lines and the passivation layer. Typical cracks can be observed by sectioning the wafer at the test chip followed by a scanning electron microscope (SEM) inspection, as shown in Figure 9.3.

Zoom-in figures of a typical top view and cross-section of failed components show that the failure begins at the vertex of a reentrant V-notch (keyhole corner), as shown in Figure 9.4.

Z. Yosibash, *Singularities in Elliptic Boundary Value Problems and Elasticity and Their Connection with Failure Initiation*, Interdisciplinary Applied Mathematics 37, DOI 10.1007/978-1-4614-1508-4_9, © Springer Science+Business Media, LLC 2012

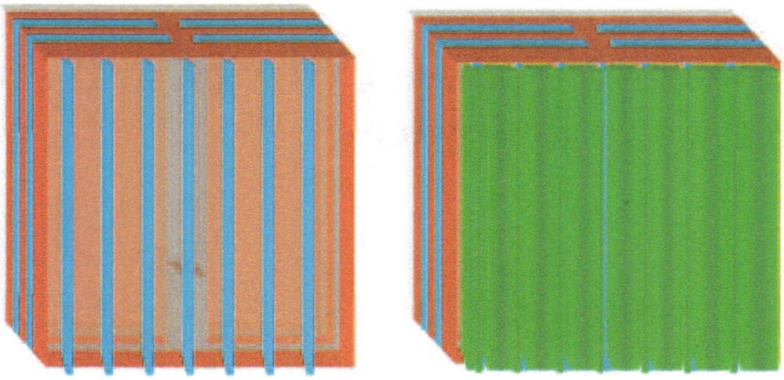

Fig. 9.1 The layered structure of a typical chip. The first off white layer is the silicon substrate, the red layers are insulators, the blue layers as well as the thin blue lines are made of metals, and the passivation layer is green.

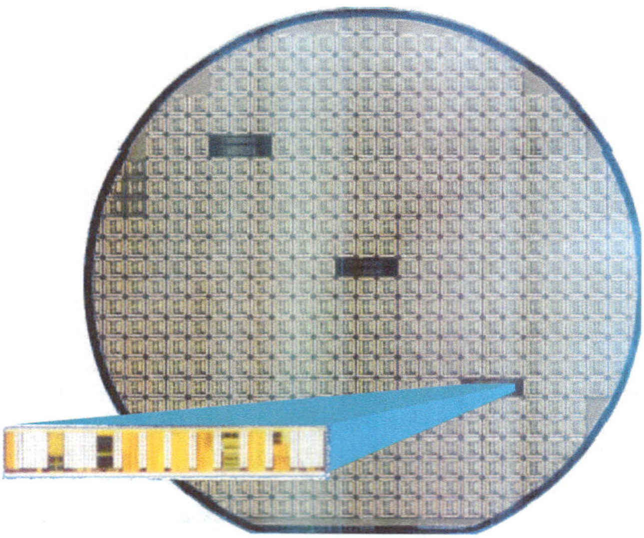

Fig. 9.2 Silicon wafer patterned with hundreds of square dies. The unpatterned areas are the scribes. The three wide rectangular dies are the test chip arrays seen in the blowup.

In an attempt to predict and eventually prevent these failures the SED criterion presented in chapter 8 is adopted. The typical feature dimension of the studied electronic devices is 0.1 to 1 μm, where the assumptions of linear elasticity still hold; see, e.g., Brandt et al. [30]. In this case, the V-notch tip, where failure begins, is a singular line at which the elastic stress tensor tends to infinity. Because failures are manifested by long planar cracks along reentrant V-notch tip lines in the passivation, a plane-strain analysis of a cross-section represents the problem well.

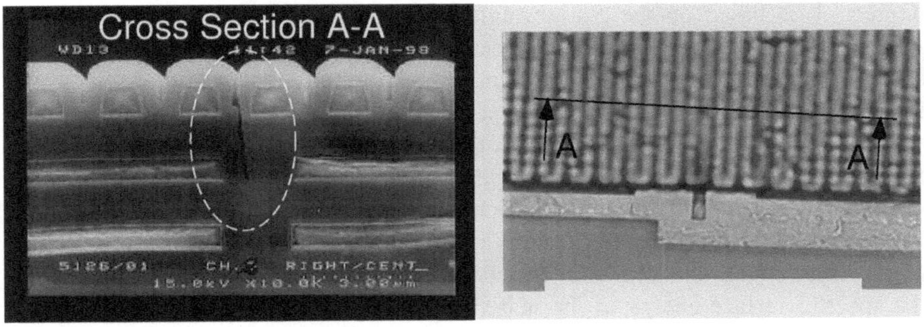

Fig. 9.3 Cracks in the passivation layer: on the right a top view of the wafer, on the left a scanning electron microscope image of the cross-section.

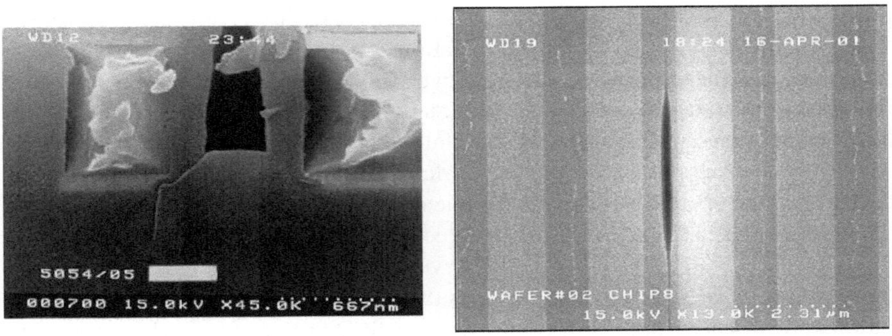

Fig. 9.4 Top view of a crack (right) and a zoom-in at a cross-section (left) of typical failure initiation sites in the passivation layer (SEM image).

The same assumptions were adopted in previous theoretical investigations of stress singularities by Michael and Hartranft [119] and Miyoshi et al. [120]. They used FEMs for the computation of the singular stress field in the vicinity of singular points under thermal loading, and concluded their work by suggesting further research for formulating a failure criterion. Sauter and Nix [155] used FEMs to investigate the thermal stresses in passivated lines bonded to substrates. Their work indicates that thermal stresses depend on the line width (increasing dramatically with decreasing aspect ratio), the passivation material and geometry (increasing with thicker and stiffer passivation). Wan et al. [187] investigated failure initiation at a $3\pi/2$ solid angle in a micromechanical silicon structure. They correlated the critical mode I stress intensity to fracture initiation, using it as the failure initiation criterion. This approach is well suited for a constant V-notch angle, but is not suited for V-notches of varying opening angles. Mazza and Dual [118] proposed a failure criterion for a silicon micromechanical structure having a reentrant corner of 135 degrees. It is based on the equilibrium of the strain energy in a radial sector of radius R and thickness h and the surface energy required to create a crack of the same length R along a thickness h, in a specific direction in silicon. Since the strain energy increases nonlinearly as a function of the radial sector R, while the surface energy increases linearly in R, there exists a radius R_{cr} (equals to 0.8 nm for silicon) for which the strain energy distribution equals the surface energy for crack creation. This critical radius (which of course is material-dependent) is chosen in their paper as the failure criterion. Mazza and Dual's criterion applies well to plane-stress situations (very thin structural layers) and requires the knowledge of the specific surface energy of the materials of interest. Since we are interested in a different geometry (plane-strain situation), and the specific surface energy is not known, a different approach is advocated.

In this chapter we demonstrate how the use of the SED failure criterion together with a carefully planned experimental program enables the prediction and eventual prevention of cracks. In Section 9.2 we identify the fabrication parameters that have the largest influence on failure initiation. As shall be shown, the mechanical properties and shape of the passivation layer and the metal lines have a dominant influence on the failure. Therefore, the material properties of the various layers have to be measured. In Section 9.3, the failure criterion (SED) is validated via an experimental program. It involves the fabrication of wafers with different values of critical parameters, followed by a numerical procedure for the computation of the SED associated with each of the fabricated wafers. To establish the critical value $(\mathrm{SED})_{cr}$ under which no failures are observed, the experimentation is carried out in three phases. This approach demonstrates that under a threshold value $(\mathrm{SED})_{cr}$, no failures are observed, highlighting the use of the SED failure criterion in engineering practice.

9.1 The SED Criterion for a Thermoelastic Problem

Consider the circular sector Ω_R shown in Figure 8.2 with traction-free boundary conditions on the faces intersecting at the singular point. A constant temperature change of $\Delta\tau =$ constant is imposed, so that the uncoupled isotropic thermoelastic problem to be solved, under the assumption of *plane-strain*, is given by

$$\mu\nabla^2\boldsymbol{u} + (\lambda+\mu)\,grad\,div\,\boldsymbol{u} = \boldsymbol{0} \quad in\ \Omega_R, \tag{9.1}$$

$$\left(2\mu\varepsilon_{\beta\gamma}(\boldsymbol{u}) + \lambda\varepsilon_{\xi\xi}(\boldsymbol{u})\delta_{\beta\gamma}\right)\mathbf{n} = \alpha(3\lambda+2\mu)\Delta\tau\mathbf{n} \quad on\ \theta = \theta_1, \theta_1+\omega. \tag{9.2}$$

The displacements in the vicinity of the singular point consist of a homogeneous part $\boldsymbol{u}^H$, as if no thermal loading were present in the *neighborhood of the singularity*, and a particular part $\boldsymbol{u}^P$, chosen so as to satisfy the inhomogeneous right-hand side of the thermoelastic system in Ω_R. The homogeneous solution is given by

$$\boldsymbol{u}^H = \sum_{i=1}^{\infty} A_i r^{\alpha_i} \begin{Bmatrix} s_1^{(i)}(\theta) \\ s_2^{(i)}(\theta) \end{Bmatrix} \Longrightarrow \boldsymbol{\sigma}^H = \sum_{i=1}^{\infty} A_i r^{\alpha_i - 1} \begin{Bmatrix} S_{11}^{(i)}(\theta) \\ S_{22}^{(i)}(\theta) \\ S_{12}^{(i)}(\theta) \end{Bmatrix}. \tag{9.3}$$

The particular solution due to a constant temperature increase is given by ([152, p. 11], [204])

$$\boldsymbol{u}^P = \beta \Delta\tau \begin{Bmatrix} x_1 \\ x_2 \end{Bmatrix}. \tag{9.4}$$

here $\beta = \frac{\alpha(3\lambda+2\mu)}{2(\lambda+\mu)}$. With this notation we decompose the strains into an elastic part, associated with the homogeneous displacements and a thermal part associated with $\boldsymbol{u}^P$

$$\varepsilon_{\gamma\zeta} = \varepsilon_{\gamma\zeta}^{El} + \varepsilon_{\gamma\zeta}^{Th}, \qquad \text{with} \quad \varepsilon_{\gamma\zeta}^{Th} = \beta\Delta\tau\delta_{\gamma\zeta}, \quad i, j = 1, 2. \tag{9.5}$$

The connection between the stress tensor and the *elastic* strain tensor is given for an isotropic elastic material via Hooke's law. For *plane-strain* conditions, the stress tensor is

$$\sigma_{\gamma\zeta} = 2\mu\varepsilon_{\gamma\zeta}^{El} + \lambda\varepsilon_{\iota\iota}^{El}\delta_{\gamma\zeta}. \tag{9.6}$$

The strain energy generated due to strain components in the x_3 direction is of no importance for a crack initiation in the x_1-x_2 plane and is not taken into consideration. Thus, the *elastic* strain energy $\mathcal{U}(\boldsymbol{u})[R]$ is

$$\mathcal{U}(\boldsymbol{u})[R] = \frac{1}{2} \iiint_{\Omega_R \times b} \sigma_{\gamma\zeta}\varepsilon_{\gamma\zeta}^{El}\, d\Omega. \tag{9.7}$$

For a domain of constant thickness b and substituting (9.6) into (9.7), one obtains

$$\mathcal{U}(\boldsymbol{u})[R] = \frac{1}{2} b \iint_{\Omega_R} \left[2\mu\varepsilon_{\gamma\zeta}^{El}\varepsilon_{\gamma\zeta}^{El} + \lambda \left(\varepsilon_{\gamma\gamma}^{El}\right)^2 \right] d\Omega. \tag{9.8}$$

Using Green's theorem, the area integral is transformed into a boundary integral, which is zero along Γ_1 and Γ_2; thus

$$\begin{aligned} \mathcal{U}^H[R] &= \frac{1}{2} b \int_{\partial\Omega_R} \left(2\mu\varepsilon_{\gamma\zeta}^{El} + \lambda\varepsilon_{\iota\iota}^{El}\delta_{\gamma\zeta}\right) n_\zeta u_\gamma^H\, dS \\ &= \frac{1}{2} b \int_{\theta_1}^{\theta_1+\omega} \left[\left(2\mu\varepsilon_{\gamma\zeta}^{El} + \lambda\varepsilon_{kk}^{El}\delta_{\gamma\zeta}\right) n_\zeta u_\gamma^H\right]_{r=R} R\, d\theta. \end{aligned} \tag{9.9}$$

Inserting (9.3) in (9.9) and using the orthogonality property for isotropic materials (8.15), the SED is given by

$$\text{SED}[R] = \sum_k A_k^2 \frac{R^{2\alpha_k-2}}{\omega} \int_{\theta_1}^{\theta_1+\omega} \left[S_{11}^{(k)} s_1^{(k)} \cos\theta + S_{12}^{(k)} \left(s_1^{(k)} \sin\theta + s_2^{(k)} \cos\theta \right) + S_{22}^{(k)} s_2^{(k)} \sin\theta \right] d\theta. \tag{9.10}$$

For the problems treated here, where a constant temperature change is imposed ($\Delta\tau$ =const), the second thermal generalized stress intensity factor is zero, $A_2 \equiv 0$. Thus, (9.10) can be written as

$$\text{SED}[R] = A_1^2 \frac{R^{2\alpha_1-2}}{\omega} I_1 + A_3^2 \frac{R^{2\alpha_3-2}}{\omega} I_3 + \mathcal{O}(R^{2\alpha_4-2}), \tag{9.11}$$

where

$$I_k \stackrel{\text{def}}{=} \int_{\theta_1}^{\theta_1+\omega} \left[S_{11}^{(k)} s_1^{(k)} \cos\theta + S_{12}^{(k)} \left(s_1^{(k)} \sin\theta + s_2^{(k)} \cos\theta \right) + S_{22}^{(k)} s_2^{(k)} \sin\theta \right] d\theta$$

is the integral of the kth eigenpair. To demonstrate that the second and further terms in (9.11) are negligible in comparison with the first term, consider the ratio of the second term to the first term in (9.11), called *Ratio*:

$$Ratio = \left(\frac{A_3}{A_1} \right)^2 R^{2(\alpha_3-\alpha_1)} \frac{I_3}{I_1} \tag{9.12}$$

For problems in which the opening angle is $3\pi/2 \leq \omega \leq 2\pi$, we have $\alpha_1 \approx 0.5$ and $\alpha_3 \approx 1$ to 1.5, so that $2(\alpha_3 - \alpha_1)$ is between 1 to 2. The ratio $\frac{I_3}{I_1}$ is close to 1 because the eigen-pairs are normalized so that the normalization factor is reflected in the coefficients A_i. The values of A_3 in all our numerical investigations are of the same order of magnitude as A_1, and in most cases are smaller, so that $\frac{A_3}{A_1} = \mathcal{O}(1)$. Thus one obtains

$$R^2 \lesssim Ratio \lesssim R^1. \tag{9.13}$$

If $R \ll 1$ (we used in our computation $R = 0.15\,\mu$m), the terms in the series (9.10) for which $\alpha_k > 1$ are orders of magnitude smaller compared to the first term, thus negligible, simplifying (9.10) to

$$\text{SED}[R] \approx \frac{A_1^2}{R^{2(1-\alpha_1)}\omega} \int_{\theta_1}^{\theta_1+\omega} \left[S_{11}^{(1)} s_1^{(1)} \cos\theta + S_{12}^{(1)} \left(s_1^{(1)} \sin\theta + s_2^{(1)} \cos\theta \right) + S_{22}^{(1)} s_2^{(1)} \sin\theta \right] d\theta. \tag{9.14}$$

SED[R] depends of course on a characteristic length size R. It should be chosen small enough so that Ω_R is within the K-dominance region, ensuring that the singular terms represent the exact solution. The small difference between SED[R] first computed by (9.14) and the values computed by (9.7) ensures that the chosen radius R is not too large. For microscopic domains considered, σ_c and K_{Ic} are unavailable at this length scale, so we chose R as a characteristic dimension of 0.15 μm and report all results for this value. The SED for any other R is obtained by (8.26).

9.2 Material Properties

The electronic device in the neighborhood of the failures shown in Figure 9.5 is a layered structure made of the passivation layer (Si_3N_4 green in the figure), the metal lines under the passivation and in the dielectric (made of aluminum, blue in

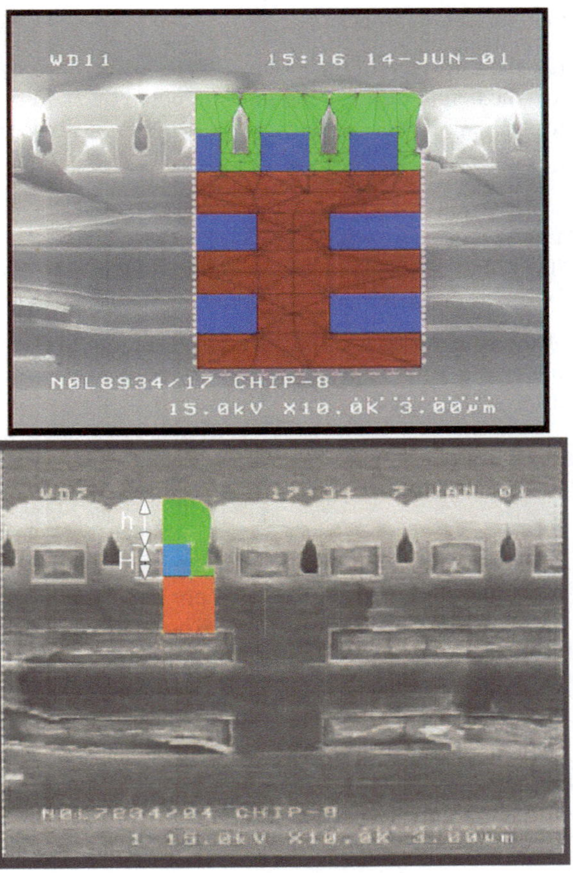

Fig. 9.5 Finite element models superimposed in color on the SEM cross-section of the test chip device. Blue, aluminum; red, SiO_2 dielectric; and green, Si_3N_4 passivation.

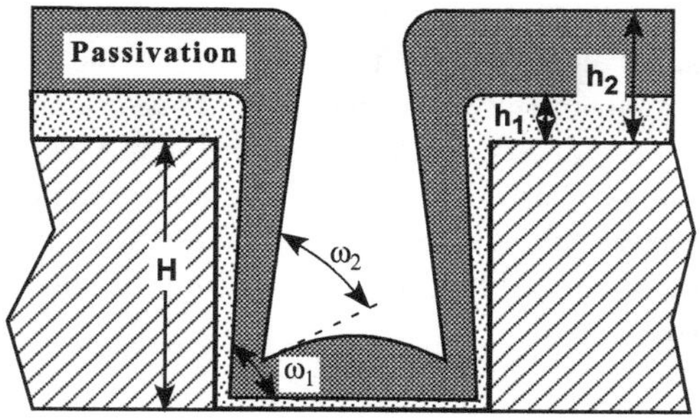

Fig. 9.6 Nonconformal step coverage of deposited passivation film (h_1 and h_2 are dimensions of two different passivation heights).

the figure) and the SiO_2 dielectric shown in red. It has been observed that failures, if they occur, begin at one of the reentrant corners above the gap in the wide metal lines. Simulating a small portion as shown in the right part of Figure 9.5 does not capture important details, and there is a need to simulate a larger portion, as shown in the left part of Figure 9.5. There are several parameters that may contribute to the failure initiation. However, the fabrication design rules allow three changes during the fabrication process:

(a) the thickness of the passivation layer (denoted by h in Figure 9.5 right),
(b) the height of the metal lines (denoted by H in Figure 9.5 right and Figure 9.6),
(c) plasma power applied during the chemical vapor deposition of the passivation layer.

Passivation thickness has two effects. First, the deposition PECVD process has a relatively poor step coverage, and therefore tends to form overhangs resulting in "keyholes" (e.g., [181, p. 95]) and singular points. Second, the reentrant angle tends to zero as the passivation thickness increases until a given thickness, and the strength of the singularity is more severe (see Figure 9.6); then, beyond $h \approx 6500$ Å, the angle increases again slightly.

9.2.1 *Material Properties of Passivation Layers*

Variation of the plasma power causes different chemical reactions (silane and ammonia) during the chemical vapor deposition of the silicon nitride. This in turn causes variation in the thermal expansion coefficient α and the Young modulus E (Poisson ratio ν is assumed to remain constant). Hence a correlation between material properties and the plasma power was necessary. These can be evaluated from measurements of residual stresses incurred in a bimaterial domain under thermal loading. A useful method to evaluate these stresses in thin films is the Stoney method described in [184, pp. 409-413]. Unpatterned layers of Si_3N_4 were deposited with different plasma powers and different thicknesses on circular Si wafers at

Power [W]	Thickness [Å]	Biaxial Stress [MPa]	E [GPa]	α [1/°C] x10^-6	Mean E [GPa]	Mean α [1/°C] x10^-6
325	5000	-39.2*			16**	8.4**
	8000	-50.8*				
350	5750	2.34	25.5	4.65		
	8000	7.04	20.0	5.22	22.7	4.94
395	6500	-46.1	27.5	1.25		
	8000	-40.8	26.0	1.48		
415	6500	-32.5	25.7	2.10	26.3	1.68
	8000	-34.3	25.9	1.90		
485	5000	+126*				
	5750	-157	72.0	0.29		
	6500	-164	75.0	0.27		
	8000	-177	79.0	0.24	79.3	0.237
505	5000	-188	83.0	0.21		
	6500	-191	85.0	0.20		
	8000	-186	82.0	0.21		

* Results ignored

** Extrapolated values

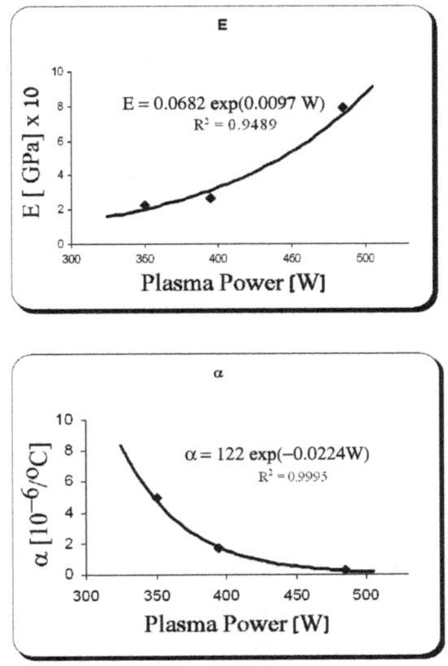

Fig. 9.7 Passivation material (Si_3N_4) properties as a function of the plasma power.

an ordinary elevated temperature, and then cooled to room temperature. The residual stress in each wafer was determined from the curvature of the bimaterial wafer.

Properties of thin layers of Si_3N_4 found in the literature indicate a large variation (see, e.g., [184]). Starting with typical values of E and α, we varied them in order to match the measured residual stresses in the wafers. A unique evaluation of the material properties could have been obtained using two different substrate materials. However, this was not available for this research. It was found that with increasing power, E increases and α decreases, obeying the empirical equations

$$E(W) = 0.682 \exp(0.0097 \times W) \quad \text{GPa}, \tag{9.15}$$

$$\alpha(W) = 1.22 \times 10^{-4} \exp(-0.0224 \times W) \quad 1/\text{C}^{\circ} \tag{9.16}$$

(where W stands for the plasma power in Watts). Figure 9.7 summarizes the results, the exponential fit equations, and the table lists the material properties.

Equations (9.15-9.16) are used for evaluating the material properties associated with each plasma power.

Because the material properties E and α are temperature-dependent, the above fit represents their averaged value between the deposition temperature (400°C) and room temperature.

A simplified finite element model as shown in the right side of Figure 9.5 was used to investigate the influence of several other fabrication parameters on the strength of the singularity. It was found that the height of the metal lines H has a large influence, while the width seemed to have little to no influence. Therefore

Table 9.1 Material properties (E, ν and α); A survey.

Property	Al line interconnect			SiO_2 diel.	Si_3N_4 pass.	Ref.
E GPa	71.5			71.7	150	[155]
	63.9*			72.9*	30	[133]
	35nm	100nm	bulk	-	-	
	24.1	16.5	62	-	-	[168]
ν	0.35			0.16	0.25	[155]
	0.36*			0.17*	0.22	[133]
α (10^{-6}/C°)	23.6*			0.55*	1	[155]
	$23.434 + 6.996\Delta\tau/10^3 + 248.1(\Delta\tau)^2/10^6$			-	1.1	[184]

* Value used in our computations.

H was chosen as the third parameter of investigation. The simplified finite element model has also been used to verify that the "plastic radius" (the maximal length measured from the V-notch tip where the equivalent stress is above yielding) is negligible compared to the passivation thickness.

9.2.2 Aluminum Lines and Dielectric Layers

For the proposed failure criterion, the Young modulus, Poisson ratio, and the coefficient of thermal expansion of the aluminum lines and SiO_2 dielectric layers are also essential. These are known to vary according to the nature of their deposition and their minor scale lengths. A literature survey shows different values for the aluminum lines. Experiments by Steinwall and Johnson [168] on aluminum fibers removed from substrates to produce free-standing fibers 8 mm long and 1 μm diameter (grain sizes of 35 and 100 nanometers) show Young's modulus in the range 16-24 GPa. However, Ohring [133, p. 426] and Tu et. al [184] list Young's moduli of evaporated thin films similar to those of bulk. In our numerical simulations we used the material properties marked by * in Table 9.1: for the SiO_2 dielectric the material properties do not vary in different references, so that $E = 72.9$ GPa, $\nu = 0.17$, and $\alpha = 10^{-6}$ 1/C°. For the aluminum lines we used the the values $E = 68.9$ GPa, $\nu = 0.36$, and $\alpha = 23.6$ 1/C°. For the Si_3N_4 passivation we used E and α from (9.15), and $\nu = 0.22$.

9.3 Experimental Validation of the Failure Criterion

Validation of the failure criterion requires the same critical value of the SED to be obtained for different configurations of the device at failure. A set of experiments was designed to test the hypothesis, and thereby to determine the failure envelope.

Table 9.2 Fabrication parameters and results of phase-1 of the experiments.

Wafer #	H [Å]	h [Å]	Plasma Power [Watt]	Cracked?
1	7000	5000	305	Cracked
2	7000	8000	305	Cracked
3	7000	5000	485	Cracked
4	7000	8000	485	Not Cracked
5	11000	5000	305	Cracked
6	11000	8000	305	Cracked
7	11000	5000	305	Cracked
8	11000	8000	305	Not Cracked
9	9000	6500	305	Cracked

Past experience showed that the failure envelope resides within the following extreme limits:

1. Al lines of height $7000\,\text{Å} \le H \le 11000\,\text{Å}$ with the standard being $H = 9000\,\text{Å}$.
2. Si_3N_4 thickness of $5000\,\text{Å} \le h \le 8000\,\text{Å}$ with the standard being $h = 6500 \pm 300\,\text{Å}$.
3. Plasma power of $305\ Watts \le p \le 485\ Watts$ with the standard being $395\ Watts$ (E and α for Si_3N_4 computed by (9.15)).

A full factorial experiment was designed in the first phase (phase-1) consisting of nine wafers fabricated (3 parameters, 2 levels + 1 center point). The precise manufacturing parameters of the nine wafers are listed in Table 9.2, and their visualization, in a form of a "test cube," is shown in Figure 9.8. The last column of Table 9.2 indicates whether a crack was detected in the passivation layer after fabrication.

Based on the results of phase-1, at high plasma power and thick passivation , failure does not occur, regardless of the metal thickness.

For refining the failure envelope another fifteen wafers were fabricated in phase 2 and 3, with parameters that lie between any pair of cracked and intact wafers from phase 1. Table 9.3 summarizes the phases 2 and 3 fabrication parameters. All 24 wafers were examined for cracks in an SEM by cross sectioning, and selected pictures for six of the wafers are shown in Figure 9.9. In order to correlate the experimental observations with the proposed failure criterion, one needs to compute the SED associated with each tested wafer. This procedure is described in the following subsection.

9.3.1 Computing SEDs by p-Version FEMs

The precise dimensions and the geometry in a neighborhood of the singular points was measured for each of the tested wafers, and a p-version parametric FE model

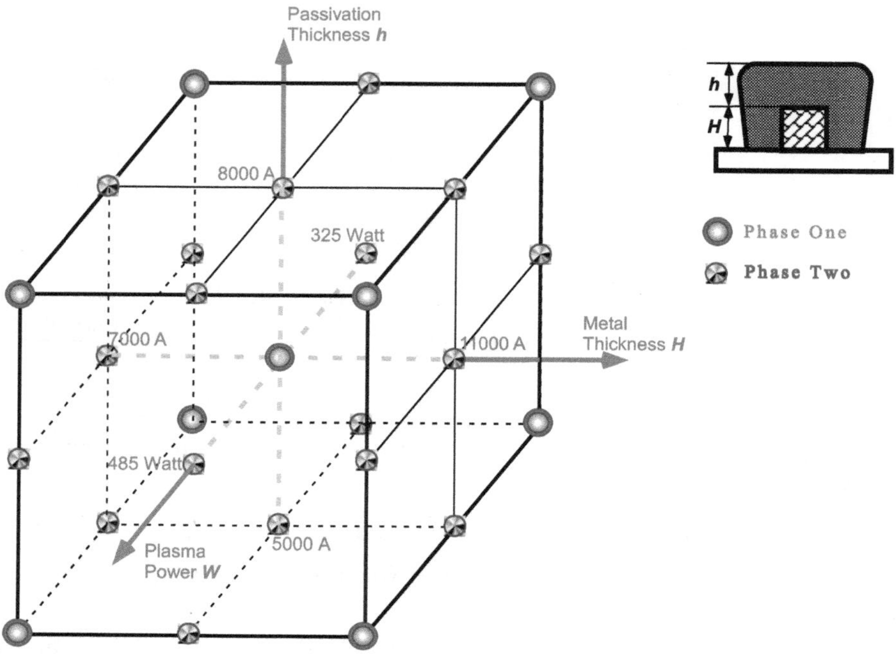

Fig. 9.8 The "test cube" illustrating the different fabrication parameters in the test plan. The "full" balls are from phase-1 experiment where the center represents the standard parameters.

Table 9.3 Fabrication parameters of phase 2&3 tests.

Wafer #	H [Å]	h [Å]	Plasma Power [Watt]	Cracked?
10	7000	6500	415	Not Cracked
11	7000	6500	515	Not Cracked
12	7000	8000	415	Not Cracked
13	11000	5750	485	Cracked
14	11000	8000	350	Cracked
15	11000	5750	350	Cracked
16	7000	6500	515	Not Cracked
17	7000	8000	415	Not Cracked
18	7000	6500	415	Not Cracked
19	9000	8000	515	Not Cracked
20	9000	8000	395	Crack roots
21	9000	6500	515	Not Cracked
22-24	9000	6500	395	Cracked

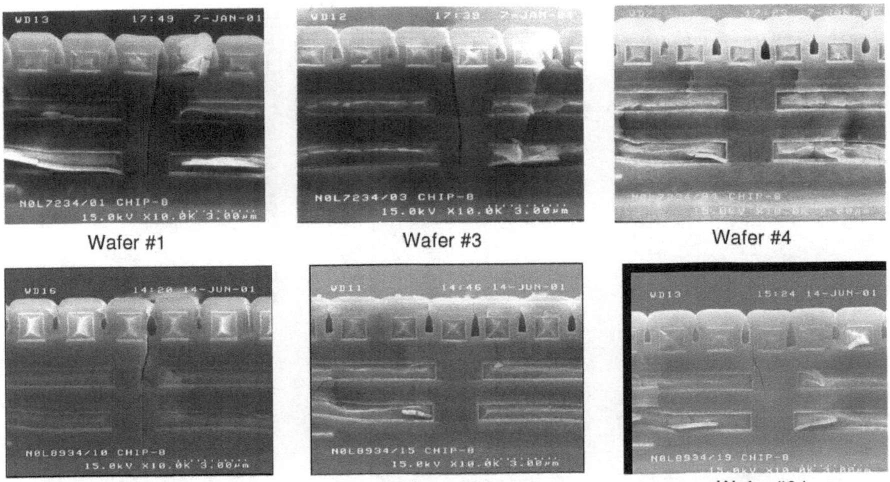

Fig. 9.9 Selected SEM cross-sections from the 24 tested wafers.

was constructed, as shown in the left picture of Figure 9.5. By varying these parameters, each of the 24 tested wafers has been represented. These models consist of two main designs:

1. Models with passivation thickness of up to 5750 Å.
2. Models with passivation thickness between 6500 Å and 8000 Å.

The reason for the two different models is that as was seen in the SEM cross-sections, the passivation tends to connect and close up on top of the keyholes, leaving the keyholes open. Another important difference is contributed by the nonconformal step coverage. The passivation is thicker on horizontal walls, and thinner on the vertical walls. During the initial stages of the passivation process (around 5000 Å), a "hill" is built in the middle of the keyhole, having sharp angles and therefore increasing the stress singularity. Continuation of the deposition and the closing of the rooftop around $h = 5750$ to 6500Å causes the sharp "crack-like" tips to become no longer sharp. Figure 9.10 presents the finite element models used for phase-1 wafers.

Using the eigenpairs, the A_i and an integration radius of $R = 0.15\ \mu$m, the SED in the vicinity of the singular points was computed for all test wafers, and is summarized in Table 9.4. To visualize the failure envelope, all wafers are shown on the test cube together with the SED values in Figure 9.11.

A semicylindrical failure envelope is observed, assessing the proposed criterion. A single value of the SED distinguishes between the cracked and intact wafers: under a threshold value of $\text{SED}_{cr}[R = 0.15\mu\text{m}] \approx 1000\ \text{J/m}^3$, all wafers manufactured are intact.

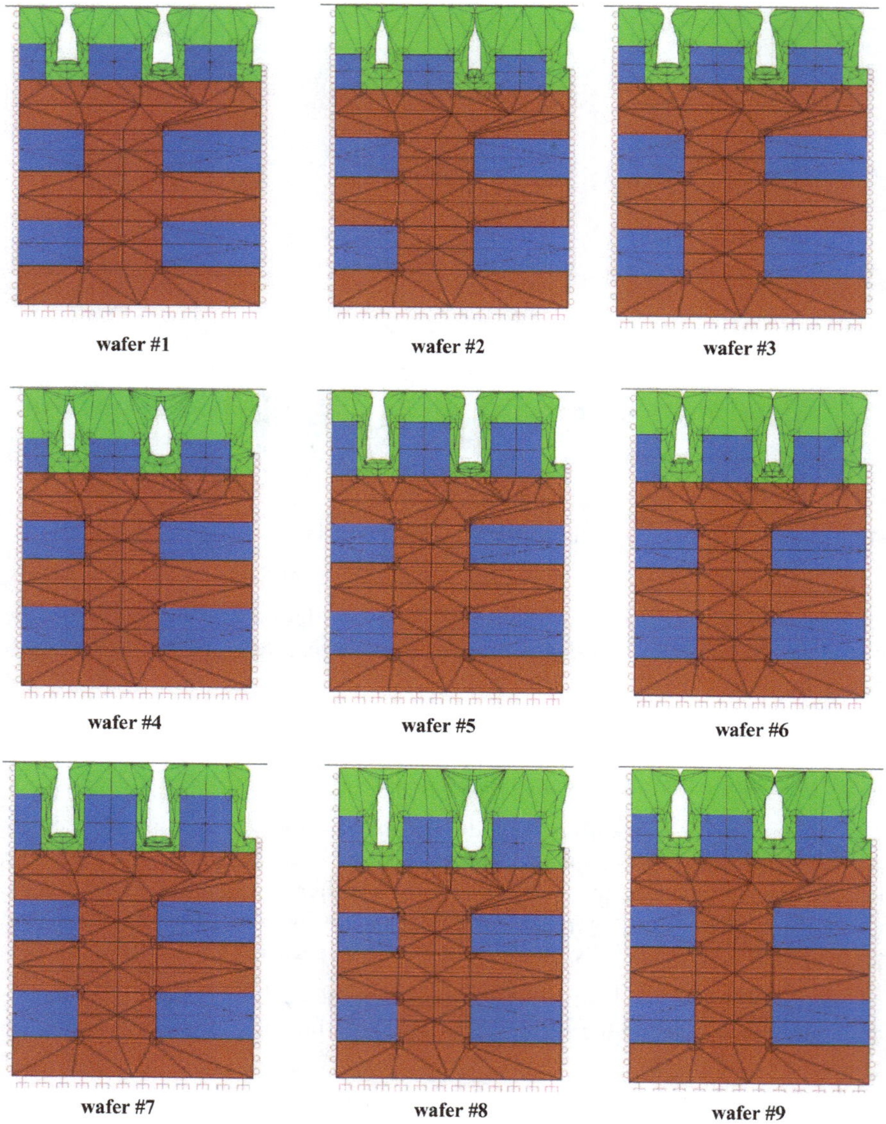

Fig. 9.10 Finite element models simulating phase-1 wafers.

It has been clearly shown that above the critical value of $SED_{cr}[R = 0.15\ \mu m] \approx 1000\ [J/m^3]$, *all wafers manufactured were cracked, for the three tested parameters.* The proposed SED criterion correlates well with empirical observations, and may be used as a standard tool for the mechanical design of failure-free electronic devices. This has major advantages because it shortens time to market by using simulation tools in place of trial-and-error fabrication processes.

Table 9.4 SED for tested wafers.

Wafer #	H [Å]	h [Å]	Plasma Power [Watts]	E [GPa]	α $[1/C°]$	ν	SED* $[R = 0.15\ \mu m]$ J/m^3
1	7000	5000	325	16.0	8.40E-06	0.22	114, 000
2	7000	8000	325	16.0	8.40E-06	0.22	124, 000
3	7000	5000	485	79.3	2.37E-07	0.22	1, 380
4	7000	8000	505	82.0	2.10E-07	0.22	985
5	9000	6500	395	25.7	1.25E-06	0.22	4, 430
6	11000	5000	325	16.0	8.40E-06	0.22	108, 000
7	11000	8000	325	16.0	8.40E-06	0.22	121, 000
8	11000	5000	485	79.3	2.37E-07	0.22	1350
9	11000	8000	505	82.0	2.10E-07	0.22	88.8
10	7000	6500	415	25.7	2.1E-06	0.22	664
11	7000	6500	515	85.0	2E-07	0.22	25.5
12	7000	8000	415	25.9	1.9E-06	0.22	985
13	11000	5750	485	79.3	2.37E-07	0.22	1, 260
14	11000	8000	350	22.7	4.94E-06	0.22	54, 400
15	11000	5750	350	22.7	4.94E-06	0.22	52, 400
16	7000	6500	515	79.3	2.4E-07	0.22	25.5
17	7000	8000	415	26.3	1.7E-06	0.22	985
18	7000	6500	415	26.3	1.7E-06	0.22	664
19	9000	8000	515	79.3	2.37E-07	0.22	13.6
20	9000	8000	395	26.3	1.68E-06	0.22	612
21	9000	6500	515	79.3	2.37E-07	0.22	39.2
22	9000	6500	395	26.3	1.68E-06	0.22	4, 430
23	9000	6500	395	26.3	1.68E-06	0.22	4, 430
24	9000	6500	395	26.3	1.68E-06	0.22	4, 430

* Boldface numbers indicate failures.

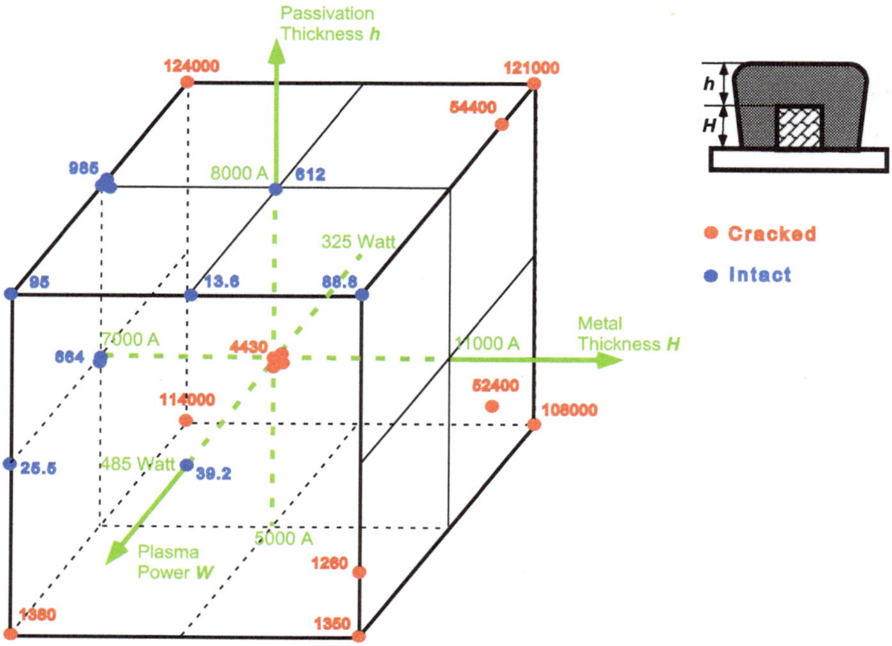

Fig. 9.11 Mapping of SEDs on the "test cube" (Units are J/m^3). The variation of the SED appears to reflect the mechanical status of the devices.

Chapter 10
Singular Solutions of the Heat Conduction (Scalar) Equation in Polyhedral Domains

Two-dimensional approximations may sometimes represent well some problems in engineering practice, but in reality, three-dimensional (3-D) domains are of interest. After obtaining in previous chapters the solution in the vicinity of singular points in 2-D, we may proceed to 3-D domains where three different types of singularities may exist.

To explain and motivate the series expansion of the elasticity solution in the vicinity of edges, vertices and vertex-edge neighborhoods, we first consider the Laplacian. This simpler elliptic problem allows more transparent analytic computations invoking all necessary characteristics of the elasticity system. Therefore, the characteristics of the solution can be more easily addressed. The solution of the Laplace equation in 3-D domains in the vicinity of singularities can be decomposed into three different forms, depending on whether it is in the neighborhood of an edge, a vertex, or the intersection of an edge and a vertex. Mathematical details on the decomposition can be found, e.g., in [6, 15, 40, 49, 73, 76] and the references therein. We consider only straight edges and assume that surfaces that intersect to form an edge are planes. A typical three-dimensional domain, denoted by Ω, containing edge singularities is shown in Figure 1.11.

The Laplace equation in 3-D reads

$$\nabla_{3D}^2 \tau \stackrel{\text{def}}{=} \Delta_{3D}\tau = 0 \qquad in\ \Omega, \qquad \text{where} \quad \Delta_{3D} = \Delta_{2D} + \partial_3^2 \stackrel{\text{def}}{=} \partial_1^2 + \partial_2^2 + \partial_3^2 \tag{10.1}$$

with the following boundary conditions:

$$\tau = g_1 \qquad on\ \Gamma_D \subset \partial\Omega, \tag{10.2}$$

$$\frac{\partial \tau}{\partial n} = g_2 \qquad on\ \Gamma_N, \subset \partial\Omega \tag{10.3}$$

$\Gamma_D \cup \Gamma_N = \partial\Omega$. In the vicinity of edges or vertices of interest, we assume that homogeneous boundary conditions are applied for clarity and simplicity of

Z. Yosibash, *Singularities in Elliptic Boundary Value Problems and Elasticity and Their Connection with Failure Initiation*, Interdisciplinary Applied Mathematics 37,
DOI 10.1007/978-1-4614-1508-4_10,

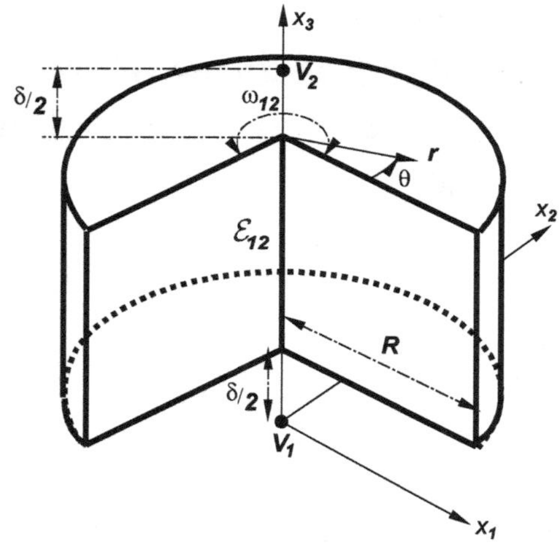

Fig. 10.1 The edge sub-domain $\mathcal{E}_{12}(R)$.

presentation. Before we explain in detail the expansion of the singular solutions, we provide here a summary for the three different singularities.

Edge Singularities: Consider one of the edges denoted by $\mathcal{E}_{ij}$ connecting the vertices V_i and V_j. Moving away from the vertex a distance $\delta/2$, we create a cylindrical sector subdomain of radius $r = R$ with the edge $\mathcal{E}_{12}$ as its axis, as shown in Figure 10.1. The solution in the edge's neighborhood can be decomposed as follows:

$$\tau(r,\theta,x_3) = \sum_{n=1}^{N}\sum_{\ell=0}^{L}\sum_{m=0}^{M} \partial_3^m A_{n\ell}(x_3)\, r^{\alpha_n+m}\, (\ln r)^{\ell}\, s^{+}_{n\ell m}(\theta) \; + \; v(r,\theta,x_3), \quad (10.4)$$

where $L \geq 0$ is an integer that is zero unless α_n is an integer, $\alpha_{n+1} \geq \alpha_n$ are called edge eigenvalues, and $A_{n\ell}(x_3)$ are analytic in x_3 and are called edge flux intensity functions (EFIFs). The $s^{+}_{n\ell m}(\theta)$ are analytic in θ, called edge eigenfunctions for $m = 0$ and shadow functions for $m > 0$. The function $v(r,\theta,x_3)$ belongs to $H^q(\mathcal{E})$, the usual Sobolev space, where q can be as large as required and depends on N and M. We shall assume that α_n for $n \leq N$ are not integers, and that no "crossing points" are of interest (see a detailed explanation in [40]). Therefore, (10.4) becomes

$$\tau(r,\theta,x_3) = \sum_{n=1}^{N}\sum_{m=0}^{M} \partial_3^m A_n(x_3)\, r^{\alpha_n+m}\; s^{+}_{nm}(\theta) \; + \; v(r,\theta,x_3). \quad (10.5)$$

Vertex Singularities: A sphere of radius $\rho = \delta$, centered at the vertex V_1 for example, is constructed and intersected with the domain Ω. Then, a cone having an opening angle θ_1 is constructed such that it intersects at V_1, and removed from the previously constructed subdomain, as shown in Figure 10.2. The resulting vertex subdomain is denoted by $V_1\rho$, and the solution τ can be decomposed in its vicinity using a spherical coordinate system by

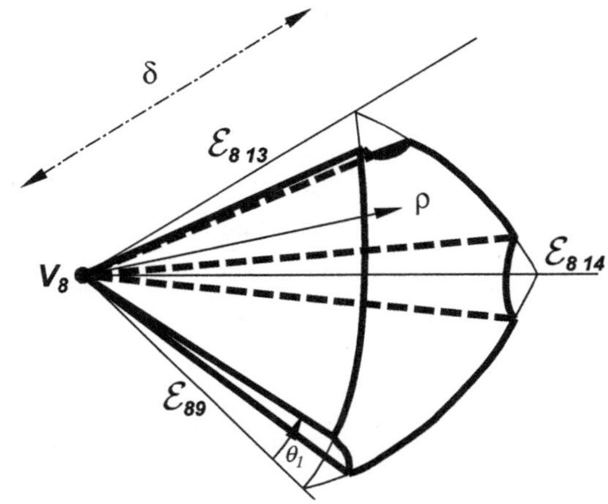

Fig. 10.2 The vertex neighborhood V_1.

$$\tau(\rho,\phi,\theta) = \sum_{\ell=1}^{L}\sum_{p=0}^{P} C_{\ell p}\, \rho^{\gamma_\ell}\, (\ln\rho)^p\, h_{\ell p}(\theta,\phi) \;+\; v(\rho,\theta,\phi), \tag{10.6}$$

where $P \geq 0$ is an integer that is zero unless γ_ℓ is an integer, $\gamma_{\ell+1} \geq \gamma_\ell$ are called vertex eigenvalues, and $h_{\ell p}(\theta,\phi)$ are analytic in ϕ and θ *away from the edges* and are called vertex eigenfunctions. The $C_{\ell p}$ are called vertex flux intensity factors (VFIF). The function $v(\rho,\theta,\phi)$ belongs to $H^q(\mathcal{V})$, where q depends on L. We shall assume that γ_ℓ for $\ell \leq L$ is not an integer. Therefore, (10.6) becomes

$$\tau(\rho,\phi,\theta) = \sum_{\ell=1}^{L} C_\ell\, \rho^{\gamma_\ell}\, h_\ell(\phi,\theta) \;+\; v(\rho,\phi,\theta). \tag{10.7}$$

Vertex-Edge Singularities: The most complicated decomposition of the solution arises in the case of a vertex-edge intersection. For example, let us consider the neighborhood where the edge $\mathcal{E}_{12}$ approaches the vertex V_1. A spherical coordinate system is located in the vertex V_1, and a cone having an opening angle θ_1 with its vertex coinciding with V_1 is constructed with $\mathcal{E}_{12}$ being its central axis. This cone is terminated by a ball-shaped basis having radius $\rho = \delta$, as shown in Figure 10.3. The resulting vertex-edge subdomain is denoted by $\mathcal{V}E_{\delta,R}(\Lambda_{11},\Lambda_{10,11})$, and the solution u can be decomposed in $\mathcal{V}E_{\delta,R}(\Lambda_{11},\Lambda_{10,11})$:

$$\tau(\rho,\phi,\theta) = \sum_{k=1}^{K}\sum_{s=0}^{S}\left(\sum_{l=1}^{L} A_{ksl}\, \rho^{\gamma_l} \;+\; m_{ks}(\rho)\right)\,(\sin\theta)^{\alpha_k}\,[\ln(\sin\theta)]^s\, g_{ks}(\phi)$$

$$+\sum_{l=1}^{L}\sum_{p=0}^{P} C_{lp}\, \rho^{\gamma_l}\,(\ln\rho)^p\, h_{lp}(\theta,\phi) \;+\; v(\rho,\theta,\phi), \tag{10.8}$$

Fig. 10.3 The vertex-edge neighborhood.

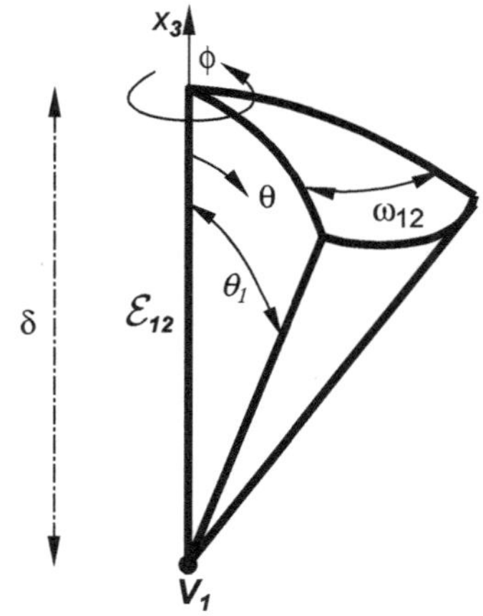

where $m_{ks}(\rho)$ are analytic in ρ, $g_{ks}(\phi)$ are analytic in ϕ, and $h_{lp}(\theta,\phi)$ are analytic in ϕ and θ. The function $v(\rho,\theta,\phi)$ belongs to $H^q(\mathcal{V}E)$, where q is as large as required depending on L and K.

The eigenvalues and eigenfunctions are associated pairs (eigenpairs) that depend on the material properties, the geometry, and the boundary conditions *in the vicinity of the singular vertex/edge only.*

10.1 Asymptotic Solution to the Laplace Equation in a Neighborhood of an Edge

Section devoted to the memory of my mentor Prof. Bernard Schiff, with whom most of this analysis was performed

Having obtained the eigenpairs for the 2-D Laplacian, we wish to use them to construct the full series expansion solution for the 3-D Laplacian. Let $r^{\alpha_n} s_n^+(\theta)$ be an eigenpair of the 2-D Laplacian (denoted by Δ_{2D}) over the x_1-x_2 plane perpendicular to an edge along the x_3-axis. Thus,

$$\Delta_{2D}\left[r^{\alpha_n} s_n^+(\theta)\right] = r^{\alpha_n-2}\left[\alpha_n^2 s_n^+(\theta) + {s^+}''_n(\theta)\right] = 0. \tag{10.9}$$

Let $A_n(x_3)$ be the edge-flux-intensity function associated with the eigenpair. It is clear that $A_n(x_3) r^{\alpha_n} s_n^+(\theta)$ does not satisfy the 3-D Laplacian unless $A_n(x_3)$ is a polynomial of degree 1 or less:

$$\begin{aligned}\Delta_{3D}\left[A_n(x_3) r^{\alpha_n} s_n^+(\theta)\right] &= (\Delta_{2D} + \partial_3^2)\left[A_n(x_3) r^{\alpha_n} s_n^+(\theta)\right] \\ &= \partial_3^2 A_n(x_3) r^{\alpha_n} s_n^+(\theta) \neq 0. \end{aligned} \tag{10.10}$$

Let us then augment the 2-D eigenfunction $A_n(x_3)r^{\alpha_n}s_n^+(\theta)$ by $\frac{-1}{4(\alpha_n+1)}\partial_3^2 A_n(x_3)$ $r^{\alpha_n+2}s_n^+(\theta)$. Then substituting in the Laplace equation, one obtains

$$\Delta_{3D}\left[A_n(x_3)r^{\alpha_n}s_n^+(\theta) - \frac{1}{4(\alpha_n+1)}\partial_3^2 A_n(x_3)r^{\alpha_n+2}s_n^+(\theta)\right]$$
$$= \frac{-1}{4(\alpha_n+1)}\partial_3^4 A_n(x_3)r^{\alpha_n+2}s_n^+(\theta) \neq 0. \tag{10.11}$$

The edge-flux-intensity function is a smooth function of the variable x_3, so it may be approximated by a basis of polynomials. Examining (10.11), one may observe that if $A_n(x_3)$ is a polynomial of degree less than or equal to three, then the two terms inside the brackets are sufficient to form the solution to the 3-D Laplacian associated with the 2-D nth eigenpair. Otherwise, one needs to add a new term, $\frac{1}{32(\alpha_n+1)(\alpha_n+2)}\partial_3^4 A_n(x_3)r^{\alpha_n+4}s_n^+(\theta)$, so that now the residual is

$$\Delta_{3D}\left[A_n(x_3)r^{\alpha_n}s_n^+(\theta) - \frac{1}{4(\alpha_n+1)}\partial_3^2 A_n(x_3)r^{\alpha_n+2}s_n^+(\theta)\right.$$
$$\left.+\frac{1}{32(\alpha_n+1)(\alpha_n+2)}\partial_3^4 A_n(x_3)r^{\alpha_n+4}s_n^+(\theta)\right]$$
$$= \frac{1}{32(\alpha_n+1)(\alpha_n+2)}\partial_3^6 A_n(x_3)r^{\alpha_n+4}s_n^+(\theta). \tag{10.12}$$

The residual now vanishes now if $A_n(x_3)$ is a polynomial of degree less than or equal to five. We may proceed in a similar fashion and obtain the following function $\tau_n^+(r,\theta,x_3)$ associated with the 2-D eigenpair $r^{\alpha_n}s_n^+(\theta)$:

$$\tau_n^+(r,\theta,x_3) = r^{\alpha_n}s_n^+(\theta)\sum_{i=0}^{\infty}\partial_3^{2i}A_n(x_3)r^{2i}\frac{(-0.25)^i}{\prod_{j=1}^{i} j(\alpha_n+j)}. \tag{10.13}$$

This function satisfies identically the 3-D Laplace equation: $\Delta_{3D}\tau_n^+ \equiv 0$. If the series (10.13) is truncated at the Nth term, the remainder, which does not satisfy the 3-D Laplace equation, is

$$\partial_3^{2N+2}A_n(x_3)r^{\alpha_n+2N}s_n^+(\theta)\frac{(-0.25)^N}{\prod_{j=1}^{N} j(\alpha_n+j)}. \tag{10.14}$$

Thus the three-dimensional edge singular solution (10.5) can be represented as

$$\tau(r,\theta,x_3) = \sum_{n=1}^{\infty}\tau_n^+(r,\theta,x_3)$$
$$= \sum_{n=1}^{\infty}A_n(x_3)r^{\alpha_n}s_n^+(\theta) + c_{n1}\partial_3^2 A_n(x_3)r^{\alpha_n+2}s_n^+(\theta)$$
$$+c_{n2}\partial_3^4 A_n(x_3)r^{\alpha_n+4}s_n^+(\theta) + \cdots \tag{10.15}$$

where $c_{n0} = 1$ and the other c_{ni}'s are given constants.

Remark 10.1 *We have shown that the 3-D solution in the vicinity of a straight edge can be obtained using the 2-D eigenfunctions of θ and the edge-flux-intensity function, complemented by additional functions called "shadow functions," which are multiplied by* derivatives *of the associated edge flux intensity function.*

Remark 10.2 *For the Laplacian, the shadow functions coincide with the 2-D eigenfunctions multiplied by a constant. As will be shown in the sequel, for a general scalar operator with different k_{ij}'s, the shadow functions are different.*

Remark 10.3 *We proved in Chapter 1 (see* (1.21)*) that the eigenfunctions $s_n^+(\theta)$ and their duals are orthogonal in the sense that*

$$\int_{\theta=0}^{\omega_{ij}} s_n^+ s_m^- \, d\theta = 0 \qquad \textit{for } m \neq n.$$

We have exploited this orthogonality condition when devising the dual function method for extracting efficiently the flux intensity factors in 2-D domains. However, in 3-D domains, since higher-order terms (r^{α_n+2}, $r^{\alpha_n+4}, \ldots$) in the expansion (10.14) *consist of the same functions $s_n^+(\theta)$ as other lower-order terms, then terms in the series expansion are no longer orthogonal. This imposes difficulties in implementing the contour integral method for 3-D domains, and remedies are provided in the next chapter.*

The asymptotic expansion of the solution in a neighborhood of an edge presented above can be brought to the classical expansion of the 3-D solution in terms of Bessel functions, as shown in the following. Let us first recall the classical solution of the Laplace equation, obtained by separation of variables. In cylindrical coordinates, the Laplace equation is

$$\frac{1}{r}\frac{\partial}{\partial r}\left(r\frac{\partial \tau}{\partial r}\right) + \frac{1}{r^2}\frac{\partial^2 \tau}{\partial \theta^2} + \frac{\partial^2 \tau}{\partial x_3^2} = 0. \tag{10.16}$$

Assume $\tau(r,\theta,x_3) = R(r)\Theta(\theta)Z(x_3)$, so that after substitution in (10.16) and division by $R(r)\Theta(\theta)Z(x_3)$, (10.16) becomes

$$\frac{1}{rR}\frac{d}{dr}(rR') + \frac{\Theta''}{r^2\Theta} + \frac{Z''}{Z} = 0. \tag{10.17}$$

The last term is independent of r and θ, so it must be a constant, denoted by $\frac{Z''}{Z} = D$. Now multiply now the equation by r^2, so that (10.17) becomes

$$\frac{r}{R}\frac{d}{dr}(rR') + \frac{\Theta''}{\Theta} + r^2 D = 0. \tag{10.18}$$

Again, the second term in (10.18) is θ-dependent while other terms are r-dependent, so that the expression $\frac{\Theta''}{\Theta}$ has to be a negative constant if an oscillatory solution in θ is sought, i.e., $\frac{\Theta''}{\Theta} = -\alpha^2$. Therefore, the solution to $\Theta(\theta)$ is of the form,

$$\Theta(\theta) = e^{i\alpha\theta}. \tag{10.19}$$

The values of α are determined by satisfying boundary conditions at $\theta = 0$ and $\theta = \omega_{12}$, i.e., these are exactly the eigenpairs for the 2-D problem, and there is an infinite number of distinct eigenpairs $\Theta_n(\theta) \equiv s_n(\theta)$. $\Theta_n(\theta)$ is given by a linear combination of $\sin(\alpha_n\theta)$ and $\cos(\alpha_n\theta)$.

The case $\frac{\Theta''}{\Theta} = +\alpha^2$ is excluded because it produces a solution that is exponential in θ, and thus cannot satisfy boundary conditions. Returning to equation (10.18),

$$r^2R'' + rR' + (Dr^2 - \alpha_n^2)R = 0, \tag{10.20}$$

there are two possibilities:

$\underline{0 > D = -\gamma^2}$:

Define $q = \gamma r$ so that $R(r) = R(q/\gamma) \stackrel{\text{def}}{=} Q(q)$, and (10.20) becomes

$$q^2Q'' + qQ' - (q^2 + \alpha_n^2)Q = 0. \tag{10.21}$$

Equation (10.21) is the modified Bessel equation, and its solution for a domain where $r = 0$ is included is the modified Bessel function of the first kind of order α_n (see [102, pp. 108-110]):

$$R(r) = I_{\alpha_n}(q) = \sum_{k=0}^{\infty} \frac{(q/2)^{2k+\alpha_n}}{\Gamma(k+1)\Gamma(\alpha_n + k + 1)}, \tag{10.22}$$

where $\Gamma(q)$ is the gamma function [102, p. 1]

$$\Gamma(q) \stackrel{\text{def}}{=} \int_0^{\infty} e^{-t}t^{q-1}dt \tag{10.23}$$

Because $\frac{Z''}{Z} = \delta = -\gamma^2$, we immediately obtain oscillatory behavior in x_3, and by imposing the boundary conditions at given x_3 we again obtain an infinite number of distinct values of γ_m, i.e., any $Z_m(x_3)$ is given by a linear combination of $\sin(\gamma_m x_3)$ and $\cos(\gamma_m x_3)$.

Summarizing, the complete solution τ, oscillatory in θ and x_3, is

$$\tau(r, \theta, x_3) = \sum_{n,m=1}^{\infty} I_{\alpha_n}(\gamma_m r)s_n^+(\theta)Z_m(x_3). \tag{10.24}$$

$0 > D = +\gamma^2$:

Again define $q = \gamma r$ so that $R(r) = R(q/\gamma) \stackrel{\text{def}}{=} Q(q)$, and (10.20) becomes

$$q^2 Q'' + qQ' + (q^2 - \alpha_n^2)Q = 0. \tag{10.25}$$

Equation (10.25) is the Bessel equation, and its solution for a domain where $r = 0$ is included is the Bessel function of the first kind of order α_n (see [102, p. 102]):

$$J_{\alpha_n}(q) \stackrel{\text{def}}{=} \sum_{k=0}^{\infty} \frac{(-1)^k (q/2)^{2k+\alpha_n}}{\Gamma(k+1)\Gamma(\alpha_n + k + 1)}. \tag{10.26}$$

Because now $\frac{Z''}{Z} = D = +\gamma^2$, we immediately obtain exponential behavior in x_3: $Z(x_3) = e^{\pm\gamma x_3}$.

Summarizing, the complete solution u, oscillatory in θ and exponential in x_3, is

$$\tau(r, \theta, x_3) = \sum_{n=1}^{\infty} J_{\alpha_n}(\gamma r) e^{\pm\gamma x_3} s_n^+(\theta). \tag{10.27}$$

The value of γ is determined by boundary conditions on r =const. For example, if $\tau = 0$ on $r = C$, then $J_{\alpha_n}(\gamma C) = 0$, so that γC is a zero of J_{α_n}.

Let us now prove that the asymptotic solution presented in (10.15) can be brought to the classical solution (10.27) if its behavior in x_3 is exponential, or to the classical solution (10.24) if its behavior in x_3 is oscillatory. To this end, we first need to introduce the following connections. Integrating (10.23) by parts, it is easily shown that the gamma function satisfies the identity

$$\Gamma(q+1) = q\Gamma(q), \tag{10.28}$$

and by recursive substitution it can be shown that

$$\begin{aligned} \Gamma(q+k+1) &= (q+k)\Gamma(q+k) = (q+k)\left[(q+k-1)\Gamma(q+k-1)\right] \\ &= \cdots = \Gamma(q+1)\prod_{\ell=1}^{k}(q+\ell), \qquad k \in \mathbb{N}. \end{aligned} \tag{10.29}$$

Having in mind that $\Gamma(j) = j!$, for any positive integer j, let us consider the following expression:

$$\begin{aligned} \Gamma(j+1)\Gamma(q+j+1) &= \left[j\Gamma(j)\right]\left[(q+j)\Gamma(q+j)\right] \\ &= \left[j(j-1)\Gamma(j-1)\right]\left[(q+j)(q+j-1)\Gamma(q+j-1\right] \\ &= \cdots = \Gamma(q+1)\prod_{k=1}^{j} k(q+k), \qquad j \in \mathbb{N}, \end{aligned} \tag{10.30}$$

or writing (10.30) somewhat differently,

$$\frac{1}{\prod_{k=1}^{j} k(q+k)} = \frac{\Gamma(q+1)}{\Gamma(j+1)\Gamma(q+j+1)}. \tag{10.31}$$

Substituting (10.31) in (10.13), the later becomes

$$\tau_n^{+}(r,\theta,x_3) = r^{\alpha_n} s_n^{+}(\theta)\Gamma(\alpha_n+1)\sum_{k=0}^{\infty} \partial_3^{2k} A_n(x_3) r^{2k} \frac{(-0.25)^k}{\Gamma(k+1)\Gamma(\alpha_n+k+1)}. \tag{10.32}$$

Assume that $A_n(x_3)$ has an exponential behavior in x_3, i.e., it may be represented as follows:

$$A_n(x_3) = e^{\pm\gamma x_3}, \Longrightarrow \partial_3^{2k} A_n(x_3) = \gamma^{2k} e^{\pm\gamma x_3}. \tag{10.33}$$

Then after substituting (10.33) in (10.32) and rearranging, one obtains

$$\tau_n^{+}(r,\theta,x_3) = \frac{2\Gamma(\alpha_n+1)}{\gamma^{\alpha_n}} e^{\pm\gamma x_3} s_n(\theta) \sum_{k=0}^{\infty} \frac{(-1)^k (r\gamma/2)^{2k+\alpha_n}}{\Gamma(k+1)\Gamma(\alpha_n+k+1)}. \tag{10.34}$$

Notice the definition of the Bessel function of the first kind of order α in (10.26); $\tau_n^{+}(r,\theta,x_3)$ can be represented in terms of the Bessel function:

$$\tau_n^{+}(r,\theta,x_3) = \frac{2\Gamma(\alpha_n+1)}{\gamma^{\alpha_n}} e^{\pm\gamma x_3} J_{\alpha_n}(\gamma r) s_n^{+}(\theta). \tag{10.35}$$

Because $\frac{2\Gamma(\alpha_n+1)}{\gamma^{\alpha_n}}$ is a constant, we can include it in the constant appearing in $s_n^{+}(\theta)$, so that (10.15) is identical to the classical solution (10.27).

If instead $A_n(x_3)$ has oscillatory behavior in x_3, i.e. it may be represented as

$$A_n(x_3) = e^{\pm i\gamma x_3}, \Longrightarrow \partial_3^{2k} a_n(x_3) = (-1)^k \gamma^{2k} e^{\pm i\gamma x_3}, \tag{10.36}$$

then after substituting (10.36) in (10.32) and rearranging, one obtains

$$\tau_n^{+}(r,\theta,x_3) = \frac{2\Gamma(\alpha_n+1)}{\gamma^{\alpha_n}} e^{\pm i\gamma x_3} s_n^{+}(\theta) \sum_{k=0}^{\infty} \frac{(r\gamma/2)^{2k+\alpha_n}}{\Gamma(k+1)\Gamma(\alpha_n+k+1)}. \tag{10.37}$$

Notice the definition of the modified Bessel function of the first kind of order α in (10.22); $\tau_n^{+}(r,\theta,x_3)$ can be represented in terms of the modified Bessel function:

$$\tau_n^{+}(r,\theta,x_3) = \frac{2\Gamma(\alpha_n+1)}{\gamma^{\alpha_n}} e^{\pm i\gamma x_3} I_{\alpha_n}(\gamma r) s_n^{+}(\theta). \tag{10.38}$$

The constant $\frac{2\Gamma(\alpha_n+1)}{\alpha_n}$ is included in the constant appearing in $s_n^+(\theta)$, and $\frac{1}{\gamma}$ is included in the constant of the oscillatory function in x_3, so that (10.15) is again identical to the classical solution (10.24).

10.2 A Systematic Mathematical Algorithm for the Edge Asymptotic Solution for a General Scalar Elliptic Equation

As shown for the simplified Laplacian the solution in the vicinity of an edge may be constructed by the 2-D eigenpairs and an infinite number of associated "shadow functions." Here a systematic mathematical algorithm for the computation of the solution in the vicinity of an edge is presented for a general scalar elliptic PDE based on five selected examples.

Consider a subdomain Ω such that only one straight edge $\mathcal{E}$ is present. The domain is generated as the product $\Omega = G \times I$, where I is the interval $[-1, 1]$, and G is a plane bounded sector of opening $\omega \in (0, 2\pi]$ and radius 1 (the case of a crack, $\omega = 2\pi$, is included); see Figure 10.4. Although any radius or interval I can be chosen, these simplified numbers have been chosen for simplicity of presentation.

The variables in G and I are (x_1, x_2) and x_3 respectively, and the two flat planes that intersect at the edge $\mathcal{E}$ are denoted by Γ_1 and Γ_2. A part of a cylindrical surface is defined as follows:

$$\Gamma_R := \left\{ \boldsymbol{x} \in \mathbb{R}^3 \mid r = R,\ \theta \in (0, \omega),\ x_3 \in I \right\}.$$

The homogeneous general scalar elliptic PDE in (1.35) is considered, i.e.,

$$\mathcal{L}(\tau) \stackrel{\text{def}}{=} -\partial_i \left(k_{ij} \partial_j \tau \right) = 0 \quad \text{in } \Omega, \tag{10.39}$$

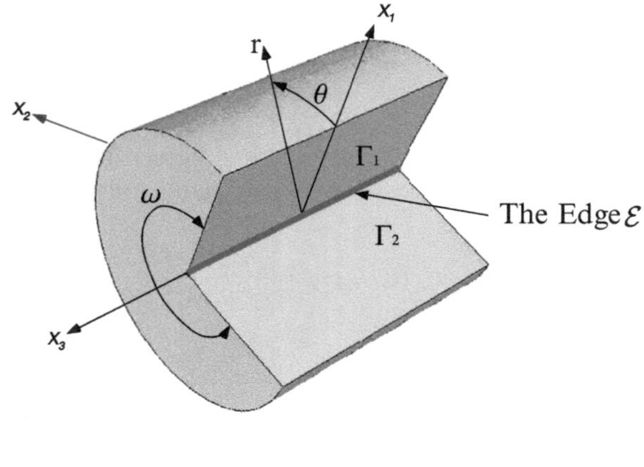

Fig. 10.4 The subdomain Ω in the vicinity of an edge.

Table 10.1 The various cases considered

		The Operator					
Case #	ω	k_{11}	k_{22}	k_{33}	k_{12}	k_{13}	k_{23}
Case 1	$3\pi/2$	1	1	1	0	0	0
Case 2	$3\pi/2$	5	4	1	-4	0	0
Case 3	2π	1	1	1	0	0	0
Case 4	2π	5	4	1	-4	0	0
Case 5	2π	1	1	1	0	$-1/2$	0

and without loss of generality, k_{33} is set as $k_{33} = 1$.In this section we consider homogeneous Dirichlet boundary conditions on Γ_1 and Γ_2, i.e.,

$$\tau(r, 0, x_3) = \tau(r, \omega, x_3) = 0. \tag{10.40}$$

All methods presented here carry over to homogeneous Neumann boundary conditions, as detailed in Section 10.3, and may also be extended to homogeneous mixed boundary conditions. For demonstration purposes three specific operators are considered: the Laplace operator $k_{ij} = \delta_{ij}$, a general operator with $k_{11} = 5, k_{22} = 4$, $k_{12} = -4$ and $k_{13} = k_{23} = 0$, and a general operator having also mixed derivatives in the x_3 direction with $k_{11} = k_{22} = 1$, $k_{13} = -1$, and $k_{12} = k_{23} = 0$. Two domains are considered, one having $\omega = 3\pi/2$ and the other a cracked domain, $\omega = 2\pi$. Combination of the two different domains and three different operators provide five specific cases according to Table 10.1.

10.2.1 The Eigenpairs and Computation of Shadow Functions

The functional representation of the exact solution for the problem $\mathcal{L}(\tau) = 0$ in a neighborhood of the edge $\mathcal{E}$ relies on splitting the operator $\mathcal{L}$ into three parts (as shown in [46]):

$$\mathcal{L} = M_0(\partial_1, \partial_2) + M_1(\partial_1, \partial_2)\partial_3 + M_2\partial_3^2, \tag{10.41}$$

where

$$M_0 \stackrel{\text{def}}{=} k_{11}\partial_1^2 + 2k_{12}\partial_1\partial_2 + k_{22}\partial_2^2, \tag{10.42}$$

$$M_1 \stackrel{\text{def}}{=} 2k_{13}\partial_1 + 2k_{23}\partial_2, \qquad M_2 \stackrel{\text{def}}{=} k_{33}. \tag{10.43}$$

The splitting allows consideration of a solution τ of the form

$$\tau = \sum_{j\geq 0} \partial_3^j A(x_3)\Phi_j(x_1, x_2). \tag{10.44}$$

Inserting (10.44) into (10.39), one obtains

$$\sum_{j\geq 0} \partial_3^j A(x_3) M_0 \Phi_j + \sum_{j\geq 0} \partial_3^{j+1} A(x_3) M_1 \Phi_j + \sum_{j\geq 0} \partial_3^{j+2} A(x_3) M_2 \Phi_j = 0, \quad (10.45)$$

and after rearranging,

$$A(x_3) M_0 \Phi_0 + \partial_3 A(x_3)(M_0 \Phi_1 + M_1 \Phi_0) + \sum_{j\geq 0} \partial_3^{j+2} A(x_3)(M_0 \Phi_{j+2} + M_1 \Phi_{j+1} + M_2 \Phi_j) = 0. \quad (10.46)$$

Equation (10.46) has to hold for any smooth function $A(x_3)$. Thus, the functions Φ_j must satisfy the three equations below:

$$\begin{cases} M_0 \Phi_0 = 0, \\ M_0 \Phi_1 + M_1 \Phi_0 = 0, \\ M_0 \Phi_{j+2} + M_1 \Phi_{j+1} + M_2 \Phi_j = 0, \quad j \geq 0, \end{cases} \qquad (x_1, x_2) \in G, \quad (10.47)$$

accompanied by homogeneous boundary conditions on the two faces $\theta = 0, \omega$.

The first PDE in (10.47) generates a solution Φ_0 of the form

$$\Phi_0 = r^\alpha \varphi_0(\theta),$$

which is nothing more than the 2-D (*called "primal"*) eigenfunction on G. Equation $(10.47)_2$ with homogeneous Dirichlet boundary conditions determines Φ_1 (which depends on Φ_0), given by

$$\Phi_1 = r^{\alpha+1} \varphi_1(\theta). \quad (10.48)$$

The terms of the sequence Φ_j ($j \geq 2$) are solutions of $(10.47)_3$ with Dirichlet boundary conditions:

$$\Phi_j = r^{\alpha+j} \varphi_j(\theta). \quad (10.49)$$

All Φ_j, where $j \geq 1$, are called the *"shadow" eigenfunctions* associated with the primal leading function Φ_0. There is an infinite number of functions Φ_j associated with positive eigenvalues α_i, and therefore

$$\Phi_j^{(\alpha_i)} = r^{\alpha_i + j} \varphi_j^{(\alpha_i)}(\theta), \quad j = 0, 1, \ldots \quad (10.50)$$

Thus, for each eigenvalue α_i, the 3-D solution associated with it is

$$\tau^{(\alpha_i)} = \sum_{j\geq 0} \partial_3^j A^{(\alpha_i)}(x_3) r^{\alpha_i + j} \varphi_j^{(\alpha_i)}(\theta), \quad (10.51)$$

and the overall solution τ is

$$\tau = \sum_{i \geq 1} \sum_{j \geq 0} \partial_3^j A^{(\alpha_i)}(x_3) r^{\alpha_i + j} \varphi_j^{(\alpha_i)}(\theta), \tag{10.52}$$

where $A^{(\alpha_i)}(x_3)$ is the edge-flux intensity function (EFIF) associated with the ith eigenvalue.

Solutions associated with the negative eigenvalues are called *dual solutions*, and are denoted by Ψ. For example,

$$\Psi_0^{(\alpha_i)} = c_0^{(\alpha_i)} r^{-\alpha_i} \psi_0^{(\alpha_i)}(\theta), \tag{10.53}$$

where $\Psi_0^{(\alpha_i)}$ is the leading dual eigen-solution and

$$\Psi_j^{(\alpha_i)} = c_0^{(\alpha_i)} r^{-\alpha_i + j} \psi_j^{(\alpha_i)}(\theta), \tag{10.54}$$

are the shadow dual eigensolutions. The constant $c_0^{(\alpha_i)}$ can be arbitrarily chosen, and is chosen to satisfy an orthonormal condition, as will be explained; see (10.65). Theoretical details and a rigorous mathematical formulation are provided in [46].

Remark 10.4 *Operators for which $M_1 = 0$ (such as the Laplacian, for example) imply that Φ_j and Ψ_j of odd rank are zero: $\Phi_j^{(\alpha_i)} = \Psi_j^{(\alpha_i)} = 0, \quad j = 1, 3, 5, \ldots$.*

10.2.2 Eigenfunctions, their Shadow Functions and Duals for Cases 1-4 (Dirichlet BCs)

For cases 1-4 the operator $\mathcal{L}$ can be split as in (10.47) with

$$M_0 = k_{11}\partial_1\partial_1 + 2k_{12}\partial_1\partial_2 + k_{22}\partial_2\partial_2, \quad M_1 = 0, \quad M_2 = 1. \tag{10.55}$$

Computing the primal and dual eigenfunctions Φ_0 and Ψ_0.

Φ_0 and Ψ_0 are the solutions of $(10.47)_1$ in the plane domain G. A change of variables is performed (details in Appendix B),

$$\xi(x_1, x_2) = \sqrt{\frac{k_{22}}{k_{11}k_{22} - k_{12}^2}} x_1 - \sqrt{\frac{k_{12}^2}{k_{22}(k_{11}k_{22} - k_{12}^2)}} x_2, \tag{10.56}$$

$$\eta(x_1, x_2) = \sqrt{\frac{1}{k_{22}}} x_2, \tag{10.57}$$

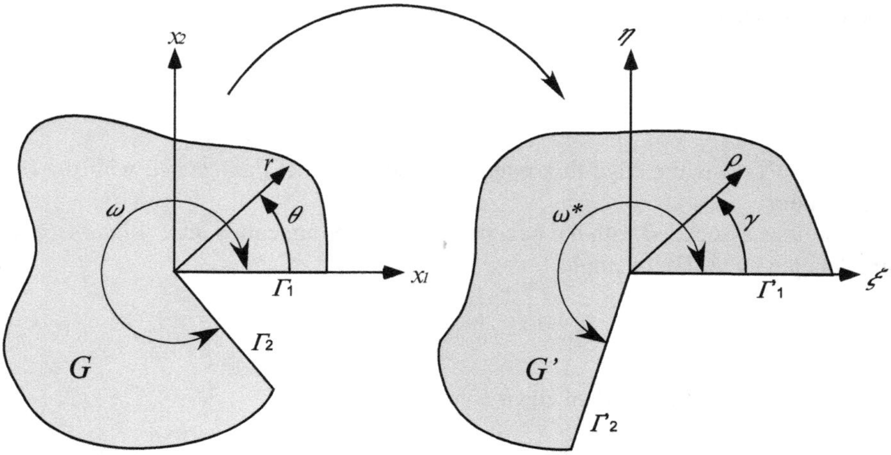

Fig. 10.5 The plane domain G before and after change of variables.

so that M_0 in the new variables is transformed into the Laplace operator

$$\frac{\partial^2}{\partial \xi^2} + \frac{\partial^2}{\partial \eta^2} \tag{10.58}$$

over a plane domain G', as illustrated in Figure 10.5. The straight lines defined by $\theta = 0$ and $\theta = \omega$ in the original domain G are transformed into the two lines defined by $\gamma = 0$ and $\gamma = \omega^*$ in the transformed domain G', where

$$\omega^* = \arctan\left(\frac{\sqrt{k_{11}k_{22} - k_{12}^2}\sin\omega}{k_{22}\cos\omega - k_{12}\sin\omega}\right). \tag{10.59}$$

Both Φ_0 and Ψ_0 have to satisfy homogeneous Dirichlet boundary conditions on $\theta = 0, \omega$ in the original domain, which become in the transformed domain

$$\Phi_0(\rho, 0) = \Phi_0(\rho, \omega^*) = \Psi_0(\rho, 0) = \Psi_0(\rho, \omega^*) = 0. \tag{10.60}$$

The solutions to the Laplace equation (by separation of variables) are

$$\begin{cases} \Phi_0(\rho, \gamma) = \rho^{\alpha}\left(A\cos(\alpha\gamma) + B\sin(\alpha\gamma)\right). \\ \Psi_0(\rho, \gamma) = c_0\rho^{-\alpha}\left(A\cos(\alpha\gamma) - B\sin(\alpha\gamma)\right). \end{cases} \tag{10.61}$$

Equation (10.60) results in (here we provide the equations for Φ_0, although the same ones are obtained for Ψ_0)

$$\begin{pmatrix} 1 & 0 \\ \cos(\alpha\omega^*) & \sin(\alpha\omega^*) \end{pmatrix}\begin{pmatrix} A \\ B \end{pmatrix} = \begin{pmatrix} 0 \\ 0 \end{pmatrix}. \tag{10.62}$$

For a nontrivial solution, the determinant of the matrix in (10.62) must vanish, i.e., α has to satisfy

$$\sin(\alpha\omega^*) = 0 \quad \Rightarrow \quad \alpha_i = \pm\frac{i\pi}{\omega^*}, \qquad i = 1, 2, \ldots, \tag{10.63}$$

There is an infinite number of distinct α_i's for which there is an associated $\Phi_0^{(\alpha_i)}$ and $\Psi_0^{(\alpha_i)}$, and distinct B_i where

$$A_i = 0.$$

The generic constant is omitted, since it is added to the EFIF in the asymptotic expansion. To obtain the solution in the original domain G, a reverse transformation of variables is performed and the functions $\Phi_0^{(\alpha_i)}$ and $\Psi_0^{(\alpha_i)}$ are obtained in the coordinates r, θ:

$$\Phi_0^{(\alpha_i)}(r,\theta) = r^{\alpha_i}\varphi_0^{(\alpha_i)}(\theta), \qquad \Psi_0^{(\alpha_i)}(r,\theta) = c_0^{(\alpha_i)} r^{-\alpha_i}\psi_0^{(\alpha_i)}(\theta),$$

where

$$\varphi_0^{(\alpha_i)}(\theta) = \left(\frac{k_{22}\cos^2\theta - k_{12}\sin(2\theta) + k_{11}\sin^2\theta}{k_{11}k_{22} - k_{12}^2}\right)^{\frac{\alpha_i}{2}} \sin\left(\alpha_i \arctan\left(\frac{\sqrt{k_{11}k_{22} - k_{12}^2}\sin\theta}{k_{22}\cos\theta - k_{12}\sin\theta}\right)\right)$$

and

$$\psi_0^{(\alpha_i)}(\theta) = \left(\frac{k_{22}\cos^2\theta - k_{12}\sin(2\theta) + k_{11}\sin^2\theta}{k_{11}k_{22} - k_{12}^2}\right)^{-\frac{\alpha_i}{2}} \sin\left(\alpha_i \arctan\left(\frac{\sqrt{k_{11}k_{22} - k_{12}^2}\sin\theta}{k_{22}\cos\theta - k_{12}\sin\theta}\right)\right).$$

One may observe that for the Laplace operator $k_{ij} = \delta_{ij}$. Then $\omega^* = \omega$, and the eigenfunctions and their duals are the well known expressions

$$\begin{cases} \Phi_0^{(\alpha_i)}(r,\theta) = r^{\alpha_i}\varphi_0^{(\alpha_i)}(\theta) = r^{\alpha_i}\sin(\alpha_i\theta), \\ \Psi_0^{(\alpha_i)}(r,\theta) = c_0^{(\alpha_i)} r^{-\alpha_i}\psi_0^{(\alpha_i)}(\theta) = c_0^{(\alpha_i)} r^{-\alpha_i}\sin(\alpha_i\theta), \end{cases} \qquad \alpha_i = \frac{i\pi}{\omega}.$$

The value of the constant $c_0^{(\alpha_i)}$ is chosen such that the primal and the dual eigenfunctions, $\Phi_0^{(\alpha_i)}$ and $\Psi_0^{(\alpha_i)}$ satisfy the orthonormal condition

$$\int_0^\omega [T(R)\Phi_0^{(\alpha_i)} \cdot \Psi_0^{(\alpha_i)} - \Phi_0^{(\alpha_i)} \cdot T(R)\Psi_0^{(\alpha_i)}]\, R\, d\theta = 1, \tag{10.64}$$

where $T(R)$ is the radial Neumann trace operator related to M_0:

$$\begin{aligned} T(R) = {} & \left(k_{11}\cos^2\theta + k_{12}\sin 2\theta + k_{22}\sin^2\theta\right)\tfrac{\partial}{\partial r} \\ & + \left(k_{12}\cos 2\theta - \tfrac{1}{2}(k_{11} - k_{22})\sin 2\theta\right)\tfrac{1}{r}\tfrac{\partial}{\partial\theta}\,. \end{aligned}$$

Table 10.2 First eigenvalue and $c_0^{(\alpha_1)}$ for cases 1-4.

Case #	ω^*	α_1	$c_0^{(\alpha_1)}$
Case 1	1.5π	2/3	0.31831
Case 2	1.1476π	0.87139	0.26903
Case 3	2π	0.5	0.31831
Case 4	2π	0.5	0.23533

Further details about (10.64) are given in [46]. The value of the constant $c_0^{(\alpha_i)}$ is extracted from equation (10.64):

$$c_0^{(\alpha_i)} = \left\{ \int_0^{\omega} \left[T(R)\left(r^{\alpha_i} \phi_0^{(\alpha_i)}\right) \cdot \left(r^{-\alpha_i} \psi_0^{(\alpha_i)}\right) - \left(r^{\alpha_i} \phi_0^{(\alpha_i)}\right) \cdot T(R)\left(r^{-\alpha_i} \psi_0^{(\alpha_i)}\right) \right] R \, d\theta \right\}^{-1}. \tag{10.65}$$

One may observe that for the Laplace operator $k_{ij} = \delta_{ij}$, the Neumann trace operator simplifies to $T = \frac{\partial}{\partial r}$, and $c_0^{(\alpha_i)} = \frac{1}{\alpha_i \omega}$, which is the known coefficient of the dual eigenfunction for a two-dimensional domain. The values of $c_0^{(\alpha_1)}$ for cases 1-4 are presented in Table 10.2.

Odd shadow functions and dual shadow functions

Once the primal eigenfunction $\Phi_0^{(\alpha_i)}$ is obtained, the first shadow function $\Phi_1^{(\alpha_i)}$ may be calculated by $(10.47)_2$. Because $M_1 \equiv 0$, the differential equation is homogeneous with homogeneous Dirichlet boundary conditions, and therefore the first shadow function vanishes:

$$\Phi_1^{(\alpha_i)} = 0. \tag{10.66}$$

The sequence of odd shadow functions $\Phi_k^{(\alpha_i)}$ (where $k = 3, 5, 7, \ldots$) are calculated as the solution of $(10.47)_3$. For $\Phi_3^{(\alpha_i)}$ we obtain

$$M_0 \Phi_3^{(\alpha_i)} = -M_2 \Phi_1^{(\alpha_i)} = 0. \tag{10.67}$$

Again, the differential equation is homogeneous with homogeneous Dirichlet boundary conditions, so that $\Phi_3^{(\alpha_i)} \equiv 0$. The same arguments hold for all odd shadow functions associated with an operator $\mathcal{L}$ having $k_{13} = k_{23} = 0$. Thus

$$\Phi_k^{(\alpha_i)} = 0, \qquad k = 3, 5, 7, \ldots \tag{10.68}$$

Computation of the dual shadow functions $\Psi_k^{(\alpha_i)}$ is along the same lines thus for any $\mathcal{L}$ with $k_{13} = k_{23} = 0$,

$$\Psi_k^{(\alpha_i)} = 0 \qquad k = 3, 5, 7, \ldots \tag{10.69}$$

The shadow function $\Phi_2^{(\alpha_i)}$ *and its dual* $\Psi_2^{(\alpha_i)}$ are the solution of $(10.47)_3$ with $j = 0$. It is an inhomogeneous differential equation with homogeneous Dirichlet boundary conditions. Its explicit form in coordinates ρ, γ is,

$$\left(\frac{\partial^2}{\partial \rho^2} + \frac{1}{\rho}\frac{\partial}{\partial \rho} + \frac{1}{\rho^2}\frac{\partial^2}{\partial \gamma^2}\right)\Phi_2^{(\alpha_i)} = -\rho^{\alpha_i}\sin(\alpha_i \gamma). \tag{10.70}$$

The homogeneous solution is

$$\Phi_2^{(\alpha_i)H} = \rho^f\left[A^H\cos(f\gamma) + B^H\sin(f\gamma)\right], \tag{10.71}$$

and the particular solution is

$$\Phi_2^{(\alpha_i)P} = \frac{-1}{4(\alpha_i+1)}\rho^{\alpha_i+2}\sin(\alpha_i\gamma). \tag{10.72}$$

The particular solution identically satisfies the homogeneous Dirichlet boundary conditions, so that the homogeneous solution must satisfy these:

$$\begin{cases} \Phi_2^{(\alpha_i)}(\rho,0) = \Phi_2^{(\alpha_i)H}(\rho,0) + \Phi_2^{(\alpha_i)P}(\rho,0) = \Phi_2^{(\alpha_i)H}(\rho,0) = 0, \\ \Phi_2^{(\alpha_i)}(\rho,\omega^*) = \Phi_2^{(\alpha_i)H}(\rho,\omega^*) + \Phi_2^{(\alpha_i)P}(\rho,\omega^*) = \Phi_2^{(\alpha_i)H}(\rho,\omega^*) = 0. \end{cases} \tag{10.73}$$

The coefficients A^H, B^H vanish, so that Φ_2 is the particular solution alone. We may conclude that $\Phi_2^{(\alpha_i)}$ is given by

$$\Phi_2^{(\alpha_i)}(r,\theta) = r^{\alpha_i+2}\varphi_2^{(\alpha_i)}(\theta), \tag{10.74}$$

where

$$\varphi_2^{(\alpha_i)}(\theta) = -\frac{1}{4(\alpha_i+1)}\left(\frac{k_{22}\cos^2\theta - k_{12}\sin(2\theta) + k_{11}\sin^2\theta}{k_{11}k_{22}-k_{12}^2}\right)^{\frac{\alpha_i+2}{2}}$$
$$\times \sin\left\{\alpha_i \arctan\left(\frac{\sqrt{k_{11}k_{22}-k_{12}^2}\sin\theta}{k_{22}\cos\theta - k_{12}\sin\theta}\right)\right\}.$$

Computing $\Psi_2^{(\alpha_i)}$ follows same arguments, and we obtain

$$\Psi_2^{(\alpha_i)}(r,\theta) = c_0^{(\alpha_i)} r^{-\alpha_i+2}\psi_2^{(\alpha_i)}(\theta), \tag{10.75}$$

where

$$\psi_2^{(\alpha_i)}(\theta) = \frac{1}{4(\alpha_i-1)}\left(\frac{k_{22}\cos^2\theta - k_{12}\sin(2\theta) + k_{11}\sin^2\theta}{k_{11}k_{22}-k_{12}^2}\right)^{\frac{-\alpha_i+2}{2}}$$
$$\times \sin\left\{\alpha_i \arctan\left(\frac{\sqrt{k_{11}k_{22}-k_{12}^2}\sin\theta}{k_{22}\cos\theta - k_{12}\sin\theta}\right)\right\}.$$

The shadow function $\Phi_4^{(\alpha_i)}$ is the solution of (10.47)$_3$ with $j = 2$. The method of extracting $\Phi_4^{(\alpha_i)}$ is very similar to the method of $\Phi_2^{(\alpha_i)}$ extraction. The explicit form of the differential equation in ρ, γ coordinates is

$$\left(\frac{\partial^2}{\partial\rho^2} + \frac{1}{\rho}\frac{\partial}{\partial\rho} + \frac{1}{\rho^2}\frac{\partial^2}{\partial\gamma^2}\right)\Phi_4^{(\alpha_i)} = \frac{-1}{4(\alpha_i + 1)}\rho^{\alpha_i+2}\sin(\alpha_i\gamma). \tag{10.76}$$

The solution of $\Phi_4^{(\alpha_i)}$ is based on the particular solution alone, since the homogeneous solution vanishes under the homogeneous Dirichlet boundary conditions. The shadow function $\Phi_4^{(\alpha_i)}$ in r, θ polar coordinates is

$$\Phi_4^{(\alpha_i)}(r,\theta) = r^{\alpha_i+4}\varphi_4^{(\alpha_i)}(\theta), \tag{10.77}$$

where

$$\varphi_4^{(\alpha_i)}(\theta) = \frac{1}{32(\alpha_i+1)(\alpha_i+2)}\left(\frac{k_{22}\cos^2\theta - k_{12}\sin(2\theta) + k_{11}\sin^2\theta}{k_{11}k_{22}-k_{12}^2}\right)^{\frac{\alpha_i+4}{2}}$$
$$\times \sin\left\{\alpha_i \arctan\left(\frac{\sqrt{k_{11}k_{22}-k_{12}^2}\sin\theta}{k_{22}\cos\theta - k_{12}\sin\theta}\right)\right\}.$$

10.2.2.1 Summary of Cases (1-4): Eigenfunctions, Shadows, and Duals (Dirichlet BCs)

The eigenfunctions and the dual eigenfunctions are defined by the eigenvalue α_i, the representative coefficient of the domain ω^* (10.59), and the representative coefficient of the dual solution $c_0^{(\alpha_i)}$. The coefficients of the selected cases associated with the first eigenvalue are presented in Table 10.2.

We provide in Figures 10.6-10.9 a graphical representation of the primal and dual eigenfunctions $\varphi_0^{(\alpha_1)}, \varphi_2^{(\alpha_1)}, \varphi_4^{(\alpha_1)}, \psi_0^{(\alpha_1)}, \psi_2^{(\alpha_1)}$ for cases 1-4. *Notice that these primal and dual eigenfunctions are determined up to an arbitrary constant. Therefore the functions plotted in Figures 10.6-10.9 correspond to the explicit functions above up to a constant.*

10.2.3 *The Primal and Dual Eigenfunctions and Shadows for Case 5 (Dirichlet BCs)*

The operator $\mathcal{L}$ for case 5 is

$$\mathcal{L} = \partial_1\partial_1 + \partial_2\partial_2 - \partial_1\partial_3 + \partial_3\partial_3$$

with

$$M_0 = \partial_1\partial_1 + \partial_2\partial_2, \quad M_1 = -\partial_1, \quad M_2 = 1.$$

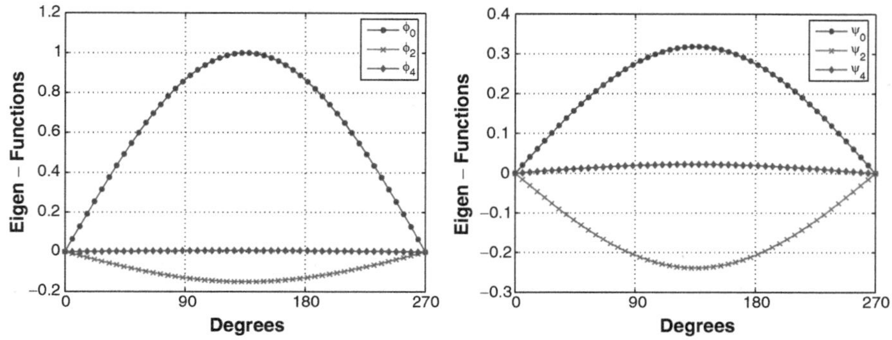

Fig. 10.6 Eigenfunctions and dual eigenfunctions associated with $\alpha_1 = 2/3$ for case 1.

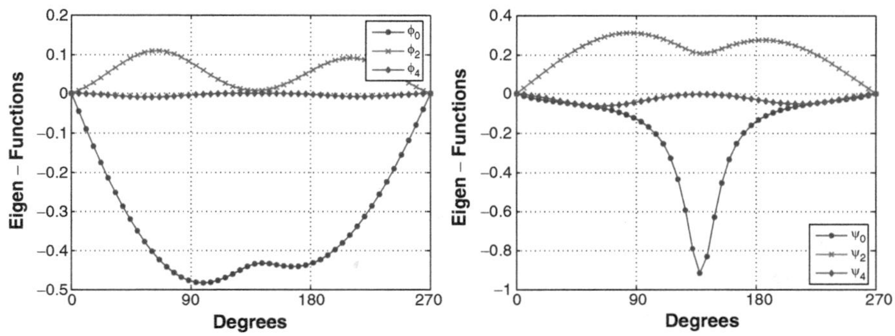

Fig. 10.7 Eigenfunctions and dual eigenfunctions associated with $\alpha_1 = 0.87139$ for case 2.

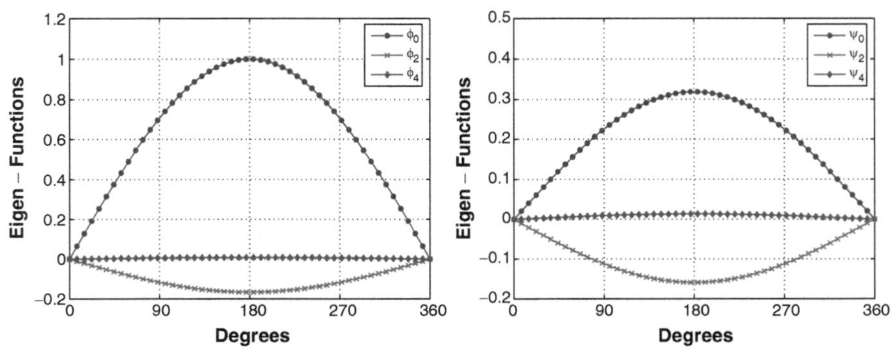

Fig. 10.8 Eigenfunctions and dual eigenfunctions associated with $\alpha_1 = 1/2$ for case 3.

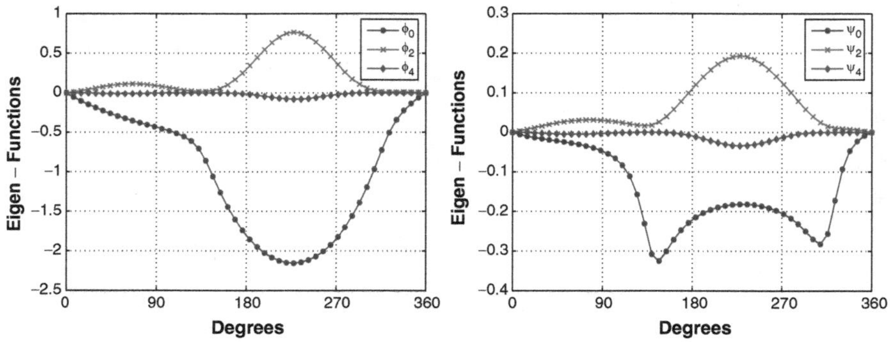

Fig. 10.9 Eigenfunctions and dual eigenfunctions associated with $\alpha_1 = 1/2$ for case 4.

For this case, $\Phi_0^{(\alpha_i)}$ and $\Psi_0^{(\alpha_i)}$ according to $(10.47)_1$ are

$$\Phi_0^{(\alpha_i)}(r,\theta) = r^{\alpha_i} \sin(\alpha_i \theta), \tag{10.78}$$

$$\Psi_0^{(\alpha_i)}(r,\theta) = c_0^{(\alpha_i)} r^{-\alpha_i} \sin(\alpha_i \theta), \tag{10.79}$$

with $\alpha_i = \frac{i\pi}{\omega}$, $i = 1, 2, \ldots$ The Neumann trace operator for case 5 is simply $T = \frac{\partial}{\partial r}$, and therefore the coefficient of the dual solution is $c_0^{(\alpha_i)} = \frac{1}{\alpha_i \omega}$. In this case, shadows of odd order are nonzero, because $M_1 \neq 0$.

The shadow function $\Phi_1^{(\alpha_i)}$ is computed by $(10.47)_2$, and the shadow functions $\Phi_2^{(\alpha_i)}$ and $\Phi_3^{(\alpha_i)}$ are computed by $(10.47)_3$ with $j = 0$ and $j = 1$ respectively:

$$\Phi_1^{(\alpha_i)}(r,\theta) = \frac{1}{4} r^{\alpha_i+1} \Big(\sin(\alpha_i - 1)\theta + \sin(\alpha_i + 1)\theta \Big),$$

$$\Phi_2^{(\alpha_i)}(r,\theta) = \frac{1}{32} r^{\alpha_i+2} \Big(\sin(\alpha_i - 2)\theta + \sin(\alpha_i + 2)\theta + \frac{2(\alpha_i - 2)}{\alpha_i + 1} \sin \alpha_i \theta \Big),$$

$$\Phi_3^{(\alpha_i)}(r,\theta) = \frac{1}{384} r^{\alpha_i+3} \Big(\sin(\alpha_i - 3)\theta + \sin(\alpha_i + 3)\theta + \frac{3(\alpha_i - 5)}{\alpha_i + 1} \Big\{ \sin(\alpha_i + 1)\theta + \sin(\alpha_i - 1)\theta \Big\} \Big).$$

The dual shadow function $\Psi_1^{(\alpha_i)}$ is the solution of $(10.47)_2$ with Dirichlet boundary conditions. The shadow functions $\Psi_2^{(\alpha_i)}$ and $\Psi_3^{(\alpha_i)}$ are the solutions of $(10.47)_3$ with $j = 0$ and $j = 1$ respectively:

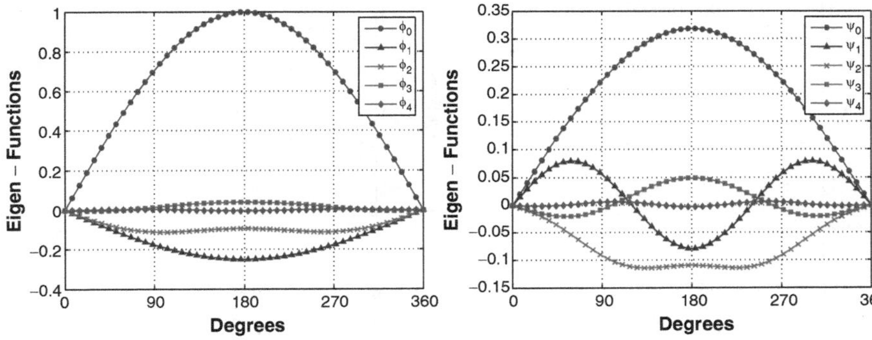

Fig. 10.10 Eigenfunctions and dual eigenfunctions associated with the first eigenvalue $\alpha_1 = 1/2$ for case 5.

$$\Psi_1^{(\alpha_i)}(r,\theta) = \frac{1}{4} c_0^{(\alpha_i)} r^{-\alpha_i+1}\Big(\sin(\alpha_i - 1)\theta + \sin(\alpha_i + 1)\theta\Big),$$

$$\Psi_2^{(\alpha_i)}(r,\theta) = \frac{1}{32} c_0^{(\alpha_i)} r^{-\alpha_i+2}\Big(\sin(\alpha_i - 2)\theta + \sin(\alpha_i + 2)\theta + \frac{2(\alpha_i + 2)}{\alpha_i - 1}\sin\alpha_i\theta\Big),$$

$$\Psi_3^{(\alpha_i)}(r,\theta) = \frac{1}{384} c_0^{(\alpha_i)} r^{-\alpha_i+3}\Big(\sin(\alpha_i - 3)\theta + \sin(\alpha_i + 3)\theta + \frac{3(\alpha_i + 5)}{\alpha_i - 1}\Big\{\sin(\alpha_i + 1)\theta + \sin(\alpha_i - 1)\theta\Big\}\Big).$$

Figure 10.10 presents the eigenfunctions, their shadows, and their duals associated with the first eigenvalue for case 5 (again, up to a constant compared to the explicit functions above).

10.3 Eigenfunctions, Shadows and Duals for Cases 1-5 with Homogeneous Neumann Boundary Conditions

The five scalar problems defined in Table 10.1 with homogeneous Neumann boundary conditions are considered:

$$k_{ij} n_i \partial_j \tau = 0, \quad \text{on } \Gamma_1, \Gamma_2. \tag{10.80}$$

The primal and dual eigenfunctions and their shadows are computed analytically following the methods in Section 10.2. Here, the derivatives in the x_1 and x_2 directions on the boundaries Γ_1 and Γ_2 have to be transformed to derivatives in terms of ξ and η on Γ_1' and Γ_2' in the transformed domain G'.

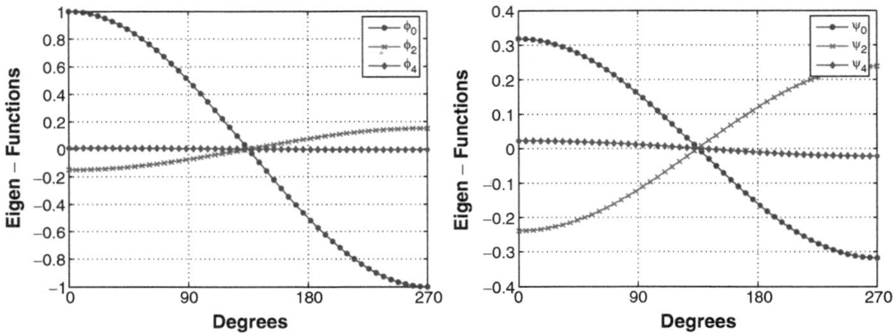

Fig. 10.11 Case 1. (Left): Eigenfunctions and shadows associated with first eigenvalue $\alpha = \frac{2}{3}$. (Right): Dual eigenfunctions and shadows associated with $\alpha = -\frac{2}{3}$.

10.3.0.1 Primal and Dual Eigenfunctions and Shadows for Case 1

Consider the differential equation

$$\begin{gathered}\left(\partial_1^2 + \partial_2^2 + \partial_3^2\right)\tau = 0 \quad \text{in } G,\\ \partial_2\tau = 0 \quad \text{on } \Gamma_1, \qquad\qquad \partial_1\tau = 0 \quad \text{on } \Gamma_2,\end{gathered} \tag{10.81}$$

where Γ_1 and Γ_2 are the surfaces defined by $\theta = 0$ and $\theta = \frac{3\pi}{2}$ respectively.

The primal and shadow functions associated with the first nonzero eigenvalue $\alpha = \frac{2}{3}$ are

$$\begin{aligned}&\Phi_0(r,\theta) = r^{\frac{2}{3}}\cos\left(\frac{2\theta}{3}\right), \quad \Phi_2(r,\theta) = -\frac{3}{20}r^{\frac{8}{3}}\cos\left(\frac{2\theta}{3}\right),\\ &\Phi_4(r,\theta) = \frac{9}{1280}r^{\frac{14}{3}}\cos\left(\frac{2\theta}{3}\right),\\ &\Phi_1(r,\theta) = \Phi_3(r,\theta) = \Phi_5(r,\theta) = 0.\end{aligned} \tag{10.82}$$

The dual and shadow functions associated with the first nonzero eigenvalue $\alpha = -\frac{2}{3}$ are (c_0 is included)

$$\begin{aligned}&\Psi_0(r,\theta) = \frac{1}{\pi}r^{-\frac{2}{3}}\cos\left(\frac{2\theta}{3}\right), \quad \Psi_2(r,\theta) = -\frac{3}{4\pi}r^{\frac{4}{3}}\cos\left(\frac{2\theta}{3}\right),\\ &\Psi_4(r,\theta) = \frac{9}{128\pi}r^{\frac{10}{3}}\cos\left(\frac{2\theta}{3}\right),\\ &\Psi_1(r,\theta) = \Psi_3(r,\theta) = \Psi_5(r,\theta) = 0.\end{aligned} \tag{10.83}$$

The graphical representation of (10.82)–(10.83) is shown in Figure 10.11.

10.3.0.2 Primal and Dual Eigenfunctions and Shadows for Case 2

The problem to solve is

$$\left(5\partial_1^2 - 8\partial_1\partial_2 + 4\partial_2^2 + \partial_3^2\right)\tau = 0, \quad \text{in } G,$$
$$4(\partial_1\tau - \partial_2\tau) = 0 \quad \text{on } \Gamma_1, \qquad 5\partial_1\tau - 4\partial_2\tau = 0 \quad \text{on } \Gamma_2, \tag{10.84}$$

where Γ_1 and Γ_2 are the surfaces defined by $\theta = 0$ and $\theta = \frac{3\pi}{2}$ respectively.

The primal and shadow functions associated with the first nonzero eigenvalue $\alpha = 0.871396$ are

$$\begin{aligned}
\Phi_0(r,\theta) &= 0.36669\left[r^{0.871396}(9-\cos(2\theta)+8*\sin(2\theta))\right]^{0.435698}\\
&\quad\times\left[\cos\left(0.871396\arctan\left(\frac{0.8\cos(\theta)+\sin(\theta)}{0.4\cos(\theta)}\right)\right)\right.\\
&\quad\left.+1.442944\sin\left(0.871396\arctan\left(\frac{0.8\cos(\theta)+\sin(\theta)}{0.4\cos(\theta)}\right)\right)\right],\\
\Phi_2(r,\theta) &= -0.006123\left[r^{2.871396}(9-\cos(2\theta)+8*\sin(2\theta))\right]^{1.435698}\\
&\quad\times\left[\cos\left(0.871396\arctan\left(\frac{0.8\cos(\theta)+\sin(\theta)}{0.4\cos(\theta)}\right)\right)\right.\\
&\quad\left.+1.442944\sin\left(0.871396\arctan\left(\frac{0.8\cos(\theta)+\sin(\theta)}{0.4\cos(\theta)}\right)\right)\right],\\
\Phi_4(r,\theta) &= 0.000033\left[r^{4.871396}(9-\cos(2\theta)+8*\sin(2\theta))\right]^{2.435698}\\
&\quad\times\left[\cos\left(0.871396\arctan\left(\frac{0.8\cos(\theta)+\sin(\theta)}{0.4\cos(\theta)}\right)\right)\right.\\
&\quad\left.+1.442944\sin\left(0.871396\arctan\left(\frac{0.8\cos(\theta)+\sin(\theta)}{0.4\cos(\theta)}\right)\right)\right],\\
\Phi_1(r,\theta) &= \Phi_3(r,\theta) = \Phi_5(r,\theta) = 0.
\end{aligned}\tag{10.85}$$

The dual and shadow functions associated with the first nonzero eigenvalue $\alpha = -0.871396$ are (c_0 is included)

$$\begin{aligned}
\Psi_0(r,\theta) &= 0.352057\left[r^{-0.871396}(9-\cos(2\theta)+8*\sin(2\theta))\right]^{-0.435698}\\
&\quad\times\left[\cos\left(0.871396\arctan\left(\frac{0.8\cos(\theta)+\sin(\theta)}{0.4\cos(\theta)}\right)\right)\right.\\
&\quad\left.+1.442944\sin\left(0.871396\arctan\left(\frac{0.8\cos(\theta)+\sin(\theta)}{0.4\cos(\theta)}\right)\right)\right],
\end{aligned}$$

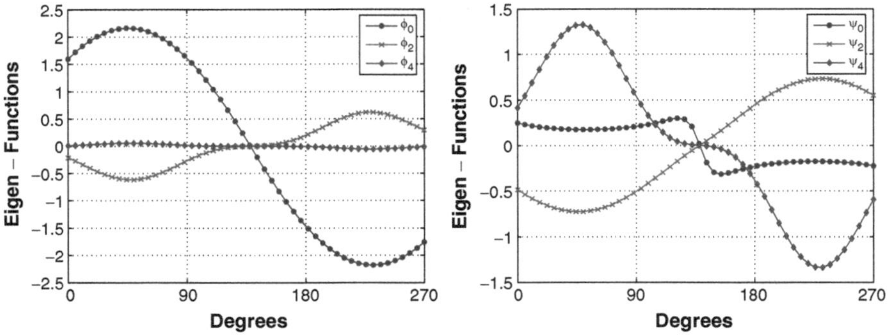

Fig. 10.12 Case 2. (Left): Eigenfunctions and shadows associated with first eigenvalue $\alpha = 0.871396$. (Right): Dual eigenfunctions and shadows associated with $\alpha = -0.871396$.

$$
\begin{aligned}
\Psi_2(r,\theta) &= -0.085548\left[r^{1.128604}(9-\cos(2\theta)+8*\sin(2\theta))\right]^{0.564302}\\
&\quad\times\left[\cos\left(0.871396\arctan\left(\frac{0.8\cos(\theta)+\sin(\theta)}{0.4\cos(\theta)}\right)\right)\right.\\
&\qquad\left.+1.442944\sin\left(0.871396\arctan\left(\frac{0.8\cos(\theta)+\sin(\theta)}{0.4\cos(\theta)}\right)\right)\right],\\
\Psi_4(r,\theta) &= 0.009174\left[r^{3.128604}(9-\cos(2\theta)+8*\sin(2\theta))\right]^{1.564302}\\
&\quad\times\left[\cos\left(0.871396\arctan\left(\frac{0.8\cos(\theta)+\sin(\theta)}{0.4\cos(\theta)}\right)\right)\right.\\
&\qquad\left.+1.442944\sin\left(0.871396\arctan\left(\frac{0.8\cos(\theta)+\sin(\theta)}{0.4\cos(\theta)}\right)\right)\right],\\
\Psi_1(r,\theta) &= \Psi_3(r,\theta) = \Psi_5(r,\theta) = 0.
\end{aligned}
\tag{10.86}
$$

The graphical representation of (10.85)–(10.86) is shown in Figure 10.12.

10.3.0.3 Primal and Dual Eigenfunctions and Shadows for Case 3

Consider the differential equation

$$
\begin{aligned}
&\left(\partial_1^2+\partial_2^2+\partial_3^2\right)\tau = 0 \quad \text{in } G,\\
&\partial_2\tau = 0 \quad \text{on } \Gamma_1, \qquad \partial_2\tau = 0 \quad \text{on } \Gamma_2,
\end{aligned}
\tag{10.87}
$$

where Γ_1 and Γ_2 are the surfaces defined by $\theta = 0$ and $\theta = 2\pi$ respectively.

We here present also the primal and shadow functions for the eigenvalue $\alpha = 0$

$$
\Phi_0(r,\theta) = 1, \quad \Phi_2(r,\theta) = -\frac{1}{4}, \quad \Phi_1(r,\theta) = \Phi_3(r,\theta) = 0. \tag{10.88}
$$

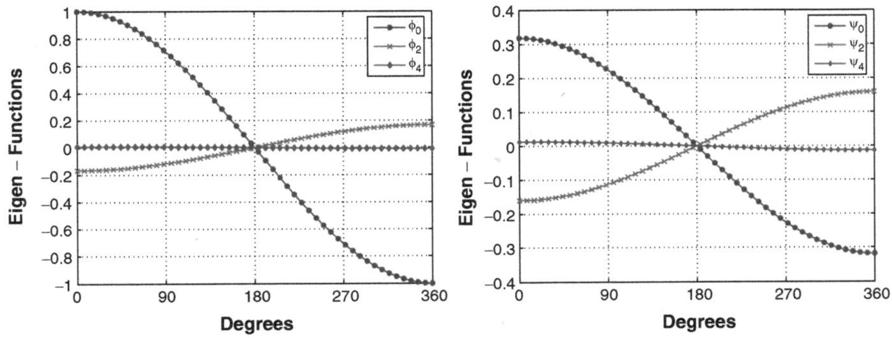

Fig. 10.13 Case 3. (Left): Eigenfunctions and shadows associated with first nonzero eigenvalue $\alpha = \frac{1}{2}$. (Right): Dual eigenfunctions and shadows associated with $\alpha = -\frac{1}{2}$.

The primal and shadow functions associated with the eigenvalue $\alpha = \frac{1}{2}$ are

$$\begin{aligned}
&\Phi_0(r,\theta) = r^{\frac{1}{2}}\cos\left(\frac{\theta}{2}\right), \quad \Phi_2(r,\theta) = -\frac{1}{6}r^{\frac{5}{2}}\cos\left(\frac{\theta}{2}\right),\\
&\Phi_4(r,\theta) = \frac{1}{120}r^{\frac{9}{2}}\cos\left(\frac{\theta}{2}\right),\\
&\Phi_1(r,\theta) = \Phi_3(r,\theta) = \Phi_5(r,\theta) = 0.
\end{aligned} \tag{10.89}$$

The dual and shadow functions associated with the eigenvalue $\alpha = -\frac{1}{2}$ are (c_0 is included)

$$\begin{aligned}
&\Psi_0(r,\theta) = \frac{1}{\pi}r^{-\frac{1}{2}}\cos\left(\frac{\theta}{2}\right), \quad \Psi_2(r,\theta) = -\frac{1}{2\pi}r^{\frac{3}{2}}\cos\left(\frac{\theta}{2}\right),\\
&\Psi_4(r,\theta) = \frac{1}{24\pi}r^{\frac{7}{2}}\cos\left(\frac{\theta}{2}\right),\\
&\Psi_1(r,\theta) = \Psi_3(r,\theta) = \Psi_5(r,\theta) = 0.
\end{aligned} \tag{10.90}$$

The graphical representation of (10.89)–(10.90) is shown in Figure 10.13.

10.3.0.4 Primal and Dual Eigenfunctions and Shadows for Case 4

Consider the differential equation

$$\begin{aligned}
&\left(5\partial_1^2 - 8\partial_1\partial_2 + 4\partial_2^2 + \partial_3^2\right)\tau = 0 \quad \text{in } G,\\
&4(\partial_1\tau - \partial_2\tau) = 0 \quad \text{on } \Gamma_1, \qquad 4(-\partial_1\tau + \partial_2\tau) = 0 \quad \text{on } \Gamma_2,
\end{aligned} \tag{10.91}$$

where Γ_1 and Γ_2 are the surfaces defined by $\theta = 0$ and $\theta = 2\pi$ respectively.

The primal and shadow functions associated with the first nonzero eigenvalue $\alpha = \frac{1}{2}$ are

$$
\begin{aligned}
\Phi_0(r,\theta) &= 0.562341\left[r^{\frac{1}{2}}(9-\cos(2\theta)+8*\sin(2\theta))\right]^{\frac{1}{4}}\\
&\quad\times\left[\cos\left(\frac{1}{2}\arctan\left(\frac{0.8\cos(\theta)+\sin(\theta)}{0.4\cos(\theta)}\right)\right)\right.\\
&\qquad\left.+0.618034\sin\left(\frac{1}{2}\arctan\left(\frac{0.8\cos(\theta)+\sin(\theta)}{0.4\cos(\theta)}\right)\right)\right],\\
\Phi_2(r,\theta) &= -0.011715\left[r^{2.5}(9-\cos(2\theta)+8*\sin(2\theta))\right]^{\frac{5}{4}}\\
&\quad\times\left[\cos\left(\frac{1}{2}\arctan\left(\frac{0.8\cos(\theta)+\sin(\theta)}{0.4\cos(\theta)}\right)\right)\right.\\
&\qquad\left.+0.618034\sin\left(\frac{1}{2}\arctan\left(\frac{0.8\cos(\theta)+\sin(\theta)}{0.4\cos(\theta)}\right)\right)\right],\\
\Phi_4(r,\theta) &= 0.000073\left[r^{4.5}(9-\cos(2\theta)+8*\sin(2\theta))\right]^{\frac{9}{4}}\\
&\quad\times\left[\cos\left(\frac{1}{2}\arctan\left(\frac{0.8\cos(\theta)+\sin(\theta)}{0.4\cos(\theta)}\right)\right)\right.\\
&\qquad\left.+0.618034\sin\left(\frac{1}{2}\arctan\left(\frac{0.8\cos(\theta)+\sin(\theta)}{0.4\cos(\theta)}\right)\right)\right],\\
\Phi_1(r,\theta) &= \Phi_3(r,\theta) = \Phi_5(r,\theta) = 0.
\end{aligned}
\tag{10.92a}
$$

The dual and shadow functions associated with the first nonzero eigenvalue $\alpha = -\frac{1}{2}$ are (c_0 is included)

$$
\begin{aligned}
\Psi_0(r,\theta) &= 1.02398\left(r^{-\frac{1}{2}}(9-\cos(2\theta)+8*\sin(2\theta))\right)^{-\frac{1}{4}}\\
&\quad\times\left[\cos\left(\frac{1}{2}\arctan\left(\frac{0.8\cos(\theta)+\sin(\theta)}{0.4\cos(\theta)}\right)\right)\right.\\
&\qquad\left.+0.618034\sin\left(\frac{1}{2}\arctan\left(\frac{0.8\cos(\theta)+\sin(\theta)}{0.4\cos(\theta)}\right)\right)\right],\\
\Psi_2(r,\theta) &= -0.063999\left[r^{1.5}(9-\cos(2\theta)+8*\sin(2\theta))\right]^{\frac{3}{4}}\\
&\quad\times\left(\cos\left(\frac{1}{2}\arctan\left(\frac{0.8\cos(\theta)+\sin(\theta)}{0.4\cos(\theta)}\right)\right)\right.\\
&\qquad\left.+0.618034\sin\left(\frac{1}{2}\arctan\left(\frac{0.8\cos(\theta)+\sin(\theta)}{0.4\cos(\theta)}\right)\right)\right),
\end{aligned}
$$

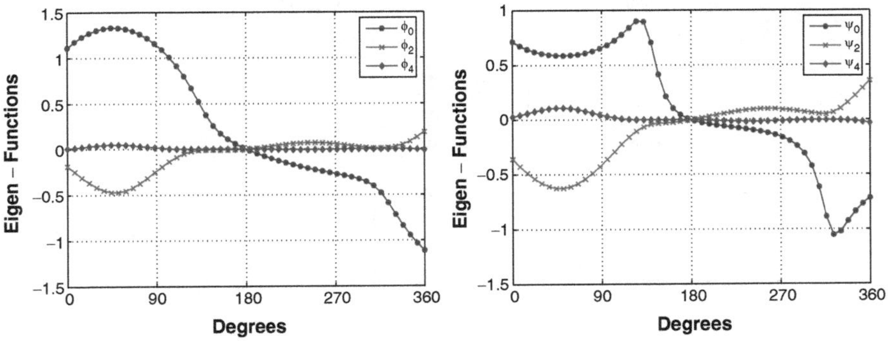

Fig. 10.14 Case 4. (Left): Eigenfunctions and shadows associated with first eigenvalue $\alpha = \frac{1}{2}$. (Right): Dual eigenfunctions and shadows associated with $\alpha = -\frac{1}{2}$.

$$
\begin{aligned}
\Psi_4(r,\theta) &= 0.000667\left[r^{3.5}(9-\cos(2\theta)+8*\sin(2\theta))\right]^{\frac{7}{4}} \\
&\quad \times\left[\cos\left(\frac{1}{2}\arctan\left(\frac{0.8\cos(\theta)+\sin(\theta)}{0.4\cos(\theta)}\right)\right)\right. \\
&\quad \left. +0.618034\sin\left(\frac{1}{2}\arctan\left(\frac{0.8\cos(\theta)+\sin(\theta)}{0.4\cos(\theta)}\right)\right)\right], \\
\Psi_1(r,\theta) &= \Psi_3(r,\theta) = \Psi_5(r,\theta) = 0. \qquad (10.92b)
\end{aligned}
$$

The graphical representation of (10.92a)–(10.92b) is shown in Figure 10.14.

10.3.0.5 Primal and Dual Eigenfunctions and Shadows for Case 5

Consider the differential equation

$$
\begin{aligned}
&\left(\partial_1^2+\partial_2^2-\partial_1\partial_3+\partial_3^2\right)\tau = 0 \quad \text{in } G, \\
&\partial_2\tau = 0 \quad \text{on } \Gamma_1, \qquad \partial_2\tau = 0 \quad \text{on } \Gamma_2, \qquad (10.93)
\end{aligned}
$$

where Γ_1 and Γ_2 are the surfaces defined by $\theta = 0$ and $\theta = 2\pi$ respectively.

The primal and shadow functions associated with the first nonzero eigenvalue $\alpha = \frac{1}{2}$ are

$$
\begin{aligned}
\Phi_0(r,\theta) &= r^{\frac{1}{2}}\cos\left(\frac{\theta}{2}\right), \qquad \Phi_1(r,\theta) = \frac{1}{4}r^{\frac{3}{2}}\cos\left(\frac{\theta}{2}\right), \\
\Phi_2(r,\theta) &= r^{\frac{5}{2}}\left(-\frac{1}{8}\cos\left(\frac{\theta}{2}\right)+\frac{1}{32}\cos\left(\frac{3\theta}{2}\right)\right),
\end{aligned}
$$

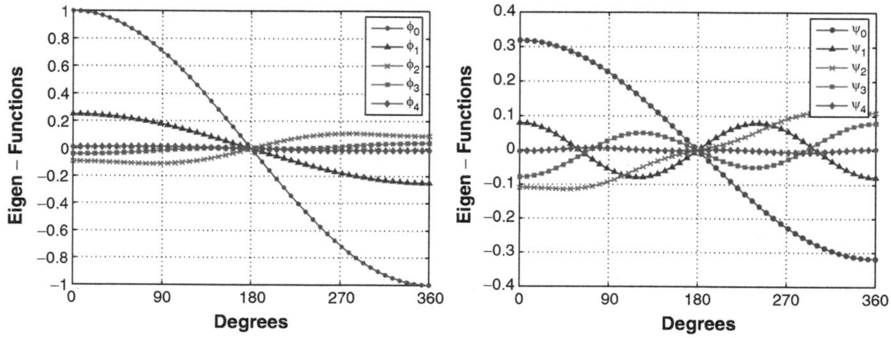

Fig. 10.15 Case 5. (Left): Eigenfunctions and shadows associated with first eigenvalue $\alpha = \frac{1}{2}$. (Right): Dual Eigenfunctions and shadows associated with $\alpha = -\frac{1}{2}$.

$$\Phi_3(r,\theta) = r^{\frac{7}{2}}\left(-\frac{1}{32}\cos\left(\frac{\theta}{2}\right) - \frac{1}{80}\cos\left(\frac{3\theta}{2}\right) + \frac{1}{384}\cos\left(\frac{5\theta}{2}\right)\right),$$

$$\Phi_4(r,\theta) = r^{\frac{9}{2}}\left(\frac{1}{64}\cos\left(\frac{\theta}{2}\right) - \frac{1}{256}\cos\left(\frac{3\theta}{2}\right) - \frac{1}{1120}\cos\left(\frac{5\theta}{2}\right) + \frac{1}{6144}\cos\left(\frac{7\theta}{2}\right)\right). \tag{10.94}$$

The dual and shadow functions associated with the first nonzero eigenvalue $\alpha = -\frac{1}{2}$ are (c_0 is included)

$$\Psi_0(r,\theta) = \frac{1}{\pi}r^{-\frac{1}{2}}\cos\left(\frac{\theta}{2}\right), \qquad \Psi_1(r,\theta) = \frac{1}{4\pi}r^{\frac{1}{2}}\cos\left(\frac{3\theta}{2}\right),$$

$$\Psi_2(r,\theta) = r^{\frac{3}{2}}\left(-\frac{12}{32\pi}\cos\left(\frac{\theta}{2}\right) + \frac{1}{32\pi}\cos\left(\frac{5\theta}{2}\right)\right),$$

$$\Psi_3(r,\theta) = r^{\frac{5}{2}}\left(-\frac{1}{16\pi}\cos\left(\frac{\theta}{2}\right) - \frac{3}{16\pi}\cos\left(\frac{3\theta}{2}\right) + \frac{1}{384\pi}\cos\left(\frac{7\theta}{2}\right)\right),$$

$$\Psi_4(r,\theta) = r^{\frac{7}{2}}\left(\frac{1}{128\pi}\cos\left(\frac{\theta}{2}\right) - \frac{1}{160\pi}\cos\left(\frac{3\theta}{2}\right) - \frac{3}{256\pi}\cos\left(\frac{5\theta}{2}\right) - \frac{1}{6144\pi}\cos\left(\frac{9\theta}{2}\right)\right). \tag{10.95}$$

The graphical representation of (10.94)–(10.95) is shown in Figure 10.15.

Chapter 11
Extracting Edge-Flux-Intensity Functions (EFIFs) Associated with Polyhedral Domains

In the previous chapter we described the asymptotic solution to heat conduction problems in the vicinity of edges, where EFIFs are functions along the edge. In this chapter we discuss two different *pointwise* extraction methods for EFIFs, and then introduce a novel method, called the quasidual function method, that extracts the functional representation of the EFIF.

11.1 Extracting Pointwise Values of the EFIFs by the L^2 Projection Method

Consider the domain in Figure 11.1 with an edge OD of interest coinciding with the x_3 coordinate. We assume that the material properties and the solid angle ω are independent of x_3. On the 2-D plane perpendicular to x_3 we define the space spanned by the 2-D Laplacian eigenpairs by $\mathfrak{S}_{2D}$:

$$\mathfrak{S}_{2D} \overset{\text{def}}{=} \Big\{ \tau_n(r,\theta) = r^{\alpha_n} s_n(\theta) \,|\, \Delta_{2D}\tau_n = 0,$$

$$\tau = 0 \text{ or } \frac{\partial \tau}{\partial n} = 0 \text{ on planes intersecting at the edge} \Big\}. \tag{11.1}$$

We wish to L^2 project the solution of the 3-D Laplace equation in the vicinity of an edge in the plane perpendicular to the edge, intersecting it at a given point x_3^* into a subspace $\mathfrak{S}_{2D}^N \subset \mathfrak{S}_{2D}$. This subspace $\mathfrak{S}_{2D}^N$ is spanned by the first N eigenpairs. Therefore, each $\tau^N \in \mathfrak{S}_{2D}^N$ is a linear combination of functions in $\mathfrak{S}_{2D}^N$:

$$\tau^N = \sum_{i=1}^{N} \tilde{A}_i(x_3^*) r^{\alpha_i} s_i(\theta) \tag{11.2}$$

Z. Yosibash, *Singularities in Elliptic Boundary Value Problems and Elasticity and Their Connection with Failure Initiation*, Interdisciplinary Applied Mathematics 37,
DOI 10.1007/978-1-4614-1508-4_11,

The projection operation is aimed at finding these $\tilde{A}_i(x_3^*)$ so that the error $\tau(x_3^*) - \tau^N(x_3^*)$ is orthogonal to the space $\mathfrak{S}_{2D}^N$ at the point x_3^*; i.e., it must be orthogonal to any $\chi^N(x_3^*) \in \mathfrak{S}_{2D}^N$:

$$\int_r \int_\theta (\tau - \tau^N)\chi^N\big|_{x_3=x_3^*} r dr d\theta = 0 \qquad \forall \chi^N(x_3^*) \in \mathfrak{S}_{2D}^N. \tag{11.3}$$

The above can be rephrased in the following form:

Find $\tau^N(x_3^*) \in \mathfrak{S}_{2D}^N$ such that

$$\int_r \int_\theta \tau\chi^N\big|_{x_3=x_3^*} r dr d\theta = \int_r \int_\theta \tau^N \chi^N\big|_{x_3=x_3^*} r dr d\theta \qquad \forall \chi^N(x_3^*) \in \mathfrak{S}_{2D}^N, \tag{11.4}$$

where χ^N being in $\mathfrak{S}_{2D}^N$ allows the representation

$$\chi^N(x_3^*) = \sum_{i=1}^{N} B_i(x_3^*) r^{\alpha_i} s_i(\theta). \tag{11.5}$$

Let us first concentrate on the right-hand-side (RHS) of (11.4), which after substitution of (11.2) and (11.5) becomes

$$\begin{aligned} \text{RHS} = {} & \tilde{A}_1(x_3^*)B_1(x_3^*) \int_r \int_\theta r^{2\alpha_1+1} s_1^2(\theta) dr d\theta + \tilde{A}_1(x_3^*)B_2(x_3^*) \\ & \times \int_r \int_\theta r^{\alpha_1+\alpha_2+1} s_1(\theta)s_2(\theta) dr d\theta + \tilde{A}_2(x_3^*)B_1(x_3^*) \\ & \times \int_r \int_\theta r^{\alpha_1+\alpha_2+1} s_1(\theta)s_2(\theta) dr d\theta + \tilde{A}_2(x_3^*)B_2(x_3^*) \\ & \times \int_r \int_\theta r^{2\alpha_2+1} s_2^2(\theta) dr d\theta + \cdots . \end{aligned} \tag{11.6}$$

The eigenfunctions of the Laplace equation with homogeneous boundary conditions in the vicinity of the edge are orthogonal (see (1.21)), so that (11.6) after integration in the radial direction from $r = 0$ to $r = R$ becomes

$$\text{RHS} = \sum_{i=1}^{N} \tilde{A}_i(x_3^*)B_i(x_3^*) \frac{R^{2\alpha_i+2}}{2\alpha_i + 2} \int_\theta s_i^2(\theta) d\theta. \tag{11.7}$$

Let us now consider the left-hand-side (LHS). Although τ is given (and may hereinafter be replaced by τ_{FE} if not known), we will use its full expansion given by (10.15). Substituting (10.15) and (11.5) in the LHS of (11.4), one obtains

$$
\begin{aligned}
\text{LHS} = & A_1(x_3^*)B_1(x_3^*) \int_r \int_\theta r^{2\alpha_1+1} s_1^2(\theta) dr d\theta \\
& + c_{11}\left(\partial_3^2 A_1(x_3^*)\right) B_1(x_3^*) \int_r \int_\theta r^{2\alpha_1+3} s_1^2(\theta) dr d\theta \\
& + c_{12}\left(\partial_3^4 A_1(x_3^*)\right) B_1(x_3^*) \int_r \int_\theta r^{2\alpha_1+5} s_1^2(\theta) dr d\theta + \cdots \\
& + A_2(x_3^*)B_2(x_3^*) \int_r \int_\theta r^{2\alpha_2+1} s_2^2(\theta) dr d\theta \\
& + c_{21}\left(\partial_3^2 A_2(x_3^*)\right) B_2(x_3^*) \int_r \int_\theta r^{2\alpha_2+3} s_2^2(\theta) dr d\theta \\
& + c_{22}\left(\partial_3^4 A_2(x_3^*)\right) B_2(x_3^*) \int_r \int_\theta r^{2\alpha_2+5} s_2^2(\theta) dr d\theta + \cdots \\
= & \sum_{i=1}^{\infty} \int_\theta s_i^2(\theta) d\theta \frac{R^{2\alpha_i+2}}{2\alpha_i+2} \Bigg\{ A_i(x_3^*)B_i(x_3^*) \\
& + R^2 \frac{c_{i1}(\alpha_i+1)}{\alpha_i+2}\left(\partial_3^2 A_i(x_3^*)\right) B_i(x_3^*) \\
& + R^4 \frac{c_{i2}(\alpha_i+1)}{\alpha_i+3}\left(\partial_3^4 A_i(x_3^*)\right) B_i(x_3^*) \\
& + R^6 \frac{c_{i3}(\alpha_i+1)}{\alpha_i+4}\left(\partial_3^6 A_i(x_3^*)\right) B_i(x_3^*) + \cdots \Bigg\}. \qquad (11.8)
\end{aligned}
$$

Since RHS has to be equal to the LHS for every $B_i(x_3^*)$, by substituting (11.7) and (11.8) in (11.4) one obtains a set of equations for the coefficients $\tilde{A}_i(x_3^*)$:

$$
\tilde{A}_i(x_3^*) = A_i(x_3^*) + R^2 \frac{c_{i1}(\alpha_i+1)}{\alpha_i+2} \partial_3^2 A_i(x_3^*) + R^4 \frac{c_{i2}(\alpha_i+1)}{\alpha_i+3} \partial_3^4 A_i(x_3^*) + \mathcal{O}(R^6). \tag{11.9}
$$

Examining (11.9) one may observe that the $\tilde{A}_i(x_3^*)$ are not equal to $A_i(x_3^*)$ as desired, but include additional terms. These are associated with $R^{2n} \partial_{x_3}^{2n} A_i(x_3^*)$, $n = 1, 2, \ldots$. To eliminate these higher order terms, one needs to compute the $\tilde{A}_i(x_3^*)$'s at decreasing values of $R < 1$, followed by Richardson's extrapolation in a manner similar to that shown in [203]. The accuracy of the method will be demonstrated by numerical examples. It is clear that when the loading is constant or depends linearly on x_3, then $A_i(x_3)$ are at most linear in x_3, so that (11.9) simplifies to $\tilde{A}_i(x_3^*) = A_i(x_3^*)$, and extracted values are independent of R. This will be demonstrated by numerical examples in Section 11.1.2.

11.1.1 Numerical Implementation

Because the exact solution τ in (11.4) is unknown, we use instead only an approximation of it, i.e., the finite element solution τ_{FE}. This approximation introduces the second source of numerical error (the first source is due to the finite R and the need of Richardson's extrapolation). Any function $\tau^N \in \mathfrak{S}_{2D}^N$ may be represented as

$$\tau^N = \left(\tilde{A}_1\ \tilde{A}_2\ \cdots\ \tilde{A}_N\right) \begin{Bmatrix} r^{\alpha_1} s_1(\theta) \\ r^{\alpha_2} s_2(\theta) \\ \cdot \\ \cdot \\ r^{\alpha_N} s_N(\theta) \end{Bmatrix} \stackrel{\text{def}}{=} \tilde{\boldsymbol{A}}^T \begin{Bmatrix} r^{\alpha_1} s_1(\theta) \\ r^{\alpha_2} s_2(\theta) \\ \cdot \\ \cdot \\ r^{\alpha_N} s_N(\theta) \end{Bmatrix}, \tag{11.10}$$

and similarly,

$$\chi^N = \{r^{\alpha_1} s_1(\theta)\ r^{\alpha_2} s_2(\theta)\ \cdots\ r^{\alpha_N} s_N(\theta)\}\, \boldsymbol{B}. \tag{11.11}$$

Substituting (11.10) and (11.11) in (11.4), with τ replaced by τ_{FE}, one obtains the following system:

$$\begin{aligned} &\text{Find } \tilde{\boldsymbol{A}} \text{ such that } \quad \forall \boldsymbol{B}(x_3^*), \\ &\int_{r=0}^{R}\int_{\theta=0}^{\omega} \tau_{FE}\, \{r^{\alpha_1} s_1(\theta)\ r^{\alpha_2} s_2(\theta)\ \cdots\ r^{\alpha_N} s_N(\theta)\}\, \boldsymbol{B}(x_3^*) r dr d\theta \\ &= \int_{r=0}^{R}\int_{\theta=0}^{\omega} \tilde{\boldsymbol{A}}^T(x_3^*) \begin{Bmatrix} r^{\alpha_1} s_1(\theta) \\ r^{\alpha_2} s_2(\theta) \\ \cdot \\ \cdot \\ r^{\alpha_N} s_N(\theta) \end{Bmatrix} \{r^{\alpha_1} s_1(\theta)\ r^{\alpha_2} s_2(\theta)\ \cdots\ r^{\alpha_N} s_N(\theta)\}\, \boldsymbol{B}(x_3^*) r dr d\theta. \end{aligned} \tag{11.12}$$

Equation (11.12) can be brought to a matrix representation

$$\mathbf{L}^T \boldsymbol{B}(x_3^*) = \tilde{\boldsymbol{A}}^T(x_3^*)[K]\boldsymbol{B}(x_3^*) \quad \forall \boldsymbol{B}(x_3^*), \tag{11.13}$$

where

$$K_{ij} = \int_{r=0}^{R}\int_{\theta=0}^{\omega} r^{\alpha_i+\alpha_j} s_i(\theta) s_j(\theta) r dr d\theta = \frac{R^{\alpha_i+\alpha_j+2}}{\alpha_i+\alpha_j+2}\int_{\theta=0}^{\omega} s_i(\theta) s_j(\theta) d\theta, \tag{11.14}$$

and in view of the orthogonality property for the Laplace equation,

$$K_{ij} = \begin{cases} \frac{R^{2\alpha_i+2}}{2\alpha_i+2}\int_{\theta=0}^{\omega} s_i^2(\theta)d\theta, & i = j, \\ 0 & i \neq j. \end{cases} \tag{11.15}$$

The elements of the vector $\mathbf{L}$ have to be computed numerically because τ_{FE} is extracted from the finite element solution:

$$L_i = \int_{r=0}^{R} \int_{\theta=0}^{\omega} \tau_{FE}(r, \theta, x_3^*) r^{\alpha_i+1} s_i(\theta) dr d\theta. \tag{11.16}$$

Since (11.13) has to hold for any $\boldsymbol{B}(x_3^*)$, it is equivalent to

$$\mathbf{L}^T = \tilde{\boldsymbol{A}}^T(x_3^*)[K]. \tag{11.17}$$

Substituting (11.15) and (11.16) in (11.17), and noticing that $[K]$ is a diagonal matrix, one obtains explicit equations for the required elements of the vector $\tilde{\boldsymbol{A}}$:

$$\tilde{A}_i(x_3^*) = L_i / \left(\frac{R^{2\alpha_i+2}}{2\alpha_i + 2} \int_{\theta=0}^{\omega} s_i^2(\theta) d\theta \right). \tag{11.18}$$

Notice that the numerical error caused by replacing τ with τ_{FE} is reflected in L_i. However, it is smaller than the pointwise error. This is due to the integration, which has a smoothing nature, and thus it decreases the relative error in L_i compared to the pointwise relative error of τ_{FE}.

Because the quality of the finite element solution τ_{FE} in the elements touching the singular edge is known to be low, and therefore the accuracy of computed $\tilde{A}_i$'s may deteriorate, the following strategy is adopted in the practical implementation. Instead of integrating on a sector from $r = 0$ to $r = R$, the integration in (11.14) and (11.16) is performed over a circular ring, $r = 0.9R$ to $r = R$. Thus instead of using (11.18) for the practical computation of the $\tilde{A}_i$'s, we use the following:

$$\tilde{A}_i(x_3^*) = \int_{0.9R}^{R} \int_{\theta=0}^{\omega} \tau_{\text{FE}}(r, \theta, x_3^*) r^{\alpha_i+1} s_i(\theta) dr d\theta / \left(\frac{(1 - 0.9^{2\alpha_i+2}) R^{2\alpha_i+2}}{2\alpha_i + 2} \int_{\theta=0}^{\omega} s_i^2(\theta) d\theta \right). \tag{11.19}$$

The numerical error in τ_{FE} can be controlled by an adaptive finite element solution using p-extension.

A different extraction procedure, based on the energy projection method shown in Section 11.2, eliminates the need of extracting τ_{FE} in the elements at the singularity.

The vector $\tilde{\boldsymbol{A}}(x_3^*)$ has to be extracted with a tight control of the numerical error using (11.18) at various R's of decreasing order. Then Richardson's extrapolating method has to be applied for obtaining the exact value at $R \to 0$. The overall algorithm is presented in the following on a model problem for which the exact solution is known.

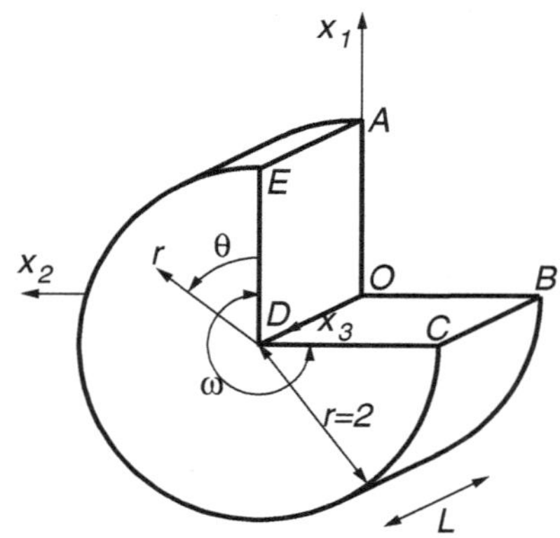

Fig. 11.1 3-D domain and notation for EFIF extraction.

11.1.2 An Example Problem and Numerical Experimentation

To test the accuracy of any numerical algorithm, we generate an example problem having an exact solution. This example problem allows one to represent constant, linear, quadratic, etc. variation of the EFIFs along an edge of interest. In view of the analytical functional representation of the solution in a neighborhood of the singular edge (10.15), we may construct a family of example problems as follows. Consider the domain in a form of a sector of a cylinder shown in Figure 11.1. On the faces $ODEA$ and $ODCB$ that intersect at the edge of interest OD, we impose homogeneous Neumann boundary conditions $\frac{\partial \tau}{\partial n} \stackrel{\text{def}}{=} q_n = 0$. On the cylindrical boundary $AECB$, at which $r = 2$, Dirichlet boundary conditions are imposed with

$$A_1 = a_{11} + a_{12}x_3 + a_{13}x_3^2 \quad \text{and} \quad A_2 = a_{21} + a_{22}x_3 + a_{23}x_3^2, \tag{11.20}$$

so that the series (10.15) is

$$\begin{aligned} \tau(r = 2, \theta, x_3) =&(a_{11} + a_{12}x_3 + a_{13}x_3^2)2^{\alpha_1}\cos(\alpha_1\theta) - a_{13}\frac{1}{2(\alpha_1 + 1)} \\ &\times 2^{\alpha_1+2}\cos(\alpha_1\theta) + (a_{21} + a_{22}x_3 + a_{23}x_3^2)2^{\alpha_2}\cos(\alpha_2\theta) \\ &- a_{13}\frac{1}{2(\alpha_2 + 1)}2^{\alpha_2+2}\cos(\alpha_2\theta), \end{aligned} \tag{11.21}$$

where $\alpha_i = i\pi/\omega$. One may observe that the flux-free boundary conditions on $ODEA$ and $ODCB$ are identically satisfied by (11.21). On the face $x_3 = 0$ of the domain we impose the Dirichlet boundary conditions

$$\begin{aligned} \tau(r, \theta, x_3 = 0) = a_{11}r^{\alpha_1}\cos(\alpha_1\theta) - a_{13}\frac{1}{2(\alpha_1 + 1)}r^{\alpha_1+2}\cos(\alpha_1\theta) \\ + a_{21}r^{\alpha_2}\cos(\alpha_2\theta) - a_{13}\frac{1}{2(\alpha_2 + 1)}r^{\alpha_2+2}\cos(\alpha_2\theta). \end{aligned} \tag{11.22}$$

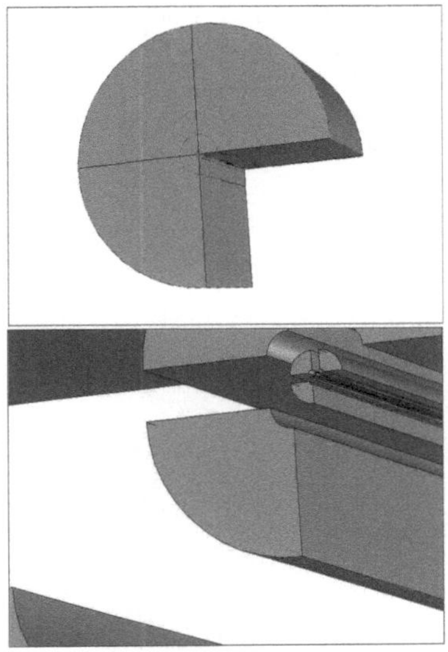

Fig. 11.2 3-D p-FEM (12 elements).

On the boundary of the domain $x_3 = L$ we may impose Dirichlet boundary conditions according to (11.21), i.e.,

$$\begin{aligned}\tau(r,\theta,x_3=L) =& (a_{11}+a_{12}L+a_{13}L^2)r^{\alpha_1}\cos(\alpha_1\theta) - a_{13}\frac{1}{2(\alpha_1+1)} \\ & \times r^{\alpha_1+2}\cos(\alpha_1\theta) + (a_{21}+a_{22}L+a_{23}L^2)r^{\alpha_2}\cos(\alpha_2\theta) \\ & - a_{13}\frac{1}{2(\alpha_2+1)}r^{\alpha_2+2}\cos(\alpha_2\theta). \end{aligned} \tag{11.23}$$

A finite element mesh containing 12 solid elements (hexahedra and pentahedra) is constructed with three refined layers in a neighborhood of the singular edge (the radius of the smallest element is $0.15^3 \times 2$). The finite element mesh with the zoomed area in the neighborhood of the singular edge is shown in Figure 11.2. We perform two analyses on the given mesh. In the first, we choose $a_{i1} = 1, a_{i2} = 0.5$, and $a_{i3} = 0, i = 1, 2$. Thus the first two EFIFs are linear in x_3:

$$A_1(x_3) = 1 + 0.5x_3, \qquad A_2(x_3) = 1 + 0.5x_3.$$

According to the mathematical analysis, the extracted EFIFs based on L^2 projection should be independent of the radius ($\partial_3^{2k} A_i(x_3) = 0 \quad \forall k = 1, 2, \dots$) and accurate.

In the second analysis we choose $a_{i1} = 1$, $a_{i2} = 0.5$, and $a_{i3} = 2$, $i = 1, 2$. Thus the first two EFIFs are parabolic with respect to x_3:

$$A_1(x_3) = 1 + 0.5x_3 + 2x_3^2, \qquad A_2(x_3) = 1 + 0.5x_3 + 2x_3^2.$$

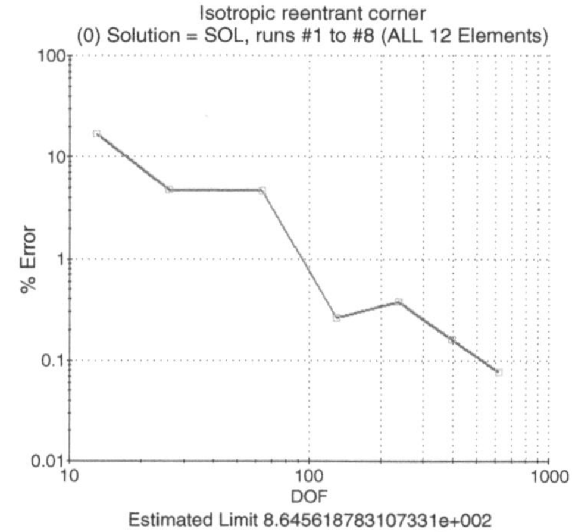

Fig. 11.3 Convergence of the error in energy-norm for the second analysis.

Table 11.1 Values of $\tilde{A}_1$ and $\tilde{A}_2$ for first analysis, where $A_1(x_3) = 1 + 0.5x_3$, $A_2(x_3) = 1 + 0.5x_3$.

	$R = 1$	$R = 0.2$	$R = 0.1$	$R = 0.01$	A_i^{EX}
$\tilde{A}_1(x_3^* = 0.5)$	1.250	1.250	1.250	1.250	1.250
$\tilde{A}_1(x_3^* = 1.0)$	1.500	1.500	1.500	1.499	1.500
$\tilde{A}_2(x_3^* = 0.5)$	1.250	1.250	1.250	1.250	1.250
$\tilde{A}_2(x_3^* = 1.0)$	1.500	1.500	1.500	1.500	1.500

For this case one should clearly see a strong dependence of the extracted EFIFs on the radius of the integration area R, and that the extracted values converge to the exact solution as $R \to 0$.

Since the finite element solution is used in the numerical procedure described in Section 11.1.1, the error of approximation must be determined before computing the EFIFs. Figure 11.3 shows the estimated relative error in energy norm as a function of the number of degrees of freedom (DOF) for the second analysis. The DOF were systematically increased by p-extension on the fixed mesh shown in Figure 11.2.

The first two nonzero EFIFs extracted using different radii at $x_3^* = 1$ and $x_3^* = 0.5$, for the first analysis, are summarized in Table 11.1

As predicted by the mathematical analysis, the extracted EFIFs are independent of the radius R (outer radius of integration for L^2 projection).

For the second analysis, with $a_{i3} \neq 0$, the extracted EFIFs are expected to be R-dependent. We summarize in Table 11.2 the first two nonzero EFIFs extracted using different radii at $x_3^* = 1$ and $x_3^* = 0.5$. It is seen that the extracted EFIFs in this case have a strong dependence on the radius of the domain on which the extraction is performed, and indeed as $R \to 0$, the extracted value approaches the exact EFIFs. However, based on the mathematical analysis, it is possible to use

Table 11.2 Values of $\tilde{A}_1$ and $\tilde{A}_2$ for the second analysis, where $A_1(x_3) = 1 + 0.5x_3 + 2x_3^2$, $A_2(x_3) = 1 + 0.5x_3 + 2x_3^2$.

	$R = 1$	$R = 0.2$	$R = 0.1$	$R = 0.01$	A_i^{EX}
$\tilde{A}_1(x_3^* = 0.5)$	1.206	1.729	1.745	1.750	1.750
$\tilde{A}_1(x_3^* = 1.0)$	2.955	3.476	3.492	3.499	3.500
$\tilde{A}_2(x_3^* = 0.5)$	1.360	1.735	1.747	1.750	1.750
$\tilde{A}_2(x_3^* = 1.0)$	3.109	3.483	3.494	3.499	3.500

Table 11.3 $\tilde{A}_1(x_3^* = 0.5)$ for various values of R, $p = 8$, and the Richardson's extrapolated values as $R \to 0$.

R	$\tilde{A}_1(x_3^* = 0.5) \stackrel{\text{def}}{=} \tilde{A}_1^{(0)}$	$\tilde{A}_1^{(1)}$	$\tilde{A}_1^{(2)}$
1.0	1.206		
		1.7477	
0.75	1.443		1.7494
		1.7490	
0.5	1.613		

Table 11.4 $\tilde{A}_2(x_3^* = 0.5)$ for various values of R, $p = 8$, and the Richardson's extrapolated values as $R \to 0$.

R	$\tilde{A}_2(x_3^* = 0.5) \stackrel{\text{def}}{=} \tilde{A}_2^{(0)}$	$\tilde{A}_2^{(1)}$	$\tilde{A}_2^{(2)}$
1.0	1.360		
		1.7485	
0.75	1.530		1.7499
		1.7496	
0.5	1.652		

Richardson's extrapolation starting with the value of R and extrapolate to $R = 0$. As an example, let us extract the values of $\tilde{A}_1$ and $\tilde{A}_2$ at $R = 1, 0.75, 0.5$, where these are known to be wrong. Using Richardson's extrapolation, with the residual error behaving like R^2 (this is known from (11.9)), we show that an excellent approximation for a_i can be obtained. For example, let us choose the point $x_3^* = 0.5$ and extract $\tilde{A}_1(x_3^* = 0.5)$. The second row in Table 11.3 represents the extracted values from the FE solution. One observes that although the extracted values at large R are off, the extrapolated value is very close to the exact solution (0.03% relative error).

The same procedure is applied to $\tilde{A}_2(x_3^* = 0.5)$ as shown in Table 11.4. The extrapolated value of $\tilde{A}_2(x_3^* = 0.5)$ is again very close to the exact solution.

11.2 The Energy Projection Method

Similarly to the L^2 projection method, the energy projection method projects τ into $\mathfrak{S}_{2D}^N$. The difference is the projection mechanism, which is based on the gradient of the function, i.e., we wish to find a member in $\mathfrak{S}_{2D}^N$ that is as close as possible to the function τ so that the error between their gradients is minimized:

$$\begin{aligned}
&\textbf{Find} \quad \tau^N(x_3^*) \in \mathfrak{S}_{2D}^N \text{ such that} \\
&\int_r \int_\theta \mathbf{grad}\tau^N \, \mathbf{grad}\chi^N \big|_{x_3=x_3^*} r dr d\theta \\
&\quad = \int_r \int_\theta \mathbf{grad}\tau \, \mathbf{grad}\chi^N \big|_{x_3=x_3^*} r dr d\theta \quad \forall \chi^N(x_3^*) \in \mathfrak{S}_{2D}^N.
\end{aligned} \tag{11.24}$$

Here **grad** should be understood as the gradient in the plane perpendicular to the edge at the point x_3^*. Using Green's theorem, (11.24) becomes

$$\begin{aligned}
&\textbf{Find} \quad \tau^N(x_3^*) \in \mathfrak{S}_{2D}^N \text{ such that} \\
&\int_\theta \left[\tau^N \frac{\partial \chi^N}{\partial r}\right]_{r=R, x_3=x_3^*} R d\theta - \int_r \int_\theta \tau^N \Delta_{2D} \chi^N \big|_{x_3=x_3^*} r dr d\theta \\
&= \int_\theta \left[\tau \frac{\partial \chi^N}{\partial r}\right]_{r=R, x_3=x_3^*} R d\theta - \int_r \int_\theta \tau \Delta_{2D} \chi^N \big|_{x_3=x_3^*} r dr d\theta \quad \forall \chi^N(x_3^*) \in \mathfrak{S}_{2D}^N.
\end{aligned} \tag{11.25}$$

Because $\chi^N(x_3^*) \in \mathfrak{S}_{2D}^N$, it satisfies identically $\Delta_{2D}\chi^N = 0$, so that (11.25) simplifies to

$$\begin{aligned}
&\textbf{Find} \quad \tau^N(x_3^*) \in \mathfrak{S}_{2D}^N \text{ such that} \\
&\int_\theta \left[\tau^N \frac{\partial \chi^N}{\partial r}\right]_{r=R, x_3=x_3^*} R d\theta = \int_\theta \left[\tau \frac{\partial \chi^N}{\partial r}\right]_{r=R, x_3=x_3^*} R d\theta, \quad \forall \chi^N(x_3^*) \in \mathfrak{S}_{2D}^N.
\end{aligned} \tag{11.26}$$

Inserting equations (10.15), (11.2), and (11.5) into (11.26), and noting the orthogonality of the eigenfunctions $s_i(\theta)$, one obtains after similar steps as in Section 11.1,

$$\tilde{A}_i(x_3^*) = A_i(x_3^*) + R^2 c_{i1} \left(\partial_3^2 A_i(x_3^*)\right) + R^4 c_{i2} \left(\partial_3^4 A_i(x_3^*)\right) + \mathcal{O}(R^6). \tag{11.27}$$

This equation is very similar to (11.9) obtained by the L^2 projection method, except that the coefficients multiplying the powers of R^{2i} are somewhat simpler. Therefore, from the theoretical point of view, the application of the energy projection method is expected to provide exactly the same behavior as the L^2 projection method. However, the practical use of the energy projection method may be more efficient compared to the previous one for two main reasons:

- The energy projection method requires integration over a 1-D circular arc, as opposed to 2-D integration required for the L^2 projection method.
- The circular arc can be taken outside the first row of elements, where numerical errors are much lower, a benefit that cannot always be realized using the L^2 projection method.

It is important to note that for a general scalar second-order boundary value problem ("anisotropic" heat transfer equation), the conclusions of the aforementioned analysis are similar. The mathematical analysis is more complicated because $\int s_i(\theta)s_j(\theta)d\theta \neq 0$ for $i \neq j$; thus an explicit equation for each $\tilde{A}_i$ cannot be obtained.

The numerical implementation of the energy projection method is along the lines outlined in Section 11.1.1, so that we provide the final formulation

$$\tilde{A}_i(x_3^*) = \frac{\alpha_i \int_{\theta=0}^{\omega} \tau_{\mathrm{FE}}(R,\theta,x_3^*)s_i(\theta)d\theta}{R^{\alpha_i} \int_{\theta=0}^{\omega} s_i^2(\theta)d\theta}. \tag{11.28}$$

11.3 A Quasidual Function Method (QDFM) for Extracting EFIFs

Utilizing the explicit structure of the solution in the vicinity of the edge, we present the quasidual function method for the extraction of the EFIFs. It can be interpreted as an extension of the dual function contour integral method in 2-D domains, and involves the computation of a surface integral $J[R]$ along a cylindrical surface of radius R away from the edge as presented in a general framework in [46]. The surface integral $J[R]$ utilizes special constructed extraction polynomials together with the dual eigenfunctions for extracting EFIFs. This accurate and efficient method provides a *polynomial approximation* of the EFIF along the edge whose order is adaptively increased so to approximate the exact EFIF. It is implemented as a post-solution operation in conjunction with the p-FEM. Numerical realization of some of the anticipated properties of $J[R]$ are provided, and it is used for extracting EFIFs associated with different scalar elliptic equations in 3-D domains, including domains having edge and vertex singularities. The numerical examples demonstrate the efficiency, robustness and high accuracy of the proposed quasidual function method.

For each eigenvalue α_i, a set of *quasidual* singular functions $K_m^{(\alpha_i)}[B_m]$ is constructed, where m is a natural integer called the *order* of the quasidual function, and $B_m(x_3)$ is a function (we choose it to be a polynomial) called the *extraction polynomial*

$$K_m^{(\alpha_i)}[B_m] \stackrel{\mathrm{def}}{=} \sum_{j=0}^{m} \partial_3^j B_m(x_3)\, \Psi_j^{(\alpha_i)}. \tag{11.29}$$

Using the quasidual functions, one can extract a scalar product of $A_i(x_3)$ with $B_m(x_3)$ on the edge. This is accomplished with the help of the *antisymmetric* boundary integral $J[R]$ over the surface Γ_R (13.1). We define $J[R](u, v)$ to be,

$$J[R](u, v) \stackrel{\text{def}}{=} \int_{\Gamma_R} (Tu \cdot v - u \cdot Tv)\, dS = \int_I \int_0^w (Tu \cdot v - u \cdot Tv)|_{r=R}\, R\, d\theta\, dx_3, \tag{11.30}$$

where $I \equiv$ the edge $\mathcal{E}$ along the x_3 axis (Figure 13.1, and T is the radial Neumann trace operator related to the operator $\mathcal{L}$:

$$T \stackrel{\text{def}}{=} \left\{ \begin{pmatrix} k_{11} & k_{12} & k_{13} \\ k_{21} & k_{22} & k_{23} \\ k_{31} & k_{32} & 1 \end{pmatrix} \begin{pmatrix} \partial_1 \\ \partial_2 \\ \partial_3 \end{pmatrix} \right\}^T \begin{pmatrix} \cos\theta \\ \sin\theta \\ 0 \end{pmatrix}. \tag{11.31}$$

With the above definition, we have the following theorem [46]

Theorem 11.1. *Take $B_m(x_3)$ such that*

$$\partial_3^j B_m(x_3) = 0 \quad \textit{for} \quad j = 0, \ldots, m-1 \quad \textit{on } \partial I. \tag{11.32}$$

Then if the EFIFs A_i in the expansion (10.52) *are smooth enough, we have*

$$J[R](\tau, K_m^{(\alpha_i)}[B_m]) = \int_I A_i(x_3)\, B_m(x_3)\, dx_3 + \mathcal{O}(R^{\alpha_1 - \alpha_i + m + 1}) \quad \textit{as} \quad R \to 0. \tag{11.33}$$

Here α_1 is the smallest of the eigenvalues α_i, $i \in \mathbb{N}$.

Theorem 11.1 allows a precise determination of $\int_I A_i(x_3)\, B_m(x_3)\, dx_3$ by computing (11.33) for two or three R values and using Richardson's extrapolation as $R \to 0$.

Remark 11.1. *For the first EFIF A_1, we obtain the highest convergence rate $\mathcal{O}(R^{m+1})$. If, moreover, the Φ_j and Ψ_j of odd rank are zero, we have the following improvement of Theorem 11.1: For any* even *integer m, condition* (11.32) *implies that the asymptotic equality* (11.33) *holds modulo a remainder in $\mathcal{O}(R^{m+2})$ instead of $\mathcal{O}(R^{m+1})$.*

11.3.0.1 The Quasidual Extraction Functions

One may consider several quasidual extraction functions of increasingly higher order:

$$K_0^{(\alpha_1)} = B_0(x_3)\Psi_0^{(\alpha_1)}(r, \theta),$$

$$K_1^{(\alpha_1)} = B_1(x_3)\Psi_0^{(\alpha_1)}(r, \theta) + \partial_3 B_1(x_3)\Psi_1^{(\alpha_1)}(r, \theta),$$

$$K_2^{(\alpha_1)} = B_2(x_3)\Psi_0^{(\alpha_1)}(r, \theta) + \partial_3 B_2(x_3)\Psi_1^{(\alpha_1)}(r, \theta) + \partial_3^2 B_2(x_3)\Psi_2^{(\alpha_1)}(r, \theta),$$

$$K_3^{(\alpha_1)} = B_3(x_3)\Psi_0^{(\alpha_1)}(r,\theta) + \partial_3 B_3(x_3)\Psi_1^{(\alpha_1)}(r,\theta) + \partial_3^2 B_3(x_3)\Psi_2^{(\alpha_1)}(r,\theta)$$
$$+ \partial_3^3 B_3(x_3)\Psi_3^{(\alpha_1)}(r,\theta).$$

According to Theorem 11.1, the difference between the integral $J[R](\tau, K_m^{(\alpha_i)}[B_m])$ and the moment $\int_I A_i(x_3)\, B_m(x_3)\, dx_3$ should be of order R^{m+1}, which is the convergence rate with respect to R.

The higher the order of the extraction function, the higher is the convergence rate with respect to R. The following conditions should be satisfied for the *extraction polynomials* B_0, B_1, B_2, and B_3 according to (11.32):

$$B_0: \quad \text{No condition required.} \tag{11.34}$$

$$B_1: \quad B_1(+1) = B_1(-1) = 0; \tag{11.35}$$

$$B_2: \quad B_2(+1) = B_2(-1) = \partial_3 B_2(+1) = \partial_3 B_2(-1) = 0; \tag{11.36}$$

$$B_3: \quad B_3(+1) = B_3(-1) = \partial_3 B_3(+1) = \partial_3 B_3(-1)$$
$$= \partial_3^2 B_3(+1) = \partial_3^2 B_3(-1) = 0. \tag{11.37}$$

The exact solution τ being unknown in general, we use instead a finite element approximation τ_{FE}, and the integral (11.30) is performed numerically using a Gaussian quadrature of order n_G:

$$J[R](\tau, K_m^{(\alpha_i)}[B_m]) = \sum_{k=1}^{n_G}\sum_{\ell=1}^{n_G} \frac{\omega}{2}\, w_k w_\ell \left(T\tau_{FE} \cdot K_m^{(\alpha_i)}[B_m] - \tau_{FE} \cdot TK_m^{(\alpha_i)}[B_m]\right)_{\xi_k,\eta_\ell}, \tag{11.38}$$

where w_k are the weights and ξ_k and η_ℓ are the abscissas of the Gaussian quadrature. The Neumann trace operator, T, operates on both τ and $K_m^{(\alpha_i)}[B_m]$. For $T\tau$ we use the numerical approximations $T\tau_{FE}$ computed by finite elements. We extract in the post-solution phase of the FE analysis τ_{FE}, $\partial_1\tau_{FE}$, $\partial_2\tau_{FE}$, and $\partial_3\tau_{FE}$, whereas $TK_m^{(\alpha_i)}[B_m]$ is computed analytically. These values are evaluated at the specific Gaussian points when the integral is computed numerically.

The numerical errors associated with the numerical integration and with replacing the exact solution by the finite element solution are negligible, as shown in [134].

11.3.1 Jacobi Polynomial Representation of the Extraction Function

We are interested in extracting the EFIF $A_i(x_3)$. Because its functional representation is unknown, a polynomial approximation is sought instead. We would like to construct an adaptive class of orthonormal polynomials with a given weight $w(x_3) = (1 - x_3^2)^m$ so to represent $B_m(x_3)$. This suggests the use of Jacobi polynomials as a natural basis. In this way, if $A_i(x_3)$ is a polynomial of degree N, it can be represented by a linear combination of Jacobi polynomials as

$$A_i(x_3) = \tilde{a}_0 J_m^{(0)} + \tilde{a}_1 J_m^{(1)}(x_3) + \cdots + \tilde{a}_N J_m^{(N)}(x_3), \tag{11.39}$$

where $J_m^{(k)}$ is the Jacobi polynomial of degree k and order m, i.e., associated with the weight $w(x_3) = (1 - x_3^2)^m$, which is denoted in the literature by $P_k^{(m,m)}$. We have the following important orthogonality property [2, pp. 773-774]

$$\int_{-1}^{1} (1 - x_3^2)^m J_m^{(n)}(x_3) J_m^{(k)}(x_3)\, dx_3 = \delta_{nk} h_k \tag{11.40}$$

with some real coefficients h_k (depending on m). The hierarchical family of extraction polynomials, denoted by $BJ_m^{(k)}(x_3)$, has to be chosen so to satisfy $BJ_m^{(k)}(\pm 1) = \partial_3 BJ_m^{(k)}(\pm 1) = \cdots = \partial_3^{m-1} BJ_m^{(k)}(\pm 1) = 0$. To accomplish this, we set

$$BJ_m^{(k)}(x_3) = (1 - x_3^2)^m \, \frac{J_m^{(k)}(x_3)}{h_k}, \tag{11.41}$$

so that according to (11.40), we retrieve the coefficients $\tilde{a}_k$ in (11.39) as a simple scalar product:

$$\int_{-1}^{1} A_i(x_3) BJ_m^{(k)}(x_3)\, dx_3 = \tilde{a}_k, \qquad k = 0, 1, \ldots, N. \tag{11.42}$$

Thus, by virtue of Theorem 11.1, the $J[R]$ integral evaluated for the quasidual functions $K_m^{(\alpha_i)}[BJ_m]$ with the extraction polynomials $BJ_m^{(k)}$, $k = 0, 1, \ldots, N$, provide approximations of the coefficients $\tilde{a}_k$.

Of course in general, $A_i(x_3)$ is an unknown function, and we wish to find a projection of it into spaces of polynomials. It is expected that as we increase the polynomial space, the approximation is better.

The EFIF $A_i(x_3)$ has an infinite Fourier expansion in the basis $J_m^{(k)}$ with a sequence of coefficients $\tilde{a}_k$,

$$A_i(x_3) = \sum_{k \geq 0} \tilde{a}_k J_m^{(k)}, \tag{11.43}$$

converging in the weighted space $L^2[w]$ with $w = (1 - x_3^2)^m$. For each fixed N, the computation of the $N + 1$ coefficients $\tilde{a}_0, \ldots, \tilde{a}_N$ provides the orthogonal projection of $A_i(x_3)$ into the space of polynomials of degree up to N in the weighted space $L^2[w]$. To accomplish this we use the $N + 1$ extraction polynomials $BJ_m^{(0)}(x_3), \ldots, BJ_m^{(N)}(x_3)$ defined in (11.41). If we want to increase the space in which $A_i(x_3)$ is projected, all that is needed is the computation of (11.42) for $k = N + 1$. In this way we obtain $A^{\text{new}}(x_3) = A^{\text{previous}}(x_3) + \tilde{a}_{N+1} J_{N+1}(x_3)$.

11.3.2 Jacobi Extraction Polynomials of Order 2

Since they satisfy (11.36), and, a fortiori, (11.35) and (11.34), the Jacobi extraction polynomials $BJ_2^{(k)}$ can be combined with the dual singular functions $K_0^{(\alpha_i)}$, $K_1^{(\alpha_i)}$, and $K_2^{(\alpha_i)}$. Then we have [2, pp. 773-774]

$$J_2^{(k)}(x_3) = \frac{1}{k^2 + 7k + 12} \sum_{l=0}^{k} \frac{(k+l+4)!}{2^l l!\,(k-l)!\,(2+l)!} (x_3 - 1)^l, \tag{11.44}$$

and the constant h_k in (11.40) is equal to

$$h_k = \frac{2^5 (k+1)(k+2)}{(2k+5)(k+3)(k+4)}. \tag{11.45}$$

Inserting (11.45) and (11.44) in (11.41), we finally obtain

$$\begin{aligned} BJ_2^{(k)}(x_3) =& \frac{(2k+5)(k+3)(k+4)}{2^5(k+1)(k+2)} \frac{(1-x_3^2)^2}{k^2+7k+12} \\ &\times \sum_{l=0}^{k} \frac{(k+l+4)!}{2^l l!\,(k-l)!\,(2+l)!} (x_3 - 1)^l. \end{aligned} \tag{11.46}$$

11.3.3 Analytical Solutions for Verifying the QDFM

We generate here analytical solutions against which numerical experiments are compared. The exact solution associated with the ith eigenpair $\tau_{\text{EX}}^{(\alpha_i)}$ is

$$\tau_{\text{EX}}^{(\alpha_i)} = \sum_{j \geq 0} \partial_3^j A_i(x_3)\, \Phi_j^{(\alpha_i)}(r, \theta). \tag{11.47}$$

So if $A_i(x_3)$ is a polynomial of order N, i.e., $A_i(x_3) = a_0 + a_1 x_3 + \cdots + a_N x_3^N$, then (11.47) has a finite number of terms in the sum, because the $N+1$ and higher derivatives are zero. Thus, (11.47) becomes

$$\tau_{\text{EX}}^{(\alpha_i)} = \sum_{j=0}^{N} \partial_3^j A_i(x_3)\, \Phi_j^{(\alpha_i)}(r, \theta). \tag{11.48}$$

Recall that by the mere construction of the $\Phi_j^{(\alpha_i)}$, we have $\mathcal{L}\tau_{\text{EX}}^{(\alpha_i)} = 0$. If we specify over the entire boundary $\partial\Omega$ the Dirichlet boundary condition as the trace of (11.48), the solution τ coincides with (11.48) at any point $\boldsymbol{x} \equiv (r, \theta, x_3)$.

We choose two examples of boundary conditions (BCs), each having a different N. The first BC, which is denoted by (BC_2), is the one for which we take $N = 2$ and

$$A_1(x_3) = 1 + x_3 + x_3^2 \quad (11.49)$$

i.e., $a_0 = a_1 = a_2 = 1$. This means that we prescribe the following Dirichlet condition on $\partial\Omega$

$$(BC_2)\ \ \tau_{EX}^{(\alpha_1)}\big|_{\partial\Omega} = (1 + x_3 + x_3^2)\Phi_0^{(\alpha_1)}(r,\theta) + (1 + 2x_3)\Phi_1^{(\alpha_1)}(r,\theta) + 2\Phi_2^{(\alpha_1)}(r,\theta).$$

The second boundary condition that we consider is for $N = 4$, denoted by (BC_4), for which we take

$$A_1(x_3) = 5 + 4x_3 + 9x_3^2 + 3x_3^3 + x_3^4, \quad (11.50)$$

i.e., $a_0 = 5$, $a_1 = 4$, $a_2 = 9$, $a_3 = 3$, and $a_4 = 1$. This means that we have the Dirichlet condition

$$\begin{aligned}(BC_4) \qquad \tau_{EX}^{(\alpha_1)}|_{\partial\Omega} =& (5 + 4x_3 + 9x_3^2 + 3x_3^3 + x_3^4)\,\Phi_0^{(\alpha_1)}(r,\theta) \\ &+ (4 + 18x_3 + 9x_3^2 + 4x_3^3)\,\Phi_1^{(\alpha_1)}(r,\theta) \\ &+ (18 + 18x_3 + 12x_3^2)\,\Phi_2^{(\alpha_1)}(r,\theta) \\ &+ (18 + 24x_3)\,\Phi_3^{(\alpha_1)}(r,\theta) + 24\Phi_4^{(\alpha_1)}(r,\theta).\end{aligned}$$

By the uniqueness of solutions, the solution of the problem with the boundary conditions (BC_2) and (BC_4) coincides with $\tau_{EX}^{(\alpha_1)}$ for the choice (11.49) and (11.50) of A_1, respectively. This means that our exact solution contains only one edge singularity (and no vertex singularities).

The domains have been discretized using a p-FEM mesh, with geometric progression toward the singular edge with a factor of 0.15, having four layers of elements. In the x_3 direction, a uniform discretization using five elements has been adopted. In Figure 11.4 we present the meshes used for opening angles of $\omega = 3\pi/2$ and $\omega = 2\pi$ (crack).

11.3.4 Numerical Results for (BC_4) Using $K_2^{(\alpha_1)}$

For the benchmark problem with boundary conditions (BC_4) for which the exact EFIF is the polynomial (11.50) of degree 4 and using the extraction polynomials $BJ_2^{(0)}(x_3), \dots, BJ_2^{(N)}(x_3)$, where $0 \le N \le 4$, we extract the EFIF for case 2 at $R = 0.05$: We performed the computation with 15 integration points and $p = 8$ in the finite element mesh, and present in Figure 11.5 the relative error as a percentage between the extracted EFIF and the exact one. As may be seen for the family of degree 4, we indeed fully recover the exact EFIF.

Fig. 11.4 The p-FE models.

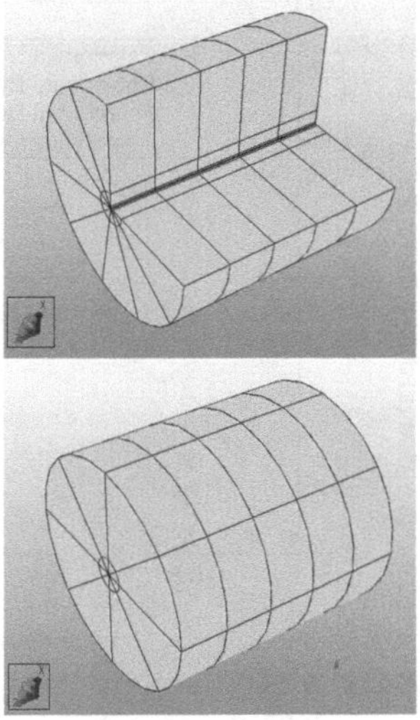

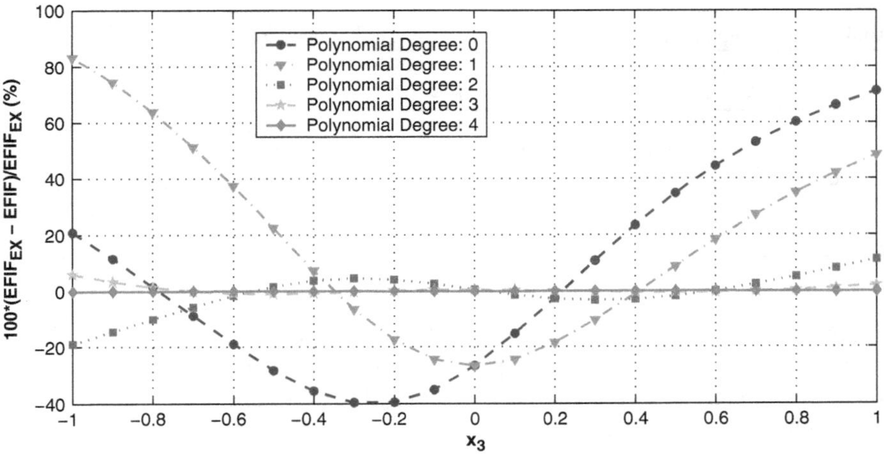

Fig. 11.5 Relative error (%) of the extracted EFIF at $R = 0.05$ using $K_2^{(\alpha_1)}$ and the hierarchical family $BJ_2^{(k)}(x_3)$, $k \leq N$, for $N = 0, 1, 2, 3, 4$.

Of course, if $N > 4$, we should fully recover the EFIF. As one increases the order of the hierarchical family, the results do not improve, but we obtain an oscillatory behavior of the solution due to numerical errors (the finite element solution is not exact), with a very small amplitude as demonstrated in Figure 11.6.

To illustrate the convergence of the extracted values as a function of R, we present in Table 11.5 the monomial coefficients of the extracted polynomial at $R = 0.9, 0.5, 0.2, 0.05$. Then we use Richardson's extrapolation, knowing that

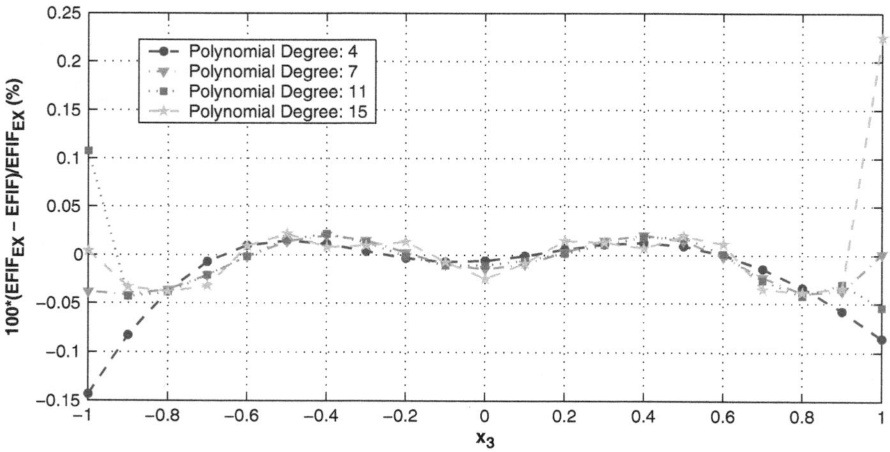

Fig. 11.6 Relative error (%) of the extracted EFIF at $R = 0.05$ using $K_2^{(\alpha_1)}$ and the hierarchical family $BJ_2^{(k)}(x_3)$, $k \le N$, for $N = 4, 7, 11, 11, 15$.

Table 11.5 Computed coefficients a_i for (BC_4), using $K_2^{(\alpha_1)}$ and $BJ_2^{(k)}(x_3)$, $k \le 4$.

	Exact	$R = 0.9$	$R = 0.5$	$R = 0.2$	$R = 0.05$	Extrapolated using $R = 0.9, 0.5$
a_0	5	5.920806968	5.089253508	5.005993235	5.000288235	5.001699446
a_1	4	4.004545148	4.002303539	4.002751475	3.998527960	4.002067521
a_2	9	9.047407703	9.008253090	9.001724824	8.989161317	9.004130510
a_3	3	2.985298783	2.995871625	3.001625541	3.005167695	2.996984837
a_4	1	0.904830390	0.983905020	1.007098452	1.025721321	0.992230769

the error behaves as $\mathcal{O}(R^4)$, cf. Remark 11.1, and the coefficients at $R = 0.9, 0.5$ to extrapolate to $R = 0$. These extrapolation results are shown in the last column of Table 11.5. The relative error in the extrapolated EFIF using the data at $R = 0.9$, 0.5 is compared with that obtained at $R = 0.5$ and 0.05 in Figure 11.7.

By extracting the EFIF from the FE solution away from the singular edge (where usually the numerical data are polluted), we demonstrate that a very good approximation is obtained by Richardson's extrapolation, taking into consideration that the error behaves as $\mathcal{O}(R^4)$. Practically, the relative error in the extrapolated EFIF is as obtained very close to the singular edge ($R = 0.05$), and much better than the values obtained when extraction is performed at $R = 0.5$.

11.3.5 A Nonpolynomial EFIF

We have demonstrated so far that the QDFM performs very well if the exact EFIF is a polynomial. A natural question is, what if the EFIF is not a polynomial? In this case we use the hierarchical algorithm for polynomial space enrichment.

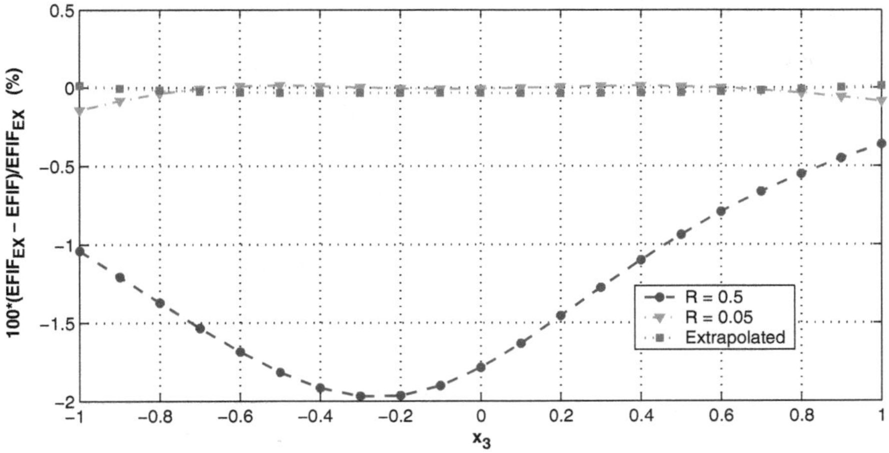

Fig. 11.7 Relative error (%) of the extracted EFIF at R =0.5, 0.05 and extrapolating from data at R =0.9, 0.5. EFIF computed using $K_2^{(\alpha_1)}$ and the hierarchical family $BJ_2^{(k)}(x_3)$, $k \le 4$.

We investigate the performance of such hierarchical space enrichment for the case that the exact EFIF is a general function, and furthermore, it contains high gradients at the ends of the edge. For example, consider case 2, where the EFIF is a function of the form

$$A_1(x_3) = \frac{\sin x_3}{(d - x_3^2)}, \tag{11.51}$$

where d is a given number. As d approaches 1, the EFIF approaches infinity at the vertices $x_3 = \pm 1$. We choose three values of $d = 2, 1.5, 1.05$.

Consider the following problem:

$$\begin{cases} \mathcal{L}(\tau) = \partial_3^2 A_1(x_3)\Phi_0^{(\alpha_1)}(r,\theta) & \text{in } \Omega, \\ \tau = A_1(x_3)\Phi_0^{(\alpha_1)}(r,\theta) & \text{on } \partial\Omega, \end{cases} \tag{11.52}$$

for which the exact solution is simply $\tau_{EX} = A_1(x_3)\Phi_0^{(\alpha_1)}(r,\theta)$.

Remark 11.2. *Theorem 11.1 does not apply stricto sensu to the solution of problem* (11.52)*. Nevertheless, it can be proved that* $J[R](\tau, K_m^{(\alpha_i)}[B])$ *yields an approximation of the moment of* A_1 *modulo a positive power of* R*.*

A refined finite element model graded toward $x_3 = \pm 1$ was generated as shown in Figure 11.8. It has 25 elements in the x_3 direction and a total of 800 solid finite elements.

To evaluate the accuracy of the extracted EFIFs, one has first to examine the numerical results, τ_{FE} and its derivatives, especially for solutions having large gradients. The graphs in Figure 11.9 present the relative error in τ_{FE} and $\partial_r \tau_{FE}$ as a

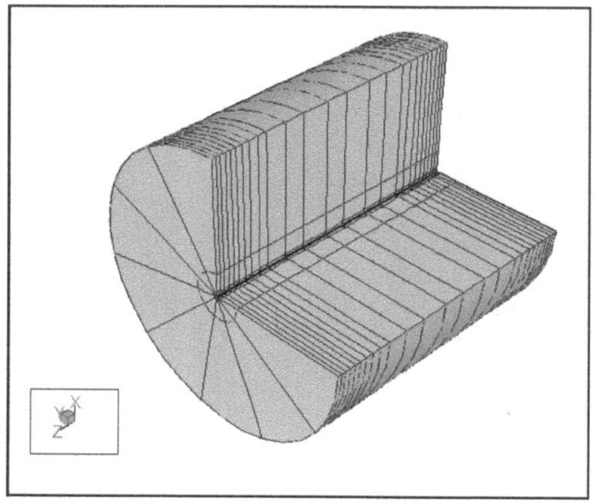

Fig. 11.8 The p-FEM model for nonpolynomial EFIFs with large gradients at $x_3 = \pm 1$.

percentage, extracted from the finite element solution at $p = 8$ for $d = 2, 1.5, 1.05$. These graphs are along the line $R = 0.05$, $\theta = 135°$, and $-1 \leq x_3 \leq 1$.

The FE results have a relative error of about 3% for $-0.8 \leq x_3 \leq 0.8$, and around 17% for $0.8 < |x_3| < 1$ for the case $d = 1.05$. This, in turn, will perturb the extraction of the EFIF by that order of magnitude when the QDFM is used, as we show in the sequel. We will also observe that the EFIFs are computed with similar accuracy and the extraction technique does not magnify the numerical error, but the opposite. For $d = 2$, 1.5, the relative error in the function and its derivatives is very small (less than 0.7%) in all the range. Therefore, the extraction of the EFIFs is expected to provide excellent results.

Using $K_2^{(\alpha_1)}$ and the hierarchical family $BJ_2^{(k)}(x_3)$, we extract the EFIFs at $R = 0.05$ using the solution at $p = 8$ and 54 Gauss integration points (due to the strong gradients of the solutions we used a higher integration scheme). We also checked with 94 Gauss integration points that the integration error in evaluating $J[R]$ is negligible.

Figure 11.10 presents the exact EFIF and the extracted EFIF using $BJ_2^{(k)}(x_3)$, $k \leq N$, of increasing order N obtained at $R = 0.05$. Notice the different ordinate scales inside the three graphs. One may easily observe the strong gradients of the EFIF at $x_3 = \pm 1$, especially for the case $d = 1.05$.

Relative errors between the extracted EFIF and the exact value are presented in Figure 11.11 (here again, the ordinate scales are different from each other). For all cases of d, the EFIF is progressively better approximated away from the large gradients ($\approx -0.85 \leq x_3 \leq -0.85$) as the order of the extraction polynomials (n) is increased.

At $N = 19$, the extracted EFIF has less than 3% relative error for the case $d = 1.05$ and less than 0.5% relative error for the cases $d = 1.5$ and $d = 2$.

The large pointwise errors in a close neighborhood of the high gradients are expected.

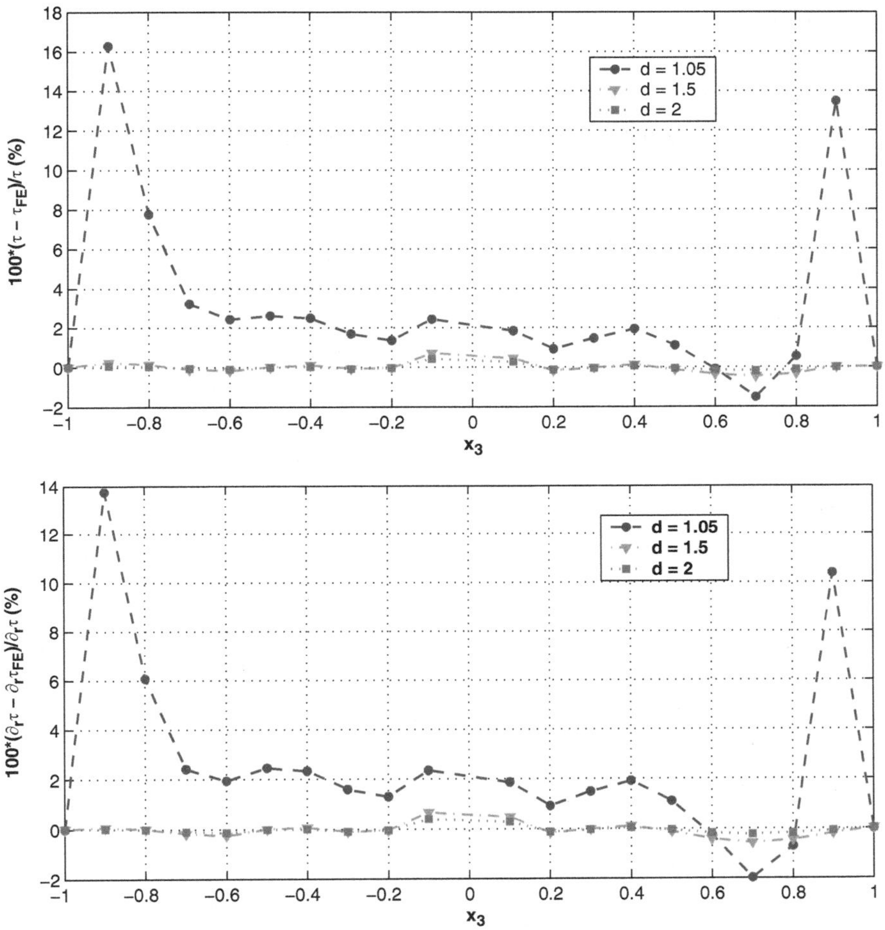

Fig. 11.9 Relative error in the FE solution and its derivatives (%) at $p = 8$ for $r = 0.05$, $\theta = 135°$, $x_3 \in [-1, 1]$ for the three problems defined by $d = 2, 1.5$, and 1.05.

11.3.6 A Domain with Edge and Vertex Singularities

To examine the vertex influence on EFIF extraction, we consider a more realistic domain constructed as an extension of the one presented in Figure 11.1 by adding two cylinders at ± 1 as shown in Figure 11.12. The added cylinders are $\Omega^{(1)} = D \times I^{(1)}$ and $\Omega^{(-1)} = D \times I^{(-1)}$, where $I^{(1)}$ is the interval $[1, 1.5]$, $I^{(-1)}$ is the interval $[-1.5, -1]$, and D is the disk of radius 1.

The domain has been discretized using a p-FEM mesh, with geometric progression toward $r = 0$ with a factor of 0.15, having four layers of elements, and

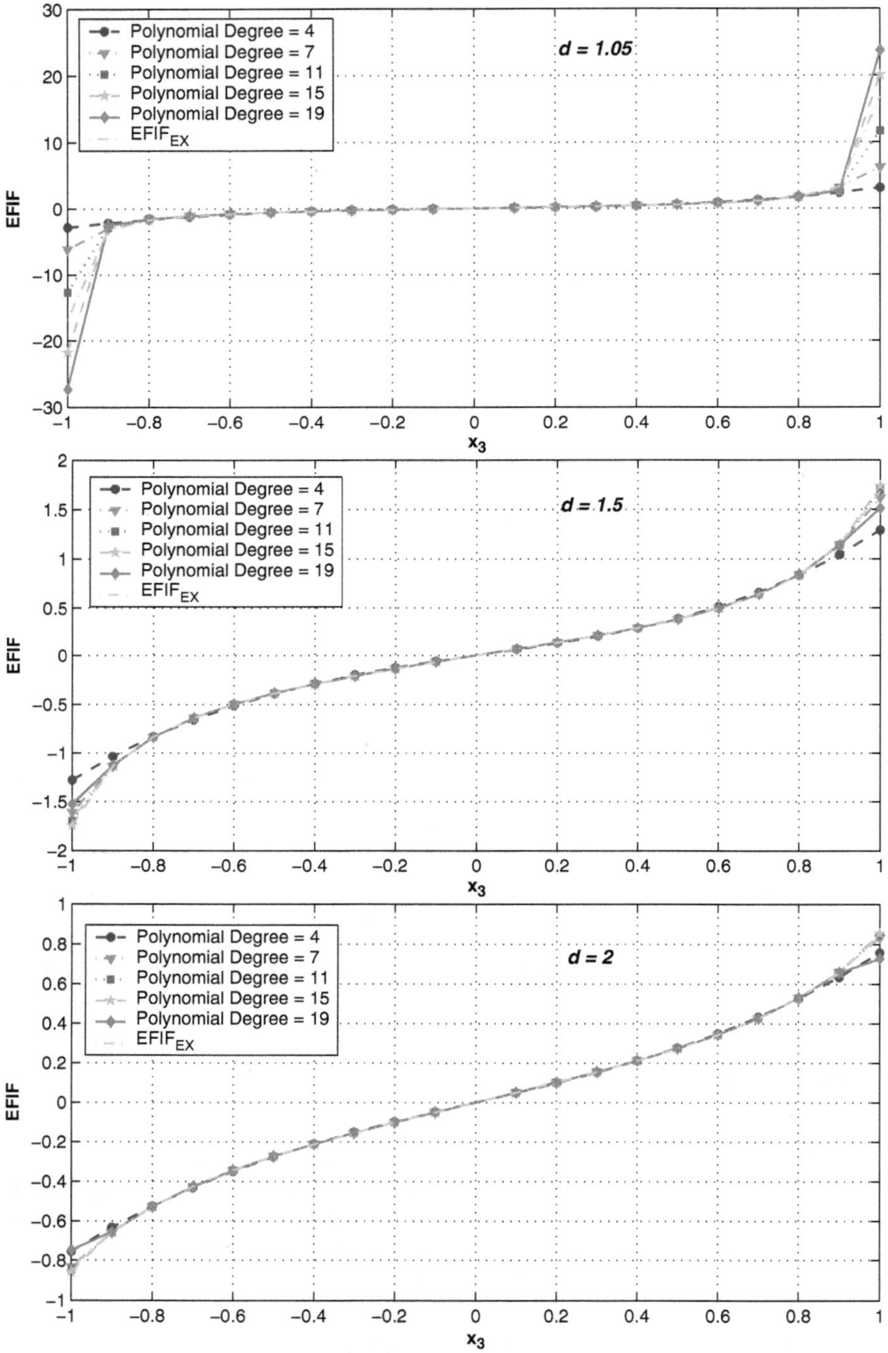

Fig. 11.10 Exact and extracted EFIF, using $K_2^{(\alpha_1)}$ and extraction polynomials of degree $\leq N$ for $N = 4, 7, 11, 15, 19$.

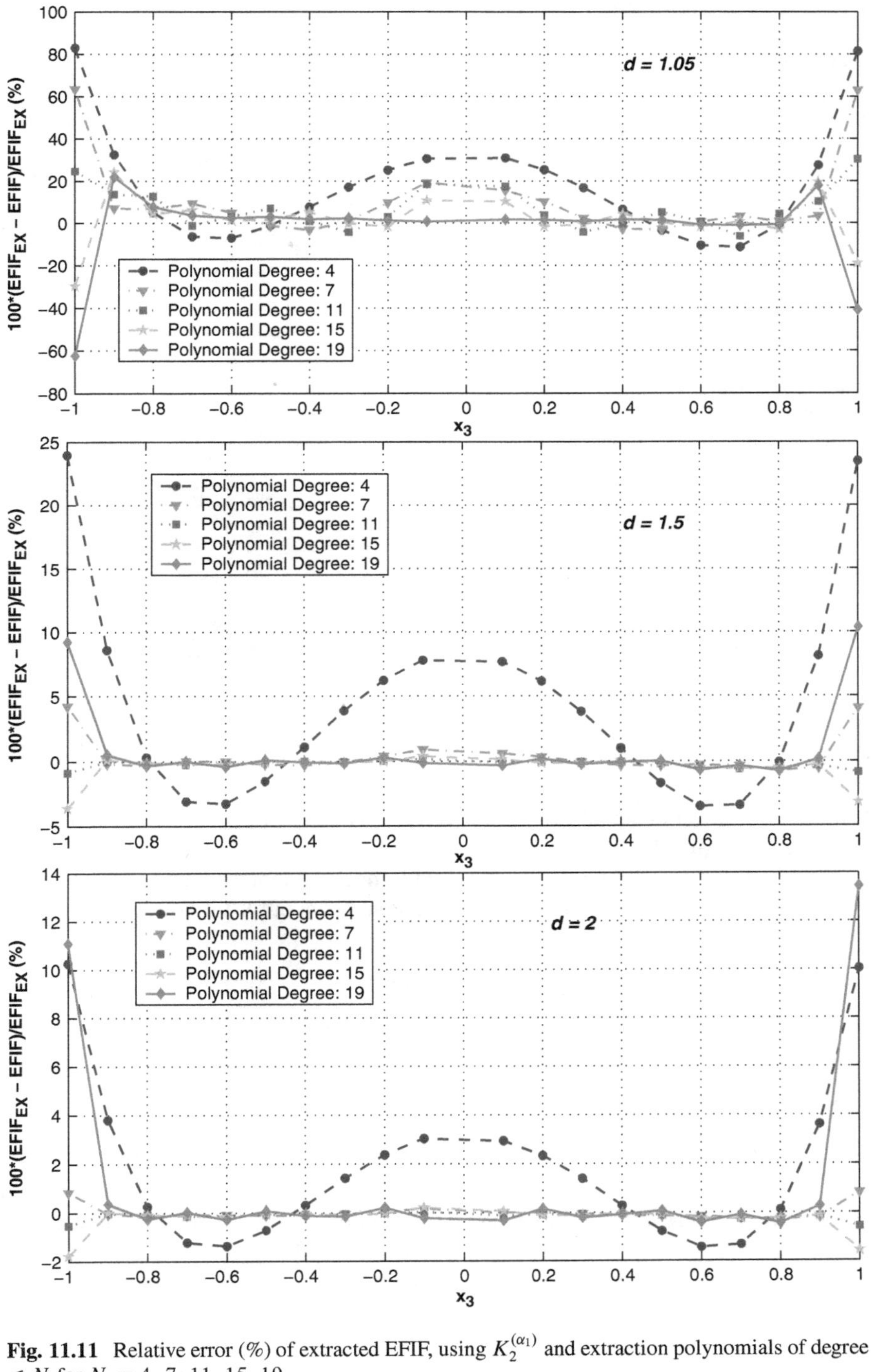

Fig. 11.11 Relative error (%) of extracted EFIF, using $K_2^{(\alpha_1)}$ and extraction polynomials of degree $\leq N$ for $N = 4, 7, 11, 15, 19$.

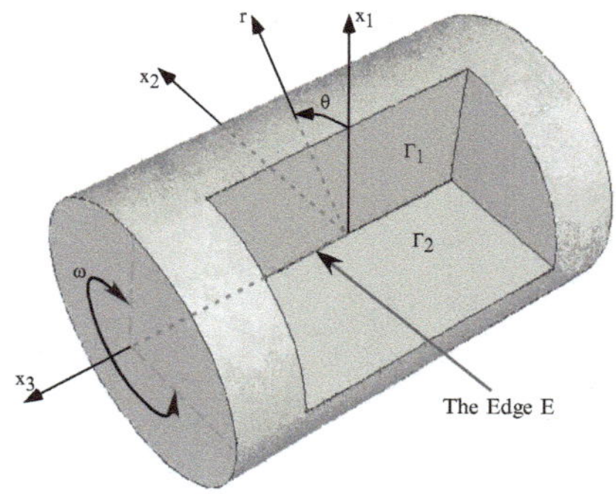

Fig. 11.12 Schematic realistic domain with two Fichera corners.

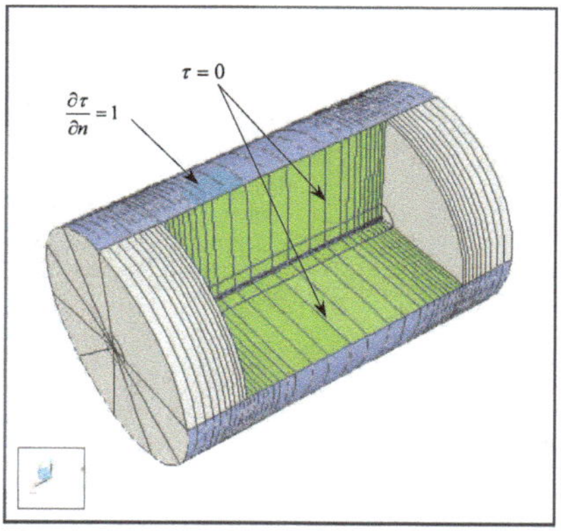

Fig. 11.13 The boundary conditions (11.53)-(11.54) applied to the FE model.

toward $x_3 = \pm 1$, having 45 layers of elements. The discretization of the domain is presented in Figure 11.13.

We consider the Laplace equation with homogeneous Neumann boundary conditions prescribed over the domain's boundary, except for the following:

$$\frac{\partial \tau}{\partial r} = 1 \quad \text{on } \Gamma, \tag{11.53}$$

$$\tau = 0 \quad \text{on } \Gamma_1 \cup \Gamma_2, \tag{11.54}$$

where

$$\Gamma := \{ \boldsymbol{x} \in \mathbb{R}^3 \mid r = 1, \ \theta \in (0, \omega), \ x_3 \in (-1.5, 1.5) \}, \tag{11.55}$$

as shown in Figure 11.13. Under these boundary conditions, vertex singularities arise at $(r, \theta, x_3) = (0, 0, -1)$ and $(r, \theta, x_3) = (0, 0, 1)$, and the exact EFIF is unknown. It can be expected that the EFIF tends to infinity at the vertices.

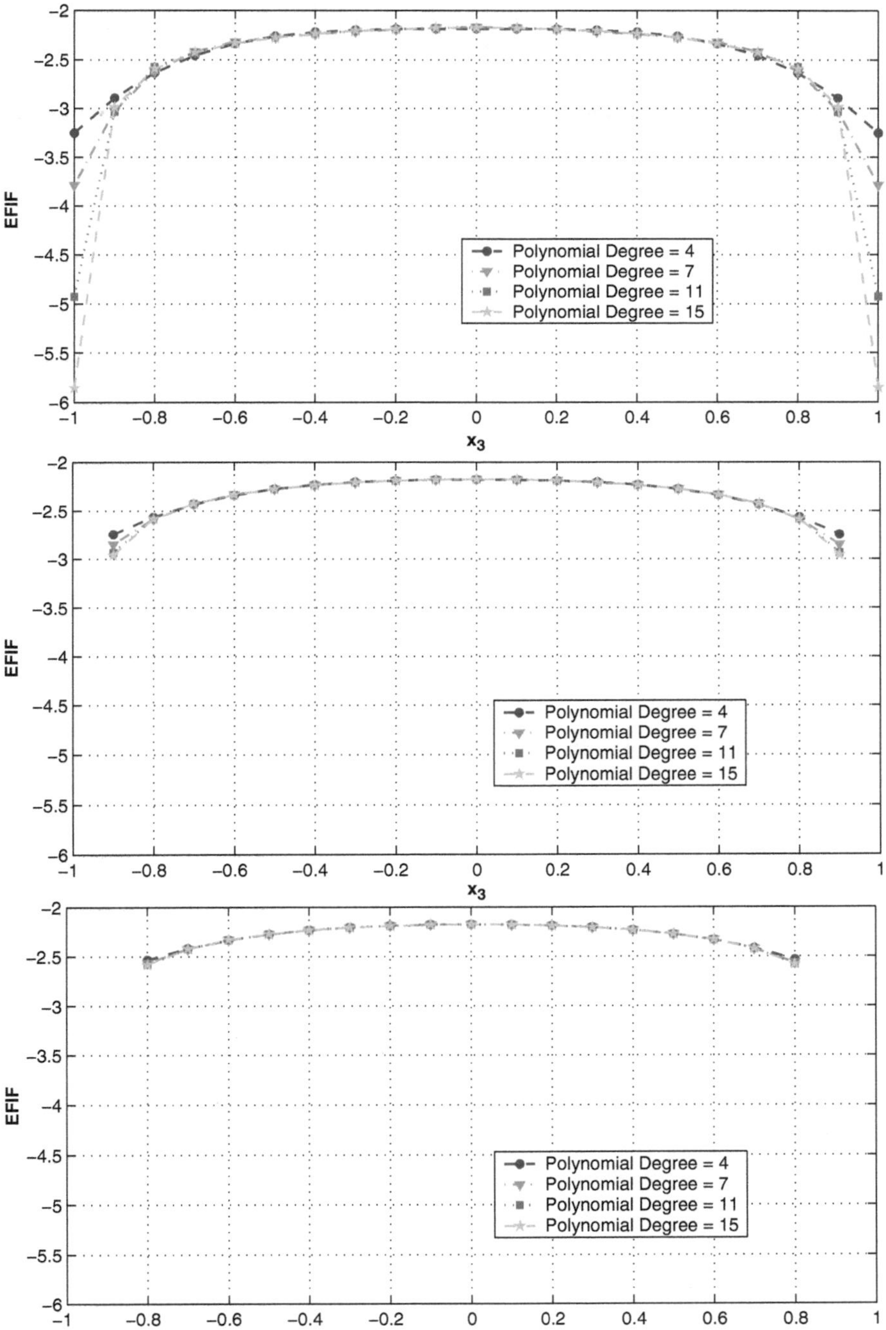

Fig. 11.14 Top: EFIF extracted on $-1 < x_3 < 1$. Middle: EFIF extracted on $-0.9 < x_3 < 0.9$. Bottom: EFIF extracted on $-0.8 < x_3 < 0.8$. All use the hierarchical extraction polynomials of degree $k = 4,\ 7,\ 11,\ 15$, with $K_2^{(\alpha_1)}$ at $R = 0.05$.

Using the extraction polynomials $BJ_2^{(0)}, \ldots, BJ_2^{(k)}$, where $4 < k < 15$, we extract the EFIF for case 1 at $R = 0.05$ on three intervals on the edge $-1 < x_3 < 1$, $-0.9 < x_3 < 0.9$, and $-0.8 < x_3 < 0.8$. These are presented in Figure 11.14. It can be observed that the EFIFs extracted on $-1 < x_3 < 1$ are influenced by the vertex singularities at $x_3 = \pm 1$.

Chapter 12
Vertex Singularities for the 3-D Laplace Equation

Although singular points in 2-D domains have been extensively investigated, the vertex singularities in 3-D domains have received scant attention due to their complexity. To the best of our knowledge, numerical methods for the investigation of vertices of conical notches, specifically the exponents of the singularity, were first introduced in [23]. Stephan and Whiteman [170] and Beagles and Whiteman [25] investigated analytically several vertices for the Laplace equation in 3-D, mainly with homogeneous Dirichlet boundary conditions, and analyzed a finite element method for the computation of eigenvalues by discretizing the Laplace-Beltrami equation (error estimates provided but no numerical results).

Analytical methods for the computation of the singularity exponents for homogeneous Dirichlet boundary conditions are provided in [25] and in [26, pp. 45-48] for axisymmetric cases. In [50] the Laplace equation in the vicinity of a conical point with Neumann boundary conditions is also discussed, with a graph describing the behavior of the eigenvalues for different opening angles ω.

In this chapter we derive explicit analytical expressions for the eigenpairs γ_i and $s_i(\theta, \varphi)$ associated with conical points and extend the modified Steklov method for the computation of eigenpairs associated with vertex singularities of the Laplace equation [213]. The analytical solutions for conical vertices for simplified problems are given in Section 12.1, against which our numerical methods are compared to demonstrate their convergence rate and accuracy. In Section 12.2 we formulate the weak eigenproblem, i.e., the modified Steklov formulation, and cast it in a form suitable for spectral/p finite element discretization. This method is aimed at computing the eigenpairs in a very efficient and accurate manner, and may be generalized to multimaterial interfaces and elasticity operators. Numerical examples are considered in Section 12.3. We first consider two problems for which analytical eigenpairs are provided in Section 12.1 to demonstrate the accuracy and efficiency of the proposed numerical methods, followed by two more-complicated example problems for which analytical results are unavailable.

Z. Yosibash, *Singularities in Elliptic Boundary Value Problems and Elasticity and Their Connection with Failure Initiation*, Interdisciplinary Applied Mathematics 37,
DOI 10.1007/978-1-4614-1508-4_12,

12.1 Analytical Solutions for Conical Vertices

Consider a three-dimensional (3-D) domain Ω having a rotationally symmetric conical vertex O on its boundary as shown in Figure 12.1 with $\omega/2 \in [0, \pi]$. Locating a spherical coordinate system in O, we aim at solving the Laplace equation with either homogeneous Dirichlet boundary conditions (BCs) in the vicinity of the conical point $\rho \to 0$,

$$\nabla^2 \tau(\rho, \theta, \varphi) = 0 \quad \text{in } \Omega, \tag{12.1}$$

$$\tau(\rho, \theta = \omega/2, \varphi) = 0 \quad \text{on } \partial\Omega_c, \tag{12.2}$$

or with homogeneous Neumann BCs,

$$\frac{\partial \tau}{\partial n}(\rho, \theta = \omega/2, \varphi) = \frac{1}{\rho}\frac{\partial \tau(\rho, \theta = \omega/2, \varphi)}{\partial \theta} = 0 \quad \text{on } \partial\Omega_c, \tag{12.3}$$

where $\partial\Omega_c = \Gamma_c$ is the surface of the cone insert. Following [102], the solution is sought by separation of variables:

$$\tau(\rho, \theta, \varphi) = R(\rho)\Theta(\theta)F(\varphi). \tag{12.4}$$

Substituting (12.4) in (12.1), one obtains a set of three ODEs as follows:

$$\rho^2 R'' + 2\rho R' - \gamma(\gamma + 1)R = 0, \tag{12.5}$$

$$F'' + \beta^2 F = 0, \tag{12.6}$$

$$-\sin^2(\theta)\Theta'' - \sin(\theta)\cos(\theta)\Theta' - (\gamma(\gamma + 1)\sin^2(\theta) - \mu^2)\Theta = 0, \tag{12.7}$$

where $\gamma(\gamma + 1)$ and β^2 are separation constants. In (12.6) we chose the separation constant as β^2 because it has to be positive if a periodic solution in φ is sought (for conical reentrant corners). The solution to (12.5) is of the form,

$$R(\rho) = A\rho^\gamma, \tag{12.8}$$

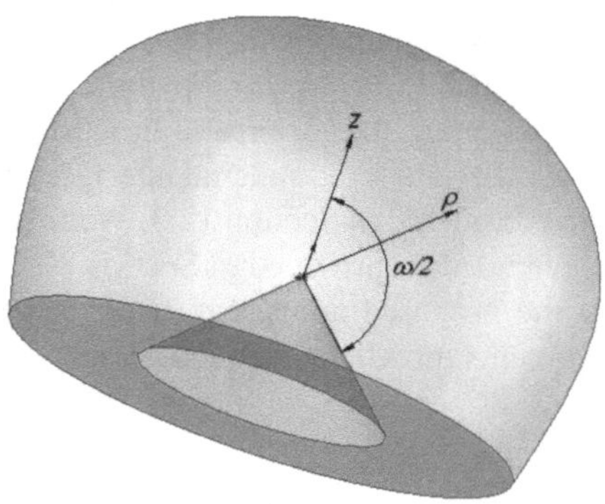

Fig. 12.1 Typical 3-D domain with a rotationally symmetric conical vertex.

where A is a generic constant. The restriction $\gamma > -1/2$ has to hold to obtain solutions that are in $H^1(\Omega)$. The solution to (12.6) has to be periodic in φ so that

$$F(\varphi) = B\sin(\beta\varphi) + C\cos(\beta\varphi), \tag{12.9}$$

where B, C are generic constants. The periodicity constraint is $F(\varphi) = F(\varphi + 2n\pi)$, and therefore β *has to be a positive integer, i.e.,* $\beta = 0, 1, 2, \ldots \stackrel{\text{def}}{=} m$. The case $m = 0$ is associated with axisymmetric solutions, independent of φ.

Changing variables $z = \cos(\theta)$, the ODE (12.7) becomes

$$(1-z^2)\frac{d^2\Theta}{dz^2} - 2z\frac{d\Theta}{dz} + \left[\gamma(\gamma+1) - \frac{m^2}{1-z^2}\right]\Theta = 0, \tag{12.10}$$

with homogeneous Dirichlet BCs,

$$\Theta(z_0) = 0 \quad \Rightarrow \quad \Theta(\cos\omega/2) = 0, \tag{12.11}$$

or homogeneous Neumann BCs,

$$\frac{1}{\rho}\frac{d\Theta(z_0)}{d\theta} = 0 \quad \Rightarrow \quad \frac{d\Theta(\cos\omega/2)}{d\theta} = 0. \tag{12.12}$$

In general z may be a complex variable, and m, γ are parameters that may take arbitrary real or complex values, called *spherical harmonics*.

The solution to (12.10) consists of a linear combination of *associated Legendre functions of degree* γ *and order* m *of the first and second kind*, denoted by $P_\gamma^m(z)$ and $Q_\gamma^m(z)$ respectively, i.e.,

$$\Theta(z) = DP_\gamma^m(z) + EQ_\gamma^m(z) \quad \Rightarrow \quad \Theta(\cos\theta) = DP_\gamma^m(\cos\theta) + EQ_\gamma^m(\cos\theta). \tag{12.13}$$

Because Legendre functions of the second kind for $m = 0$ tend to ∞ along the axis of symmetry of the domain, then $E \equiv 0$. Furthermore, for $m > 0$ the leading term of $Q_\gamma^m(z)$ is

$$2^{m/2-1}\Gamma(m)\cos(m\pi)(1-z)^{-m/2}.$$

Then at $\theta = 0$, i.e., $z = 1$, $Q_\gamma^m(\cos 0)$ is unbounded, and thus one must choose $E = 0$, which reduces the solution (12.13) to

$$\Theta(\cos\theta) = DP_\gamma^m(\cos\theta), \tag{12.14}$$

where $P_\gamma^m(\cos(\omega/2))$ is the associated Legendre function of the first kind. For example, for the case $m = 0$, the Legendre function P_γ can be computed using the Mehler-Dirichlet formula [102, (7.4.10)]:

$$P_\gamma(\cos\omega/2) = \frac{\sqrt{2}}{\pi}\int_0^{\omega/2} \frac{\cos\left(\gamma+\frac{1}{2}\right)t}{\sqrt{\cos t - \cos(\omega/2)}}dt. \tag{12.15}$$

It is important to notice that [102]

$$P_\gamma^m(\cos\theta) = P_{-1-\gamma}^m(\cos\theta), \tag{12.16}$$

which has an important implication on the solution, i.e., if a given P_γ^m is a solution, then also $P_{-1-\gamma}^m$, i.e., if a given γ is found to satisfy the BCs, so will $-1-\gamma$.

Because there is an infinite number of γ's that are determined by the boundary conditions (detailed in the next subsection), each being a root of the Legendre function $P_{\gamma_\ell}^m$, we denote them by two indices $\gamma_\ell^{(m)}$, so the overall solution can be represented by

$$\tau(\rho,\theta,\phi) = \sum_{m=0}\sum_{\ell=1} \rho^{\gamma_\ell^{(m)}} \left[A_{m,\ell}\sin(m\varphi) + B_{m,\ell}\cos(m\varphi)\right] P_{\gamma_\ell}^m(\cos\theta). \tag{12.17}$$

12.1.1 Homogeneous Dirichlet BCs

Consider, for example, the domain in Figure 12.1 with the conical point at the apex of a cone insert having a solid angle $\omega = 6\pi/4$. There is an infinite number of γ's for which the homogeneous Dirichlet BC (12.2) holds. These γ's are found by the root of (12.2):

$$P_\gamma^m(\cos 3\pi/4) = 0, \tag{12.18}$$

$m = 0$: *axi-symmetric solution.* For the case $m = 0$, (12.18) reads $P_\gamma(\cos 3\pi/4) = 0$. Using Mathematica, [192] one may easily obtain, e.g., the following first four nonnegative $\gamma^{(0)}$'s for which (12.18) holds:

$$\gamma_1^{(0)} = 0.463098561780106, \quad \gamma_2^{(0)} = 1.81322787311022,$$

$$\gamma_3^{(0)} = 3.153048711303707, \quad \gamma_4^{(0)} = 4.48976080342872.$$

The associated Legendre functions of the first kind are shown in Figure 12.2.

For each of the $\gamma_i^{(0)}$'s a solution is obtained of the form

$$\tau_i(\rho,\theta,\varphi) = A_i \rho^{\gamma_i^{(0)}} P_{\gamma_i^{(0)}}(\cos\theta),$$

so that the overall solution is a linear combination:

$$\tau(\rho,\theta,\varphi) = \sum_i A_i \rho^{\gamma_i^{(0)}} P_{\gamma_i^{(0)}}(\cos\theta). \tag{12.19}$$

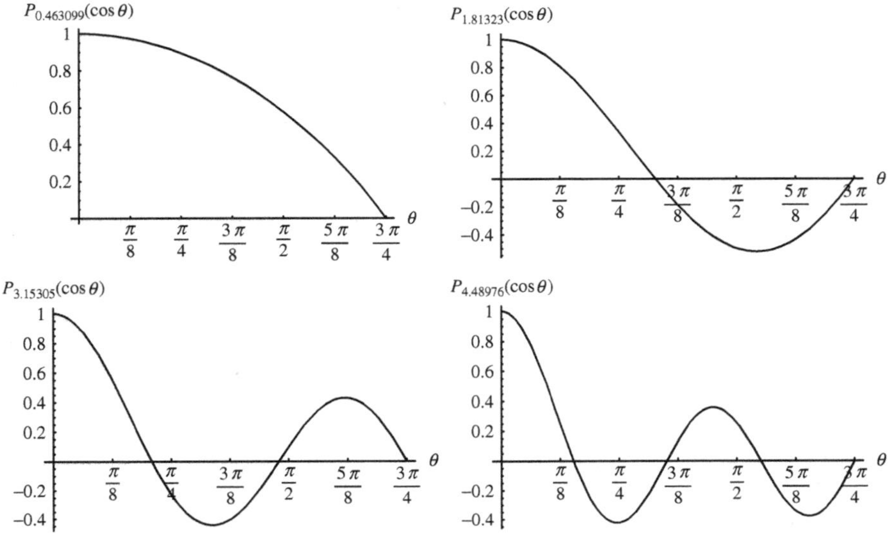

Fig. 12.2 First four eigenfunctions, Dirichlet BCs for a conical point having a solid angle $3\pi/4$.

Table 12.1 First four γ's for $m = 0, 1, 2, 3$ for Dirichlet BCs associated with $\omega = 6\pi/4$.

	$\gamma_1^{(m)}$	$\gamma_2^{(m)}$	$\gamma_3^{(m)}$	$\gamma_4^{(m)}$
$m = 0$	0.46309856178010	1.81322787311022	3.153048711303707	4.48976080342872
$m = 1$	1.24507709100149	2.54898557133218	3.868541068328044	5.19403335518201
$m = 2$	2.13656665895361	3.37380855301073	4.655359106556064	5.95715662710399
$m = 3$	3.07712950983885	4.25338593246190	5.492126885263152	6.76456426448560

Remark 12.1 *Notice that since $\gamma_1^{(0)} < 1$, the first derivative is unbounded as $\rho \to 0$.*

$m = 1, 2, 3, \ldots$. For an arbitrary m, the solution of $P_\gamma^m(\cos 3\pi/4) = 0$ can be obtained. We summarize in Table 12.1 the first four γ's for $m = 0, 1, 2, 3$.

In Figure 12.3 we plot the variation of the smallest eigenvalues $\gamma_1^{(0)}$, $\gamma_1^{(1)}$, and $\gamma_1^{(2)}$ as a function of ω starting from a flat plate $\omega/2 = \pi/2$ up to a reentrant line $\omega/2 = \pi$.

12.1.2 Homogeneous Neumann BCs

For the same domain as that in Section 12.1.1 with $\omega = 6\pi/4$, the homogeneous Neumann BC (12.3) reads

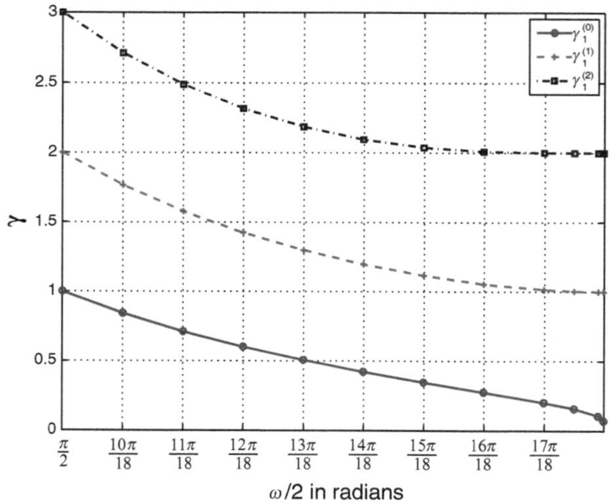

Fig. 12.3 $\gamma_1^{(0)}$, $\gamma_1^{(1)}$, and $\gamma_1^{(1)}$ as a function of the cone reentrant angle ω, Dirichlet BCs.

$$\frac{1}{\rho}\frac{dP_\gamma^m(\cos\theta)}{d\theta}\bigg|_{\theta=3\pi/4} = 0 \quad \Rightarrow \quad -\sin\theta \frac{dP_\gamma^m(\cos\theta)}{d\cos\theta}\bigg|_{\theta=3\pi/4} = 0. \qquad (12.20)$$

Using the recurrence [102, (7.12.16) on p. 195]

$$\left(z^2-1\right)\frac{dP_\gamma^m(z)}{dz} = \gamma z P_\gamma^m(z) - (\gamma+m)P_{\gamma-1}^m(z), \qquad (12.21)$$

the BC (12.20) becomes

$$\cos(3\pi/4)\gamma P_\gamma^m(\cos 3\pi/4) - (\gamma+m)P_{\gamma-1}^m(\cos 3\pi/4) = 0, \qquad (12.22)$$

for which there exists an infinite number of γ's.

The smallest nonnegative eigenvalue is 0, associated with the so-called rigid body motion (known to exist for homogeneous Neumann BCs), and is of no interest, since it describes a constant solution.

$m = 0$*: axisymmetric solution* The first four nonnegative (and nonzero) γ's for which (12.22) holds are

$$\gamma_1^{(0)} = 1.24507709100149, \quad \gamma_2^{(0)} = 2.548985521168983,$$
$$\gamma_3^{(0)} = 3.86854093155942, \quad \gamma_4^{(0)} = 5.194033355182022.$$

The associated Legendre eigenfunctions of the first kind are shown in Figure 12.4.

$m = 1, 2, 3, \ldots$ For an arbitrary m we summarize in Table 12.2 the first four (nonzero) γ's for $m = 0, 1, 2, 3$.

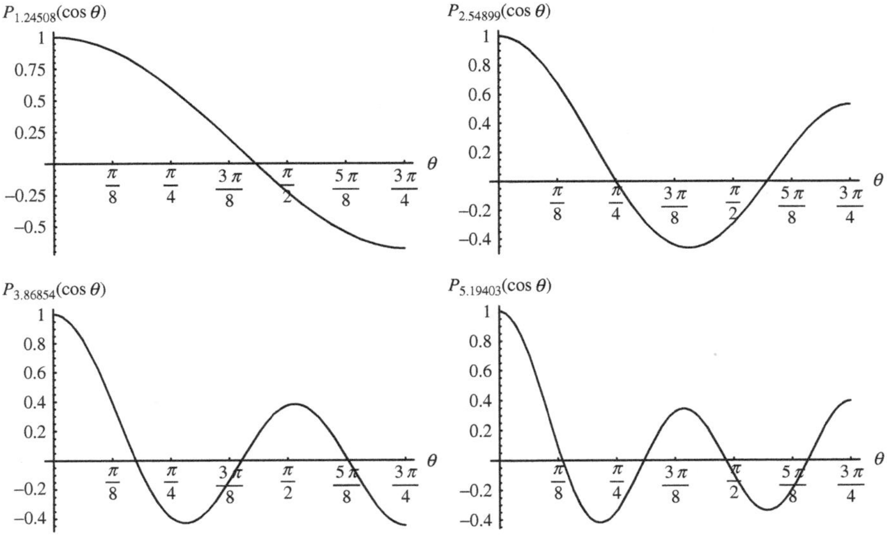

Fig. 12.4 First four eigenfunctions associated with $m = 0$, Neumann BCs, $\omega = 6\pi/4$.

Table 12.2 First four γ's for $m = 0, 1, 2, 3$ for Neumann BCs associated with $\omega = 6\pi/4$.

	$\gamma_1^{(m)}$	$\gamma_2^{(m)}$	$\gamma_3^{(m)}$	$\gamma_4^{(m)}$
$m = 0$	1.2450770910	2.54898552117	3.86854093156	5.19403335518
$m = 1$	0.8571676765	2.00000000000	3.27090467124	4.57561722130
$m = 2^*$	1.8742536963	2.88678057132	4.08498821549	5.35319993628
$m = 3^{**}$	2.9130094418	3.84636536605	4.96120003163	6.18222599268

* For $m = 2$, $\gamma = 1$ is also an eigensolution, but $P_1^2(\cos\theta) \equiv 0$.
** For $m = 3$, $\gamma = 1, 2$ are also eigensolutions, but $P_1^3(\cos\theta) = P_2^3(\cos\theta) \equiv 0$.

12.2 The Modified Steklov Weak Form and Finite Element Discretization

Here we develop the formulation and numerical procedures that will efficiently and reliably compute approximations for the singular solutions (eigenpairs) for such problems when used in conjunction with the spectral/p-version of the finite element method. The "modified Steklov method" is general, that is, applicable to singularities associated with corners, anisotropic multimaterial interfaces, and abrupt changes in boundary conditions.

Consider (12.1) with either (12.2) or (12.3) boundary conditions in the vicinity of the vertex, in an artificial subdomain Ω_R created by the intersection of Ω with two spheres of radii $R_1 < R_2$ as shown in Figure 12.5. Since the solution in the

Fig. 12.5 The subdomain Ω_R in the vicinity of the vertex.

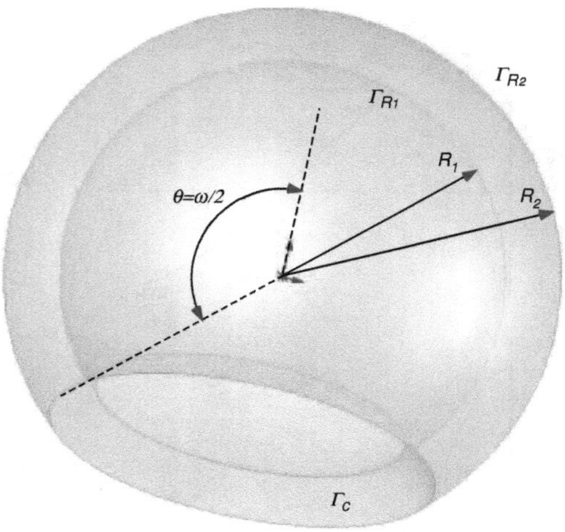

vicinity of the vertex is of the form $\tau = \rho^\gamma s(\theta, \varphi)$, on the surface of the sphere R_1 one obtains

$$\frac{\partial \tau}{\partial n}(\rho = R_1) = -\frac{\partial \tau}{\partial \rho}(\rho = R_1) = -\gamma R_1^{\gamma-1} s(\theta, \varphi) = -\frac{\gamma}{R_1}\tau(R_1, \theta, \varphi). \quad (12.23)$$

Similarly, on the surface of the sphere R_2, one obtains

$$\frac{\partial \tau}{\partial n}(\rho = R_2) = \frac{\partial \tau}{\partial \rho}(\rho = R_2) = \frac{\gamma}{R_2}\tau(R_2, \theta, \varphi). \quad (12.24)$$

Thus, the strong (classical) modified Steklov formulation in Ω_R (see [210]) is obtained:

$$\nabla^2 \tau(\rho, \theta, \varphi) = 0 \quad \text{in } \Omega_R, \quad (12.25)$$

$$\frac{1}{\rho}\frac{\partial \tau(\rho, \theta = \omega/2, \varphi)}{\partial \theta} = 0 \text{ or } \tau(\rho, \theta = \omega/2, \varphi) = 0 \quad \text{on } \Gamma_c \stackrel{\text{def}}{=} \partial \Omega_c, \quad (12.26)$$

$$\frac{\partial \tau}{\partial n}(\rho = R_1) = -\frac{\gamma}{R_1}\tau(R_1, \theta, \varphi) \quad \text{on } \Gamma_{R_1}, \quad (12.27)$$

$$\frac{\partial \tau}{\partial n}(\rho = R_2) = \frac{\gamma}{R_2}\tau(R_2, \theta, \varphi) \quad \text{on } \Gamma_{R_2}. \quad (12.28)$$

The strong modified Steklov formulation may be brought to a weak form by multiplying (12.25) by a test function $\chi(\rho, \theta, \varphi)$, integrating over Ω_R, and using Green's theorem to obtain

$$\iiint_{\Omega_R} (\nabla \tau) \cdot (\nabla \chi) d\Omega = \iint_{\partial \Omega_R} \frac{\partial \tau}{\partial n} \chi d\Gamma, \quad (12.29)$$

For homogeneous Neumann or Dirichlet boundary conditions on $\partial\Omega_c$, that part of the boundary diminishes in the RHS of (12.29), and considering (12.27)-(12.28), one finally obtains the weak modified Steklov eigenformulation:

$$\begin{aligned}&\text{Seek } \gamma \in \mathbb{R} \text{ and } 0 \neq \tau \in \mathcal{E}(\Omega_R) \text{ such that } \forall \chi \in \mathcal{E}(\Omega_R),\\ &\mathcal{B}(\tau,\chi) = \gamma \left[\mathcal{M}_{R_2}(\tau,\chi) - \mathcal{M}_{R_1}(\tau,\chi)\right],\end{aligned} \tag{12.30}$$

where

$$\begin{aligned}\mathcal{B}(\tau,\chi) \stackrel{\text{def}}{=} \int_{\rho=R_1}^{R_2}\int_{\theta=0}^{\frac{\omega}{2}}\int_{\varphi=0}^{2\pi}&\left(\rho^2\sin\theta\frac{\partial\tau}{\partial\rho}\frac{\partial\chi}{\partial\rho} + \sin\theta\frac{\partial\tau}{\partial\theta}\frac{\partial\chi}{\partial\theta}\right.\\ &\left.+\frac{1}{\sin\theta}\frac{\partial\tau}{\partial\varphi}\frac{\partial\chi}{\partial\varphi}\right)d\rho\, d\theta\, d\varphi,\end{aligned} \tag{12.31}$$

$$\begin{aligned}\gamma\left[\mathcal{M}_{R_2}(\tau,\chi) - \mathcal{M}_{R_1}(\tau,\chi)\right] \stackrel{\text{def}}{=} \gamma&\left[R_2\int_{\theta=0}^{\frac{\omega}{2}}\int_{\varphi=0}^{2\pi}\tau\,\chi|_{R_2}\sin\theta\, d\theta\, d\varphi\right.\\ &\left.-R_1\int_{\theta=0}^{\frac{\omega}{2}}\int_{\varphi=0}^{2\pi}\tau\,\chi|_{R_1}\sin\theta\, d\theta\, d\varphi\right].\end{aligned} \tag{12.32}$$

Remark 12.2 *Notice that in 2-D [210], the RHS of the weak form is independent of the radius of the circular domain, whereas in 3-D there is an explicit dependency on R_1 and R_2.*

Remark 12.3 *For homogeneous Dirichlet boundary conditions, the energy space $\mathcal{E}(\Omega_R)$ is restricted to functions that automatically satisfy these, i.e., to $\mathcal{E}_o(\Omega_R)$.*

Remark 12.4 *The third term in the integrand in (12.31) is singular $\frac{1}{\sin\theta}$. A remedy for this difficulty may be obtained if an asymmetric weak formulation is considered.*

Remark 12.5 *The weak formulation (12.30) may be generalized to cases that involve a V-notch or a crack front (see e.g. Figures 12.8-12.10). In these cases the integral on the variable φ is to be performed from 0 to φ_2 (the solid angle of the V-notch opening; a crack is a V-notch for which $\varphi_2 = 2\pi$). We generalize the formulation for these latter cases in the following.*

12.2.0.1 An Asymmetric Weak Eigenform

In view of Remark 12.4, we multiply (12.25) by a special test function $\sin\theta w(\rho,\theta,\varphi)$ and follow the steps described above to obtain an asymmetric weak modified Steklov eigenformulation that does not contain singular terms:

$$\begin{aligned}&\text{Seek } \gamma \in \mathbb{R} \text{ and } 0 \neq \tau \in \mathcal{E}(\Omega_R) \text{ such that } \forall \chi \in \mathcal{E}(\Omega_R),\\ &\tilde{\mathcal{B}}(\tau,\chi) = \gamma\left[\tilde{\mathcal{M}}_{R_2}(\tau,\chi) - \tilde{\mathcal{M}}_{R_1}(\tau,\chi)\right],\end{aligned} \tag{12.33}$$

where

$$\tilde{\mathcal{B}}(\tau,\chi) \stackrel{\text{def}}{=} \int_{\rho=R_1}^{R_2}\int_{\theta=0}^{\frac{\omega}{2}}\int_{\varphi=0}^{\varphi_2}\left(\rho^2\sin^2\theta\frac{\partial\tau}{\partial\rho}\frac{\partial\chi}{\partial\rho}+\sin^2\theta\frac{\partial\tau}{\partial\theta}\frac{\partial\chi}{\partial\theta}+\frac{\partial\tau}{\partial\varphi}\frac{\partial\chi}{\partial\varphi}\right.$$
$$\left.+\frac{\sin(2\theta)}{2}\frac{\partial\tau}{\partial\theta}\chi\right)d\rho\,d\theta\,d\varphi, \qquad (12.34)$$

$$\gamma\left[\tilde{\mathcal{M}}_{R_2}(\tau,\nu)-\tilde{\mathcal{M}}_{R_1}(\tau,\nu)\right] \stackrel{\text{def}}{=} \gamma\left[R_2\int_{\theta=0}^{\frac{\omega}{2}}\int_{\varphi=0}^{\varphi_2}\tau\,\chi|_{R_2}\sin^2\theta\,d\theta\,d\varphi\right.$$
$$\left.-R_1\int_{\theta=0}^{\frac{\omega}{2}}\int_{\varphi=0}^{\varphi_2}\tau\,\chi|_{R_1}\sin^2\theta\,d\theta\,d\varphi\right]. \qquad (12.35)$$

For convenience of numerical application (and for future use of the p-FE method), we perform a change of variables in (12.34-12.35) as follows:

$$\rho=\frac{1-\xi}{2}R_1+\frac{1+\xi}{2}R_1 \quad\rightarrow\quad d\rho=\frac{R_2-R_1}{2}d\xi, \qquad (12.36)$$

$$\theta=\frac{1+\eta}{2}\frac{\omega}{2} \quad\rightarrow\quad d\theta=\frac{\omega}{4}d\eta, \qquad (12.37)$$

$$\varphi=\frac{1+\zeta}{2}\varphi_2 \quad\rightarrow\quad d\varphi=\frac{\varphi_2}{2}d\zeta, \qquad (12.38)$$

so that (12.34-12.35) become

$$\tilde{\mathcal{B}}(\tau,\chi)=\frac{\omega\varphi_2}{4(R_2-R_1)}\iiint_{-1}^{1}\rho^2(\xi)\sin^2\theta(\eta)\frac{\partial\tau}{\partial\xi}\frac{\partial\chi}{\partial\xi}\,d\xi\,d\eta\,d\zeta \qquad (12.39)$$
$$+\frac{(R_2-R_1)\varphi_2}{\omega}\iiint_{-1}^{1}\sin^2\theta(\eta)\frac{\partial\tau}{\partial\eta}\frac{\partial\chi}{\partial\eta}\,d\xi\,d\eta\,d\zeta$$
$$+\frac{(R_2-R_1)}{4}\left[\frac{\omega}{\varphi_2}\iiint_{-1}^{1}\frac{\partial\tau}{\partial\zeta}\frac{\partial\chi}{\partial\zeta}\,d\xi\,d\eta\,d\zeta\right.$$
$$\left.+\varphi_2\iiint_{-1}^{1}\frac{\sin(2\theta(\eta))}{2}\frac{\partial\tau}{\partial\eta}\chi\,d\xi\,d\eta\,d\zeta\right]$$

$$\gamma\left[\tilde{\mathcal{M}}_{R_2}(\tau,\nu)-\tilde{\mathcal{M}}_{R_1}(\tau,\nu)\right]=\gamma\frac{\omega\varphi_2}{8}\left[\iint_{-1}^{1}\left(R_2\,(\tau\,\chi)|_{\xi=1}-R_1\,(\tau\,\chi)|_{\xi=-1}\right)\right.$$
$$\left.\sin^2\theta(\eta)\,d\eta\,d\zeta\right] \qquad (12.40)$$

12.2.1 Application of p/Spectral Finite Element Methods

The weak form (12.33) may be represented in terms of a matrix formulation using the p-version or spectral finite element methods. The finite-dimensional space corresponding to the weak form is spanned by a set of shape functions $\Phi_i(\xi, \eta, \zeta)$, $i = 1, \dots, (p+1)(q+1)(s+1)$, where $(p+1)$ represents the number of basis functions that span the functional space in ξ (and $(q+1)$, $(s+1)$ correspond to the number of basis functions in η and ζ respectively). In terms of the shape functions and their coefficients, one has $\tau = \sum_{i=1}^{(p+1)(q+1)(s+1)} a_i \Phi_i(\xi, \eta, \zeta) = \boldsymbol{a}_{tot}^T \boldsymbol{\Phi}$ and similarly $\chi = \boldsymbol{\Phi}^T \boldsymbol{b}_{tot}$. Denoting by $\boldsymbol{a}_R$ and $\boldsymbol{b}_R$ the coefficients that multiply basis functions *that are nonzero* on Γ_{R_1} and Γ_{R_2}, (12.30) or (12.33) becomes

$$[K]\boldsymbol{a}_{tot} = \gamma \begin{bmatrix} [M_{R_1}] & [0] \\ [0] & [M_{R_2}] \end{bmatrix} \boldsymbol{a}_R, \tag{12.41}$$

where $[K]$ is the stiffness matrix, and $[M_{R_i}]$ are the generalized mass matrices corresponding to the terms in $\mathbf{a}_R$ on the boundaries Γ_{R_i}.

We may partition $\boldsymbol{a}_{tot} = \{\boldsymbol{a}_R \ , \ \boldsymbol{a}_{in}\}$. By partitioning $[K]$, we may represent the eigenproblem (12.41) in the form:

$$\begin{bmatrix} [K_R] & [K_{R\text{-}in}] \\ [K_{in\text{-}R}] & [K_{in}] \end{bmatrix} \{\boldsymbol{a}_R \ , \ \boldsymbol{a}_{in}\} = \gamma \begin{bmatrix} [M_{R_1}] & [0] \\ [0] & [M_{R_2}] \end{bmatrix} \boldsymbol{a}_R. \tag{12.42}$$

The relation in (12.42) can be used for eliminating $\mathbf{a}_{in}$ using *static condensation* (named also Schur-complement), thus obtaining the reduced eigenproblem

$$[K_S]\boldsymbol{a}_R = \gamma \begin{bmatrix} [M_{R_1}] & [0] \\ [0] & [M_{R_2}] \end{bmatrix} \boldsymbol{a}_R, \tag{12.43}$$

where:

$$[K_S] = [K_R] - [K_{R\text{-}in}][K_{in}]^{-1}[K_{in\text{-}R}].$$

For the solution of (12.43), it is important to note that $[K_S]$ is, in general, a full matrix. However, since the order of the matrices is relatively small, the solution is inexpensive.

Remark 12.6 *For conical vertices the solution in Ω_R is regular and the p/spectral FEM will converge exponentially [178], and furthermore the dual eigenpairs are obtained, since solutions of the form $\rho^{-\gamma-1}$ belong to $\mathcal{E}(\Omega_R)$.*

Remark 12.7 *Implementation of homogeneous Dirichlet boundary conditions on one or more of the boundaries is realized by the substitution of* 0 *in the rows* *and columns of the matrices* $[K], [M_{R_i}]$ *that correspond to unknown values on the boundaries, except the diagonal term which is set equal to one in* $[K]$ *and* 0.01 *in* $[M_{R_i}]$. *This is numerically equivalent to restricting the space in which the functions belong to* $\mathcal{E}_o(\Omega_R)$, *and produces artificial eigenvalues of* 100.

12.2.1.1 The Basis Functions

We construct the basis functions so that the first $2(q+1)(s+1)$ are nonzero on the two boundaries $\rho = R_1$ and $\rho = R_2$, whereas all the others are zero on these two boundaries. A polynomial basis (in terms of the variable $-1 \leq t \leq 1$, based on the Legendre polynomials [178]) is chosen to represent the solution in $\rho(\xi)$, $\theta(\eta)$, or $\varphi(\zeta)$ (t is replaced by ξ or η or ζ):

$$
\begin{aligned}
P_1(t) &= (1-t)/2,\\
P_2(t) &= (1+t)/2,\\
P_3(t) &= \sqrt{\frac{3}{8}}(t^2-1),\\
P_4(t) &= \sqrt{\frac{5}{8}}t(t^2-1),\\
P_5(t) &= \sqrt{\frac{7}{128}}(5t^4-6t^2+1),\\
P_6(t) &= \sqrt{\frac{9}{128}}t(7t^4-10t^2+3),\\
P_7(t) &= \sqrt{\frac{11}{512}}(21t^6-35t^4+15t^2-1),\\
P_8(t) &= \sqrt{\frac{13}{512}}t(33t^6-63t^4+35t^2-5),\\
P_9(t) &= \sqrt{\frac{15}{32768}}(429t^8-924t^6+630t^4-140t^2+5).
\end{aligned}
$$

If the domain of interest has conical vertices as shown in Figure 12.1, the basis functions have to be periodic in φ with period 2π. Therefore, in this case a sin and cos basis is chosen as the basis functions in φ:

$$
Q_k(\zeta) = \begin{cases} \cos(k\frac{1+\zeta}{4}\varphi_2) & k = 0, 2, 4, 6, \ldots, \\ \sin((k+1)\frac{1+\zeta}{4}\varphi_2) & k = 1, 3, 5, 7, \ldots. \end{cases} \tag{12.44}
$$

Otherwise, for aperiodic solutions such as the vertices in the domains shown in Figures 12.8 and 12.10, the polynomial basis is chosen to represent the solution in the φ variable.

Therefore, the basis functions are defined as

$$\Phi_{i+(s+1)(j-1)+(s+1)(q+1)(k-1)}(\xi,\eta,\zeta) = \begin{cases} P_i(\xi)P_j(\eta)Q_k(\zeta) & \text{periodic solutions,} \\ P_i(\xi)P_j(\eta)P_k(\zeta) & \text{aperiodic solutions,} \end{cases} \tag{12.45}$$

$$i = 1,\ldots,p+1, \quad j = 1,\ldots,q+1, \quad k = 1,\ldots,s+1$$

resulting in a $(p+1)(q+1)(s+1) \times (p+1)(q+1)(s+1)$ stiffness matrix $[K]$, which after static condensation is reduced to a $2(q+1)(s+1) \times 2(q+1)(s+1)$ eigenproblem. The formulation described here was implemented utilizing the Mathematica package [192] for the generation of the required matrices and the computation of the eigenvalues and eigenvectors.

12.3 Numerical Examples

Four example problems are considered. The first two involve a conical vertex with either homogeneous Neumann or homogeneous Dirichlet BCs for which analytical solutions are available, so that the convergence rate of the modified Steklov asymmetric method can be assessed. The third example problem involves a vertex generated at the intersection of a crack front and a flat plane with homogeneous Neumann BCs. This case is considered because the artificial subdomain contains a singular edge (along the crack front); therefore the convergence rate is slower. This example problem does not have an analytical solution, and numerical approximations are provided. The last example problem involves a vertex at the intersection of a V-notch front with a conical reentrant corner, with homogeneous Neumann BCs where a singular edge also exists in the subdomain along the V-notch front.

12.3.1 *Conical Vertex, $\omega/2 = 3\pi/4$, Homogeneous Neumann BCs*

We demonstrate the accuracy and efficiency of the modified Steklov asymmetric eigenformulation by considering a conical vertex with $\omega/2 = 3\pi/4$. We choose $R_1 = 0.95$ and $R_2 = 1$ because the eigenvalues are insensitive to R_1 (keeping $R_2 = 1$) for $R_1 > 0.9$. This is because the exact solution is of the form ρ^γ, which may be well represented by polynomials for $0.9 < \rho < 1$ (see also the 2-D case in

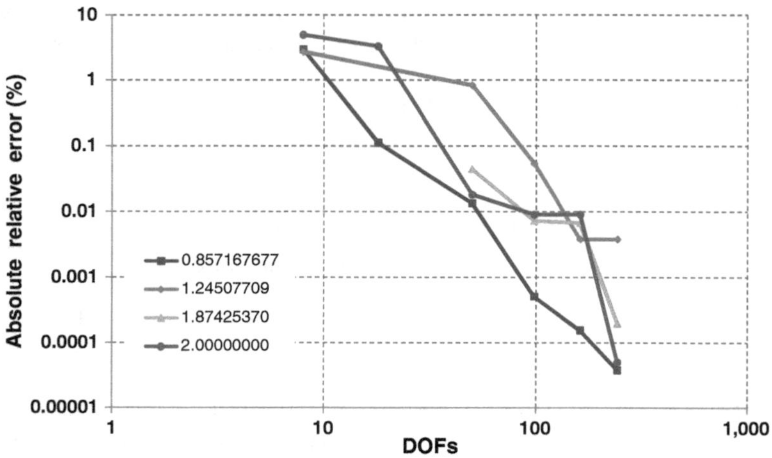

Fig. 12.6 Convergence of the first four nonzero eigenvalues γ for the conical vertex with $\omega/2 = 3\pi/4$ and homogeneous Neumann BCs.

[210]). We first consider homogeneous Neumann boundary conditions on Γ_c, and summarize the first seven computed eigenvalues in Table 12.3, together with the relative error in % defined as

$$\text{Relative error } \% = \frac{100(\gamma_i^{\text{FE}} - \gamma_i^{\text{EX}})}{\gamma_i^{\text{EX}}}.$$

To better demonstrate the accuracy and fast convergence rate of the modified asymmetric Steklov method, we plot in Figure 12.6 the relative error as a percentage of the first five eigenvalues as the number of DOFs is increased. A clear high rate of convergence for the first eigenvalues is observed, yielding an accuracy of order 10^{-4}% relative error with fewer than 300 DOFs but only 200 DOFs in the condensed eigenproblem.

12.3.2 Conical Vertex, $\omega/2 = 3\pi/4$, Homogeneous Dirichlet BCs

In this section we consider the previous conical vertex with $\omega/2 = 3\pi/4$, $R_1 = 0.95$, and $R_2 = 1$, but homogeneous Dirichlet boundary conditions are applied on Γ_c. The first five computed eigenvalues are summarized in Table 12.4, together with the relative error. The convergence rate of the first four eigenvalues is shown in the plot in Figure 12.7. One may observe the fast rate of convergence in this example problem also.

Table 12.3 Conical point with $\omega/2 = 3\pi/4$ and homogeneous Neumann BCs. Convergence of first seven eigenvalues (except first eigenvalue which is 0) for $R_1 = 0.95$ as the approximation functional space is enriched.

p, q, s DOFs	1,1,1 8 DOFs	1,2,2 18 DOFs	1,4,4 50 DOFs	1,6,6 98 DOFs	1,8,8 162 DOFs	2,8,8 243 DOFs
$\gamma_1^{(1)} = 0.85716767$	0.83175 (-2.96 %)	0.85812 (0.11 %)	0.85728 (1.3E-2 %)	0.857172 (1.0E-3 %)	0.857169 (1.5E-4 %)	0.857168 (4.0E-5 %)
$\gamma_1^{(0)} = 1.24507709$	1.21102 (-2.73 %)		1.25547 (8.3E-1 %)	1.24441 (-5.4E-2 %)	1.24503 (-3.8E-3 %)	1.24503 (-3.8E-3 %)
$\gamma_1^{(2)} = 1.87425369$			1.87508 (4.4E-2 %)	1.87439 (7.0E-3 %)	1.87438 (6.7E-3 %)	1.87425 (-2E-4 %)
$\gamma_2^{(1)} = 2.00000000$	2.09971 (4.99 %)	2.0664 (3.32 %)	2.00036 (1.8E-2 %)	2.00018 (9.0E-3 %)	2.00018 (9.0E-3 %)	2.000001 (5E-5 %)
$\gamma_2^{(0)} = 2.54898552$		2.72871 (7.05 %)	2.57948 (1.19 %)	2.54832 (-2.6E-2 %)	2.54971 (2.8E-2 %)	2.54915 (6.4E-3 %)
$\gamma_2^{(2)} = 2.88678057$			2.89257 (2.0E-1%)	2.88773 (3.3E-2 %)	2.88774 (3.3E-2 %)	2.88678 (-2E-5 %)
$\gamma_1^{(3)} = 2.91300944$				2.914 (3.4E-2 %)	2.91401 (3.4E-2 %)	2.91301 (2E-5 %)

In parentheses the relative error in % is reported.

Table 12.4 Conical point with $\omega/2 = 3\pi/4$ and homogeneous Dirichlet BCs. Convergence of first five eigenvalues for $R_1 = 0.95$ as the approximation functional space is enriched.

p, q, s DOFs	1,2,2 18 DOFs	1,3,3 32 DOFs	1,4,4 50 DOFs	1,6,6 98 DOFs	1,8,8 162 DOFs
$\gamma_1^{(0)} = 0.463098562$	0.483284 (4.4E0 %)	0.452786 (-2.2E0 %)	0.46808 (1.1E0 %)	0.463774 (1.5E-1 %)	0.463204 (2.3E-2 %)
$\gamma_1^{(1)} = 1.24507709$	1.24489 (-1.5E-2 %)	1.24374 (-1.1E-1 %)	1.2453 (1.8E-2 %)	1.24509 (1.0E-3 %)	1.24508 (2.3E-4 %)
$\gamma_2^{(0)} = 1.813227873$		1.69437 (-6.6E0 %)	1.87592 (3.5E0 %)	1.8105 (-1.5E-1 %)	1.81333 (5.6E-3%)
$\gamma_1^{(2)} = 2.136566659$			2.13683 (1.2E-2 %)	2.13683 (1.2E-2 %)	2.13681 (1.1E-2%)
$\gamma_2^{(1)} = 2.54551186$	2.97811 (1.7E1 %)	2.55054 (2.0E-1%)	2.55029 (1.9E-1 %)	2.54961 (1.6E-1 %)	2.54954 (1.6E-1 %)

In parentheses the relative error in % is reported.

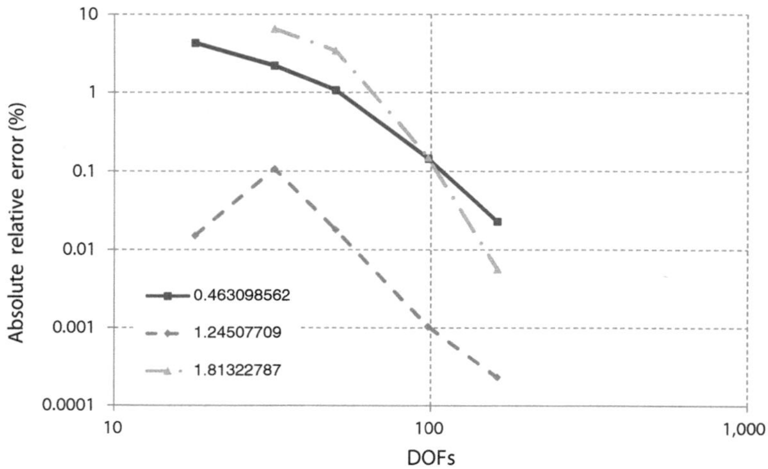

Fig. 12.7 Convergence of the first four nonzero eigenvalues γ for the conical vertex with $\omega/2 = 3\pi/4$ and homogeneous Dirichlet BCs.

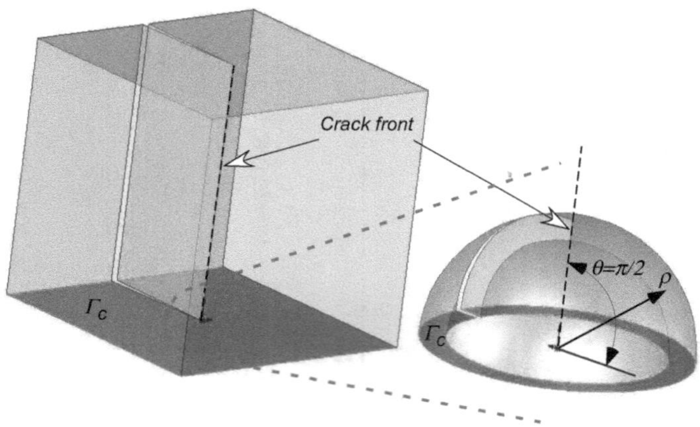

Fig. 12.8 A crack front intersecting a free face. Right: The 3-D domain with the crack. Left: The artificial subdomain used for the computation of the eigenpairs.

12.3.3 Vertex at the Intersection of a Crack Front with a Flat Face, Homogeneous Neumann BCs

In many practical applications, cracks are present in 3-D domains, and a vertex singularity exists at the intersection of the crack face with the boundary of the domain. Such a situation is described in Figure 12.8, where a crack front intersects a flat free face. Homogeneous Neumann boundary conditions are prescribed on the crack surfaces and the flat face.

Taking $R_1 = 0.95$ and $R_2 = 1$, we summarize the first 5 computed eigenvalues in Table 12.5. For this example problem the analytical eigenvalues are unknown, but estimated to be 0, 0.5, 1, 1.5, 2.0, and the relative error is computed relative to the estimated values.

The convergence rate of the first four eigenvalues is shown in the plot in Figure 12.9. Because the computational domain contains a singular edge, along the crack front, the convergence of the first (most singular) eigenvalue is much slower compared to the previous two example problems. A remedy to this situation is the use of a p-FE method and to refine the computational mesh in the vicinity of the singular edge. However, this necessitates an assembly procedure. Nevertheless, high accuracy is achieved with the presented method with a moderate number of degrees of freedom.

12.3.4 *Vertex at the Intersection of a V-Notch Front with a Conical Reentrant Corner, Homogeneous Neumann BCs*

The last example problem has a vertex at the intersection of a conical insert with a reentrant corner as described in Figure 12.10. Homogeneous Neumann boundary conditions are prescribed on all surfaces and the flat face.

Taking $R_1 = 0.95$ and $R_2 = 1$, we summarize the first four (nonzero) computed eigenvalues in Table 12.6 (first eigenvalue is zero so is not considered). For this example problem the analytical eigenvalues are unknown. Thus, the relative error cannot be computed. One may easily observe the clear convergence of the eigenvalues as the number of DOFs is increased.

Remark 12.8 *In all four considered example problems, for each positive eigenvalue γ_i computed, the modified Steklov problem provided the negative eigenvalue $-1-\gamma_i$ with high accuracy. This eigenvalue with the corresponding eigenfunction may be used in future studies for the extraction of the vertex stress-intensity factor.*

12.4 Other Methods for the Computation of the Vertex eigenpairs, and Extensions to the Elasticity System

Another popular method for the computation of eigenpairs by Leguillon [105, 109] leads to a quadratic weak eigenproblem and is briefly described here on the basis of the Laplace equation. The same quadratic eigenproblem can also be deduced by a Mellin transform (see, for example, [98, 99]). The advantage of this method compared to the modified Steklov method is a formulation over the two-dimensional manifold $0 < \theta < \omega/2,\ 0 < \varphi < 2\pi$. However, it has the disadvantage of being a quadratic eigenproblem.

Table 12.5 Vertex at the intersection of a crack front with a free face. Convergence of first five eigenvalues (except first eigenvalue which is 0) for $R_1 = 0.95$ as the approximation functional space is enriched.

p, q, s DOFs	1,1,1 8 DOFs	2,3,3 48 DOFs	2,4,4 75 DOFs	2,6,6 147 DOFs	2,9,9 300 DOFs
$\gamma_{Est} = 0.5$	0.475752 (-4.8E0 %)	0.48559691 (-2.9E0 %)	0.49063114 (-1.9E0 %)	0.49505940 (-9.9E-1 %)	0.49757218 (-4.9E-1 %)
$\gamma_{Est} = 1.0$		1.23250596 (2.3E1 %)	1.00362152 (3.6E-1 %)	1.00001671 (1.7E-1 %)	1.00000002 (2.5E-6 %)
$\gamma_{Est} = 1.5$				1.49506022 (-3.3E-1 %)	1.50000103 (6.9E-5 %)
$\gamma_{Est} = 2.0$	2.31019 (1.6E1 %)	2.07592021 (3.8E0 %)	2.03081227 (1.5E0 %)	1.99928949 (-3.6E-2 %)	1.99998322 (-8.4E-4 %)

In parentheses the relative error in % is reported.

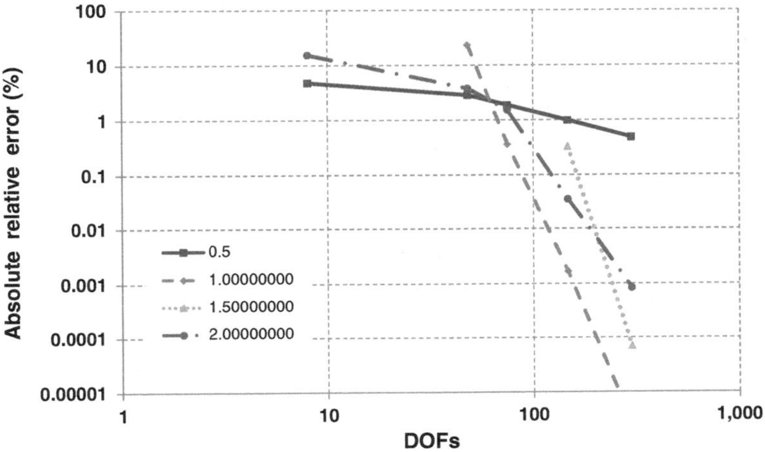

Fig. 12.9 Convergence of the first five non-zero eigenvalues γ for the crack front with a free face.

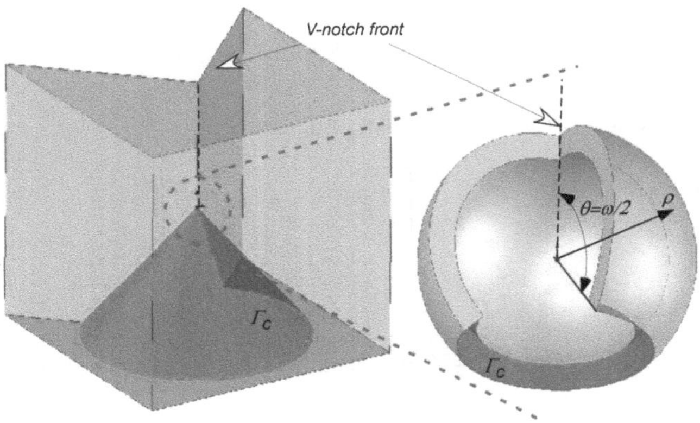

Fig. 12.10 A V-notch intersecting a conical reentrant corner. Right: The 3-D domain. Left: The artificial subdomain used for the computation of the eigenpairs.

Table 12.6 Vertex at the intersection of a V-notch front with a conical reentrant corner, homogeneous Neumann BCs, $\omega/2 = 3\pi/4$, $\varphi_2 = 6\pi/4$. Convergence of first four eigenvalues (except first eigenvalue, which is 0) for $R_1 = 0.95$ as the approximation functional space is enriched.

$p, q, s = 1, 2, 2$ 18 DOFs	$p, q, s = 1, 4, 4$ 50 DOFs	$p, q, s = 1, 6, 6$ 98 DOFs	$p, q, s = 1, 8, 8$ 162 DOFs
0.600442	0.535591	0.536327	0.536642
1.068311	1.195313	1.19021	1.190185
1.503901	1.255467	1.24441	1.245032
1.741498	1.730020	1.72647	1.727616

Consider the weak form of the Laplace equation given in (12.29), being defined over the infinite cone $0 < \rho < \infty,\ 0 < \theta < \omega/2,\ 0 < \varphi < 2\pi$:

$$\int_{\rho=0}^{\infty} \iint_{\theta\varphi} (\nabla \tau) \cdot (\nabla \chi) d\Omega = \iint_{\rho,\varphi} \left[\frac{\partial \tau}{\partial \theta} \chi \right]_{\theta=\omega/2} d\Gamma + \iint_{\theta,\varphi} \left[\frac{\partial \tau}{\partial \rho} \chi \right]_{\rho=\infty} d\Gamma. \tag{12.46}$$

By choosing the special test function

$$\chi(\rho, \theta, \varphi) = F(\rho)\tilde{\chi}(\theta, \varphi)$$

such that $F(\rho)$ has finite support on $\rho \in (0, \infty)$, the last integral in the RHS vanishes, and because homogeneous boundary conditions are prescribed on $\theta = \omega/2$, the first integral in the RHS also vanishes, so that (12.46) is simplified to

$$\int_{\rho=0}^{\infty} \iint_{\theta\varphi} (\nabla \tau) \cdot (\nabla \chi) d\Omega = 0. \tag{12.47}$$

The sought function τ in the vicinity of the vertex allows the representation $\tau(\rho, \theta, \varphi) = r^{\gamma} \tilde{\tau}(\theta, \varphi)$.

The gradient operator in a spherical coordinate system is

$$\nabla \stackrel{\text{def}}{=} \partial_\rho \hat{\boldsymbol{\rho}} + \frac{1}{\rho} \partial_\theta \hat{\boldsymbol{\theta}} + \frac{1}{\rho \sin\theta} \partial_\varphi \hat{\boldsymbol{\varphi}}. \tag{12.48}$$

Applying the ∇ operator on χ and τ, one obtains

$$\nabla \chi = F' \tilde{\chi} \hat{\boldsymbol{\rho}} + \frac{F}{\rho} \left[\partial_\theta \tilde{\chi} \hat{\boldsymbol{\theta}} + \frac{1}{\sin\theta} \partial_\varphi \tilde{\chi} \hat{\boldsymbol{\varphi}} \right], \tag{12.49}$$

$$\nabla \tau = \gamma \rho^{\gamma-1} \tilde{\tau} \hat{\boldsymbol{\rho}} + \rho^{\gamma-1} \tilde{\tau} \left[\partial_\theta \tilde{\tau} \hat{\boldsymbol{\theta}} + \frac{1}{\sin\theta} \partial_\varphi \tilde{\tau} \hat{\boldsymbol{\varphi}} \right]. \tag{12.50}$$

Inserting (12.49-12.50) into (12.47), and recalling that $d\Omega = \rho^2 \sin\theta \, d\rho d\theta d\varphi$, one obtains

$$\int_{\rho=0}^{\infty} \iint_{\theta\varphi} \gamma \rho^{\gamma+1} \tilde{\tau} F' \tilde{\chi} \sin\theta \, d\rho d\theta d\varphi \tag{12.51}$$

$$+ \int_{\rho=0}^{\infty} \iint_{\theta\varphi} F \rho^{\gamma} \left[\partial_\theta \tilde{\chi} \partial_\theta \tilde{\tau} + \frac{1}{\sin^2\theta} \partial_\varphi \tilde{\chi} \partial_\varphi \tilde{\tau} \right] \sin\theta d\rho d\theta d\varphi = 0.$$

Integrating the first integral in (12.51) by parts (in the ρ coordinate) leads to

$$\int_{\rho=0}^{\infty} \gamma \rho^{\gamma+1} F' d\rho \iint_{\theta\varphi} \tilde{\tau} \tilde{\chi} \sin\theta \, d\theta d\varphi = \gamma \rho^{\gamma+1} F(\rho) \big|_{\rho=0}^{\rho=\infty} \iint_{\theta\varphi} \tilde{\tau} \tilde{\chi} \sin\theta \, d\theta d\varphi$$

$$- \int_{\rho=0}^{\infty} \gamma(\gamma+1) \rho^{\gamma} F d\rho \iint_{\theta\varphi} \tilde{\tau} \tilde{\chi} \sin\theta \, d\theta d\varphi. \tag{12.52}$$

The first term in the RHS of (12.52) vanishes because $\rho^{\gamma+1} = 0$ at $\rho = 0$ and $F(\rho) \to 0$ as $\rho \to \infty$ due to its compact support. Therefore, after substituting (12.52) into (12.51), one obtains

$$\int_{\rho=0}^{\infty} \rho^{\gamma} F(\rho) d\rho \iint_{\theta\varphi} \left\{ \left[\gamma(\gamma+1)\tilde{\tau}\tilde{\chi} + \partial_\theta \tilde{\chi} \partial_\theta \tilde{\tau} + \frac{1}{\sin^2\theta} \partial_\varphi \tilde{\chi} \partial_\varphi \tilde{\tau} \right] \sin\theta d\theta d\varphi \right\} = 0. \quad (12.53)$$

The weak formulation above has to hold for any $F(\rho)$. Therefore, (12.53) reduces to a quadratic eigenproblem over the two-dimensional manifold spanned by θ and φ:

$$\int_{\theta=0}^{\omega/2} \int_{\varphi=0}^{2\pi} \left[\partial_\theta \tilde{\chi} \partial_\theta \tilde{\tau} + \frac{1}{\sin^2\theta} \partial_\varphi \tilde{\chi} \partial_\varphi \tilde{\tau} \right] \sin\theta d\varphi d\theta = -\gamma(\gamma+1) \int_{\theta=0}^{\omega/2} \int_{\varphi=0}^{2\pi} \tilde{\tau}\tilde{\chi} \sin\theta d\varphi d\theta. \quad (12.54)$$

For the Laplace equation it is possible to perform the substitution $\beta = \gamma(\gamma+1)$ so to reduce the quadratic eigenproblem to a usual eigenproblem. However, this is not possible for a general scalar elliptic equation or the elasticity system.

The weak eigenproblem can be solved using the finite element method, discretizing the trial and test functions $\tilde{\tau}(\theta, \varphi)$ and $\tilde{\chi}(\theta, \varphi)$ by 2-D shape functions in θ and φ. Furthermore, considering vertices created by intersection of edges such as those in Figure 12.10, the integration over φ has to be performed from 0 to φ_2, the solid angle of the reentrant V-notch.

Remark 12.9 *When homogeneous Dirichlet boundary conditions are prescribed on the faces intersecting at the vertex, the trial and test functions are to be restricted to satisfy these conditions identically.*

12.4.1 Extension of the Method to the Elasticity System

The quadratic eigenproblem presented in detail for the computation of vertex eigenpairs associated with the Laplace equation was applied to the elasticity system in [105]. Denoting the Cartesian displacement eigenvector by $\boldsymbol{s} \stackrel{\text{def}}{=} (s_1(\theta,\varphi), s_2(\theta,\varphi), s_3(\theta,\varphi))^T$, and the trial function by $\boldsymbol{v} \stackrel{\text{def}}{=} (v_1(\theta,\varphi), v_2(\theta,\varphi), v_3(\theta,\varphi))^T$ the following eigenproblem is obtained:

$$-\gamma(\gamma+1)a(\boldsymbol{s},\boldsymbol{v}) - (\gamma+1)b(\boldsymbol{s},\boldsymbol{v}) + \gamma c(\boldsymbol{s},\boldsymbol{v}) + d(\boldsymbol{s},\boldsymbol{v}) = 0 \quad (12.55)$$

with

$$a(\boldsymbol{s},\boldsymbol{v}) = \iint_{\theta\varphi} C_{ijk\ell} B_j B_\ell s_i \bar{v}_k \sin\theta \; d\theta d\varphi,$$

$$b(\boldsymbol{s},\boldsymbol{v}) = \iint_{\theta\varphi} C_{ijk\ell} \left(D_j B_\ell \, \partial_\theta s_i \, \bar{v}_k + G_j B_\ell \, \partial_\varphi s_i \, \bar{v}_k\right) \sin\theta \; d\theta d\varphi,$$

$$c(\boldsymbol{s},\boldsymbol{v}) = \iint_{\theta\varphi} C_{ijk\ell} \left(B_j D_\ell s_i \, \partial_\theta \bar{v}_k + B_j G_\ell s_i \, \partial_\varphi \bar{v}_k\right) \sin\theta \; d\theta d\varphi,$$

$$d(\boldsymbol{s},\boldsymbol{v}) = \iint_{\theta\varphi} C_{ijk\ell} \left(D_j D_\ell \, \partial_\theta s_i \, \partial_\theta \bar{v}_k + D_j G_\ell \, \partial_\theta s_i \, \partial_\varphi \bar{v}_k + G_j D_\ell \, \partial_\varphi s_i \, \partial_\theta \bar{v}_k \right.$$
$$\left. + G_j G_\ell \, \partial_\varphi s_i \, \partial_\varphi \bar{v}_k\right) \sin\theta \; d\theta d\varphi,$$

where $C_{ijk\ell}$ is the elasticity tensor defined for an isotropic material in (1.43), $\bar{\cdot}$ is the complex conjugate (eigenfunctions may be complex), and

$$\mathbf{B} = \begin{Bmatrix} \cos\varphi \sin\theta \\ \sin\varphi \sin\theta \\ \cos\theta \end{Bmatrix}, \quad \mathbf{D} = \begin{Bmatrix} \cos\varphi \cos\theta \\ \sin\varphi \cos\theta \\ -\sin\theta \end{Bmatrix}, \quad \mathbf{G} = \begin{Bmatrix} -\sin\varphi/\sin\theta \\ \cos\varphi/\sin\theta \\ 0 \end{Bmatrix}. \tag{12.56}$$

The quadratic weak eigenproblem can be cast in a finite element framework and numerically solved; details are provided in Section 13.3.1.

Vertex eigenvalues for isotropic as well as anisotropic materials have been widely studied numerically by methods similar to that presented above in several works in addition to those already cited [7,22,56,64].

Some exact vertex eigenvalues.

Computation of exact vertex eigenpairs is in most cases a tedious, if not an impossible task. However, there are some special cases for which these can be explicitly given. For example, the eigenvalues for a rotationally symmetric conical vertex in an isotropic material can be computed "almost" analytically out of a transcendental equation.

For Dirichlet homogeneous boundary conditions ($\boldsymbol{u} = \mathbf{0}$), the exact eigenvalues are the solutions of the equation [24]

$$D_1\left(\frac{\omega}{2},\gamma\right) = \frac{-(\gamma+1)}{\sin\frac{\omega}{2}} \left[P_\gamma^2 \cos\frac{\omega}{2}(\gamma + 4\nu - 3)\right.$$
$$+ P_\gamma P_{\gamma+1}\left(3 - 4\nu - (2\gamma+1)\cos^2\frac{\omega}{2}\right)$$
$$\left. + P_{\gamma+1}^2(\gamma+1)\cos\frac{\omega}{2}\right] = 0, \tag{12.57}$$

and for traction-free boundary conditions,

$$D_2\left(\frac{\omega}{2},\gamma\right) = 2\gamma(\gamma+1)\left[D_1\left(\frac{\omega}{2},\gamma\right) - 4(1-\nu)\sin\frac{\omega}{2}P_\gamma P'_\gamma\right] \\ +2((1-\nu)\sin 2\frac{\omega}{2}(P'_\gamma)^2 = 0. \qquad (12.58)$$

Here $P_\gamma \equiv P_\gamma(\cos\frac{\omega}{2})$ are Legendre functions of the first kind, which can be calculated using the Mehler-Dirichlet formula (12.15). The integrand is singular for $t = \frac{\omega}{2}$, so that special techniques have to be applied for the numerical treatment. The integrand denominator is expressed as

$$\sqrt{-\sin\left(\frac{t+\frac{\omega}{2}}{2}\right)\sin\left(\frac{t-\frac{\omega}{2}}{2}\right)\left(\frac{2}{t-\frac{\omega}{2}}\right)\left(t-\frac{\omega}{2}\right)}, \qquad (12.59)$$

so that

$$P_\gamma(\cos\frac{\omega}{2}) = \frac{\sqrt{2}}{\pi}\int_0^{\frac{\omega}{2}} \frac{1}{\sqrt{\frac{\omega}{2}-t}} \frac{\cos\left(\gamma+\frac{1}{2}\right)t}{\sqrt{\sin\left(\frac{t+\frac{\omega}{2}}{2}\right)\left(\frac{2}{\frac{\omega}{2}-t}\right)\sin\left(\frac{\frac{\omega}{2}-t}{2}\right)}}dt. \qquad (12.60)$$

The product of the last two terms in the denominator can be computed for $(\frac{\omega}{2}-t) \ll 1$ by expanding the sin in a Taylor series. If we now perform a change of variables

$$\frac{\omega}{2}x^2 = t$$

and define

$$G(x) = \frac{\sqrt{t}\cos(\gamma+1/2)t}{\sqrt{\sin\left(\frac{t+\frac{\omega}{2}}{2}\right)\left(\frac{2}{\frac{\omega}{2}-t}\right)\sin\left(\frac{\frac{\omega}{2}-t}{2}\right)}},$$

then we may compute numerically the integral given by (12.15):

$$P_\gamma(\cos\frac{\omega}{2}) = \frac{2\sqrt{2}}{\pi}\sum_{j=1}^{\left[\frac{n+1}{2}\right]} G(x_j), \qquad (12.61)$$

where $x_j = \cos\frac{(2j-1)\pi}{2n}$. The expression P'_γ in (12.58) can be computed using the following relationship:

$$\sin^2\frac{\omega}{2}P'_\gamma\left(\cos\frac{\omega}{2}\right) = \gamma P_{\gamma-1}\left(\cos\frac{\omega}{2}\right) - \gamma\cos\frac{\omega}{2}P_\gamma\left(\cos\frac{\omega}{2}\right) \qquad (12.62)$$

Table 12.7 γ_1 for clamped and traction-free boundary conditions, in isotropic materials.

ν	$\frac{\omega}{2} = 0.51\pi$	$\frac{\omega}{2} = \frac{2}{3}\pi$	$\frac{\omega}{2} = \frac{5}{6}\pi$	$\frac{\omega}{2} = 0.97\pi$
	Clamped BCs.			
0.0	0.9793	0.6886	0.4014	0.1760
0.3	0.9861	0.7528	0.4334	0.1827
0.49	0.9988	0.9309	0.4846	0.1916
	Traction Free BCs.			
0.0	0.9706	0.8334	0.9411	0.9983
0.3	0.9614	0.7456	0.8978	0.9973
0.4	0.9584	0.7093	0.8755	0.9971

The above equations were solved numerically in [23], and in Table 12.7 the values for γ_1 (smallest eigenvalue) are reported from Table 1 in [23].

Chapter 13
Edge EigenPairs and ESIFs of 3-D Elastic Problems

We proceed to elasticity problems in three-dimensional (3-D) polyhedral domains in the vicinity of an edge and provide the solution in an explicit form. It involves a family of eigenfunctions with their shadows, and the associated "edge-stress-intensity functions" (ESIFs), which are functions along the edges. Utilizing the explicit structure of the solution in the vicinity of the edge, we extend the use of the *quasidual function method* (QDFM) presented in Section 11.3 for EFIFs [46, 134] to the extraction of ESIFs. It provides a polynomial approximation of the ESIF along the edge whose order is adaptively increased so to approximate the exact ESIF. The QDFM is implemented as a post-solution operation in conjunction with the p -version finite element method. Numerical examples are provided in which we extract ESIFs associated with traction-free or homogeneous Dirichlet boundary conditions in 3-D cracked domains or 3-D V-notched domains. These demonstrate the efficiency, robustness, and high accuracy of the proposed QDFM.

The three-dimensional elastic solution in the vicinity of an *edge* is characterized:

- by an *exponent* α that belongs to a discrete set $\{\alpha_i,\ i \in \mathbb{N}\}$ of eigenvalues depending only on the geometry, material properties, and boundary conditions in the vicinity of the edge, and which determines the level of nonsmoothness of the singularity. Any *eigenvalue* α_i is computed by solving a 2-D problem.
- by an associated *eigenfunction* $\boldsymbol{\varphi}_0^{(\alpha)}(\theta)$ that depends on the geometry of the domain, material properties, and boundary conditions. These eigenfunctions are computed by solving a set of 2-D problems.
- by an ESIF along the edge, denoted by $A_i(x_3)$ (x_3 is a coordinate along the edge). $A_i(x_3)$ is associated with the i th eigenvalue, determining the "amount of energy" residing in each singularity.

From the engineering perspective, $A_i(x_3)$ for $\alpha_i < 1$ are of major importance because these are correlated to failure initiation. In many situations, $\alpha_i < 1$ when the opening at the edge is nonconvex. For example, α_i is equal to $\frac{1}{2}$ in the presence of cracks.

Z. Yosibash, *Singularities in Elliptic Boundary Value Problems and Elasticity and Their Connection with Failure Initiation*, Interdisciplinary Applied Mathematics 37,
DOI 10.1007/978-1-4614-1508-4_13, © Springer Science+Business Media, LLC 2012

First we provide a mathematical algorithm for the construction of the asymptotic elastic solution in the vicinity of an edge (which is an extension of the two-dimensional case), and then we compute a polynomial approximation of the edge stress intensity function by the QDFM.

The eigenpairs of the three-dimensional cracked or notched domain in the vicinity of the edge were first addressed by Hartranft and Sih in [79], shown to be computed by a recursive procedure. At the time, however, these were not presented explicitly, and the general structure of the asymptotic expansion was not observed. We explicitly express the elasticity solution in the vicinity of an edge as a combination of eigenfunctions and their shadows. These shadows are "new functions" appearing in 3-D domains, having no counterparts in 2-D domains as far as homogeneous operators with constant coefficients are concerned. The dual eigenfunctions and their dual shadows are computed also, which are required subsequently for the quasidual function method. Using the eigenfunctions and their shadows, the functional $J[R]$ is used (see Theorem 11.1 in Section 11.3), which can be viewed as an extension of the 2-D contour integral to 3-D domains. It is a surface integral along a cylindrical surface, enabling us to present the edge-stress-intensity function explicitly as a function of x_3 (the coordinate along the edge). The method presented is implemented as a post-processing step in a p-version finite element code, and the numerical performance is documented on several example problems. Using the $J[R]$ functional, and newly constructed extraction polynomials, one may extract the ESIFs in the vicinity of any edge (including crack front) in any polyhedron. The *functional representation of the ESIFs along* x_3 is obtained (as opposed to other methods providing pointwise values of the ESIFs along the edge) and is very accurate, efficient, and robust. Most importantly, *the method is adaptive*, providing a better polynomial representation of the ESIF as the special hierarchical family of extraction polynomials is increased.

- We start with notation followed by a mathematical algorithm for obtaining the asymptotic expansion of the solution in a neighborhood of an edge in terms of eigenfunctions, their shadows, and the structure of the ESIFs. The dual eigenfunctions, and their shadows, which are associated with the primal eigenfunctions, are addressed as well. For TF/TF crack and clamped V-notch domains we provide explicit formulas for the eigenfunctions, duals, and shadows.
- The $J[R]$ integral (introduced in Chapter 11 in the context of scalar elliptic problems) is then extended to the elasticity system [46]. It requires the construction of extracting polynomials, denoted by $BJ(x_3)$, and the data on a cylindrical surface of radius R around the edge. A short explanation on its application in conjunction with the finite element method is given, and a hierarchical family of extraction polynomials is constructed.
- The hierarchical family of extraction polynomials is used in several numerical tests to extract the ESIFs associated with:
  - A cracked domain with traction-free boundary conditions.
  - V-notched domain with clamped boundary conditions.

  - A problem of engineering relevance, i.e., a compact tension specimen subjected to tension load such that only mode I is excited along the crack front. We compare the extracted ESIF by our method with a pointwise extraction method. This example problem demonstrates the efficiency and robustness of the quasidual function method in handling realistic geometries in engineering practice.

- Finally, we address edges in anisotropic materials and multimaterial interfaces.
  - We introduce a new method for the computation of eigenpairs [109] and their shadows.
  - A pathological case is then discussed with a remedy suggested and proven to overcome the difficulties.
  - We then extend the QDFM to complex eigenpairs and provide two example problems for extracting ESIFs along a crack at a bimaterial interface and a crack at the interface of two anisotropic materials.

13.1 The Elastic Solution for an Isotropic Material in the Vicinity of an Edge

In this section we derive the asymptotic solution in the neighborhood of an edge in an isotropic elastic domain. It can be presented as an asymptotic series of eigenpairs (the well-known eigenpairs of the 2-D cross-section) and the associated edge-stress-intensity functions. However, as opposed to planar elastic problems, each of the eigenpairs is accompanied by an infinite number of shadow functions with an increasing exponential order.

13.1.1 Differential Equations for 3-D Eigenpairs

Consider a domain Ω in which one straight edge $\mathcal{E}$ of interest is present. The domain is generated as the product $\Omega = G \times I$, where I is the interval $[-1, 1]$, and G is a plane-bounded sector of opening $\omega \in (0, 2\pi]$, and for simplicity assume that it has radius 1 as shown in Figure 13.1 (the case of a crack, $\omega = 2\pi$, is included). Although any G or I can be chosen, these simplified ones have been chosen for simplicity of presentation.

The variables in G and I are (x_1, x_2) and x_3 respectively, and the coordinates (x_1, x_2, x_3) are denoted by $\boldsymbol{x}$. Let (r, θ) be the polar coordinates centered at the vertex of G so that G coincides with $\{(x_1, x_2) \in \mathbb{R}^2 \mid r \in (0, 1),\ \theta \in (0, \omega)\}$. The *edge* $\mathcal{E}$ of interest is the set $\{\boldsymbol{x} \in \mathbb{R}^3 \mid r = 0,\ x_3 \in I\}$. The two flat planes

Fig. 13.1 Domain of interest Ω.

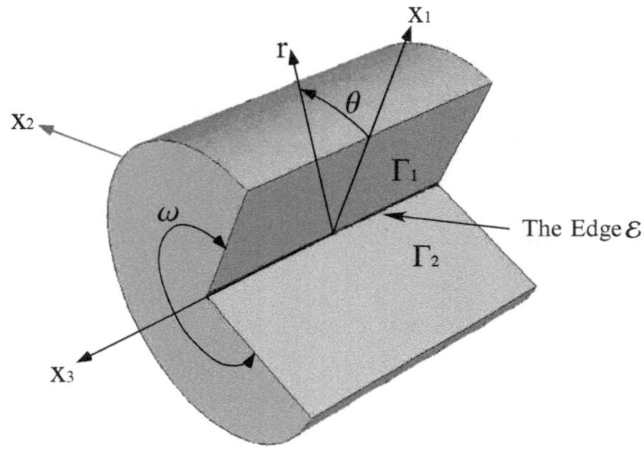

that intersect at the edge $\mathcal{E}$ are denoted by Γ_1 and Γ_2. For any R, $0 < R < 1$, the cylindrical surface Γ_R is defined as follows:

$$\Gamma_R := \left\{ \boldsymbol{x} \in \mathbb{R}^3 \mid r = R,\ \theta \in (0,\omega),\ x_3 \in I \right\}. \tag{13.1}$$

Remark 13.1. The methods presented are restricted to geometries for which the edges are straight lines and the angle ω is fixed along x_3 .

Remark 13.2. In general, the eigenpairs associated with the elasticity operator may be complex. However, in most practical cases the eigenvalues smaller than 1 are of interest, and these are usually real. Here we address real eigenpairs only, whereas the general case will be addressed in Section 13.3.

As usual, $\boldsymbol{u} = \{u_1, u_2, u_3\}^T$ and $\tilde{\boldsymbol{u}} = \{u_r, u_\theta, u_3\}^T$, and we use either of them when convenient.

The Navier-Lamé (N-L) equations that describe the elastic isotropic problem in cylindrical coordinates are

$$(\lambda + 2\mu)\partial_r^2 u_r + (\lambda + 2\mu)\tfrac{1}{r}\partial_r u_r - (\lambda + 2\mu)\tfrac{1}{r^2} u_r + \mu \tfrac{1}{r^2}\partial_\theta^2 u_r + \mu\partial_3^2 u_r$$
$$-(\lambda + 3\mu)\tfrac{1}{r^2}\partial_\theta u_\theta + (\lambda + \mu)\tfrac{1}{r}\partial_r\partial_\theta u_\theta + (\lambda + \mu)\partial_r\partial_3 u_3 = 0, \tag{13.2}$$

$$(\lambda + \mu)\tfrac{1}{r}\partial_r\partial_\theta u_r + (\lambda + 3\mu)\tfrac{1}{r^2}\partial_\theta u_r + (\lambda + 2\mu)\tfrac{1}{r^2}\partial_\theta^2 u_\theta + \mu\partial_r^2 u_\theta$$
$$+\mu\tfrac{1}{r}\partial_r u_\theta - \mu\tfrac{1}{r^2}u_\theta + \mu\partial_3^2 u_\theta + (\lambda + \mu)\tfrac{1}{r}\partial_3\partial_\theta u_3 = 0, \tag{13.3}$$

$$(\lambda + \mu)\partial_r\partial_3 u_r + (\lambda + \mu)\tfrac{1}{r}\partial_3 u_r + (\lambda + \mu)\tfrac{1}{r}\partial_3\partial_\theta u_\theta + \mu\partial_r^2 u_3$$
$$+\mu\tfrac{1}{r}\partial_r u_3 + \mu\tfrac{1}{r^2}\partial_\theta^2 u_3 + (\lambda + 2\mu)\partial_3^2 u_3 = 0. \tag{13.4}$$

The system (13.2)-(13.4) can be split into three operators:

$$\mathcal{L}(\tilde{\boldsymbol{u}}) = [M_0(\partial_r, \partial_\theta)]\tilde{\boldsymbol{u}} + [M_1(\partial_r, \partial_\theta)]\partial_3\tilde{\boldsymbol{u}} + [M_2(\partial_r, \partial_\theta)]\partial_3^2\tilde{\boldsymbol{u}} = 0, \tag{13.5}$$

with

$$[M_0] = \begin{pmatrix} (\lambda+2\mu)\left(\partial_r^2 + \frac{1}{r}\partial_r - \frac{1}{r^2}\right) + \mu\frac{1}{r^2}\partial_\theta^2 & -(\lambda+3\mu)\frac{1}{r^2}\partial_\theta + (\lambda+\mu)\frac{1}{r}\partial_r\partial_\theta & 0 \\ (\lambda+\mu)\frac{1}{r}\partial_r\partial_\theta + (\lambda+3\mu)\frac{1}{r^2}\partial_\theta & (\lambda+2\mu)\frac{1}{r^2}\partial_\theta^2 + \mu\left(\partial_r^2 + \frac{1}{r}\partial_r - \frac{1}{r^2}\right) & 0 \\ 0 & 0 & \mu\left(\partial_r^2 + \frac{1}{r}\partial_r + \frac{1}{r^2}\partial_\theta^2\right) \end{pmatrix}, \quad (13.6)$$

$$[M_1] = \begin{pmatrix} 0 & 0 & (\lambda+\mu)\partial_r \\ 0 & 0 & (\lambda+\mu)\frac{1}{r}\partial_\theta \\ (\lambda+\mu)\left(\partial_r + \frac{1}{r}\right) & (\lambda+\mu)\frac{1}{r}\partial_\theta & 0 \end{pmatrix}, \quad (13.7)$$

$$[M_2] = \begin{pmatrix} \mu & 0 & 0 \\ 0 & \mu & 0 \\ 0 & 0 & (\lambda+2\mu) \end{pmatrix}. \quad (13.8)$$

The splitting allows the consideration of a solution $\tilde{\boldsymbol{u}}$ of the form

$$\tilde{\boldsymbol{u}} = \sum_{j\geq 0} \partial_3^j A(x_3)\boldsymbol{\Phi}_j(r,\theta). \quad (13.9)$$

The N-L system in view of (13.9) becomes

$$\sum_{j\geq 0} \partial_3^j A(x_3)[M_0]\boldsymbol{\Phi}_j + \sum_{j\geq 0} \partial_3^{j+1} A(x_3)[M_1]\boldsymbol{\Phi}_j + \sum_{j\geq 0} \partial_3^{j+2} A(x_3)[M_2]\boldsymbol{\Phi}_j = 0, \quad (13.10)$$

and after rearranging,

$$A(x_3)[M_0]\boldsymbol{\Phi}_0 + \partial_3 A(x_3)([M_0]\boldsymbol{\Phi}_1 + [M_1]\boldsymbol{\Phi}_0) + \sum_{j\geq 0} \partial_3^{j+2} A(x_3)([M_0]\boldsymbol{\Phi}_{j+2} + [M_1]\boldsymbol{\Phi}_{j+1} + [M_2]\boldsymbol{\Phi}_j) = 0. \quad (13.11)$$

Equation (13.11) has to hold for any smooth function $A(x_3)$. Thus, the functions $\boldsymbol{\Phi}_j$ must satisfy the equations below, each defined on a two-dimensional domain G:

$$\begin{cases} [M_0]\boldsymbol{\Phi}_0 = 0, \\ [M_0]\boldsymbol{\Phi}_1 + [M_1]\boldsymbol{\Phi}_0 = 0, & (r,\theta) \in G, \\ [M_0]\boldsymbol{\Phi}_{j+2} + [M_1]\boldsymbol{\Phi}_{j+1} + [M_2]\boldsymbol{\Phi}_j = 0, \quad j \geq 0, \end{cases} \quad (13.12)$$

accompanied by homogeneous boundary conditions on the two surfaces Γ_1 and Γ_2, discussed in the sequel.

The first partial differential equation in (13.12) generates the solution $\boldsymbol{\Phi}_0$ associated with the eigenvalue α, called a *primal eigenfunction*, which is the well-known two-dimensional eigenfunction

$$\boldsymbol{\Phi}_0 = r^{\alpha}\boldsymbol{\varphi}_0(\theta). \tag{13.13}$$

The second PDE in (13.12) generates the function $\boldsymbol{\Phi}_1$, which depends on $\boldsymbol{\Phi}_0$:

$$\boldsymbol{\Phi}_1 = r^{\alpha+1}\boldsymbol{\varphi}_1(\theta). \tag{13.14}$$

The terms of the sequence $\boldsymbol{\Phi}_j$ (where $j \geq 2$) are the solutions of the third equation of (13.12). These are of the form

$$\boldsymbol{\Phi}_j = r^{\alpha+j}\boldsymbol{\varphi}_j(\theta). \tag{13.15}$$

All $\boldsymbol{\Phi}_j$, $j \geq 1$ are called *shadow eigenfunctions* of the primal eigenfunction $\boldsymbol{\Phi}_0$. There exists an infinite number of shadow functions $\boldsymbol{\Phi}_j$ for each eigenvalue α_i (these are obtained by applying boundary conditions, as will be discussed in Section 13.1.2):

$$\boldsymbol{\Phi}_j^{(\alpha_i)} = r^{\alpha_i+j}\boldsymbol{\varphi}_j^{(\alpha_i)}(\theta), \quad j = 0, 1, \ldots \tag{13.16}$$

Thus, for each eigenvalue α_i, the 3-D solution in the vicinity of an edge is

$$\tilde{\boldsymbol{u}}^{(\alpha_i)} = \sum_{j\geq 0} \partial_3^j A_i(x_3) r^{\alpha_i+j}\boldsymbol{\varphi}_j^{(\alpha_i)}(\theta), \tag{13.17}$$

and the overall solution $\tilde{\boldsymbol{u}}$ is

$$\tilde{\boldsymbol{u}} = \sum_{i\geq 1} \tilde{\boldsymbol{u}}^{(\alpha_i)} = \sum_{i\geq 1}\sum_{j\geq 0} \partial_3^j A_i(x_3) r^{\alpha_i+j}\boldsymbol{\varphi}_j^{(\alpha_i)}(\theta), \tag{13.18}$$

where $A_i(x_3)$ is the edge-stress-intensity-function (ESIF) of the i th eigenvalue.

Because the operator $\mathcal{L}$ is self-adjoint, for any real eigenvalue α_i the number $-\alpha_i$ is also an eigenvalue. It is associated with an eigenfunction $\boldsymbol{\Phi}_0^{(-\alpha_i)}$ and its shadows $\boldsymbol{\Phi}_j^{(-\alpha_i)}$ by similar formulas as in (13.16). Solutions of (13.12) for the *negative eigenvalues* $-\alpha_i$ are called the *dual singular solutions*, and are denoted by $\boldsymbol{\Psi}_j^{(\alpha_i)}$. For normalization purposes, a real coefficient $c_0^{(\alpha_i)}$ is chosen as in (10.65), linking $\boldsymbol{\Phi}_j^{(\alpha_i)}$ with $\boldsymbol{\Psi}_j^{(\alpha_i)}$:

$$\boldsymbol{\Psi}_0^{(\alpha_i)} = r^{-\alpha_i}\boldsymbol{\psi}_0^{(\alpha_i)}(\theta) = c_0^{(\alpha_i)} r^{-\alpha_i}\boldsymbol{\varphi}_0^{(-\alpha_i)}(\theta) \tag{13.19}$$

and

$$\boldsymbol{\Psi}_j^{(\alpha_i)} = r^{-\alpha_i+j}\boldsymbol{\psi}_j^{(\alpha_i)}(\theta) = c_0^{(\alpha_i)} r^{-\alpha_i+j}\boldsymbol{\varphi}_j^{(-\alpha_i)}(\theta). \tag{13.20}$$

13.1.2 Boundary Conditions for the Primal, Dual and Shadow Functions

Two types of homogeneous boundary conditions are considered on Γ_1 and Γ_2 surfaces, traction-free and clamped.

13.1.2.1 Traction-Free Boundary Conditions

The traction-free boundary conditions on Γ_1, Γ_2 may be expressed as

$$\begin{cases} (\sigma_{r\theta})\,|_{\theta=0,\omega} = 0, \\ (\sigma_{\theta\theta})\,|_{\theta=0,\omega} = 0, \\ (\sigma_{\theta 3})\,|_{\theta=0,\omega} = 0. \end{cases} \Rightarrow \begin{cases} \left(\mu\left(\frac{1}{r}\partial_\theta u_r + \partial_r u_\theta - \frac{1}{r}u_\theta\right)\right)|_{\theta=0,\omega} = 0 \\ \Big((\lambda+2\mu)\frac{1}{r}u_r + \lambda\partial_r u_r + (\lambda+2\mu)\frac{1}{r}\partial_\theta u_\theta \\ \qquad\qquad + \lambda\partial_3 u_3\Big)|_{\theta=0,\omega} = 0 \\ \left(\mu\left(\partial_3 u_\theta + \frac{1}{r}\partial_\theta u_3\right)\right)|_{\theta=0,\omega} = 0 \end{cases}$$

By defining the operator matrices $[T_0]$ and $[T_1]$

$$[T_0] = \begin{pmatrix} \mu\frac{1}{r}\partial_\theta & \mu\partial_r - \mu\frac{1}{r} & 0 \\ (\lambda+2\mu)\frac{1}{r} + \lambda\partial_r & (\lambda+2\mu)\frac{1}{r}\partial_\theta & 0 \\ 0 & 0 & \mu\frac{1}{r}\partial_\theta \end{pmatrix}, \quad [T_1] = \begin{pmatrix} 0 & 0 & 0 \\ 0 & 0 & \lambda \\ 0 & \mu & 0 \end{pmatrix}, \tag{13.21}$$

the traction-free BCs (13.21) are cast in a simpler form:

$$[T](\tilde{\boldsymbol{u}})|_{\Gamma_1,\Gamma_2} = ([T_0(\partial_r,\partial_\theta)]\tilde{\boldsymbol{u}} + [T_1(\partial_r,\partial_\theta)]\partial_3\tilde{\boldsymbol{u}})\,|_{\Gamma_1,\Gamma_2} = 0. \tag{13.22}$$

Inserting (13.9) in (13.22), one obtains

$$A(x_3)[T_0]\boldsymbol{\Phi}_0|_{\Gamma_1,\Gamma_2} + \sum_{j\ge 0}\partial_3^{j+1}A(x_3)\left([T_0]\boldsymbol{\Phi}_{j+1} + [T_1]\boldsymbol{\Phi}_j\right)|_{\Gamma_1,\Gamma_2} = 0. \tag{13.23}$$

Equation (13.23) has to hold for any smooth function $A(x_3)$, and therefore the boundary conditions for the primal and shadow eigenfunctions are

$$\begin{cases} [T_0]\boldsymbol{\Phi}_0 = 0, \\ [T_0]\boldsymbol{\Phi}_{j+1} = -[T_1]\boldsymbol{\Phi}_j, \quad j \ge 0, \end{cases} \qquad \text{on} \quad \Gamma_1, \Gamma_2. \tag{13.24}$$

The first equation in (13.24) expresses the boundary conditions for $\boldsymbol{\Phi}_0$, which are identical to the two-dimensional boundary conditions. The second equation in (13.24) contains the boundary conditions for each $\boldsymbol{\Phi}_j$, for $j \geq 1$.

13.1.2.2 Clamped Boundary Conditions

Clamped boundary conditions on Γ_1, Γ_2 are

$$\tilde{\boldsymbol{u}}|_{\Gamma_1,\Gamma_2} = \sum_{j\geq 0} \partial_3^j A(x_3)\boldsymbol{\Phi}_j(r,\theta)|_{\Gamma_1,\Gamma_2} = 0. \tag{13.25}$$

Equation (13.25) has to hold for any smooth function $A(x_3)$, and therefore the clamped boundary conditions for the eigenfunctions are

$$\boldsymbol{\Phi}_j(r,\theta) = 0 \qquad \text{on} \quad \Gamma_1, \Gamma_2. \tag{13.26}$$

Explicit expressions for the primal and dual eigenfunctions and their shadows for a traction-free crack and a clamped $3\pi/2$ V-notch are presented in the following sections.

13.1.3 Primal and Dual Eigenfunctions and Shadow Functions for a Traction-Free Crack

The displacements $\tilde{\boldsymbol{u}}$ (13.18) in the case of a cracked domain ($0 \leq \theta \leq \omega = 2\pi$) with traction-free boundary conditions on the crack surfaces Γ_1 and Γ_2 are constructed by the primal and shadow functions $\boldsymbol{\Phi}_j$, $j \geq 0$. Here $\boldsymbol{\Phi}_0$ and $\boldsymbol{\Psi}_0$ are the solutions of the first differential equation of (13.12). The boundary conditions applied to $\boldsymbol{\Phi}_0$ and $\boldsymbol{\Psi}_0$ are prescribed in the first equation of (13.24). There is an infinite number of eigenvalues α_i for which there is an associated $\boldsymbol{\Phi}_0^{(\alpha_i)}$ and $\boldsymbol{\Psi}_0^{(\alpha_i)}$, where the positive α_i 's are associated with $\boldsymbol{\Phi}_0^{(\alpha_i)}$ and the negative α_i 's are associated with $\boldsymbol{\Psi}_0^{(\alpha_i)}$. We consider the first three eigenvalues only ($\alpha_1 = \alpha_2 = \alpha_3 = \frac{1}{2}$). The dual eigenfunction $\boldsymbol{\Psi}_0^{(\alpha_i)}$ includes the normalization factor $c_0^{(\alpha_i)}$ chosen such that the primal and dual eigenfunctions satisfy the orthonormal condition

$$\int_0^{\omega} \left([T]\boldsymbol{\Phi}_0^{(\alpha_i)} \cdot \boldsymbol{\Psi}_0^{(\alpha_i)} - \boldsymbol{\Phi}_0^{(\alpha_i)} \cdot [T]\boldsymbol{\Psi}_0^{(\alpha_i)} \right) R\, d\theta = 1. \tag{13.27}$$

After the primal eigenfunction $\boldsymbol{\Phi}_0^{(\alpha_i)}$ and the dual eigenfunction $\boldsymbol{\Psi}_0^{(\alpha_i)}$ are computed, the first shadow function $\boldsymbol{\Phi}_1^{(\alpha_i)}$ and the first dual shadow function $\boldsymbol{\Psi}_1^{(\alpha_i)}$ may be computed by the second differential equation in (13.12), with the second equation of (13.24) as the boundary conditions. The boundary conditions contain the operators $[T_0]$ and $[T_1]$, as defined in (13.21).

The shadow function $\boldsymbol{\Phi}_2^{(\alpha_i)}$ and the dual shadow function $\boldsymbol{\Psi}_2^{(\alpha_i)}$ are the solution of the third equation in (13.12), with the second equation of (13.24) as the boundary conditions.

The primal solution $\boldsymbol{\Phi}_0^{(\alpha_1)}$ in the case of a crack is known as a mode I solution. The eigenvalue in this case is $\alpha_1 = 1/2$, and the primal and shadow functions for $\lambda = 0.5769$ and $\mu = 0.3846$, for example (corresponding to the engineering material properties $E = 1$ and $\nu = 0.3$), are

$$\boldsymbol{\Phi}_0^{(\alpha_1)}(r,\theta) = r^{\frac{1}{2}} \begin{pmatrix} 2.6\sin(\frac{1}{2}\theta) + \sin(\frac{3}{2}\theta) \\ 4.6\cos(\frac{1}{2}\theta) + \cos(\frac{3}{2}\theta) \\ 0 \end{pmatrix},$$

$$\boldsymbol{\Phi}_1^{(\alpha_1)}(r,\theta) = r^{\frac{3}{2}} \begin{pmatrix} 0 \\ 0 \\ -2\sin(\frac{1}{2}\theta) - 3.06667\sin(\frac{3}{2}\theta) \end{pmatrix},$$

$$\boldsymbol{\Phi}_2^{(\alpha_1)}(r,\theta) = r^{\frac{5}{2}} \begin{pmatrix} 0.23333\sin(\frac{1}{2}\theta) + 0.65644\sin(\frac{3}{2}\theta) \\ -0.76667\cos(\frac{1}{2}\theta) + 0.03244\cos(\frac{3}{2}\theta) \\ 0 \end{pmatrix}, \tag{13.28}$$

and the dual shadow functions are

$$\boldsymbol{\Psi}_0^{(\alpha_1)}(r,\theta) = 0.05542 r^{-\frac{1}{2}} \begin{pmatrix} \sin(\frac{1}{2}\theta) + 1.53333\sin(\frac{3}{2}\theta) \\ \cos(\frac{1}{2}\theta) + 0.86667\cos(\frac{3}{2}\theta) \\ 0 \end{pmatrix},$$

$$\boldsymbol{\Psi}_1^{(\alpha_1)}(r,\theta) = 0.05542 r^{\frac{1}{2}} \begin{pmatrix} 0 \\ 0 \\ -1.73333\sin(\frac{1}{2}\theta) - 0.66667\sin(\frac{3}{2}\theta) \end{pmatrix},$$

$$\boldsymbol{\Psi}_2^{(\alpha_1)}(r,\theta) = 0.05542 r^{\frac{3}{2}} \begin{pmatrix} 0.23778\sin(\frac{1}{2}\theta) - 0.1\sin(\frac{3}{2}\theta) \\ 0.495556\cos(\frac{1}{2}\theta) - 0.43333\cos(\frac{3}{2}\theta) \\ 0 \end{pmatrix}. \tag{13.29}$$

Problem 13.1. For $\lambda = 0.5769$ and $\mu = 0.3846$ and a state of plane-strain show that the eigenfunctions u_r and u_θ for mode I in 2-D (see Table 5.3) are identical to the r and θ components of $\boldsymbol{\Phi}_0^{(\alpha_1)}(r,\theta)$ in (13.28). Notice that the angle θ here is measured from the crack surface, whereas in Chapter 5 it is measured from the bisector of the crack face.

The primal and dual eigenfunctions and shadow functions associated with $\alpha_1 = 1/2$ are presented in Figure 13.2.

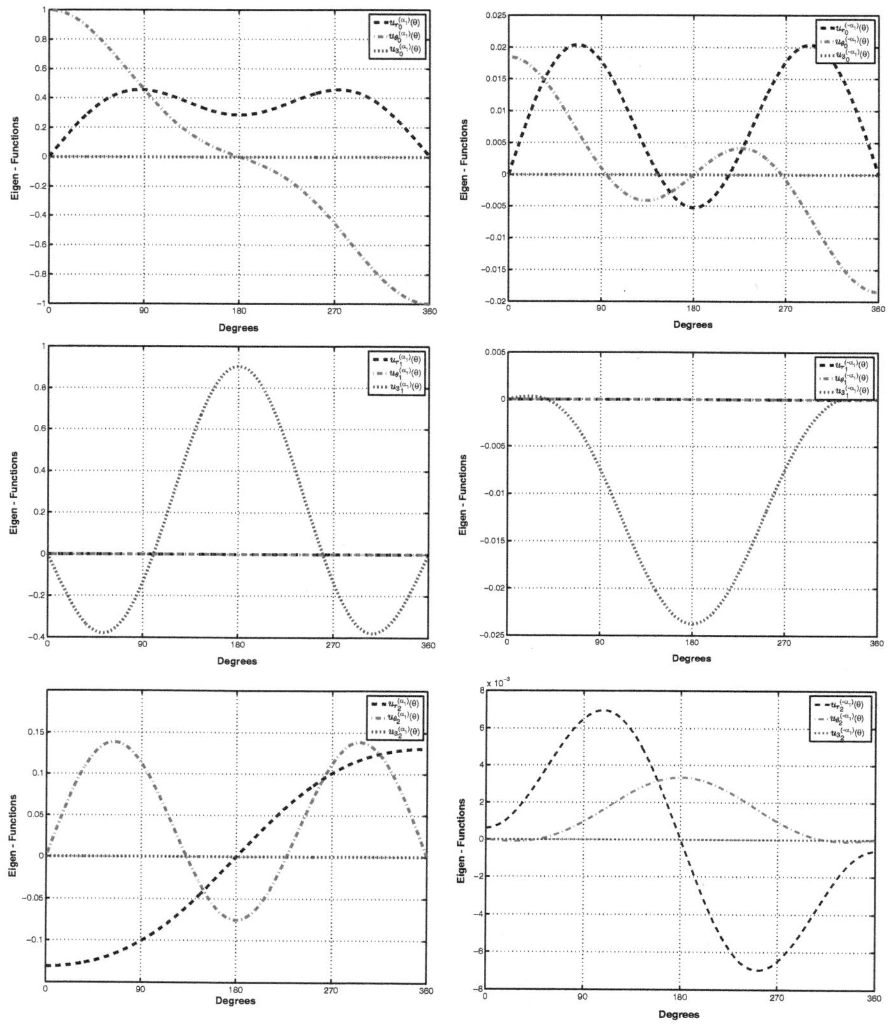

Fig. 13.2 The eigenfunctions (left) and the dual eigenfunctions (right) associated with $\alpha_1 = \frac{1}{2}$ in the case of a cracked domain ($\omega = 2\pi$), $\lambda = 0.5769$, and $\mu = 0.3846$.

The primal solution $\boldsymbol{\Phi}_0^{(\alpha_2)}$ in the case of a crack is known as a mode II solution. The eigenvalue in the case is $\alpha_2 = 1/2$, and the primal and shadow functions for $\lambda = 0.5769$ and $\mu = 0.3846$ are

$$\boldsymbol{\Phi}_0^{(\alpha_2)}(r,\theta) = r^{\frac{1}{2}} \begin{pmatrix} 0.86667\cos(\frac{1}{2}\theta) + \cos(\frac{3}{2}\theta) \\ -1.53333\sin(\frac{1}{2}\theta) - \sin(\frac{3}{2}\theta) \\ 0 \end{pmatrix},$$

$$\boldsymbol{\Phi}_1^{(\alpha_2)}(r,\theta) = r^{\frac{3}{2}} \begin{pmatrix} 0 \\ 0 \\ -0.66667\cos(\frac{1}{2}\theta) \end{pmatrix},$$

$$\boldsymbol{\Phi}_2^{(\alpha_2)}(r,\theta) = r^{\frac{5}{2}} \begin{pmatrix} 0.07778\cos(\frac{1}{2}\theta) - 0.07956\cos(\frac{3}{2}\theta) \\ 0.25556\sin(\frac{1}{2}\theta) + 0.10775\sin(\frac{3}{2}\theta) \\ 0 \end{pmatrix}, \tag{13.30}$$

and the dual shadow functions are

$$\boldsymbol{\Psi}_0^{(\alpha_2)}(r,\theta) = 0.05542 r^{-\frac{1}{2}} \begin{pmatrix} \cos(\frac{1}{2}\theta) + 4.6\cos(\frac{3}{2}\theta) \\ \sin(\frac{1}{2}\theta) - 2.6\sin(\frac{3}{2}\theta) \\ 0 \end{pmatrix},$$

$$\boldsymbol{\Psi}_1^{(\alpha_2)}(r,\theta) = 0.05542 r^{\frac{1}{2}} \begin{pmatrix} 0 \\ 0 \\ -2\cos(\frac{3}{2}\theta) \end{pmatrix},$$

$$\boldsymbol{\Psi}_2^{(\alpha_2)}(r,\theta) = 0.05542 r^{\frac{3}{2}} \begin{pmatrix} -0.27067\cos(\frac{1}{2}\theta) - 0.3\cos(\frac{3}{2}\theta) \\ -0.31067\sin(\frac{1}{2}\theta) + 1.3\sin(\frac{3}{2}\theta) \\ 0 \end{pmatrix}. \tag{13.31}$$

Problem 13.2. For $\lambda = 0.5769$ and $\mu = 0.3846$ and a state of plane-strain show that the eigenfunctions u_r and u_θ for mode II in 2-D (see Table 5.3) are identical to the r and θ components of $\boldsymbol{\Phi}_0^{(\alpha_2)}(r,\theta)$ in (13.30). Notice that the angle θ here is measured from the crack surface, whereas in Chapter 5 it is measured from the bisector of the crack face.

The primal and dual eigenfunctions and shadow functions associated with $\alpha_2 = 1/2$ are presented in Figure 13.3.

The third eigenvalue in the case of a cracked domain with traction free boundary conditions is $\alpha_3 = 1/2$, and the primal and shadow functions for $\lambda = 0.5769$ and $\mu = 0.3846$ are

$$\boldsymbol{\Phi}_0^{(\alpha_3)}(r,\theta) = r^{\frac{1}{2}} \begin{pmatrix} 0 \\ 0 \\ \cos(\frac{1}{2}\theta) \end{pmatrix},$$

$$\boldsymbol{\Phi}_1^{(\alpha_3)}(r,\theta) = r^{\frac{3}{2}} \begin{pmatrix} -0.29333\cos(\frac{1}{2}\theta) \\ 0.10667\sin(\frac{1}{2}\theta) \\ 0 \end{pmatrix},$$

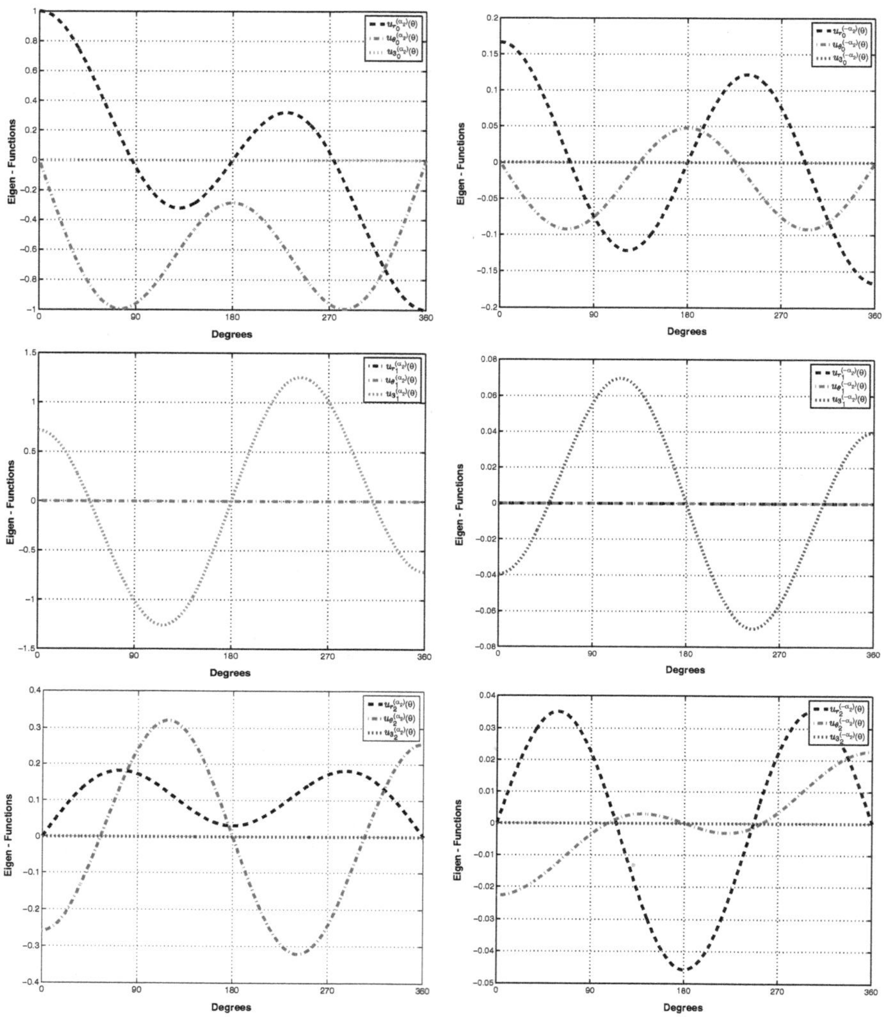

Fig. 13.3 The eigenfunctions (left) and the dual eigenfunctions (right) associated with $\alpha_2 = \frac{1}{2}$ in the case of a cracked domain ($\omega = 2\pi$), $\lambda = 0.5769$, and $\mu = 0.3846$.

$$\boldsymbol{\Phi}_2^{(\alpha_3)}(r,\theta) = r^{\frac{5}{2}} \begin{pmatrix} 0 \\ 0 \\ -0.3\cos(\frac{1}{2}\theta) \end{pmatrix}, \tag{13.32}$$

and the dual shadow functions are

$$\boldsymbol{\Psi}_0^{(\alpha_3)}(r,\theta) = 0.82760 r^{-\frac{1}{2}} \begin{pmatrix} 0 \\ 0 \\ \cos(\frac{1}{2}\theta) \end{pmatrix},$$

$$\boldsymbol{\Psi}_1^{(\alpha_3)}(r,\theta) = 0.82760 r^{\frac{1}{2}} \begin{pmatrix} -0.13333\cos(\frac{1}{2}\theta) \\ -0.53333\sin(\frac{1}{2}\theta) \\ 0 \end{pmatrix},$$

$$\boldsymbol{\Psi}_2^{(\alpha_3)}(r,\theta) = 0.82760 r^{\frac{3}{2}} \begin{pmatrix} 0 \\ 0 \\ -1.16667\cos(\frac{1}{2}\theta) \end{pmatrix}. \tag{13.33}$$

The primal and dual eigenfunctions and shadowfunctions associated with $\alpha_3 = 1/2$, are presented in Figure 13.4.

Similarly to the first three $\frac{1}{2}$ eigenvalues, the fourth to sixth eigenvalues are $\alpha_4 = \alpha_5 = \alpha_6 = 1$, and the corresponding primal and shadow eigenfunctions are

$$\boldsymbol{\Phi}_0^{(\alpha_4)}(r,\theta) = r \begin{pmatrix} \cos(2\theta) + \frac{\mu}{\mu+\lambda} \\ -\sin(2\theta) \\ 0 \end{pmatrix}, \quad \boldsymbol{\Phi}_1^{(\alpha_4)}(r,\theta) = r^2 \begin{pmatrix} 0 \\ 0 \\ \cos(2\theta) - \frac{1}{2} \end{pmatrix},$$

$$\boldsymbol{\Phi}_2^{(\alpha_4)}(r,\theta) = r^3 \begin{pmatrix} -\frac{15\lambda^2+26\lambda\mu+12\mu^2}{48(\lambda+\mu)^2}\cos(2\theta) + \frac{\lambda}{8(\mu+\lambda)} \\ \frac{(3\lambda+2\mu)(2\lambda+3\mu)}{48(\lambda+\mu)^2}\sin(2\theta) \\ 0 \end{pmatrix}, \tag{13.34}$$

$$\boldsymbol{\Phi}_0^{(\alpha_5)}(r,\theta) = r \begin{pmatrix} 0 \\ 1 \\ 0 \end{pmatrix}, \quad \boldsymbol{\Phi}_1^{(\alpha_5)}(r,\theta) = r^2 \begin{pmatrix} 0 \\ 0 \\ -\frac{1}{2}\sin(2\theta) \end{pmatrix},$$

$$\boldsymbol{\Phi}_2^{(\alpha_5)}(r,\theta) = r^3 \begin{pmatrix} \frac{3\lambda+2\mu}{24(\lambda+\mu)}\sin(2\theta) \\ \frac{\mu}{24(\lambda+\mu)}\cos(2\theta) - \frac{1}{8} \\ 0 \end{pmatrix}, \tag{13.35}$$

$$\boldsymbol{\Phi}_0^{(\alpha_6)}(r,\theta) = r \begin{pmatrix} 0 \\ 0 \\ \cos\theta \end{pmatrix}, \quad \boldsymbol{\Phi}_1^{(\alpha_6)}(r,\theta) = r^2 \begin{pmatrix} -\frac{\lambda}{3\lambda+2\mu}\cos\theta \\ -\frac{2\mu^2}{(\lambda-\mu)(3\lambda+2\mu)}\sin\theta \\ 0 \end{pmatrix},$$

$$\boldsymbol{\Phi}_2^{(\alpha_6)}(r,\theta) = r^3 \begin{pmatrix} 0 \\ 0 \\ \left(-\frac{3}{8} + \frac{\lambda^2}{6\lambda^2-2\lambda\mu-4\mu^2}\right)\cos\theta \end{pmatrix}. \tag{13.36}$$

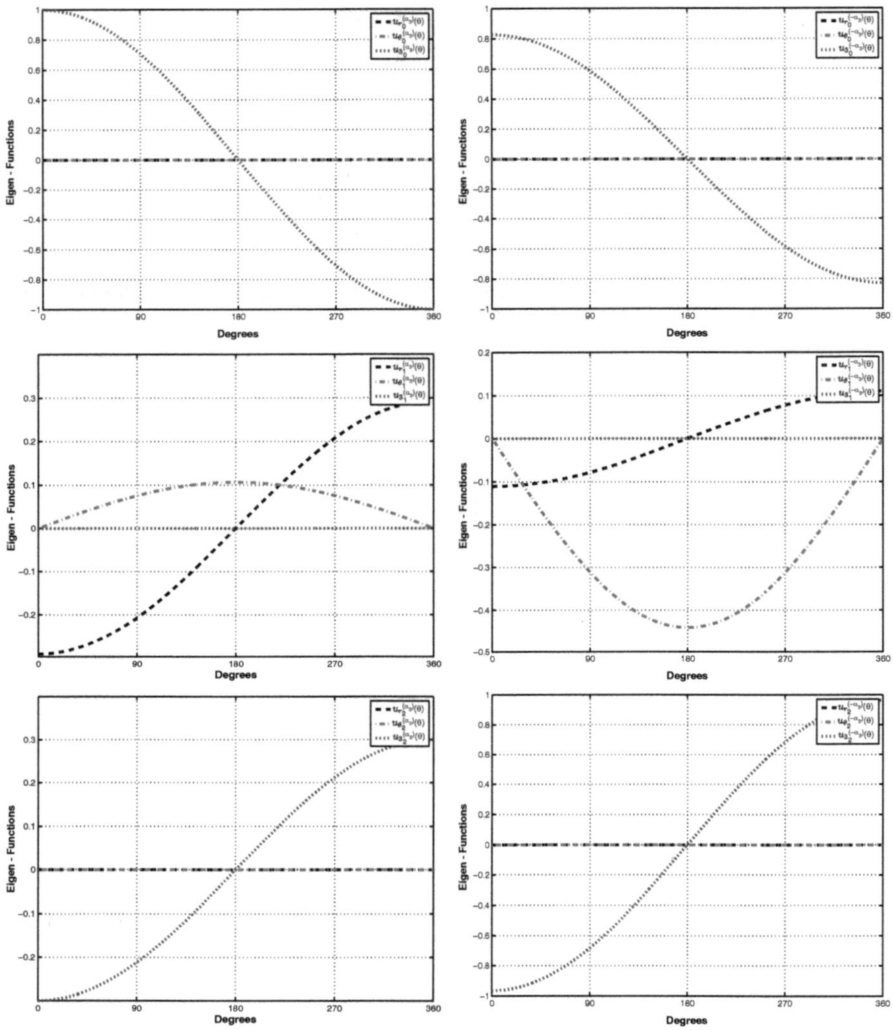

Fig. 13.4 The eigenfunctions (left) and the dual eigenfunctions (right) associated with $\alpha_3 = \frac{1}{2}$ in the case of a cracked domain ($\omega = 2\pi$), $\lambda = 0.5769$, and $\mu = 0.3846$.

The stresses in the vicinity of the crack edge can be easily obtained after the eigenfunctions and shadows are computed. Considering the first six nonzero eigenvalues, the primal eigenfunctions and their shadows (13.28), (13.30), (13.32), (13.34), (13.35), (13.36), and using the kinematic conditions (1.49) and Hooke's law (1.43), the stress tensor is computed (we present terms up to $\sqrt{r}$). To be consistent with engineering notation we also define

$$\mu A_1(z) \stackrel{\text{def}}{=} \frac{K_{\rm I}(z)}{4\sqrt{2\pi}}, \quad \frac{\mu}{3} A_2(z) \stackrel{\text{def}}{=} \frac{K_{\rm II}(z)}{4\sqrt{2\pi}}, \quad \frac{\mu}{2} A_3(z) \stackrel{\text{def}}{=} \frac{K_{\rm III}(z)}{\sqrt{2\pi}},$$

$$2\mu A_4(z) \stackrel{\text{def}}{=} T_2, \quad \mu A_6(z) \stackrel{\text{def}}{=} T_6. \tag{13.37}$$

To compare with the classical 2-D stresses we use the angle denoted here by $\theta^* = \theta - \pi$, which is measured as in Chapter 5 from the bisector of the crack. The stress tensor in polar coordinates reads:

$$\begin{aligned}
\sigma_{rr} &= \frac{K_{\rm I}(z)}{\sqrt{2\pi r}} \frac{1}{4} \left[5\cos\left(\frac{\theta^*}{2}\right) - \cos\left(\frac{3\theta^*}{2}\right) \right] + \frac{K_{\rm II}(z)}{\sqrt{2\pi r}} \frac{1}{4} \left[-5\sin\left(\frac{\theta^*}{2}\right) + 3\sin\left(\frac{3\theta^*}{2}\right) \right] \\
&\quad + T_4(z)\left[1 + \cos(2\theta^*)\right] + \mathcal{O}(\sqrt{r}), \\
\sigma_{\theta\theta} &= \frac{K_{\rm I}(z)}{\sqrt{2\pi r}} \frac{1}{4} \left[3\cos\left(\frac{\theta^*}{2}\right) + \cos\left(\frac{3\theta^*}{2}\right) \right] - \frac{K_{\rm II}(z)}{\sqrt{2\pi r}} \frac{3}{4} \left[\sin\left(\frac{\theta^*}{2}\right) + \sin\left(\frac{3\theta^*}{2}\right) \right] \\
&\quad + T_4(z)\left[1 - \cos(2\theta^*)\right] + \mathcal{O}(\sqrt{r}), \\
\sigma_{zz} &= \frac{2\lambda}{5\mu} \left[\frac{K_{\rm I}(z)}{\sqrt{2\pi r}} \cos\left(\frac{\theta^*}{2}\right) - \frac{K_{\rm II}(z)}{\sqrt{2\pi r}} \sin\left(\frac{\theta^*}{2}\right) + T_4(z) \right] + \mathcal{O}(\sqrt{r}), \qquad\qquad (13.38) \\
\sigma_{\theta z} &= -\frac{K_{\rm III}(z)}{\sqrt{2\pi r}} \cos\left(\frac{\theta^*}{2}\right) + T_6(z)\sin\theta^* + \mathcal{O}(\sqrt{r}), \\
\sigma_{rz} &= -\frac{K_{\rm III}(z)}{\sqrt{2\pi r}} \sin\left(\frac{\theta^*}{2}\right) - T_6(z)\cos\theta^* + \mathcal{O}(\sqrt{r}), \\
\sigma_{r\theta} &= \frac{K_{\rm I}(z)}{\sqrt{2\pi r}} \frac{1}{4} \left[\sin\left(\frac{\theta^*}{2}\right) + \sin\left(\frac{3\theta^*}{2}\right) \right] + \frac{K_{\rm II}(z)}{\sqrt{2\pi r}} \frac{1}{4} \left[\cos\left(\frac{\theta^*}{2}\right) + 3\cos\left(\frac{3\theta^*}{2}\right) \right] \\
&\quad - T_4(z)\sin(2\theta^*) + \mathcal{O}(\sqrt{r}).
\end{aligned}$$

13.1.4 Primal and Dual Eigenfunctions and Shadow Functions for a Clamped $3\pi/2$ V-notch

The displacements $\tilde{\boldsymbol{u}}$ in the case of a V-notched domain ($0 \le \theta \le \omega = \frac{3\pi}{2}$) with clamped boundary conditions (13.26) on the surfaces Γ_1 and Γ_2 are constructed by the primal and shadow functions $\boldsymbol{\Phi}_j$, $j \ge 0$.

There is an infinite number of eigenvalues α_i for which there is an associated $\boldsymbol{\Phi}_0^{(\alpha_i)}$ and $\boldsymbol{\Psi}_0^{(\alpha_i)}$, where the positive α_i 's are associated with $\boldsymbol{\Phi}_0^{(\alpha_i)}$ and the negative α_i 's are associated with $\boldsymbol{\Psi}_0^{(\alpha_i)}$. We consider only the first three eigenvalues of the $3\pi/2$ V-notched domain, $\alpha_1 = 0.595156$, $\alpha_2 = 0.759042$, $\alpha_3 = 0.66667$. The dual eigenfunction $\boldsymbol{\Psi}_0^{(\alpha_i)}$ includes the normalization factor $c_0^{(\alpha_i)}$, chosen such that the primal and dual eigenfunctions satisfy the orthonormal condition as defined in (13.27).

The primal and shadow functions associated with $\alpha_1 = 0.595156$ for $\lambda = 0.5769$ and $\mu = 0.3846$ are

$$\boldsymbol{\Phi}_0^{(\alpha_1)}(r,\theta) = r^{0.59516}\begin{pmatrix} -1.40993\cos(0.404840\theta) + 1.40993\cos(1.595160\theta) \\ +\sin(0.404840\theta) - 1.98793\sin(1.595160\theta) \\ 1.98794\cos(0.404840\theta) - 1.98794\cos(1.595160\theta) \\ +2.80286\sin(0.404840\theta) - 1.40993\sin(1.595160\theta) \\ 0 \end{pmatrix},$$

$$\boldsymbol{\Phi}_1^{(\alpha_1)}(r,\theta) = r^{1.59516}\begin{pmatrix} 0 \\ 0 \\ 1.17022\cos(0.404840\theta) - 1.17022\cos(1.595160\theta) \\ -0.82998\sin(0.404840\theta) + 1.64996\sin(1.595160\theta) \end{pmatrix},$$

$$\boldsymbol{\Phi}_2^{(\alpha_1)}(r,\theta) = r^{2.59516}\begin{pmatrix} -0.14583\cos(0.404840\theta) + 0.14583\cos(1.595160\theta) \\ +0.10343\sin(0.404840\theta) - 0.20562\sin(1.595160\theta) \\ -0.31156\cos(0.404840\theta) + 0.31156\cos(1.595160\theta) \\ -0.43928\sin(0.404840\theta) + 0.22097\sin(1.595160\theta) \\ 0 \end{pmatrix},$$

(13.39)

and the dual shadow functions are

$$\boldsymbol{\Psi}_0^{(\alpha_1)}(r,\theta) = -0.05898 r^{-0.59516}\begin{pmatrix} 0.70924\cos(0.404840\theta) - 0.70924\cos(1.595160\theta) \\ -0.50303\sin(0.404840\theta) + \sin(1.595160\theta) \\ -0.50303\cos(0.4048440\theta) + 0.50303\cos(1.595160\theta) \\ -0.70924\sin(0.404840\theta) + 0.35677\sin(1.595160\theta) \\ 0 \end{pmatrix},$$

$$\boldsymbol{\Psi}_1^{(\alpha_1)}(r,\theta) = -0.05898 r^{0.40484}\begin{pmatrix} 0 \\ 0 \\ -0.296125\cos(0.404840\theta) + 0.29612\cos(1.595160\theta) \\ +0.21002\sin(0.404840\theta) - 0.41751\sin(1.595160\theta) \end{pmatrix},$$

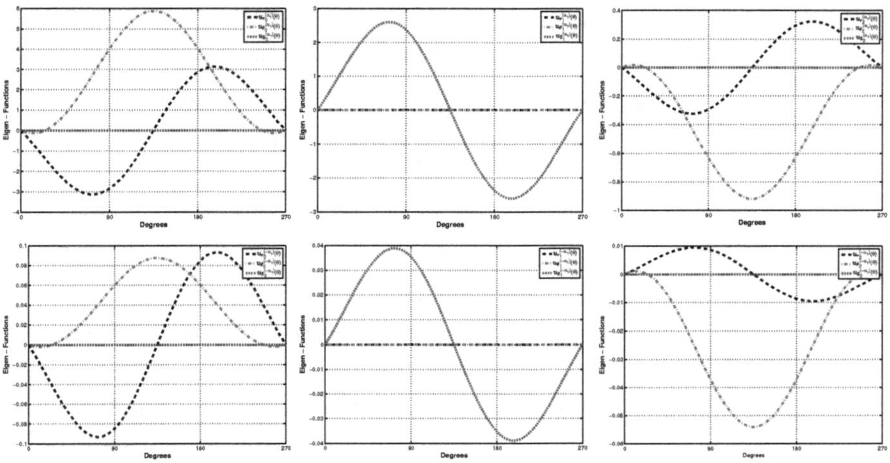

Fig. 13.5 Eigenfunctions (top) and the dual eigenfunctions (bottom) associated with $\alpha_1 = 0.595156$ for a clamped V-notched domain ($\omega = \frac{3\pi}{2}$), $\lambda = 0.5769$, and $\mu = 0.3846$.

$$\boldsymbol{\Psi}_2^{(\alpha_1)}(r,\theta) = -0.05898 r^{1.40484} \begin{pmatrix} 0.07044\cos(0.404840\theta) - 0.07044\cos(1.595160\theta) \\ +0.15980\sin(0.404840\theta) - 0.11044\sin(1.595160\theta) \\ 0.32312\cos(0.404840\theta) - 0.32312\cos(1.595160\theta) \\ +0.02033\sin(0.404840\theta) + 0.20608\sin(1.595160\theta) \\ 0 \end{pmatrix}. \tag{13.40}$$

The primal and dual eigenfunction and shadow functions associated with $\alpha_1 = 0.595156$ are presented in Figure 13.5.

The primal and shadow functions $\boldsymbol{\Phi}_0^{(\alpha_i)}$ for $\alpha_2 = 0.759042$, where $\lambda = 0.5769$ and $\mu = 0.3846$ are

$$\boldsymbol{\Phi}_0^{(\alpha_2)}(r,\theta) = r^{0.75904} \begin{pmatrix} 1.56791\cos(0.240960\theta) - 1.56791\cos(1.759040\theta) \\ +\sin(0.240960\theta) - 2.45835\sin(1.759040\theta) \\ 2.45835\cos(0.240960\theta) - 2.45835\cos(1.759040\theta) \\ -3.85448\sin(0.240960\theta) + 1.56791\sin(1.759040\theta) \\ 0 \end{pmatrix},$$

$$\boldsymbol{\Phi}_1^{(\alpha_2)}(r,\theta) = r^{1.75904} \begin{pmatrix} 0 \\ 0 \\ -1.50622\cos(0.240960\theta) + 1.50622\cos(1.759040\theta) \\ -0.96065\sin(0.240960\theta) + 2.36163\sin(1.759040\theta) \end{pmatrix},$$

$$\boldsymbol{\Phi}_2^{(\alpha_2)}(r,\theta) = r^{2.75904}\begin{pmatrix} -0.15202\cos(0.24096\theta) + 0.15202\cos(1.75904\theta) \\ +0.10782\sin(0.24096\theta) + 0.03358\sin(1.75904\theta) \\ -0.27837\cos(0.24096\theta) + 0.27837\cos(1.75904\theta) \\ -0.39248\sin(0.24096\theta) + 0.65140\sin(1.75904\theta) \\ 0 \end{pmatrix}. \tag{13.41}$$

and the dual shadow functions are

$$\boldsymbol{\Psi}_0^{(\alpha_2)}(r,\theta) = -0.05520 r^{-0.75904}\begin{pmatrix} -0.63779\cos(0.2409556\theta) + 0.63779\cos(1.75904\theta) \\ -0.40678\sin(0.240960\theta) + Sin(1.75904\theta) \\ -0.40678\cos(0.240960\theta) + 0.40678\cos(1.75904\theta) \\ +0.63779\sin(0.240960\theta) - 0.25944\sin(1.759040\theta) \\ 0 \end{pmatrix},$$

$$\boldsymbol{\Psi}_1^{(\alpha_2)}(r,\theta) = -0.05520 r^{0.24096}\begin{pmatrix} 0 \\ 0 \\ 0.24923\cos(0.240960\theta) - 0.24923\cos(1.759040\theta) \\ +0.15896\sin(0.240960\theta) - 0.39077\sin(1.759040\theta) \end{pmatrix},$$

$$\boldsymbol{\Psi}_2^{(\alpha_2)}(r,\theta) = -0.05520 r^{1.24096}\begin{pmatrix} -0.00053\cos(0.240960\theta) + 0.00053\cos(1.759040\theta) \\ +0.00016\sin(0.240960\theta) + 0.00034\sin(1.759040\theta) \\ 0.17120\cos(0.240960\theta) - 0.17120\cos(1.759040\theta) \\ -0.42767\sin(0.240960\theta) + 0.26843\sin(1.759040\theta) \\ 0 \end{pmatrix}. \tag{13.42}$$

The primal and dual eigenfunction and shadow functions associated with $\alpha_2 = 0.759042$ are presented in Figure 13.6.

The primal and shadow functions for $\alpha_3 = 0.666667$, where $\lambda = 0.5769$ and $\mu = 0.3846$ are

$$\boldsymbol{\Phi}_0^{(\alpha_3)}(r,\theta) = r^{0.66667}\begin{pmatrix} 0 \\ 0 \\ \sin(0.66667\theta) \end{pmatrix},$$

$$\boldsymbol{\Phi}_1^{(\alpha_3)}(r,\theta) = r^{1.66667}\begin{pmatrix} -0.28846\sin(0.66667\theta) \\ 0 \\ 0 \end{pmatrix},$$

$$\boldsymbol{\Phi}_2^{(\alpha_3)}(r,\theta) = r^{2.66667}\begin{pmatrix} 0 \\ 0 \\ -0.23654\sin(0.66667\theta) \end{pmatrix}, \tag{13.43}$$

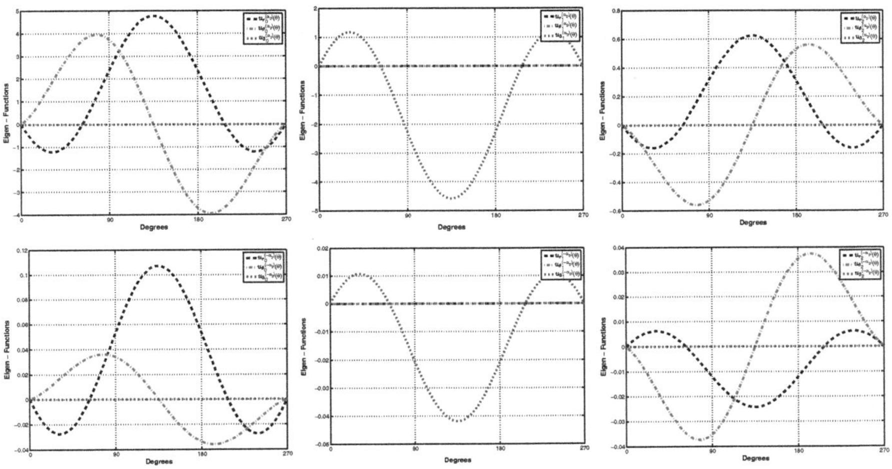

Fig. 13.6 Eigenfunctions (top) and dual eigenfunctions (bottom) associated with $\alpha_2 = 0.759042$ for a clamped V-notched domain ($\omega = \frac{3\pi}{2}$), $\lambda = 0.5769$, and $\mu = 0.3846$.

and the dual shadow functions are

$$\boldsymbol{\Psi}_0^{(\alpha_3)}(r,\theta) = -0.82760r^{-0.66667}\begin{pmatrix} 0 \\ 0 \\ -\sin(0.66667\theta) \end{pmatrix},$$

$$\boldsymbol{\Psi}_1^{(\alpha_3)}(r,\theta) = -0.82760r^{0.33333}\begin{pmatrix} 0.46875\sin(0.66667\theta) \\ 0 \\ 0 \end{pmatrix},$$

$$\boldsymbol{\Psi}_2^{(\alpha_3)}(r,\theta) = -0.82760r^{1.33333}\begin{pmatrix} 0 \\ 0 \\ 1.45313\sin(0.66667\theta) \end{pmatrix}. \tag{13.44}$$

The primal and dual eigenfunction and shadow functions associated with $\alpha_3 = 0.666667$ are presented in Figure 13.7.

13.2 Extracting ESIFs by the $J[R]$ -Integral

Once the asymptotic series representing the elastic solution in the vicinity of an edge is available, we proceed to extraction of ESIFs by recalling the $J[R]$ -integral introduced in (11.30). Here we show an improvement of the method and apply it to the elasticity equations.

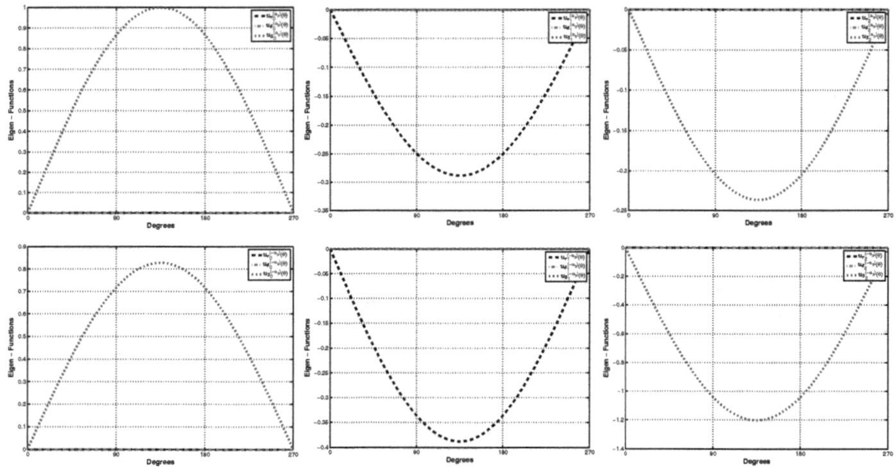

Fig. 13.7 Eigenfunctions (top) and dual eigenfunctions (bottom) associated with $\alpha_3 = 0.666667$ for a clamped V-notched domain ($\omega = \frac{3\pi}{2}$), $\lambda = 0.5769$, and $\mu = 0.3846$.

We construct the *quasidual-singularfunctions* $\mathbf{K}_m^{(\alpha_i)}[BJ]$ for each eigenvalue α_i where m is a natural integer called the *order* of the quasidual function, and $BJ(x_3)$ is a function related to the Jacobi polynomials called an *extraction polynomial*. Each $\mathbf{K}_m^{(\alpha_i)}[BJ]$ is characterized by the number of dual singular functions m needed to construct it and the extraction polynomial BJ :

$$\mathbf{K}_m^{(\alpha_i)}[BJ] \stackrel{\text{def}}{=} \sum_{j=0}^{m} \partial_3^j\, BJ(x_3)\, \boldsymbol{\Psi}_j^{(\alpha_i)}. \tag{13.45}$$

Using the quasidual functions, we have shown that we can extract a scalar product of $A_i(x_3)$ with $BJ(x_3)$ on $\mathcal{E}$. This is accomplished with the help of the *antisymmetric* boundary integral $J[R]$, over the surface Γ_R (13.1) defined for the elasticity system as

$$\begin{aligned} J[R](\boldsymbol{u},\boldsymbol{v}) &\stackrel{\text{def}}{=} \int_{\Gamma_R} ([T]|_{\Gamma_R}\boldsymbol{u}\cdot\boldsymbol{v} - \boldsymbol{u}\cdot[T]|_{\Gamma_R}\boldsymbol{v})\, d\Gamma \\ &= \int_I \int_0^{\omega} ([T]|_{\Gamma_R}\boldsymbol{u}\cdot\boldsymbol{v} - \boldsymbol{u}\cdot[T]|_{\Gamma_R}\boldsymbol{v})|_{r=R}\, R\, d\theta\, dx_3, \end{aligned} \tag{13.46}$$

where $I \equiv \mathcal{E}$ (the edge) along the x_3 axis (Figure 13.1) and $[T]|_{\Gamma_R}$ is the radial Neumann trace operator related to the operator $\mathcal{L}$ on the surface Γ_R :

$$[T]|_{\Gamma_R}\tilde{\boldsymbol{u}} \stackrel{\text{def}}{=} \begin{pmatrix} \sigma_{rr} \\ \sigma_{r\theta} \\ \sigma_{r3} \end{pmatrix} = \begin{pmatrix} (\lambda+2\mu)\partial_r + \lambda\frac{1}{r} & \lambda\frac{1}{r}\partial_\theta & \lambda\partial_3 \\ \mu\frac{1}{r}\partial_\theta & -\mu\frac{1}{r} + \mu\partial_r & 0 \\ \mu\partial_3 & 0 & \mu\partial_r \end{pmatrix} \begin{pmatrix} u_r \\ u_\theta \\ u_3 \end{pmatrix}. \tag{13.47}$$

With the above definitions we have the following theorem [46]:

Theorem 13.1. *Take* $BJ(x_3)$ *such that*

$$\partial_3^j BJ(x_3) = 0 \quad for \quad j = 0, \ldots, m-1 \quad on\ \partial I. \tag{13.48}$$

Then if the ESIFs A_i *in the expansion* (13.18) *are smooth enough,*

$$J[R](\tilde{\boldsymbol{u}}, \mathbf{K}_m^{(\alpha_i)}[BJ]) = \int_I A_i(x_3)\, BJ(x_3)\, dx_3 + \mathcal{O}(R^{\alpha_1 - \alpha_i + m + 1}) \quad as \quad R \to 0. \tag{13.49}$$

Here α_1 *is the smallest of the positive real eigenvalues* α_i , $i \in \mathbb{N}$, *and we assume that any other complex eigenvalue* α *with positive real part satisfies* $\Re\alpha \geq \alpha_1$, *as mentioned in Remark 13.2.*

Theorem 13.1 allows a precise determination of $\int_I A_i(x_3)\, BJ(x_3)\, dx_3$ by computing (13.49) for two or three R values and using Richardson's extrapolation as $R \to 0$.

The construction of the $BJ(x_3)$ extraction functions based on the Jacobi polynomials is explained in Section 11.3.1, so if $A_i(x_3)$ is a polynomial of degree N represented by a linear combination of Jacobi polynomials as

$$A_i(x_3) = \tilde{a}_0 J_n^{(0)} + \tilde{a}_1 J_n^{(1)}(x_3) + \cdots + \tilde{a}_N J_n^{(N)}(x_3), \tag{13.50}$$

then

$$BJ_m^{(k)}(x_3) = (1 - x_3^2)^m\, \frac{J_m^{(k)}(x_3)}{h_k}\,, \tag{13.51}$$

so that

$$\int_{-1}^{1} A_i(x_3) BJ_m^{(k)}(x_3)\, dx_3 = \tilde{a}_k, \qquad k = 0, 1, \ldots, N. \tag{13.52}$$

13.2.1 Jacobi Extraction Polynomials of Order 4

For the sake of simplicity, the first three dual singular functions $\mathbf{K}_0^{(\alpha_i)}$, $\mathbf{K}_1^{(\alpha_i)}$, and $\mathbf{K}_2^{(\alpha_i)}$ are considered here. Thus, according to Theorem 13.1, it is necessary that the Jacobi extraction polynomials satisfy the conditions in (13.48) at least to $m = 2$. In [134] it was observed that if the minimal condition is satisfied, one does indeed recover the expected rate of convergence with respect to R . However, poor results are evident at the two ends of the edge (this behavior was observed also if the edge portion along which EFIFs was extracted was entirely within the domain and away from the vertices, i.e., $-0.6 < x_3 < 0.6$). This phenomenon is attributed to the large values of the derivatives of the Jacobi polynomials at the endpoints (interested readers are referred to [134, Appendix C]. Therefore we select the Jacobi extraction polynomials $BJ_4^{(k)}$, which satisfy (13.48) up to $m = 4$. The Jacobi extraction polynomials $BJ_4^{(k)}$ are used for the construction of the dual singular functions $\mathbf{K}_0^{(\alpha_i)}$, $\mathbf{K}_1^{(\alpha_i)}$, and $\mathbf{K}_2^{(\alpha_i)}$. We have [2, pp. 773-774]

$$J_4^{(k)}(x_3) = \frac{(k+4)!}{(k+8)!} \sum_{l=0}^{k} \frac{(k+l+8)!}{2^l l!\,(k-l)!\,(4+l)!} (x_3 - 1)^l, \tag{13.53}$$

and the constant h_k in (13.51) is equal to

$$h_k = \frac{2^9 (k+4)!(k+4)!}{(2k+9)(k+8)!}. \tag{13.54}$$

Inserting (13.54) and (13.53) in (13.51), we finally obtain

$$BJ_4^{(k)}(x_3) = \frac{(2k+9)(1-x_3^2)^4}{2^9(k+4)!} \sum_{l=0}^{k} \frac{(k+l+8)!}{2^l l!\,(k-l)!\,(4+l)!} (x_3 - 1)^l. \tag{13.55}$$

13.2.1.1 Numerical Computation of the $J[R]$ Integral

The exact solution $\tilde{\boldsymbol{u}}$ is in general unknown, so we use instead a finite element approximation $\tilde{\boldsymbol{u}}_{\mathrm{FE}}$, and the integral (13.46) is computed numerically using a Gaussian quadrature of order n_G

$$\begin{aligned} &J[R](\tilde{\boldsymbol{u}}, \mathbf{K}_m^{(\alpha_i)}[BJ]) \\ &= \sum_{k=1}^{n_G} \sum_{\ell=1}^{n_G} \frac{\omega}{2} w_k w_\ell \left([T]\tilde{\boldsymbol{u}}_{\mathrm{FE}} \cdot \mathbf{K}_m^{(\alpha_i)}[BJ_n^{(k)}] - \tilde{\boldsymbol{u}}_{\mathrm{FE}} \cdot [T]\mathbf{K}_m^{(\alpha_i)}[BJ_n^{(k)}] \right)_{\xi_k, \eta_\ell}, \end{aligned} \tag{13.56}$$

where w_k are the weights and ξ_k and η_ℓ are the abscissas of the Gaussian quadrature. The Neumann trace operator, $[T]$, operates on both $\tilde{\boldsymbol{u}}$ and $\mathbf{K}_m^{(\alpha_i)}[BJ_n^{(k)}]$. For $[T]\tilde{\boldsymbol{u}}$ we use the numerical approximations $[T]\tilde{\boldsymbol{u}}_{\mathrm{FE}}$ computed by finite elements (notice that such extractions are easily computed by the p-version of the FEM at any point within an element), whereas $[T]\mathbf{K}_m^{(\alpha_i)}[BJ_n^{(k)}]$ is computed analytically. These values are evaluated at the specific Gaussian points at which the integral is computed numerically.

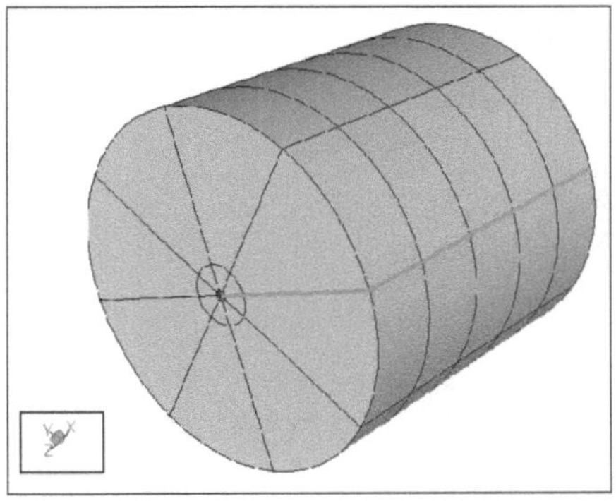

Fig. 13.8 The p-FEM model of the cracked domain.

13.2.2 Numerical Example: A Cracked Domain ($\omega = 2\pi$) with Traction-Tree Boundary Conditions

We can generate an exact solution to a crack in a three-dimensional isotropic domain with traction-free boundary conditions by computing analytically the primal and shadow eigenfunctions $\boldsymbol{\Phi}_0$, $\boldsymbol{\Phi}_1$, $\boldsymbol{\Phi}_2$. Their formulas are presented in Section 13.1.3. We refer to the first three eigenvalues only for a cracked domain, $\alpha_1 = \alpha_2 = \alpha_3 = \frac{1}{2}$, and they are the only eigenvalues that are smaller than 1. Next we choose the ESIFs $A_i(x_3)$, $i = 1, 2, 3$, to be, for example, polynomials of order 3 at most, i.e., $A_i(x_3) = a_0^{(i)} + a_1^{(i)} x_3 + a_2^{(i)} x_3^2$. We obtain therefore an exact solution (13.17) with a finite number of terms in the sum, because the third and higher derivatives of $A_i(x_3)$ are zero. The exact i th eigensolution is

$$\tilde{\boldsymbol{u}}^{(\alpha_i)} = \sum_{j\geq 0}^{2} \partial_3^j A_i(x_3) r^{\alpha_i + j} \boldsymbol{\varphi}_j^{(\alpha_i)}(\theta), \tag{13.57}$$

Let us consider the following ESIFs (polynomials of order 3):

$$A_1(x_3) = 3 + 4x_3 + 5x_3^2, \quad A_2(x_3) = 2 + 3x_3 + 4x_3^2, \quad A_3(x_3) = 5 + 4x_3 + 2x_3^2. \tag{13.58}$$

Then the corresponding exact solution is

$$\tilde{\boldsymbol{u}} = \sum_{i=1}^{3} \tilde{\boldsymbol{u}}^{(\alpha_i)} = \sum_{i=1}^{3} \sum_{j\geq 0}^{2} \partial_3^j A_i(x_3) r^{\alpha_i + j} \boldsymbol{\varphi}_j^{(\alpha_i)}(\theta). \tag{13.59}$$

The domain has been discretized using a p-FEM mesh, with geometric progression toward the singular edge with a factor of 0.15, having four layers of elements. In the x_3 direction, a uniform discretization using five elements has been adopted. In Figure 13.8 we present the mesh used for the cracked domain.

We specify on the entire boundary $\partial\Omega$ Dirichlet boundary conditions according to the exact solution $\tilde{\boldsymbol{u}}$ (13.59). In this way, the exact solution at any point

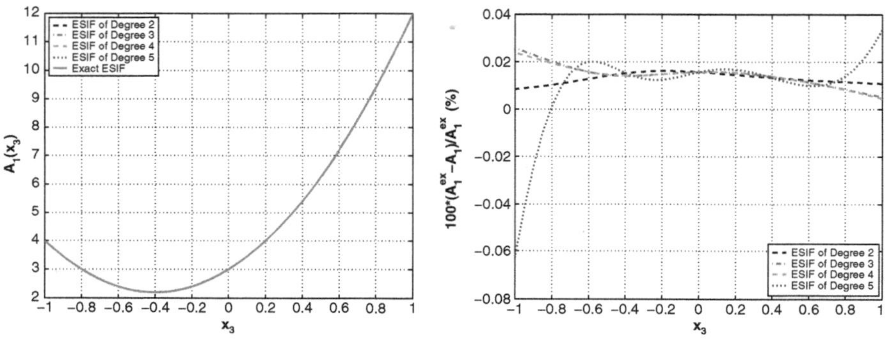

Fig. 13.9 $A_1(x_3)$ (left) and its relative error (right) at $R = 0.05$. Computations done with $BJ_4^{(k)}$, $k = 2, 3, 4, 5$, where $A_1^{ex}(x_3) = 3 + 4x_3 + 5x_3^2$, $\omega = 2\pi$, $\lambda = 0.5769$, and $\mu = 0.3846$.

$\boldsymbol{x} \equiv (r, \theta, x_3)$ is therefore (13.59). In all numerical examples the Young modulus is taken to be 1 and the Poisson ratio 0.3, so the Lamé constants are $\lambda = 0.5769$ and $\mu = 0.3846$.

When $J[R]$ is computed with the quasidual function $\mathbf{K}_m^{(\alpha_i)}$, $BJ_4^{(k)}(x_3)$, and the dual and shadow eigenfunctions $\boldsymbol{\Psi}_0$, $\boldsymbol{\Psi}_1$, $\boldsymbol{\Psi}_2$ given in Section 13.1.3, we obtain according to (13.49) the coefficient $\tilde{a}_j^{(\alpha_i)}$:

$$J[0] = \int_{-1}^{1} A_i(x_3) BJ_4^{(j)}(x_3) dx_3 = \tilde{a}_j^{(\alpha_i)}, \qquad j = 0, 1, \ldots, n. \tag{13.60}$$

The ESIF is then easily represented by a linear combination of the Jacobi polynomials as

$$A_i(x_3) = \tilde{a}_0 J_4^{(0)}(x_3) + \tilde{a}_1 J_4^{(1)}(x_3) + \tilde{a}_2 J_4^{(2)}(x_3) + \cdots . \tag{13.61}$$

The advantage of the hierarchical family of polynomials is that one can adaptively increase the polynomial order of the ESIF. For example, if one is interested in projecting $A_i(x_3)$ into the space of polynomials of degree up to n , the $n+1$ coefficients $\tilde{a}_0, \ldots, \tilde{a}_n$ are computed using the $n+1$ extraction polynomials $BJ_4^{(0)}(x_3), \ldots, BJ_4^{(n)}(x_3)$ defined in (13.51).

To increase the space in which $A_i(x_3)$ is projected, all that is needed is the computation of (13.56) for $n+1$. In this way, the new A_i^{new} equals $A_i^{\text{previous}} + \tilde{a}_{n+1} J_4^{(n+1)}(x_3)$. We illustrate the extracted polynomial representation of the ESIF, $A_1(x_3)$, $A_2(x_3)$, $A_3(x_3)$, of order $2, 3, 4, 5$, and its relative error using the data at $R = 0.05$ in Figures 13.9, 13.10 and 13.11 respectively and using $\mathbf{K}_2^{(\alpha_i)}$.

Notice that the relative error of the extracted ESIFs is lower than 0.1% . The results show an accurate and efficient method.

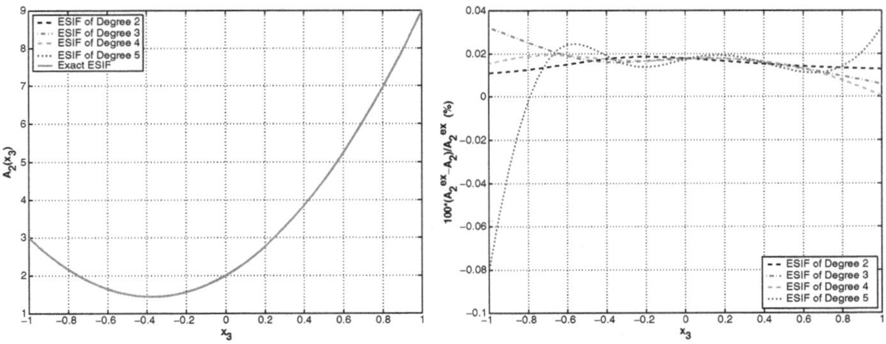

Fig. 13.10 $A_2(x_3)$ (left) and its relative error (right) at $R = 0.05$. Computations done with $BJ_4^{(k)}$, $k = 2, 3, 4, 5$, where $A_2^{ex}(x_3) = 2 + 3x_3 + 4x_3^2$, $\omega = 2\pi$, $\lambda = 0.5769$ and $\mu = 0.3846$.

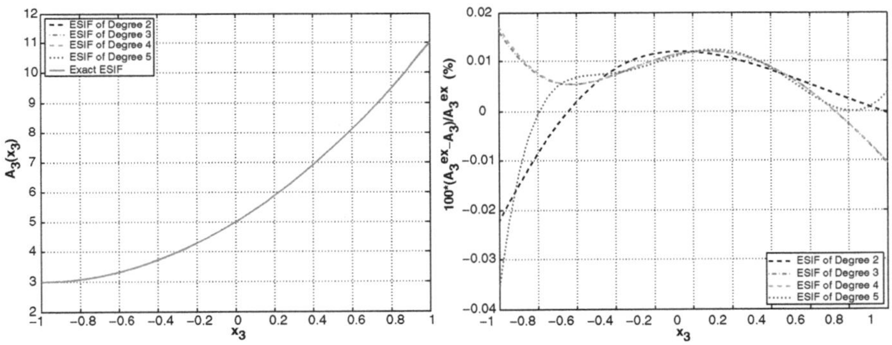

Fig. 13.11 $A_3(x_3)$ (left) and its relative error (right) at $R = 0.05$. Computations done with $BJ_4^{(k)}$, $k = 2, 3, 4, 5$, where $A_3^{ex}(x_3) = 5 + 4x_3 + 2x_3^2$, $\omega = 2\pi$, $\lambda = 0.5769$ and $\mu = 0.3846$.

13.2.3 Numerical Example: A Clamped V-notched Domain $\left(\omega = \frac{3\pi}{2}\right)$

As in the previous section, we generate an exact solution to a V-notched domain ($\omega = \frac{3\pi}{2}$) with clamped boundary conditions on the surfaces Γ_1 and Γ_2 by computing analytically the primal and shadow eigenfunctions $\boldsymbol{\Phi}_0$, $\boldsymbol{\Phi}_1$, $\boldsymbol{\Phi}_2$. Their formulas are presented in Section 13.1.4.

We select the ESIF to be polynomials of order 2 as presented in (13.58), such that the the exact solution (13.17) contains only three terms in the sum, (13.57).

The domain $\omega = \frac{3\pi}{2}$ has been discretized using a p-FEM mesh, with geometric progression toward the singular edge with a factor of 0.15, having four layers of elements. In the x_3 direction, a uniform discretization using five elements

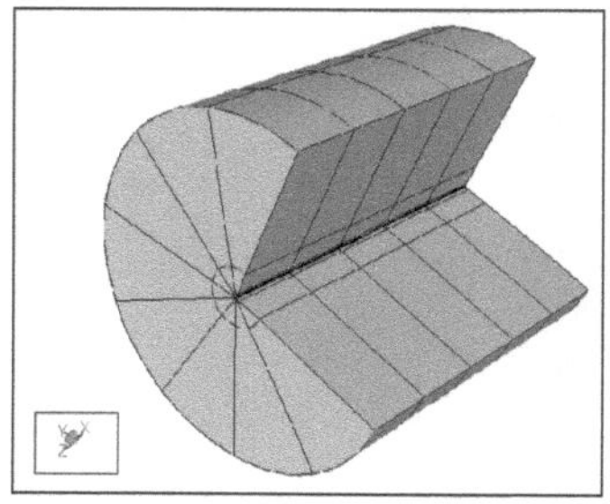

Fig. 13.12 The p -FEM model of the $\omega = \frac{3\pi}{2}$ V-notched domain.

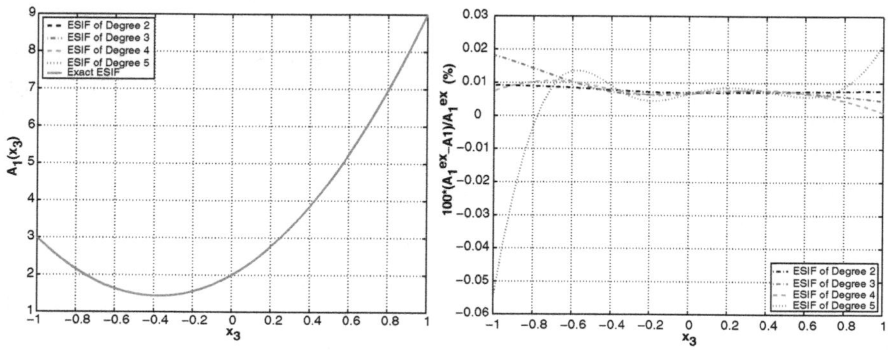

Fig. 13.13 $A_1(x_3)$ (left) and its relative error (right) at $R = 0.05$ using $BJ_4^{(k)}$, $k = 2, 3, 4, 5$, where $A_1^{ex}(x_3) = 3 + 4x_3 + 5x_3^2$, $\omega = \frac{3\pi}{2}$, $\lambda = 0.5769$, and $\mu = 0.3846$.

has been adopted, as presented in Figure 13.12. We specify over the entire boundary $\partial\Omega$ displacement boundary conditions according to the exact solution $\tilde{\boldsymbol{u}}$ (13.59). The FE solution at any point $\boldsymbol{x} \equiv (r, \theta, x_3)$ is therefore the exact solution (13.59).

After computing the $J[R]$ integrals, the computation of the polynomial representation of the ESIF is simple, using a linear combination of the Jacobi polynomials (13.61). We illustrate the extracted polynomial representation of the ESIF, $A_1(x_3)$, $A_2(x_3)$, $A_3(x_3)$, and their relative errors using the data at $R = 0.05$ in Figures 13.13, 13.14, and 13.15 respectively, using $\mathbf{K}_2^{(\alpha_i)}$.

The relative error of the extracted ESIF is less than 0.1% .

13.2.4 Numerical Example of Engineering Importance: Compact Tension Specimen

In this section we compare the ESIFs computed by the quasidual function method with a pointwise extraction method of stress intensity factors (SIFs $K_{\rm I}$ and $K_{\rm II}$)

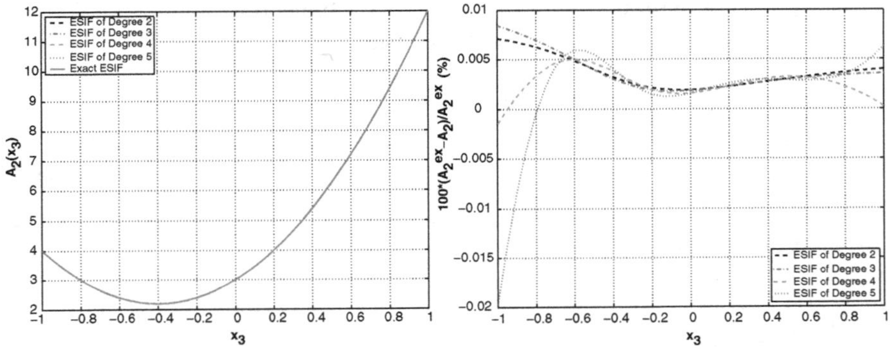

Fig. 13.14 $A_2(x_3)$ (left) and its relative error (right) at $R = 0.05$ using $BJ_4^{(k)}$, $k = 2, 3, 4, 5$, where $A_1^{ex}(x_3) = 2 + 3x_3 + 4x_3^2$, $\omega = \frac{3\pi}{2}$, $\lambda = 0.5769$, and $\mu = 0.3846$.

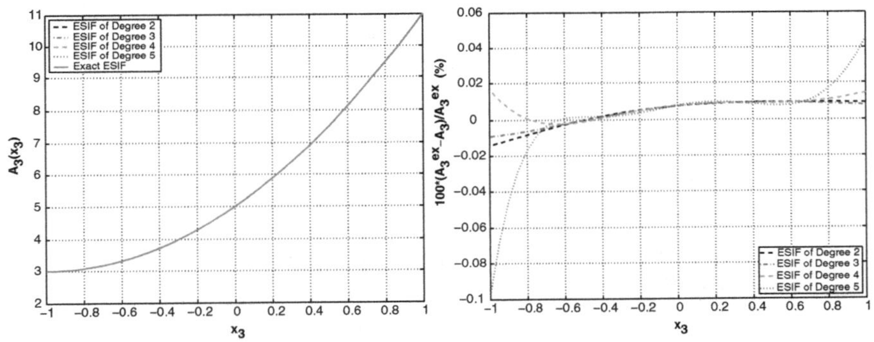

Fig. 13.15 $A_3(x_3)$ (left) and its relative error (right) at $R = 0.05$ using $BJ_4^{(k)}$, $k = 2, 3, 4, 5$, where $A_3^{ex}(x_3) = 5 + 4x_3 + 2x_3^2$, $\omega = \frac{3\pi}{2}$, $\lambda = 0.5769$, and $\mu = 0.3846$.

available in [1]. In the classical fracture-mechanics literature the plane-strain SIFs are reported, which multiply a specific mode I or mode II eigenfunction. To compare the ESIFs and the SIFs, we first present the relationship between the functions A_1 , A_2 and the SIFs K_{I} and K_{II} . We then describe the compact tension specimen (CTS) used for determination of fracture toughness. For the CTS we extract the ESIF using the quasidual function method and pointwise values of SIFs and compare them.

13.2.4.1 The Relation Between the SIFs K_{I} , K_{II} and the ESIF

Under the assumption of plane-strain and mode I loading, the classical solution $\boldsymbol{u}$ in the vicinity of a crack edge is (see Table 5.3; notice that one has to substitute

$\theta + \pi$ instead of θ in the expressions of Table 5.3 due to the different definition of the angle in Chapter 5 compared to this chapter)

$$\begin{Bmatrix} u_1 \\ u_2 \end{Bmatrix} = \frac{K_{\rm I}(x_3)}{2\mu}\sqrt{\frac{r}{2\pi}} \begin{Bmatrix} \cos((\theta+\pi)/2)\left[\kappa - 1 + 2\sin^2((\theta+\pi)/2)\right] \\ \sin((\theta+\pi)/2)\left[\kappa + 1 - 2\cos^2((\theta+\pi)/2)\right] \end{Bmatrix}, \tag{13.62}$$

where $\kappa = 3 - 4\nu$. In the case of the plane-strain assumption and mode II loading, the classical solution $\boldsymbol{u}$ in the vicinity of a crack edge is

$$\begin{Bmatrix} u_1 \\ u_2 \end{Bmatrix} = \frac{K_{\rm II}(x_3)}{2\mu}\sqrt{\frac{r}{2\pi}} \begin{Bmatrix} \sin((\theta+\pi)/2)\left[\kappa + 1 + 2\cos^2((\theta+\pi)/2)\right] \\ \cos((\theta+\pi)/2)\left[\kappa - 1 - 2\sin^2((\theta+\pi)/2)\right] \end{Bmatrix}. \tag{13.63}$$

Comparing the displacements expressed above with those expressed in terms of the ESIFs (for $\lambda = 0.5769$ and $\mu = 0.3846$, see Section 13.1.3), the relation between A_1 and $K_{\rm I}$ and the relation between A_2 and $K_{\rm II}$ in the case of plane strain is

$$\begin{aligned} &\frac{K_{\rm I}}{0.7692\sqrt{2\pi}}\cos((\theta+\pi)/2)\left[0.8 + 2\sin^2((\theta+\pi)/2)\right] \\ &\quad = A_1\left[-2.6\sin\left(\frac{1}{2}\theta\right) - \sin\left(\frac{3}{2}\theta\right)\right], \end{aligned} \tag{13.64}$$

$$\begin{aligned} &\frac{K_{\rm II}}{0.7692\sqrt{2\pi}}\sin((\theta+\pi)/2)\left[2.8 + 2\cos^2((\theta+\pi)/2)\right] \\ &\quad = A_2\left[2.2\cos\left(\frac{1}{2}\theta\right) - \frac{1}{3}\cos\left(\frac{3}{2}\theta\right)\right], \end{aligned} \tag{13.65}$$

which after algebraic manipulation is shown to be independent of θ:

$$A_1 = 0.259312 K_{\rm I}, \qquad A_2 = 0.777938 K_{\rm II}. \tag{13.66}$$

Remark 13.3. The strain component ε_{33} computed using the displacements in (13.28), for the case A_1, is a constant

$$\varepsilon_{33} = \frac{\partial^2 u_3}{\partial x_3^2} = 0. \tag{13.67}$$

On the other hand, if the plane-stress condition is assumed, ε_{33} is given by

$$\begin{aligned} \varepsilon_{33} &= \frac{\sigma_{11}}{E} - \frac{\nu}{E}(\sigma_{11} + \sigma_{22}) \\ \Rightarrow \varepsilon_{33} &= -\frac{\nu}{E}(\sigma_{11} + \sigma_{22}) = -0.923076 r^{-\frac{1}{2}}\sin\left(\frac{1}{2}\theta\right), \end{aligned} \tag{13.68}$$

and therefore in $3-D$ the plane-stress condition cannot be represented in the vicinity of a singular edge.

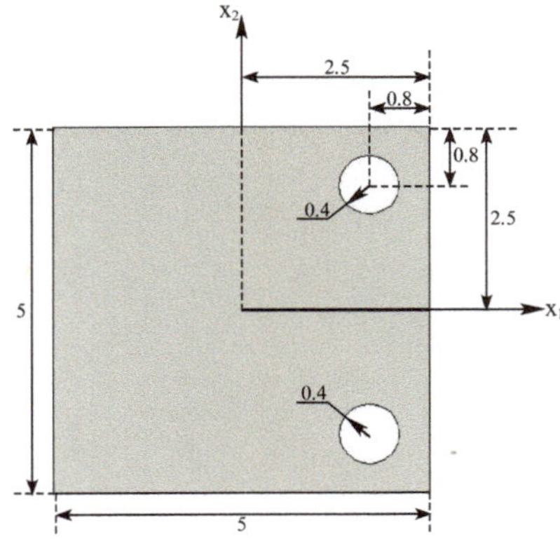

Fig. 13.16 Dimensions of CTS. The thickness of the specimen is 2 ranging over $-1 < x_3 < 1$.

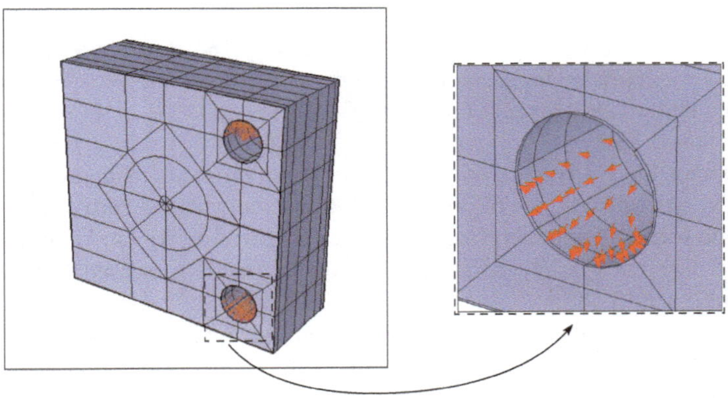

Fig. 13.17 The p -FEM model of the CTS with a constant loading in the x_3 direction (the loading at the upper hole is as in the shown lower hole, in the opposite direction).

13.2.4.2 Compact Tension Specimen (CTS) Under a Constant Tension Along x_3

The classical compact tension specimen (see 2-D view in Figure 13.16) under bearing loads at the tearing holes having an equivalent force in the x_2 direction and being independent of x_3 is presented in Figure 13.17. All other faces are traction free. The thickness of the specimen is 2, ranging over $-1 < x_3 < 1$. The specimen is subjected to a tension load of 100 Newton such that only mode I is excited along the crack front. Although the loading is independent of x_3 , because of the vertex singularities at $x_3 = \pm 1$ we anticipate a variation in A_1 as the vertices are approached. The domain is discretized using a p -FEM mesh, with geometric progression toward the singular edge with a factor of 0.15 where the

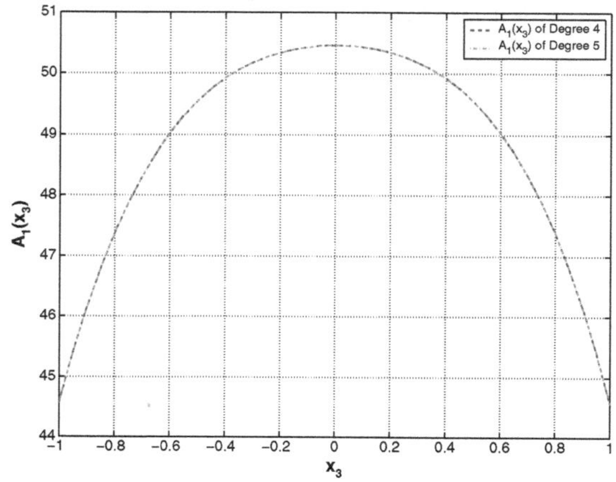

Fig. 13.18 A_1 extracted at $R = 0.05$ using polynomials of degree up to 4 and up to 5 for the CTS.

smallest layer in the vicinity of the edge is at $r = 0.15^3$. In the x_3 direction, we also used a mesh graded in a geometric progression close to the vertex singularity at $x_3 = \pm 1$. The smallest layer in the vicinity of the vertex is $-1 < x_3 < -1+0.15^2$, $1 < x_3 < 1 - 0.15^2$.

We extract the ESIF A_1 , A_2 , and A_3 as polynomials of degree 4 and 5 at $R = 0.05$; A_2 and A_3 are of order 10^{-3} (the exact value is zero except perhaps close to the vertices), and therefore negligible compared to A_1 , and thus not plotted here. The difference in A_1 as the polynomial degree is increased from 4 to 5 is shown in Figure 13.18.

It may be noticed that the difference between the approximation of 4th and 5th order polynomials is negligible and we use in the sequel polynomial degree 5 for approximating A_1 . Next we compute A_1 and K_I (extracted by the pointwise contour integral method, see [1], at several points along the edge) at $R = 0.5$, $R = 0.3$, $R = 0.2$, and $R = 0.05$ and plot these in Figure 13.19.

One may observe the good convergence of the ESIF as $R \to 0$ compared to the pointwise SIFs.

Next, we wish to demonstrate that the ESIFs can be used away from the singular edge, so a coarse mesh is sufficient. We use the same model with a coarse mesh in the vicinity of the edge where the smallest layer in the vicinity of the edge is at $r = 0.15$. In the x_3 direction the same discretization as in the fine mesh is employed, and we perform an FE analysis, using the trunk space up to $p = 7$, having 125,442 DOF. The computed function $A_1(x_3)$ and the pointwise values of K_I at $R = 0.5$, $R = 0.3$, and $R = 0.2$ are presented in Figure 13.20. Although the loading is constant in x_3 , the vertex singularities influence the ESIF, and as seen usually in practice, the crack propagation in the middle of the specimen is usually faster than at the outer surfaces. The results obtained using the ESIF extraction method are generated faster than pointwise extraction methods (K_I extraction) and do not require plane-stress or plane-strain assumptions.

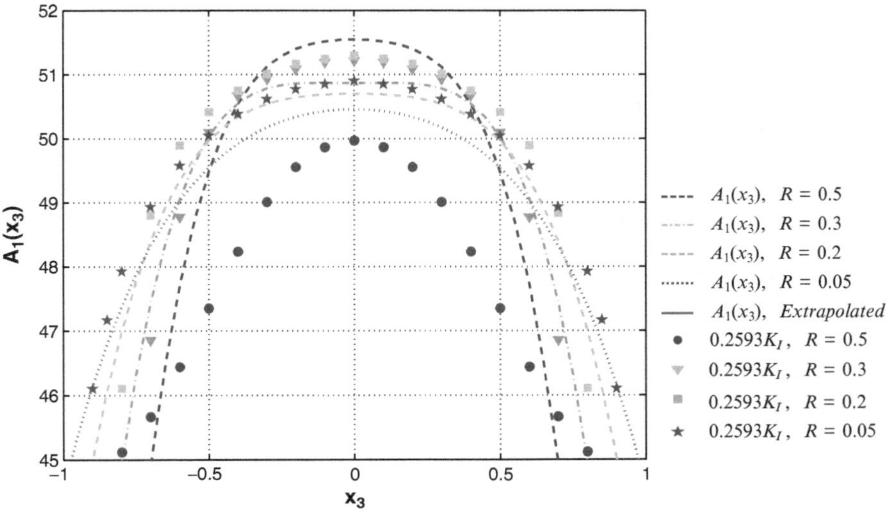

Fig. 13.19 $A_1(x_3)$ and K_{I} extracted using different R 's for the compact tension specimen.

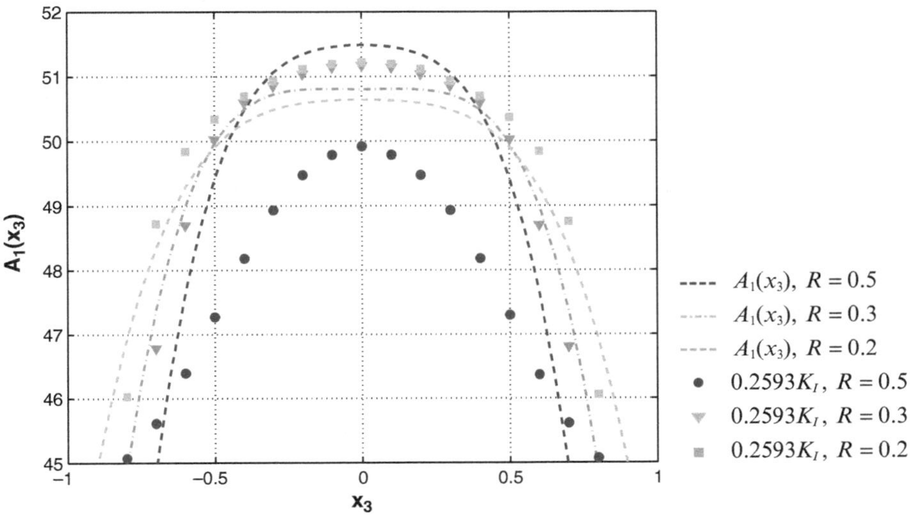

Fig. 13.20 $A_1(x_3)$ and K_{I} extracted using different R 's for the compact tension specimen using coarse grid with 125442 DOF.

It is easy to see that the results of the extracted ESIF using the coarse mesh with 125,442 DOFs are similar to the results obtained using the refined mesh with 150,726 DOFs.

13.3 Eigenpairs and ESIFs for Anisotropic and Multimaterial Interfaces

In anisotropic materials and/or multimaterial interfaces the computation of eigenpairs and the application of the QDFM for the extraction of ESIFs is technically more entangled. For a 3-D interface crack between dissimilar anisotropic materials the energy release rate was extended for the computation of pointwise K_{I}, K_{II}, and K_{III} in [84]. SIF extraction for anisotropic cracked domains is reported in [19,137]. Employing the M-integral, good pointwise approximations of SIFs are reported in [19], and in [137] anisotropic penny-shaped cracked configurations are investigated based on both the traditional displacement boundary element method and the displacement discontinuity method. Transversely isotropic bimaterial cracked domains are investigated in [211] by the dual boundary element method. The fundamental solution for the bimaterial solid occupying an infinite region is incorporated into the dual boundary integral equations, and modes I, II, and III SIFs are computed by the crack opening displacements method.

In anisotropic materials and multimaterial interfaces in the vicinity of edges, difficulties are encountered due to the possible existence of complex eigenpairs on the one hand and intractable analytical derivation on the other. For specific pathological cases (one of which is the cracked configuration, of major importance in fracture mechanics), the numerical methods for computing shadow functions break down because of conceptual difficulties. An explanation of these difficulties is provided in Section 13.3.3. A numerical example is provided that illustrates the problems and the remedy. Finally, the QDFM is extended to extract complex ESIFs in subsection 13.3.4. Numerical examples for multimaterial interfaces involving anisotropic materials are provided for which the complex eigenpairs and shadow functions are numerically computed and complex ESIFs extracted. These examples show the efficiency and high accuracy of the numerical approximations.

Usually, for a general anisotropic domain, Hooke's law is given in a Cartesian coordinate system and may be represented also in a cylindrical coordinate system:

$$\boldsymbol{\sigma} = [E]\boldsymbol{\varepsilon}, \qquad \text{or} \qquad \tilde{\boldsymbol{\sigma}} = [\tilde{E}]\tilde{\boldsymbol{\varepsilon}}, \tag{13.69}$$

where $[E]$ (respectively $[\tilde{E}]$) is symmetric and $[\tilde{E}]$ depends on $[E]$ and θ,

$$[E] = \begin{pmatrix} E_{11} & E_{12} & E_{13} & E_{14} & E_{15} & E_{16} \\ & E_{22} & E_{23} & E_{24} & E_{25} & E_{26} \\ & & E_{33} & E_{34} & E_{35} & E_{36} \\ & & & E_{44} & E_{45} & E_{46} \\ & & & & E_{55} & E_{56} \\ & & & & & E_{66} \end{pmatrix}, \tag{13.70}$$

whereas $[\tilde{E}]$ is associated with $[E]$ as follows:

$$\tilde{E}_{11} = \frac{1}{8}\Big(3E_{11} + 2E_{12} + 3E_{22} + 4E_{66} + 4(E_{11} - E_{22})\cos(2\theta)$$
$$+ (E_{11} - 2E_{12} + E_{22} - 4E_{66})\cos(4\theta)$$
$$+ 8(E_{16} + E_{26})\sin(2\theta) + 4(E_{16} - E_{26})\sin(4\theta)\Big),$$

$$\tilde{E}_{12} = \frac{1}{8}\Big(E_{11} + 6E_{12} + E_{22} - 4E_{66} - (E_{11} - 2E_{12} + E_{22} - 4E_{66})\cos(4\theta)$$
$$+ 4(-E_{16} + E_{26})\sin(4\theta)\Big),$$

$$\tilde{E}_{13} = \frac{1}{2}\Big(E_{13} + E_{23} + (E_{13} - E_{23})\cos(2\theta) + 2E_{36}\sin(2\theta)\Big),$$

$$\tilde{E}_{14} = \frac{1}{4}\Big[(3E_{14} + E_{24} - 2E_{56})\cos(\theta) + (E_{14} - E_{24} + 2E_{56})\cos(3\theta),$$
$$-2\Big(E_{15} + E_{25} - 2E_{46} + (E_{15} - E_{25} - 2E_{46})\cos(2\theta)\Big)\sin(\theta)\Big]$$

$$\tilde{E}_{15} = \frac{1}{4}\Big[(3E_{15} + E_{25} + 2E_{46})\cos(\theta) + (E_{15} - E_{25} - 2E_{46})\cos(3\theta)$$
$$+2\Big(E_{14} + E_{24} + 2E_{56} + (E_{14} - E_{24} + 2E_{56})\cos(2\theta)\Big)\sin(\theta)\Big],$$

$$\tilde{E}_{16} = \frac{1}{8}\Big[4(E_{16} + E_{26})\cos(2\theta) + 4(E_{16} - E_{26})\cos(4\theta)$$
$$-2\Big(E_{11} - E_{22} + (E_{11} - 2E_{12} + E_{22} - 4E_{66})\cos(2\theta)\Big)\sin(2\theta)\Big],$$

$$\tilde{E}_{22} = \frac{1}{8}\Big(3E_{11} + 2E_{12} + 3E_{22} + 4E_{66} + 4(-E_{11} + E_{22})\cos(2\theta)$$
$$+ (E_{11} - 2E_{12} + E_{22} - 4E_{66})\cos(4\theta)$$
$$- 8(E_{16} + E_{26})\sin(2\theta) + 4(E_{16} - E_{26})\sin(4\theta)\Big),$$

$$\tilde{E}_{23} = \frac{1}{2}\Big(E_{13} + E_{23} + (-E_{13} + E_{23})\cos(2\theta) - 2E_{36}\sin(2\theta)\Big),$$

$$\tilde{E}_{24} = \frac{1}{4}\Big((E_{14} + 3E_{24} + 2E_{56})\cos(\theta)$$
$$+ (-E_{14} + E_{24} - 2E_{56})\cos(3\theta) - (3E_{15} + E_{25} + 2E_{46})\sin(\theta)$$
$$+ (E_{15} - E_{25} - 2E_{46})\sin(3\theta)\Big),$$

$$\begin{aligned}
\tilde{E}_{25} &= \frac{1}{4}\Bigg((E_{15} + 3E_{25} - 2E_{46})\cos(\theta) \\
&\quad + (-E_{15} + E_{25} + 2E_{46})\cos(3\theta) + (3E_{14} + E_{24} - 2E_{56})\sin(\theta) \\
&\quad + (-E_{14} + E_{24} - 2E_{56})\sin(3\theta)\Bigg), \\
\tilde{E}_{26} &= \frac{1}{8}\Bigg(4(E_{16} + E_{26})\cos(2\theta) + 4(-E_{16} + E_{26})\cos(4\theta) \\
&\quad + 2(-E_{11} + E_{22})\sin(2\theta) + (E_{11} - 2E_{12} + E_{22} - 4E_{66})\sin(4\theta)\Bigg), \\
\tilde{E}_{33} &= E_{33}, \\
\tilde{E}_{34} &= E_{34}\cos(\theta) - E_{35}\sin(\theta), \\
\tilde{E}_{35} &= E_{35}\cos(\theta) + E_{34}\sin(\theta), \\
\tilde{E}_{36} &= E_{36}\cos(2\theta) + (-E_{13} + E_{23})\cos(\theta)\sin(\theta), \\
\tilde{E}_{44} &= \frac{1}{2}\Bigg(E_{44} + E_{55} + (E_{44} - E_{55})\cos(2\theta) - 2E_{45}\sin(2\theta)\Bigg), \\
\tilde{E}_{45} &= E_{45}\cos(2\theta) + (E_{44} - E_{55})\cos(\theta)\sin(\theta), \\
\tilde{E}_{46} &= \frac{1}{2}\Bigg[2\cos(2\theta)(E_{46}\cos(\theta) - E_{56}\sin(\theta)) \\
&\quad + \Bigg((-E_{14} + E_{24})\cos(\theta) + (E_{15} - E_{25})\sin(\theta)\Bigg)\sin(2\theta)\Bigg], \\
\tilde{E}_{55} &= E_{55}\cos(\theta)^2 + E_{44}\sin(\theta)^2 + E_{45}\sin(2\theta), \\
\tilde{E}_{56} &= \sin(\theta)\Bigg(E_{46}\cos(2\theta) + (-E_{14} + E_{24})\cos(\theta)\sin(\theta)\Bigg) \\
&\quad + \cos(\theta)\Bigg(E_{56}\cos(2\theta) + (-E_{15} + E_{25})\cos(\theta)\sin(\theta)\Bigg), \\
\tilde{E}_{66} &= \frac{1}{8}\Bigg(E_{11} - 2E_{12} + E_{22} + 4E_{66} - (E_{11} - 2E_{12} + E_{22} - 4E_{66})\cos(4\theta) \\
&\quad + 4(-E_{16} + E_{26})\sin(4\theta)\Bigg).
\end{aligned} \tag{13.71}$$

The Navier-Lamé (N-L) equations for an elastic anisotropic domain without body forces in cylindrical coordinates are

$$\begin{aligned}
&\Bigg(-\tilde{E}_{22}\frac{1}{r} + \tilde{E}_{15}\partial_3 + \tilde{E}_{55} r\partial_3^2 + 2\tilde{E}_{56}\partial_\theta\partial_3 + \tilde{E}_{66}\frac{1}{r}\partial_\theta^2 + \tilde{E}_{11}\partial_r \\
&\quad + 2\tilde{E}_{15} r\partial_r\partial_3 + 2\tilde{E}_{16}\partial_r\partial_\theta + \tilde{E}_{11} r\partial_r^2\Bigg) u_r
\end{aligned}$$

$$+\left(\tilde{E}_{26}\frac{1}{r}+(\tilde{E}_{14}-\tilde{E}_{24}-\tilde{E}_{56})\partial_3+\tilde{E}_{45}r\partial_3^2-(\tilde{E}_{22}+\tilde{E}_{66})\frac{1}{r}\partial_\theta\right.$$
$$+(\tilde{E}_{25}+\tilde{E}_{46})\partial_\theta\partial_3+\tilde{E}_{26}\frac{1}{r}\partial_\theta^2-\tilde{E}_{26}\partial_r$$
$$\left.+(\tilde{E}_{14}+\tilde{E}_{56})r\partial_r\partial_3+(\tilde{E}_{12}+\tilde{E}_{66})\partial_r\partial_\theta+\tilde{E}_{16}r\partial_r^2\right)u_\theta$$
$$+\left((\tilde{E}_{13}-\tilde{E}_{23})\partial_3+\tilde{E}_{35}r\partial_3^2-\tilde{E}_{24}\frac{1}{r}\partial_\theta\right. \tag{13.72}$$
$$+(\tilde{E}_{36}+\tilde{E}_{45})\partial_\theta\partial_3+\tilde{E}_{46}\frac{1}{r}\partial_\theta^2+(\tilde{E}_{15}-\tilde{E}_{25})\partial_r+(\tilde{E}_{13}+\tilde{E}_{55})r\partial_r\partial_3$$
$$\left.+(\tilde{E}_{14}+\tilde{E}_{56})\partial_r\partial_\theta+\tilde{E}_{15}r\partial_r^2\right)u_3=0,$$

$$\left(\tilde{E}_{26}\frac{1}{r}+(\tilde{E}_{24}+2\tilde{E}_{56})\partial_3+\tilde{E}_{45}r\partial_3^2+(\tilde{E}_{22}+\tilde{E}_{66})\frac{1}{r}\partial_\theta\right.$$
$$+(\tilde{E}_{25}+\tilde{E}_{46})\partial_\theta\partial_3+\tilde{E}_{26}\frac{1}{r}\partial_\theta^2+(2\tilde{E}_{16}+\tilde{E}_{26})\partial_r$$
$$\left.+(\tilde{E}_{14}+\tilde{E}_{56})r\partial_r\partial_3+(\tilde{E}_{12}+\tilde{E}_{66})\partial_r\partial_\theta+\tilde{E}_{16}r\partial_r^2\right)u_r$$
$$+\left(-\tilde{E}_{66}\frac{1}{r}+\tilde{E}_{46}\partial_3+\tilde{E}_{44}r\partial_3^2+2\tilde{E}_{24}\partial_\theta\partial_3\right.$$
$$\left.+\tilde{E}_{22}\frac{1}{r}\partial_\theta^2+\tilde{E}_{66}\partial_r+2\tilde{E}_{46}r\partial_r\partial_3+2\tilde{E}_{26}\partial_r\partial_\theta+\tilde{E}_{66}r\partial_r^2\right)u_\theta$$
$$+\left(2\tilde{E}_{36}\partial_3+\tilde{E}_{34}r\partial_3^2+\tilde{E}_{46}\frac{1}{r}\partial_\theta+(\tilde{E}_{23}+\tilde{E}_{44})\partial_\theta\partial_3+\tilde{E}_{24}\frac{1}{r}\partial_\theta^2+2\tilde{E}_{56}\partial_r\right.$$
$$\left.+(\tilde{E}_{36}+\tilde{E}_{45})r\partial_r\partial_3+(\tilde{E}_{25}+\tilde{E}_{46})\partial_r\partial_\theta+\tilde{E}_{56}r\partial_r^2\right)u_3=0. \tag{13.73}$$

$$\left((\tilde{E}_{23}+\tilde{E}_{55})\partial_3+\tilde{E}_{35}r\partial_3^2+\tilde{E}_{24}\frac{1}{r}\partial_\theta+(\tilde{E}_{36}+\tilde{E}_{45})\partial_\theta\partial_3\right.$$
$$+\tilde{E}_{46}\frac{1}{r}\partial_\theta^2+(\tilde{E}_{15}+\tilde{E}_{25})\partial_r+(\tilde{E}_{13}+\tilde{E}_{55})r\partial_r\partial_3$$
$$\left.+(\tilde{E}_{14}+\tilde{E}_{56})\partial_r\partial_\theta+\tilde{E}_{15}r\partial_r^2\right)u_r+\left((\tilde{E}_{45}-\tilde{E}_{36})\partial_3+\tilde{E}_{34}r\partial_3^2\right.$$
$$-\tilde{E}_{46}\frac{1}{r}\partial_\theta+(\tilde{E}_{23}+\tilde{E}_{44})\partial_\theta\partial_3$$
$$\left.+\tilde{E}_{24}\frac{1}{r}\partial_\theta^2+(\tilde{E}_{36}+\tilde{E}_{45})r\partial_r\partial_3+(C25+\tilde{E}_{46})\partial_r\partial_\theta+\tilde{E}_{56}r\partial_r^2\right)u_\theta$$

$$+\left(\tilde{E}_{35}\partial_3+\tilde{E}_{33}r\partial_3^2+2\tilde{E}_{34}\partial_\theta\partial_3+\tilde{E}_{44}\frac{1}{r}\partial_\theta^2+\tilde{E}_{55}\partial_r\right.$$

$$\left.+2\tilde{E}_{35}r\partial_r\partial_3+2\tilde{E}_{45}\partial_r\partial_\theta+\tilde{E}_{55}r\partial_r^2\right)u_3=0. \tag{13.74}$$

We split the N-L equations into three operators as follows:

$$\begin{aligned}\mathcal{L}(\tilde{\boldsymbol{u}}) &= [M_0(\partial_r,\partial_\theta,\theta,\tilde{E}_{ij})]\tilde{\boldsymbol{u}}+[M_1(\partial_r,\partial_\theta,\theta,\tilde{E}_{ij})]\partial_3\tilde{\boldsymbol{u}}\\ &\quad+[M_2(\partial_r,\partial_\theta,\theta,\tilde{E}_{ij})]\partial_3^2\tilde{\boldsymbol{u}}=0.\end{aligned} \tag{13.75}$$

The splitting (13.75) allows an expression of the solution $\tilde{\boldsymbol{u}}$ as the series (13.18) in which the *shadow functions* are determined by the recurrence relations (13.12) accompanied by homogeneous boundary conditions (BCs) on the two surfaces Γ_1 and Γ_2. The PDE system (13.12) results in an ODE system for the computation of $\tilde{\boldsymbol{\varphi}}_j^{(\alpha_i)}$ after the substitution of $[M_i]$ and $\tilde{\boldsymbol{\Phi}}_j^{(\alpha_i)}(r,\theta)=r^{\alpha_i+j}\tilde{\boldsymbol{\varphi}}_j^{(\alpha_i)}(\theta)$:

$$\begin{cases}[\mathcal{M}_0]\tilde{\boldsymbol{\varphi}}_0^{(\alpha_i)}=0,\\ [\mathcal{M}_0]\tilde{\boldsymbol{\varphi}}_1^{(\alpha_i)}=-[\mathcal{M}_1]\tilde{\boldsymbol{\varphi}}_0^{(\alpha_i)}, & 0<\theta<\omega,\\ [\mathcal{M}_0]\tilde{\boldsymbol{\varphi}}_{j+2}^{(\alpha_i)}=-[\mathcal{M}_1]\tilde{\boldsymbol{\varphi}}_{j+1}^{(\alpha_i)}-[\mathcal{M}_2]\tilde{\boldsymbol{\varphi}}_j^{(\alpha_i)},\quad j\geq 0,\end{cases} \tag{13.76}$$

where

$$\begin{aligned}[\mathcal{M}_0]\tilde{\boldsymbol{\varphi}}_\gamma^{(\alpha_i)} &= \left([A_1]\partial_\theta^2+((\alpha_i+\gamma)[A_2]+[A_3])\,\partial_\theta\right.\\ &\quad\left.+\left((\alpha_i+\gamma)^2[A_4]+(\alpha_i+\gamma)[A_5]+[A_6]\right)\right)\tilde{\boldsymbol{\varphi}}_\gamma^{(\alpha_i)},\\ [\mathcal{M}_1]\tilde{\boldsymbol{\varphi}}_\gamma^{(\alpha_i)} &= ([A_7]\partial_\theta+((\alpha_i+\gamma))[A_8]+[A_9]))\,\tilde{\boldsymbol{\varphi}}_\gamma^{(\alpha_i)},\\ [\mathcal{M}_2]\tilde{\boldsymbol{\varphi}}_\gamma^{(\alpha_i)} &= [A_{10}]\tilde{\boldsymbol{\varphi}}_\gamma^{(\alpha_i)},\end{aligned} \tag{13.77}$$

and

$$[A_1]=\begin{pmatrix}\tilde{E}_{66} & \tilde{E}_{26} & \tilde{E}_{46}\\ \tilde{E}_{26} & \tilde{E}_{22} & \tilde{E}_{24}\\ \tilde{E}_{46} & \tilde{E}_{24} & \tilde{E}_{44}\end{pmatrix},\quad [A_2]=\begin{pmatrix}2\tilde{E}_{16} & \tilde{E}_{12}+\tilde{E}_{66} & \tilde{E}_{14}+\tilde{E}_{56}\\ \tilde{E}_{12}+\tilde{E}_{66} & 2\tilde{E}_{26} & \tilde{E}_{25}+\tilde{E}_{46}\\ \tilde{E}_{14}+\tilde{E}_{56} & \tilde{E}_{25}+\tilde{E}_{46} & 2\tilde{E}_{45}\end{pmatrix},$$

$$[A_3]=\begin{pmatrix}0 & -\tilde{E}_{22}-\tilde{E}_{66} & -\tilde{E}_{24}\\ \tilde{E}_{22}+\tilde{E}_{66} & 0 & \tilde{E}_{46}\\ \tilde{E}_{24} & -\tilde{E}_{46} & 0\end{pmatrix},\quad [A_4]=\begin{pmatrix}\tilde{E}_{11} & \tilde{E}_{16} & \tilde{E}_{15}\\ \tilde{E}_{16} & \tilde{E}_{66} & \tilde{E}_{56}\\ \tilde{E}_{15} & \tilde{E}_{56} & \tilde{E}_{55}\end{pmatrix},$$

$$[A_5]=\begin{pmatrix}0 & -\tilde{E}_{16}-\tilde{E}_{26} & -\tilde{E}_{25}\\ \tilde{E}_{16}+\tilde{E}_{26} & 0 & \tilde{E}_{56}\\ \tilde{E}_{25} & -\tilde{E}_{56} & 0\end{pmatrix},\quad [A_6]=\begin{pmatrix}-\tilde{E}_{22} & \tilde{E}_{26} & 0\\ \tilde{E}_{26} & -\tilde{E}_{66} & 0\\ 0 & 0 & 0\end{pmatrix},$$

$$[A_7] = \begin{pmatrix} 2\tilde{E}_{56} & (\tilde{E}_{25} + \tilde{E}_{46}) & (\tilde{E}_{36} + \tilde{E}_{45}) \\ (\tilde{E}_{25} + \tilde{E}_{46}) & 2\tilde{E}_{24} & (\tilde{E}_{23} + \tilde{E}_{44}) \\ (\tilde{E}_{36} + \tilde{E}_{45}) & (\tilde{E}_{23} + \tilde{E}_{44}) & 2\tilde{E}_{34} \end{pmatrix},$$

$$[A_8] = \begin{pmatrix} 2\tilde{E}_{15} & (\tilde{E}_{14} + \tilde{E}_{56}) & (\tilde{E}_{13} + \tilde{E}_{55}) \\ (\tilde{E}_{14} + \tilde{E}_{56}) & 2\tilde{E}_{46} & (\tilde{E}_{36} + \tilde{E}_{45}) \\ (\tilde{E}_{13} + \tilde{E}_{55}) & (\tilde{E}_{36} + \tilde{E}_{45}) & 2\tilde{E}_{35} \end{pmatrix},$$

$$[A_9] = \begin{pmatrix} \tilde{E}_{15} & (\tilde{E}_{14} - \tilde{E}_{24} - \tilde{E}_{56}) & (\tilde{E}_{13} - \tilde{E}_{23}) \\ (\tilde{E}_{24} + 2\tilde{E}_{56}) & \tilde{E}_{46} & 2\tilde{E}_{36} \\ (\tilde{E}_{23} + \tilde{E}_{55}) & -(\tilde{E}_{36} - \tilde{E}_{45}) & \tilde{E}_{35} \end{pmatrix},$$

$$[A_{10}] = \begin{pmatrix} \tilde{E}_{55} & \tilde{E}_{45} & \tilde{E}_{35} \\ \tilde{E}_{45} & \tilde{E}_{44} & \tilde{E}_{34} \\ \tilde{E}_{35} & \tilde{E}_{34} & \tilde{E}_{33} \end{pmatrix}. \tag{13.78}$$

Notice that $\gamma = 0, 1, 2, \ldots$ correspond to $\tilde{\boldsymbol{\varphi}}_0$, $\tilde{\boldsymbol{\varphi}}_1$, $\tilde{\boldsymbol{\varphi}}_2$, ...

For any eigenvalue α_i also $-\bar{\alpha}_i$ is an eigenvalue with an associated *dual function* that is the solution of $(13.76)_1$ [46]:

$$\tilde{\boldsymbol{\Psi}}_0^{(\alpha_i)} = c_0^{(\alpha_i)} r^{-\bar{\alpha}_i} \tilde{\boldsymbol{\psi}}_0^{(\alpha_i)}(\theta), \tag{13.79}$$

where $c_0^{(\alpha_i)}$ is a real coefficient chosen for normalization purposes (to be discussed in the following). The *shadow dual functions* are obtained from $(13.76)_{2,3}$:

$$\tilde{\boldsymbol{\Psi}}_j^{(\alpha_i)} = r^{-\bar{\alpha}_i + j} \tilde{\boldsymbol{\psi}}_j^{(\alpha_i)}(\theta), \quad j = 1, 2, \ldots \tag{13.80}$$

The ODE system (13.76) is complemented by either homogeneous Dirichlet boundary conditions (clamped-BCs) on Γ_1 and Γ_2 ,

$$\tilde{\boldsymbol{\varphi}}_j^{(\alpha_i)}(\theta = 0, \omega) = \mathbf{0}, \qquad j = 0, 1, \ldots \tag{13.81}$$

or traction free BCs,

$$\begin{cases} [\mathcal{T}_0]\tilde{\boldsymbol{\varphi}}_0 = 0, \\ [\mathcal{T}_0]\tilde{\boldsymbol{\varphi}}_{j+1} + [\mathcal{T}_1]\tilde{\boldsymbol{\varphi}}_j = 0,, \quad j \geq 0, \end{cases} \qquad for\ \theta = 0, \omega, \tag{13.82}$$

where

$$[\mathcal{T}_0]\tilde{\boldsymbol{\varphi}}_\gamma^{(\alpha_i)} = ([B_1]\partial_\theta + ((\alpha_i + \gamma)[B_2] + [B_3]))\, \tilde{\boldsymbol{\varphi}}_\gamma^{(\alpha_i)},$$
$$[\mathcal{T}_1]\tilde{\boldsymbol{\varphi}}_\gamma^{(\alpha_i)} = [B_4]\tilde{\boldsymbol{\varphi}}_\gamma^{(\alpha_i)}, \tag{13.83}$$

and

$$[B_1] = [A_1], \quad [B_2] = \begin{pmatrix} \tilde{E}_{16} & \tilde{E}_{66} & \tilde{E}_{56} \\ \tilde{E}_{12} & \tilde{E}_{26} & \tilde{E}_{25} \\ \tilde{E}_{14} & \tilde{E}_{46} & \tilde{E}_{45} \end{pmatrix}, \quad [B_3] = \begin{pmatrix} \tilde{E}_{26} & \tilde{E}_{66} & 0 \\ \tilde{E}_{22} & -\tilde{E}_{26} & 0 \\ \tilde{E}_{24} & -\tilde{E}_{46} & 0 \end{pmatrix},$$

$$[B_4] = \begin{pmatrix} \tilde{E}_{56} & \tilde{E}_{46} & \tilde{E}_{36} \\ \tilde{E}_{25} & \tilde{E}_{24} & \tilde{E}_{23} \\ \tilde{E}_{45} & \tilde{E}_{44} & \tilde{E}_{34} \end{pmatrix}. \tag{13.84}$$

Remark 13.4. The matrices $[A_i]$ and $[B_i]$ are $\tilde{E}$ dependent. If material properties are given in a Cartesian coordinate system, $\tilde{E}$ is represented in terms of $[E]$, and each of the matrices is a combination of nine independent matrices, multiplying trigonometric functions: $[A_i] + [A_i]_{c1}\cos(\theta) + [A_i]_{c2}\cos(2\theta) + [A_i]_{c3}\cos(3\theta) + [A_i]_{c4}\cos(4\theta) + [A_i]_{s1}\sin(\theta) + [A_i]_{s2}\sin(2\theta) + [A_i]_{s3}\sin(3\theta) + [A_i]_{s4}\sin(4\theta)$. Although in the practical computational scheme we use the decomposition above, here we condense our notation by using $[\tilde{E}]$.

13.3.1 Computing Eigenpairs

In this subsection we introduce a new method (in this book) for the computation of eigenpairs first introduced by Leguillon and Sanchez-Palencia in [109], resulting in a quadratic weak eigenproblem. Any eigenvalue α and primal eigenfunctions $r^\alpha \tilde{\boldsymbol{\varphi}}_0^{(\alpha)}$ (and dual eigenfunctions $r^{-\alpha}\tilde{\boldsymbol{\psi}}_0^{(\alpha)}$) are the solution of $(13.76)_1$ and $(13.77)_1$ with $\gamma = 0$, resulting in a quadratic eigenproblem

$$[A_1]\tilde{\boldsymbol{\varphi}}_0'' + (\alpha[A_2] + [A_3])\,\tilde{\boldsymbol{\varphi}}_0' + \left(\alpha^2[A_4] + \alpha[A_5] + [A_6]\right)\tilde{\boldsymbol{\varphi}}_0 = 0, \qquad \theta \in (0,\omega). \tag{13.85}$$

The above equation is augmented by either homogeneous Dirichlet BCs, or traction free BCs according to $(13.82)_1$:

$$\begin{cases} \{[B_1]\tilde{\boldsymbol{\varphi}}_0' + (\alpha[B_2] + [B_3])\,\tilde{\boldsymbol{\varphi}}_0\}_{\theta=0,\omega} = 0, & \text{traction-free}, \\ \{\tilde{\boldsymbol{\varphi}}_0\}\,|_{\theta=0,\omega} = 0, & \text{homogeneous Dirichlet}. \end{cases} \tag{13.86}$$

Since the eigenpairs may be complex, we formulate the *sesquilinear* form corresponding to (13.85) on the element level, followed by an assembly procedure. Multiplying (13.85) by a test function $\bar{\boldsymbol{v}}^{e_i}$ then integrating over the 1-D element (from ω_{i-1} to ω_i) and integrating by parts the second derivative term ($\tilde{\boldsymbol{\varphi}}_0^{e_i} = \tilde{\boldsymbol{\varphi}}_0(\omega_{i-1} \leq \theta \leq \omega_i)$), one obtains

$$\left\{\{[A_1]\tilde{\boldsymbol{\varphi}}_0^{e_i\,\prime}\}^T \bar{\boldsymbol{v}}^{e_i}\right\}\Big|_{\omega_{i-1}}^{\omega_i} - \int_{\omega_{i-1}}^{\omega_i} \{[A_1]\tilde{\boldsymbol{\varphi}}_0^{e_i\,\prime}\}^T \bar{\boldsymbol{v}}^{e_i\,\prime} d\theta \tag{13.87}$$
$$+ \int_{\omega_{i-1}}^{\omega_i} \{(\alpha[A_2] + [A_3])\, \tilde{\boldsymbol{\varphi}}_0^{e_i\,\prime}\}^T \bar{\boldsymbol{v}}^{e_i} d\theta$$
$$+ \int_{\omega_{i-1}}^{\omega_i} \{(\alpha^2[A_4] + \alpha[A_5] + [A_6])\, \tilde{\boldsymbol{\varphi}}_0^{e_i}\}^T \bar{\boldsymbol{v}}^{e_i} d\theta = 0.$$

After enforcing traction-free BCs (13.86)

$$\left\{\{[A_1]\tilde{\boldsymbol{\varphi}}_0^{e_i\,\prime}\}^T \bar{\boldsymbol{v}}^{e_i}\right\}\Big|_{\omega_{i-1}}^{\omega_i} = \left\{\{[B_1]\tilde{\boldsymbol{\varphi}}_0^{e_i\,\prime}\}^T \bar{\boldsymbol{v}}^{e_i}\right\}\Big|_{\omega_{i-1}}^{\omega_i}$$
$$= -\left\{\{(\alpha[B_2] + [B_3])\, \tilde{\boldsymbol{\varphi}}_0^{e_i}\}^T \bar{\boldsymbol{v}}^{e_i}\right\}\Big|_{\omega_{i-1}}^{\omega_i}, \tag{13.88}$$

we define the elemental sesquilinear forms

$$\mathcal{B}_0^0(\tilde{\boldsymbol{\varphi}}_0^{e_i}, \bar{\boldsymbol{v}}^{e_i}) = -\int_{\omega_{i-1}}^{\omega_i} \{[A_1]\tilde{\boldsymbol{\varphi}}_0^{e_i\,\prime}\}^T \bar{\boldsymbol{v}}^{e_i\,\prime} d\theta + \int_{\omega_{i-1}}^{\omega_i} \{[A_3]\tilde{\boldsymbol{\varphi}}_0^{e_i\,\prime}\}^T \bar{\boldsymbol{v}}^{e_i} d\theta$$
$$+ \int_{\omega_{i-1}}^{\omega_i} \{[A_6]\tilde{\boldsymbol{\varphi}}_0^{e_i}\}^T \bar{\boldsymbol{v}}^{e_i} d\theta - \left\{\{[B_3]\tilde{\boldsymbol{\varphi}}_0^{e_i}\}^T \bar{\boldsymbol{v}}^{e_i}\right\}\Big|_{\omega_{i-1}}^{\omega_i},$$
$$\mathcal{B}_1^0(\tilde{\boldsymbol{\varphi}}_0^{e_i}, \bar{\boldsymbol{v}}^{e_i}) = \int_{\omega_{i-1}}^{\omega_i} \{[A_2]\tilde{\boldsymbol{\varphi}}_0^{e_i\,\prime}\}^T \bar{\boldsymbol{v}}^{e_i} d\theta + \int_{\omega_{i-1}}^{\omega_i} \{[A_5]\tilde{\boldsymbol{\varphi}}_0^{e_i}\}^T \bar{\boldsymbol{v}}^{e_i} d\theta$$
$$-\left\{\{[B_2]\tilde{\boldsymbol{\varphi}}_0^{e_i}\}^T \bar{\boldsymbol{v}}^{e_i}\right\}\Big|_{\omega_{i-1}}^{\omega_i},$$
$$\mathcal{B}_2^0(\tilde{\boldsymbol{\varphi}}_0^{e_i}, \bar{\boldsymbol{v}}^{e_i}) = \int_{\omega_{i-1}}^{\omega_i} \{[A_4]\tilde{\boldsymbol{\varphi}}_0^{e_i}\}^T \bar{\boldsymbol{v}}^{e_i} d\theta. \tag{13.89}$$

Finally, assembling all elements, the quadratic sesquilinear eigenform for the evaluation of the primal and dual eigenpairs is obtained:

$$\text{Seek} \quad \alpha \in \mathbb{C}, \quad 0 \neq \tilde{\boldsymbol{\varphi}}_0 \in \mathcal{E}(0,\omega), \quad \text{s.t.} \quad \forall \boldsymbol{v} \in \mathcal{E}(0,\omega),$$
$$\mathcal{B}_0^0(\tilde{\boldsymbol{\varphi}}_0, \bar{\boldsymbol{v}}) + \alpha\mathcal{B}_1^0(\tilde{\boldsymbol{\varphi}}_0, \bar{\boldsymbol{v}}) + \alpha^2\mathcal{B}_2^0(\tilde{\boldsymbol{\varphi}}_0, \bar{\boldsymbol{v}}) = 0, \tag{13.90}$$

where $\mathcal{B}_0^0(\tilde{\boldsymbol{\varphi}}_0, \bar{\boldsymbol{v}})$, $\mathcal{B}_1^0(\tilde{\boldsymbol{\varphi}}_0, \bar{\boldsymbol{v}})$, and $\mathcal{B}_2^0(\tilde{\boldsymbol{\varphi}}_0, \bar{\boldsymbol{v}})$ are the assembled sesquilinear forms resulting from $\mathcal{B}_0^0(\tilde{\boldsymbol{\varphi}}_0^{e_i}, \boldsymbol{v}^{e_i})$, $\mathcal{B}_1^0(\tilde{\boldsymbol{\varphi}}_0^{e_i}, \boldsymbol{v}^{e_i})$, and $\mathcal{B}_2^0(\tilde{\boldsymbol{\varphi}}_0^{e_i}, \boldsymbol{v}^{e_i})$ respectively. In the assembly procedure continuity of the displacement is enforced which insures continuity of traction between two materials in case of multimaterial interfaces. The two surfaces $\theta = 0, \omega$ are therefore under a traction-free condition. In the case of Dirichlet BCs, the $\mathcal{E}$ space is replaced by $\mathcal{E}_o$.

13.3.1.1 p-FEMs for the Solution of the Weak Eigenformulation

The eigenfunctions are smooth functions, and thus the application of p-FEMs for the solution of (13.90) should result in exponential convergence rates. To this end, $\tilde{\varphi}_0 = \left(u_{r0}\ u_{\theta 0}\ u_{30}\right)^T$ is expressed in terms of the basis functions $\Phi_k(\xi)$ (integrals of Legendre polynomials) in the standard element:

$$u_{r0}^{e_i}(\xi) = \sum_{k=1}^{p+1} a_k \Phi_k(\xi), \quad u_{\theta 0}^{e_i}(\xi) = \sum_{k=1}^{p+1} a_{p+1+k} \Phi_k(\xi),$$
$$u_{30}^{e_i}(\xi) = \sum_{k=1}^{p+1} a_{2p+2+k} \Phi_k(\xi), \tag{13.91}$$

or

$$\tilde{\varphi}_0^{e_i} = \begin{pmatrix} \Phi_1(\xi)\cdots\Phi_{p+1}(\xi) & 0\cdots 0 & 0\cdots 0 \\ 0\cdots 0 & \Phi_1(\xi)\cdots\Phi_{p+1}(\xi) & 0\cdots 0 \\ 0\cdots 0 & 0\cdots 0 & \Phi_1(\xi)\cdots\Phi_{p+1}(\xi) \end{pmatrix} \begin{pmatrix} a_1 \\ \vdots \\ a_{3(p+1)} \end{pmatrix}$$
$$\stackrel{\text{def}}{=} [\Phi]\boldsymbol{a}_0^{e_i}.$$

Similarly, $\tilde{\boldsymbol{v}}^{e_i} \stackrel{\text{def}}{=} [\Phi]\boldsymbol{b}_0^{e_i}$ and $d\theta = \frac{\omega}{2} d\xi$. Substituting (13.92) in (13.90), one obtains the FE formulation of the weak eigenform:

$$\boldsymbol{a}_0^T \left(\alpha^2 [K_2^0] + \alpha [K_1^0] + [K_0^0]\right) = \boldsymbol{0}, \tag{13.92}$$

where $[K_0^0]$, $[K_1^0]$, $[K_2^0]$ are the assembled matrices corresponding to $[K_0^0]^{e_i}$, $[K_1^0]^{e_i}$, $[K_2^0]^{e_i}$ respectively and $\omega^{e_i} = \omega_i - \omega_{i-1}$:

$$[K_2^0]^{e_i} = \frac{\omega^{e_i}}{2} \int_{-1}^{1} [\Phi]^T [A_4]^T [\Phi] d\xi,$$
$$[K_1^0]^{e_i} = -\left\{[\Phi]^T [B_2]^T [\Phi]\right\} \Big|_{\omega_{i-1}}^{\omega_i}$$
$$+ \int_{-1}^{1} [\Phi']^T [A_2]^T [\Phi] d\xi + \frac{\omega^{e_i}}{2} \int_{-1}^{1} [\Phi]^T [A_5]^T [\Phi] d\xi,$$

$$[K_0^0]^{e_i} = -\left\{[\Phi]^T[B_3]^T[\Phi]\right\}\Big|_{\omega_{i-1}}^{\omega_i} - \frac{2}{\omega^{e_i}}\int_{-1}^{1}[\Phi']^T[A_1]^T[\Phi']d\xi$$
$$+\int_{-1}^{1}[\Phi']^T[A_3]^T[\Phi]d\xi + \frac{\omega^{e_i}}{2}\int_{-1}^{1}[\Phi]^T[A_6]^T[\Phi]d\xi. \quad (13.93)$$

Here $\boldsymbol{a}_0$ is the vector of assembled coefficients of $\boldsymbol{a}_0^{e_i}$. For clamped BCs, $[B_2]|_0 = [B_2]|_\omega = [B_3]|_0 = [B_3]|_\omega = [0]$. The quadratic eigenproblem (13.92) is solved by a linearization process according to [7]. Setting $\boldsymbol{d}_0 = \alpha\boldsymbol{a}_0$, the $(3pQ+3)\times(3pQ+3)$ (Q is the number of elements) quadratic eigenproblem is transformed into a linear $(6pQ+6)\times(6pQ+6)$ "standard eigenproblem":

$$\begin{pmatrix}\boldsymbol{a}_0\\ \boldsymbol{d}_0\end{pmatrix}^T\begin{pmatrix}0 & [K_0^0]\\ I & [K_1^0]\end{pmatrix} = \alpha\begin{pmatrix}\boldsymbol{a}_0\\ \boldsymbol{d}_0\end{pmatrix}^T\begin{pmatrix}I & 0\\ 0 & -[K_2^0]\end{pmatrix}. \quad (13.94)$$

Because the eigenpairs may be complex, the complex α and $\boldsymbol{a}_0$ are denoted by $\alpha = \alpha_{\Re} + i\alpha_{\Im}$ and $\boldsymbol{a}_0 = \boldsymbol{a}_{0\Re} + i\boldsymbol{a}_{0\Im}$.

The normalization factor c_0. The dual eigenfunctions $\tilde{\boldsymbol{\psi}}_0$ are the solutions of (13.94) associated with negative eigenvalues. The normalization factor $c_0^{(\alpha)}$ is determined so that the primal and dual eigenfunctions satisfy an orthonormal condition (10.64) under integration along a circular curve with the edge at its center:

$$\int_0^\omega\left\{[T]\tilde{\boldsymbol{\Phi}}_0^{(\alpha)}\cdot\bar{\tilde{\boldsymbol{\Psi}}}_0^{(\alpha)} - \tilde{\boldsymbol{\Phi}}_0^{(\alpha)}\cdot[T]\bar{\tilde{\boldsymbol{\Psi}}}_0^{(\alpha)}\right\}Rd\theta = 1, \quad (13.95)$$

where $[T]$ is the Neumann trace operator (related to $\mathcal{L}$) on a circular surface around the edge:

$$[T]\tilde{\boldsymbol{u}} \stackrel{\text{def}}{=} \begin{pmatrix}\sigma_{rr}\\ \sigma_{r\theta}\\ \sigma_{r3}\end{pmatrix}$$
$$= \begin{pmatrix}
\frac{1}{r}\tilde{E}_{12}+\tilde{E}_{15}\partial_3+\frac{1}{r}\tilde{E}_{16}\partial_\theta+\tilde{E}_{11}\partial_r & & \\
-\frac{1}{r}\tilde{E}_{16}+\tilde{E}_{14}\partial_3+\frac{1}{r}\tilde{E}_{12}\partial_\theta+\tilde{E}_{16}\partial_r & \tilde{E}_{13}\partial_3+\frac{1}{r}\tilde{E}_{14}\partial_\theta+\tilde{E}_{15}\partial_r & \\
\frac{1}{r}\tilde{E}_{26}+\tilde{E}_{56}\partial_3+\frac{1}{r}\tilde{E}_{66}\partial_\theta+\tilde{E}_{16}\partial_r & & \\
-\frac{1}{r}\tilde{E}_{66}+\tilde{E}_{46}\partial_3+\frac{1}{r}\tilde{E}_{26}\partial_\theta+\tilde{E}_{66}\partial_r & \tilde{E}_{36}\partial_3+\frac{1}{r}\tilde{E}_{46}\partial_\theta+\tilde{E}_{56}\partial_r & \\
\frac{1}{r}\tilde{E}_{25}+\tilde{E}_{55}\partial_3+\frac{1}{r}\tilde{E}_{56}\partial_\theta+\tilde{E}_{15}\partial_r & & \\
-\frac{1}{r}\tilde{E}_{56}+\tilde{E}_{45}\partial_3+\frac{1}{r}\tilde{E}_{25}\partial_\theta+\tilde{E}_{56}\partial_r & \tilde{E}_{35}\partial_3+\frac{1}{r}\tilde{E}_{45}\partial_\theta+\tilde{E}_{55}\partial_r &
\end{pmatrix}\begin{pmatrix}u_r\\ u_\theta\\ u_3\end{pmatrix}. \quad (13.96)$$

The operator $[T]$ is split according to $[T] = [T_0(\partial_r,\partial_\theta)] + [T_1(\partial_r,\partial_\theta)]\partial_3$ with

$$[T_0(\partial_r,\partial_\theta)]\tilde{\boldsymbol{\Phi}}_j = \left([T_a]\frac{1}{r}\partial_\theta + [T_b]\partial_r + ([T_c]+j[T_b])\frac{1}{r}\right)\tilde{\boldsymbol{\Phi}}_j,$$
$$\left[T_1(\partial_r,\partial_\theta)\right]\tilde{\boldsymbol{\Phi}}_j = \left[T_d\right]\tilde{\boldsymbol{\Phi}}_j, \tag{13.97}$$

$$[T_a] = \begin{pmatrix} \tilde{E}_{16} & \tilde{E}_{12} & \tilde{E}_{14} \\ \tilde{E}_{66} & \tilde{E}_{26} & \tilde{E}_{46} \\ \tilde{E}_{56} & \tilde{E}_{25} & \tilde{E}_{45} \end{pmatrix}, \quad [T_b] = \begin{pmatrix} \tilde{E}_{11} & \tilde{E}_{16} & \tilde{E}_{15} \\ \tilde{E}_{16} & \tilde{E}_{66} & \tilde{E}_{56} \\ \tilde{E}_{15} & \tilde{E}_{56} & \tilde{E}_{55} \end{pmatrix},$$

$$[T_c] = \begin{pmatrix} \tilde{E}_{12} & -\tilde{E}_{16} & 0 \\ \tilde{E}_{26} & -\tilde{E}_{66} & 0 \\ \tilde{E}_{25} & -\tilde{E}_{56} & 0 \end{pmatrix}, \quad [T_d] = \begin{pmatrix} \tilde{E}_{15} & \tilde{E}_{14} & \tilde{E}_{13} \\ \tilde{E}_{56} & \tilde{E}_{46} & \tilde{E}_{36} \\ \tilde{E}_{55} & \tilde{E}_{45} & \tilde{E}_{35} \end{pmatrix}. \tag{13.98}$$

Because the eigenpairs and their duals are independent of x_3, one obtains

$$[T]\tilde{\boldsymbol{\Phi}}_0^{(\alpha)} = [T_0]\tilde{\boldsymbol{\Phi}}_0^{(\alpha)} = R^{\alpha-1}\left\{[T_a]\tilde{\boldsymbol{\varphi}}_0' + \alpha[T_b]\tilde{\boldsymbol{\varphi}}_0 + [T_c]\tilde{\boldsymbol{\varphi}}_0\right\},$$
$$[T]\bar{\tilde{\boldsymbol{\Psi}}}_0^{(\alpha)} = [T_0]\bar{\tilde{\boldsymbol{\Psi}}}_0^{(\alpha)} = R^{-\alpha-1}\left\{[T_a]\bar{\tilde{\boldsymbol{\psi}}}_0' - \alpha[T_b]\bar{\tilde{\boldsymbol{\psi}}}_0 + [T_c]\bar{\tilde{\boldsymbol{\psi}}}_0\right\}. \tag{13.99}$$

Inserting (13.79) and (13.99) in (13.95), the expression for the normalization factor c_0 is obtained:

$$c_0^{(\alpha)}\left(\int_0^\omega \left\{\left([T_a]\tilde{\boldsymbol{\varphi}}_0' + \alpha[T_b]\tilde{\boldsymbol{\varphi}}_0 + [T_c]\tilde{\boldsymbol{\varphi}}_0\right)\cdot\bar{\tilde{\boldsymbol{\psi}}} \right.\right.$$
$$\left.\left. -\tilde{\boldsymbol{\varphi}}_0\cdot\left([T_a]\bar{\tilde{\boldsymbol{\psi}}}_0' - \alpha[T_b]\bar{\tilde{\boldsymbol{\psi}}}_0 + [T_c]\bar{\tilde{\boldsymbol{\psi}}}\right)\right\} d\theta\right) = 1 \tag{13.100}$$

Substituting $c_0 = c_{0_\Re} + i c_{0_\Im}$, $\tilde{\boldsymbol{\varphi}}_0 = \tilde{\boldsymbol{\varphi}}_{0_\Re} + i\tilde{\boldsymbol{\varphi}}_{0_\Im}$, and $\bar{\tilde{\boldsymbol{\psi}}}_0 = \tilde{\boldsymbol{\psi}}_{0_\Re} - i\tilde{\boldsymbol{\psi}}_{0_\Im}$ into (13.100), the following system is obtained:

$$\begin{cases} c_{0_\Re}^{(\alpha)} I_{c0}^{\Re} - c_{0_\Im}^{(\alpha)} I_{c0}^{\Im} = 1, \\ c_{0_\Re}^{(\alpha)} I_{c0}^{\Im} + c_{0_\Im}^{(\alpha)} I_{c0}^{\Re} = 0, \end{cases} \Rightarrow \begin{cases} c_{0_\Re}^{(\alpha)} = \frac{I_{c0}^{\Re}}{I_{c0}^{\Re 2} + I_{c0}^{\Im 2}}, \\ c_{0_\Im}^{(\alpha)} = \frac{-I_{c0}^{\Im}}{I_{c0}^{\Re 2} + I_{c0}^{\Im 2}}, \end{cases} \tag{13.101}$$

where

$$I_{c0}^{\Re} = \int_0^\omega \left([T_a]\tilde{\boldsymbol{\varphi}}_{0_\Re}' + \alpha_\Re[T_b]\tilde{\boldsymbol{\varphi}}_{0_\Re} - \alpha_\Im[T_b]\tilde{\boldsymbol{\varphi}}_{0_\Im} + [T_c]\tilde{\boldsymbol{\varphi}}_{0_\Re}\right)\cdot\tilde{\boldsymbol{\psi}}_{0_\Re}\, d\theta$$
$$+\int_0^\omega \left([T_a]\tilde{\boldsymbol{\varphi}}_{0_\Im}' + \alpha_\Re[T_b]\tilde{\boldsymbol{\varphi}}_{0_\Im} + \alpha_\Im[T_b]\tilde{\boldsymbol{\varphi}}_{0_\Re} + [T_c]\tilde{\boldsymbol{\varphi}}_{0_\Im}\right)\cdot\tilde{\boldsymbol{\psi}}_{0_\Im}\, d\theta$$
$$+\int_0^\omega \tilde{\boldsymbol{\varphi}}_{0_\Re}\cdot\left([T_a]\tilde{\boldsymbol{\psi}}_{0_\Re}' - \alpha_\Re[T_b]\tilde{\boldsymbol{\psi}}_{0_\Re} + \alpha_\Im[T_b]\tilde{\boldsymbol{\psi}}_{0_\Im} + [T_c]\tilde{\boldsymbol{\psi}}_{0_\Re}\right) d\theta$$
$$+\int_0^\omega \tilde{\boldsymbol{\varphi}}_{0_\Im}\cdot\left([T_a]\tilde{\boldsymbol{\psi}}_{0_\Im}' - \alpha_\Re[T_b]\tilde{\boldsymbol{\psi}}_{0_\Im} - \alpha_\Im[T_b]\tilde{\boldsymbol{\psi}}_{0_\Re} + [T_c]\tilde{\boldsymbol{\psi}}_{0_\Im}\right) d\theta,$$

$$
\begin{aligned}
I_{c0}^{\Im} = &- \int_0^\omega \left([T_a]\tilde{\boldsymbol{\varphi}}'_{0\Re} + \alpha_{\Re}[T_b]\tilde{\boldsymbol{\varphi}}_{0\Re} - \alpha_{\Im}[T_b]\tilde{\boldsymbol{\varphi}}_{0\Im} + [T_c]\tilde{\boldsymbol{\varphi}}_{0\Re}\right) \cdot \tilde{\boldsymbol{\psi}}_{0\Im}\, d\theta \\
&+ \int_0^\omega \left([T_a]\tilde{\boldsymbol{\varphi}}'_{0\Im} + \alpha_{\Re}[T_b]\tilde{\boldsymbol{\varphi}}_{0\Im} + \alpha_{\Im}[T_b]\tilde{\boldsymbol{\varphi}}_{0\Re} + [T_c]\tilde{\boldsymbol{\varphi}}_{0\Im}\right) \cdot \tilde{\boldsymbol{\psi}}_{0\Re}\, d\theta \\
&+ \int_0^\omega \tilde{\boldsymbol{\varphi}}_{0\Im} \cdot \left([T_a]\tilde{\boldsymbol{\psi}}'_{0\Re} - \alpha_{\Re}[T_b]\tilde{\boldsymbol{\psi}}_{0\Re} + \alpha_{\Im}[T_b]\tilde{\boldsymbol{\psi}}_{0\Im} + [T_c]\tilde{\boldsymbol{\psi}}_{0\Re}\right) d\theta \\
&- \int_0^\omega \tilde{\boldsymbol{\varphi}}_{0\Re} \cdot \left([T_a]\tilde{\boldsymbol{\psi}}'_{0\Im} - \alpha_{\Re}[T_b]\tilde{\boldsymbol{\psi}}_{0\Im} - \alpha_{\Im}[T_b]\tilde{\boldsymbol{\psi}}_{0\Re} + [T_c]\tilde{\boldsymbol{\psi}}_{0\Im}\right) d\theta.
\end{aligned}
\tag{13.102}
$$

13.3.2 Computing Complex Primal and Dual Shadow Functions

13.3.2.1 The Weak Form for the Computation of Primal and Dual Shadow Functions

The primal and dual shadow functions $\tilde{\boldsymbol{\varphi}}_\gamma$ and $\tilde{\boldsymbol{\psi}}_\gamma$ are the solutions of system (13.76) ($\tilde{\boldsymbol{\varphi}}_1$, $\tilde{\boldsymbol{\psi}}_1$ are the solutions of $(13.76)_2$ whereas $\tilde{\boldsymbol{\varphi}}_\gamma$, $\tilde{\boldsymbol{\psi}}_\gamma$, $\gamma \geq 2$, are the solutions of $(13.76)_3$). Here $\tilde{\boldsymbol{\psi}}_\gamma$ is computed by replacing $(\tilde{\boldsymbol{\varphi}}^{(\alpha)}_{\gamma\Re} + i\tilde{\boldsymbol{\varphi}}^{(\alpha)}_{\gamma\Im})$ and $(\alpha_{\Re} + i\alpha_{\Im})$ by $(\tilde{\boldsymbol{\psi}}^{(\alpha)}_{\gamma\Re} + i\tilde{\boldsymbol{\psi}}^{(\alpha)}_{\gamma\Im})$ and $(-\alpha_{\Re} + i\alpha_{\Im})$ in the relevant equation of system (13.76). Notice that $\alpha_{\Re} + i\alpha_{\Im}$ is known, obtained by solving the eigenvalue problem in the previous subsection. The weak formulation for $\tilde{\boldsymbol{\varphi}}^{e_i}_\gamma$, on the element level, is obtained by multiplying the appropriate equation in (13.76) by a test function $\bar{\boldsymbol{v}}^{e_i}$ and integrating over ω^{e_i}. Applying integration by parts to the second-derivative term, one obtains

$$
\begin{aligned}
&\left\{\left([A_1]\tilde{\boldsymbol{\varphi}}^{e_i\,\prime}_\gamma\right)^T \bar{\boldsymbol{v}}^{e_i}\right\} \Big|_{\omega_{i-1}}^{\omega_i} - \int_{\omega_{i-1}}^{\omega_i} \left([A_1]\tilde{\boldsymbol{\varphi}}^{e_i\,\prime}_\gamma\right)^T \bar{\boldsymbol{v}}^{e_i\,\prime} d\theta \\
&+ \int_{\omega_{i-1}}^{\omega_i} \left[((\alpha+\gamma)[A_2] + [A_3])\,\tilde{\boldsymbol{\varphi}}^{e_i\,\prime}_\gamma\right]^T \bar{\boldsymbol{v}}^{e_i}\, d\theta \\
&+ \int_{\omega_{i-1}}^{\omega_i} \left[((\alpha+\gamma)^2[A_4] + (\alpha+\gamma)[A_5] + [A_6])\,\tilde{\boldsymbol{\varphi}}^{e_i}_\gamma\right]^T \bar{\boldsymbol{v}}^{e_i}\, d\theta \\
&+ \int_{\omega_{i-1}}^{\omega_i} \left([A_7]\tilde{\boldsymbol{\varphi}}^{e_i\,\prime}_{\gamma-1}\right)^T \bar{\boldsymbol{v}}^{e_i}\, d\theta + \int_{\omega_{i-1}}^{\omega_i} \left[((\alpha+\gamma-1)[A_8] + [A_9])\,\tilde{\boldsymbol{\varphi}}^{e_i}_{\gamma-1}\right]^T \bar{\boldsymbol{v}}^{e_i}\, d\theta \\
&+ \varrho \int_{\omega_{i-1}}^{\omega_i} \left[([A_{10}])\,\tilde{\boldsymbol{\varphi}}^{e_i}_{\gamma-2}\right]^T \bar{\boldsymbol{v}}^{e_i}\, d\theta = 0,
\end{aligned}
\tag{13.103}
$$

where

$$\varrho = \begin{cases} 0, & \gamma = 1, \\ 1, & \gamma \geq 2. \end{cases} \tag{13.104}$$

Traction-free boundary conditions are applied to each element and using (13.83), we represent the first term in (13.103):

$$\begin{aligned} \left\{ \left([A_1] \tilde{\boldsymbol{\varphi}}_{\gamma}^{e_i \prime} \right)^T \bar{\boldsymbol{v}}^{e_i} \right\} \Big|_{\omega_{i-1}}^{\omega_i} &= \left\{ \left([B_1] \tilde{\boldsymbol{\varphi}}_{\gamma}^{e_i \prime} \right)^T \bar{\boldsymbol{v}}^{e_i} \right\} \Big|_{\omega_{i-1}}^{\omega_i} \\ &= - \left\{ \left[((\alpha + \gamma)[B_2] + [B_3]) \, \tilde{\boldsymbol{\varphi}}_{\gamma}^{e_i} \right]^T \bar{\boldsymbol{v}}^{e_i} \right\} \Big|_{\omega_{i-1}}^{\omega_i} \\ &\quad - \left\{ \left([B_4] \tilde{\boldsymbol{\varphi}}_{\gamma-1}^{e_i} \right)^T \bar{\boldsymbol{v}}^{e_i} \right\} \Big|_{\omega_{i-1}}^{\omega_i}. \end{aligned} \tag{13.105}$$

The sesquilinear form for the computation of the shadow function $\tilde{\boldsymbol{\varphi}}_{\gamma}$ is

$$\text{Seek} \quad \tilde{\boldsymbol{\varphi}}_{\gamma} \in \mathcal{E}(0, \omega) \quad \text{s.t.} \quad \mathcal{B}^{\gamma}(\tilde{\boldsymbol{\varphi}}_{\gamma}, \bar{\boldsymbol{v}}) = \mathcal{F}^{\gamma}(\bar{\boldsymbol{v}}), \quad \forall \bar{\boldsymbol{v}} \in \mathcal{E}(0, \omega), \tag{13.106}$$

where $\mathcal{B}^{\gamma}(\tilde{\boldsymbol{\varphi}}_{\gamma}, \bar{\boldsymbol{v}})$ and $\mathcal{F}^{\gamma}(\bar{\boldsymbol{v}})$ are the assembled forms of $\mathcal{B}^{\gamma}(\tilde{\boldsymbol{\varphi}}_{\gamma}^{e_i}, \bar{\boldsymbol{v}}^{e_i})$ and $\mathcal{F}^{\gamma}(\bar{\boldsymbol{v}}^{e_i})$:

$$\begin{aligned} \mathcal{B}^{\gamma}(\tilde{\boldsymbol{\varphi}}_{\gamma}^{e_i}, \bar{\boldsymbol{v}}^{e_i}) = &- \int_{\omega_{i-1}}^{\omega_i} \left([A_1] \tilde{\boldsymbol{\varphi}}_{\gamma}^{e_i \prime} \right)^T \bar{\boldsymbol{v}}^{e_i \prime} d\theta \\ &+ \int_{\omega_{i-1}}^{\omega_i} \left[((\alpha + \gamma)[A_2] + [A_3]) \, \tilde{\boldsymbol{\varphi}}_{\gamma}^{e_i \prime} \right]^T \bar{\boldsymbol{v}}^{e_i} d\theta \\ &+ \int_{\omega_{i-1}}^{\omega_i} \left[((\alpha + \gamma)^2 [A_4] + (\alpha + \gamma)[A_5] + [A_6]) \, \tilde{\boldsymbol{\varphi}}_{\gamma}^{e_i} \right]^T \bar{\boldsymbol{v}}^{e_i} d\theta \\ &- \left\{ \left[((\alpha + \gamma)[B_2] + [B_3]) \, \tilde{\boldsymbol{\varphi}}_{\gamma}^{e_i} \right]^T \bar{\boldsymbol{v}}^{e_i} \right\} \Bigg|_{\omega_{i-1}}^{\omega_i}, \end{aligned} \tag{13.107}$$

$$\begin{aligned} \mathcal{F}^{\gamma}(\boldsymbol{v}^{e_i}) = &- \int_{\omega_{i-1}}^{\omega_i} \left([A_7] \tilde{\boldsymbol{\varphi}}_{\gamma-1}^{e_i \prime} \right)^T \bar{\boldsymbol{v}}^{e_i} d\theta \\ &- \int_{\omega_{i-1}}^{\omega_i} \left[((\alpha + \gamma - 1)[A_8] + [A_9]) \, \tilde{\boldsymbol{\varphi}}_{\gamma-1}^{e_i} \right]^T \bar{\boldsymbol{v}}^{e_i} d\theta \\ &- \varrho \int_{\omega_{i-1}}^{\omega_i} \left[([A_{10}]) \, \tilde{\boldsymbol{\varphi}}_{\gamma-2}^{e_i} \right]^T \bar{\boldsymbol{v}}^{e_i} d\theta + \left\{ \left([B_4] \tilde{\boldsymbol{\varphi}}_{\gamma-1}^{e_i} \right)^T \bar{\boldsymbol{v}}^{e_i} \right\} \Bigg|_{\omega_{i-1}}^{\omega_i}. \end{aligned} \tag{13.108}$$

In the assembly procedure, continuity of the displacement is enforced. The two surfaces $\theta = 0, \omega$ are therefore under traction-free condition. In the case of Dirichlet boundary conditions the energy space $\mathcal{E}$ is replaced by $\mathcal{E}_o$.

13.3.2.2 p -FEMs for the Solution of (13.106)

We apply p -FEMs for the solution of (13.106) as in Section 13.3.1.1. To this end, $\tilde{\boldsymbol{\varphi}}_\gamma = \left(u_{r\gamma}\ u_{\theta\gamma}\ u_{3\gamma}\right)^T = [\Phi]\boldsymbol{a}_\gamma$ and $\bar{\boldsymbol{v}} \stackrel{\text{def}}{=} [\Phi]\boldsymbol{b}_\gamma$.

The resulting FE formulation is

$$\boldsymbol{a}_\gamma^T [K^\gamma] = \boldsymbol{F}^{\gamma T}, \tag{13.109}$$

where

$$\begin{aligned}
[K^\gamma]^{e_i} &= -\left\{[\Phi]^T\left((\alpha+\gamma)[B_2]^T + [B_3]^T\right)[\Phi]\right\}\Big|_{\omega_{i-1}}^{\omega_i} - \frac{2}{\omega^{e_i}}\int_{-1}^{1}[\Phi']^T[A_1]^T[\Phi']d\xi \\
&\quad + \int_{-1}^{1}[\Phi']^T\left((\alpha+\gamma)[A_2]^T + [A_3]^T\right)[\Phi]d\xi \\
&\quad + \frac{\omega^{e_i}}{2}\int_{-1}^{1}[\Phi]^T\left((\alpha+\gamma)^2[A_4]^T + (\alpha+\gamma)[A_5]^T + [A_6]^T\right)[\Phi]d\xi, \\
\boldsymbol{F}^{\gamma e_i} &= \left\{\boldsymbol{a}_{\gamma-1}^{e_i T}[\Phi]^T[B_4]^T[\Phi]\right\}\Big|_{\omega_{i-1}}^{\omega_i} - \int_{-1}^{1}\boldsymbol{a}_{\gamma-1}^{e_i T}[\Phi']^T[A_7]^T[\Phi]d\xi \\
&\quad - \frac{\omega^{e_i}}{2}\int_{-1}^{1}\boldsymbol{a}_{\gamma-1}^{e_i T}[\Phi]^T\left((\alpha+\gamma-1)[A_8]^T + [A_9]^T\right)[\Phi]d\xi \\
&\quad - \varrho\frac{\omega^{e_i}}{2}\int_{-1}^{1}\boldsymbol{a}_{\gamma-2}^{e_i T}[\Phi]^T\left([A_{10}]^T\right)[\Phi]d\xi,
\end{aligned} \tag{13.110}$$

where $[K^\gamma]$ are the assembled matrices formed of $[K^\gamma]^{e_i}$, and $\boldsymbol{F}^\gamma$ are the assembled vectors formed of $\boldsymbol{F}^{\gamma e_i}$. For clamped BCs, $[B_2]|_0 = [B_2]|_\omega = [B_3]|_0 = [B_3]|_\omega = [0]$.

Substituting $\alpha = \alpha_{\Re} + i\alpha_{\Im}$, $\boldsymbol{a}_\gamma = \boldsymbol{a}_\gamma^{\Re} + i\boldsymbol{a}_\gamma^{\Im}$, $\boldsymbol{a}_{\gamma-1} = \boldsymbol{a}_{\gamma-1}^{\Re} + i\boldsymbol{a}_{\gamma-1}^{\Im}$, and $\boldsymbol{a}_{\gamma-2} = \boldsymbol{a}_{\gamma-2}^{\Re} + i\boldsymbol{a}_{\gamma-2}^{\Im}$ into (13.110), we obtain the FE formulation

$$\begin{pmatrix}\boldsymbol{a}_\gamma^{\Re}\\ \boldsymbol{a}_\gamma^{\Im}\end{pmatrix}^T \begin{pmatrix}[K^\gamma]_{\Re} & [K^\gamma]_{\Im}\\ -[K^\gamma]_{\Im} & [K^\gamma]_{\Re}\end{pmatrix} = \begin{pmatrix}\boldsymbol{F}_{\Re}^{\gamma}\\ \boldsymbol{F}_{\Im}^{\gamma}\end{pmatrix}^T, \tag{13.111}$$

where

$$
\begin{aligned}
[K^{\gamma}]^{e_i}_{\Re} &= -\left\{[\Phi]^T\left((\alpha_{\Re}+\gamma)[B_2]^T+[B_3]^T\right)[\Phi]\right\}|^{\omega_i}_{\omega_{i-1}} - \frac{2}{\omega^{e_i}}\int_{-1}^{1}[\Phi']^T[A_1]^T[\Phi']d\xi \\
&+\int_{-1}^{1}[\Phi']^T\left((\alpha_{\Re}+\gamma)[A_2]^T+[A_3]^T\right)[\Phi]d\xi \\
&+\frac{\omega^{e_i}}{2}\int_{-1}^{1}[\Phi]^T\left(((\alpha_{\Re}+\gamma)^2-\alpha_{\Im}^2)[A_4]^T+(\alpha_{\Re}+\gamma)[A_5]^T+[A_6]^T\right)[\Phi]d\xi, \\
[K^{\gamma}]^{e_i}_{\Im} &= -\left\{[\Phi]^T\left(\alpha_{\Im}[B_2]^T\right)[\Phi]\right\}|^{\omega_i}_{\omega_{i-1}} + \int_{-1}^{1}[\Phi']^T\left(\alpha_{\Im}[A_2]^T\right)[\Phi]d\xi \\
&+\frac{\omega^{e_i}}{2}\int_{-1}^{1}[\Phi]^T\left(2(\alpha_{\Re}+\gamma)\alpha_{\Im}[A_4]^T+\alpha_{\Im}[A_5]^T\right)[\Phi]d\xi, \\
\boldsymbol{F}^{\gamma e_i}_{\Re} &= \left\{\boldsymbol{a}^{\Re e_i^T}_{\gamma-1}[\Phi]^T[B_4]^T[\Phi]\right\}|^{\omega_i}_{\omega_{i-1}} - \int_{-1}^{1}\boldsymbol{a}^{\Re e_i^T}_{\gamma-1}[\Phi']^T[A_7]^T[\Phi]d\xi \\
&-\frac{\omega^{e_i}}{2}\int_{-1}^{1}((\alpha_{\Re}+\gamma-1)\boldsymbol{a}^{\Re e_i}_{\gamma-1}-\alpha_{\Im}\boldsymbol{a}^{\Im e_i}_{\gamma-1})^T[\Phi]^T[A_8]^T[\Phi]d\xi \\
&-\frac{\omega^{e_i}}{2}\int_{-1}^{1}\boldsymbol{a}^{\Re e_i^T}_{\gamma-1}[\Phi]^T[A_9]^T[\Phi]d\xi - \frac{\omega}{2}\int_{-1}^{1}\boldsymbol{a}^{\Re e_i^T}_{\gamma-2}[\Phi]^T[A_{10}]^T[\Phi]d\xi, \\
\boldsymbol{F}^{\gamma e_i}_{\Im} &= \left\{\boldsymbol{a}^{\Im e_i^T}_{\gamma-1}[\Phi]^T[B_4]^T[\Phi]\right\}|^{\omega_i}_{\omega_{i-1}} - \int_{-1}^{1}\boldsymbol{a}^{\Im e_i^T}_{\gamma-1}[\Phi']^T[A_7]^T[\Phi]d\xi \\
&-\frac{\omega^{e_i}}{2}\int_{-1}^{1}(\alpha_{\Im}\boldsymbol{a}^{\Re e_i}_{\gamma-1}+(\alpha_{\Re}+\gamma-1)\boldsymbol{a}^{\Im e_i}_{\gamma-1})^T[\Phi]^T[A_8]^T[\Phi]d\xi \\
&-\frac{\omega^{e_i}}{2}\int_{-1}^{1}\boldsymbol{a}^{\Im e_i^T}_{\gamma-1}[\Phi]^T[A_9]^T[\Phi]d\xi - \frac{\omega^{e_i}}{2}\int_{-1}^{1}\boldsymbol{a}^{\Im e_i^T}_{\gamma-2}[\Phi]^T[A_{10}]^T[\Phi]d\xi,
\end{aligned}
\tag{13.112}
$$

where $[K^{\gamma}]_{\Re}$, $[K^{\gamma}]_{\Im}$ are the assembled matrices from $[K^{\gamma}]^{e_i}_{\Re}$ and $[K^{\gamma}]^{e_i}_{\Im}$, and $\boldsymbol{F}^{\gamma}_{\Re}$, $\boldsymbol{F}^{\gamma}_{\Im}$ are the assembled vectors from $\boldsymbol{F}^{\gamma e_i}_{\Re}$ and $\boldsymbol{F}^{\gamma e_i}_{\Im}$.

13.3.3 Difficulties in Computing Shadows and Remedies for Several Pathological Cases

There are several pathological cases, among which one is of major importance, the cracked case, where the numerical methods presented fail, and remedies are to be implemented. These pathological cases occur when $\alpha_i = \alpha_j - n$ for n an integer. For a cracked configuration, for example, $\alpha_1 = \alpha_2 = \alpha_3 = 1/2$, $\alpha_4 = \alpha_5 = \alpha_6 = 1$, $\alpha_7 = \alpha_8 = \alpha_9 = 3/2$, etc. In this case consider for example the first shadow function associated with $\alpha_1 = 1/2$, which has to satisfy the *inhomogeneous* ODE (13.76)$_2$:

$$[\mathcal{M}_0]\tilde{\boldsymbol{\varphi}}_1^{(\alpha_1)} = \left([A_1]\partial_\theta^2 + \left(\frac{3}{2}[A_2] + [A_3]\right)\partial_\theta + \left((\frac{3}{2})^2[A_4] + \frac{3}{2}[A_5] + [A_6]\right)\right)\tilde{\boldsymbol{\varphi}}_1^{(\alpha_1)} = -[\mathcal{M}_1]\tilde{\boldsymbol{\varphi}}_0^{(\alpha_1)} \quad (13.113)$$

Formally, the solution (13.113) may be obtained by the inverse of the operator $[\mathcal{M}_0]$ applied to the RHS. Practically, when FE discretization is applied, the operator $[\mathcal{M}_0]$ results in the matrix $[K^1]$, which has to be inverted and must not be singular. This is equivalent to requiring a particular solution without the homogeneous part of the solution. However, the LHS of (13.113) is exactly the ODE for the computation of $\tilde{\boldsymbol{\varphi}}_0^{(\alpha_1+1)}$, except that for $\tilde{\boldsymbol{\varphi}}_0^{(\alpha_1+1)}$, the ODE is *homegeneous*:

$$[\mathcal{M}_0]\tilde{\boldsymbol{\varphi}}_0^{(\alpha_1+1)} = \left([A_1]\partial_\theta^2 + \left(\frac{3}{2}[A_2] + [A_3]\right)\partial_\theta + \left(\left(\frac{3}{2}\right)^2[A_4] + \frac{3}{2}[A_5] + [A_6]\right)\right)\tilde{\boldsymbol{\varphi}}_1^{(\alpha_1)} = \mathbf{0}. \quad (13.114)$$

In the continuum case (i.e., theoretically as the number of degrees of freedom tends to infinity), $[K^1]$ is singular and may not be inverted. Practically, because the eigenvalues are computed numerically, the larger the eigenvalue, the worse is the approximation, so $[K^1]$ is not identically singular, but ill-conditioned, and as the polynomial degree is increased (resulting in better approximation of eigenvalues), the more ill-conditioned $[K^1]$ becomes. Of course this situation occurs with any of the dual shadow functions computed numerically.

A remedy to this problem is achieved if one notices that only a particular solution of (13.113) is sought, therefore a constraint can be added that the sought solution be orthogonal to the homogeneous solution for the operator $[\mathcal{M}_0]$. Practically, in the FE formulation one has to enforce the additional condition that the scalar product between $\boldsymbol{a}_1^{(\alpha_1)}$ and $\bar{\boldsymbol{a}}_0^{(\alpha_1+1)}$, for example, be zero. Or in general, we add the scalar product between $\boldsymbol{a}_\gamma^{(\alpha_i)}$ and $\bar{\boldsymbol{a}}_0^{(\alpha_j)}$, where $\alpha_j = \alpha_i + \gamma$ (if the α_j exist) to ensure that $\boldsymbol{a}_\gamma^{(\alpha_i)}$ is not dependent on $\boldsymbol{a}_0^{(\alpha_j)}, \bar{\boldsymbol{a}}_0^{(\alpha_j)}$.

The system (13.111) for these pathological cases therefore becomes:

$$\underbrace{\begin{pmatrix} \boldsymbol{a}_\gamma^{\Re(\alpha_i)} \\ \boldsymbol{a}_\gamma^{\Im(\alpha_i)} \end{pmatrix}^T}_{(1\times S)} \underbrace{\begin{pmatrix} [K^\gamma]_\Re & [K^\gamma]_\Im & \boldsymbol{a}_0^{\Re(\alpha_j)} & -\boldsymbol{a}_0^{\Im(\alpha_j)} \\ -[K^\gamma]_\Im & [K^\gamma]_\Re & \boldsymbol{a}_0^{\Im(\alpha_j)} & \boldsymbol{a}_0^{\Re(\alpha_j)} \end{pmatrix}}_{(S\times(S+2))} = \underbrace{\begin{pmatrix} \boldsymbol{F}_\Re^\gamma \\ \boldsymbol{F}_\Im^\gamma \\ 0 \\ 0 \end{pmatrix}^T}_{(1\times(S+2))}, \quad (13.115)$$

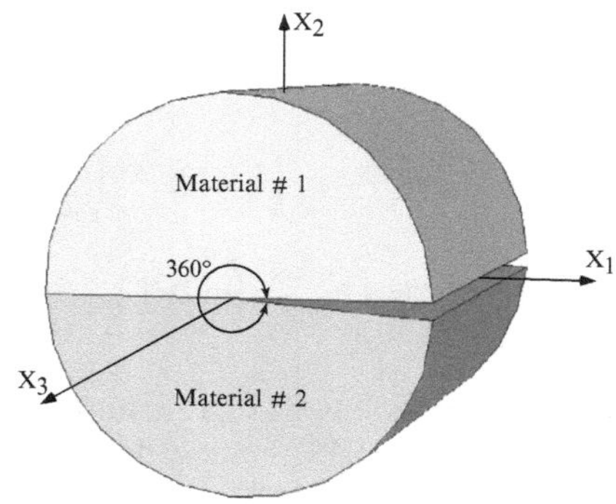

Fig. 13.21 Bimaterial cracked domain.

where $S = 3pQ + 3$, where Q is the number of elements. Since system (13.115) is now an overdetermined system of equations, we use the least squares solution to determine $\boldsymbol{a}_\gamma^{\Re(\alpha_i)}$ and $\boldsymbol{a}_\gamma^{\Im(\alpha_i)}$.

To demonstrate the pathological case discussed, as an example, we compute the eigenpairs and first two shadow functions of an orthotropic bimaterial cracked domain, shown in Figure 13.21. Both materials are made of the same high-modulus graphite-epoxy system with different fiber orientations. Referring to the principal direction of the fibers, the material properties are

$$E_L = 1.38 \times 10^5 \text{MPa}, \quad E_T = E_z = 1.45 \times 10^4 \text{MPa}, \qquad (13.116)$$
$$G_{LT} = G_{Lz} = G_{Tz} = 0.586 \times 10^4 \text{MPa}, \quad \nu_{LT} = \nu_{Lz} = \nu_{Tz} = 0.21,$$

where the subscripts L, T, z refer to fiber, transverse, and thickness directions of individual materials. The orientation of fibers of the upper material (Material #1) is in the x_1 direction, whereas the orientation at the lower material (Material #2) is in the x_3 direction. The first three eigenvalues for this example problem, computed using eight elements at $p = 15$, are

$$\alpha_{1,2} = 0.5 \pm i0.05106124425, \qquad \alpha_3 = 0.5, \qquad (13.117)$$

In this example one obtains also the eigenvalues $\alpha = 2.5 \pm i0.05106124425$, that satisfy $\alpha = \alpha_{1,2} + 2$, therefore cause the $[K^2]$ matrix, associated with $\tilde{\boldsymbol{\varphi}}_2^{(\alpha_1)}$, to become singular. Figure 13.22 shows the condition number of the matrix $[K^2]$ associated with $\tilde{\boldsymbol{\varphi}}_2^{(\alpha_1)}$, computed using increasing p-level.

It may be observed that the condition number of $[K^2]$ increases continuously as p is increased. The condition number of $[K^2]$ after incorporating the constraint of the scalar product remains constant.

The functions u_{r2} and $u_{\theta 2}$ of $\tilde{\boldsymbol{\varphi}}_2^{(\alpha_{1,2})}$ computed with and without the scalar product condition are presented in Figures 13.23 and 13.24 respectively.

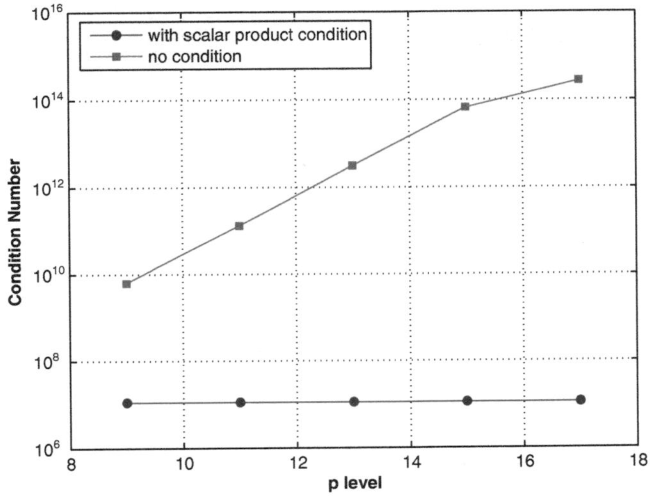

Fig. 13.22 Condition number of $[K^2]$ associated with $\tilde{\varphi}_2^{(\alpha_{1,2})}$, $\alpha_{1,2} = 0.5 \pm i0.0510612$.

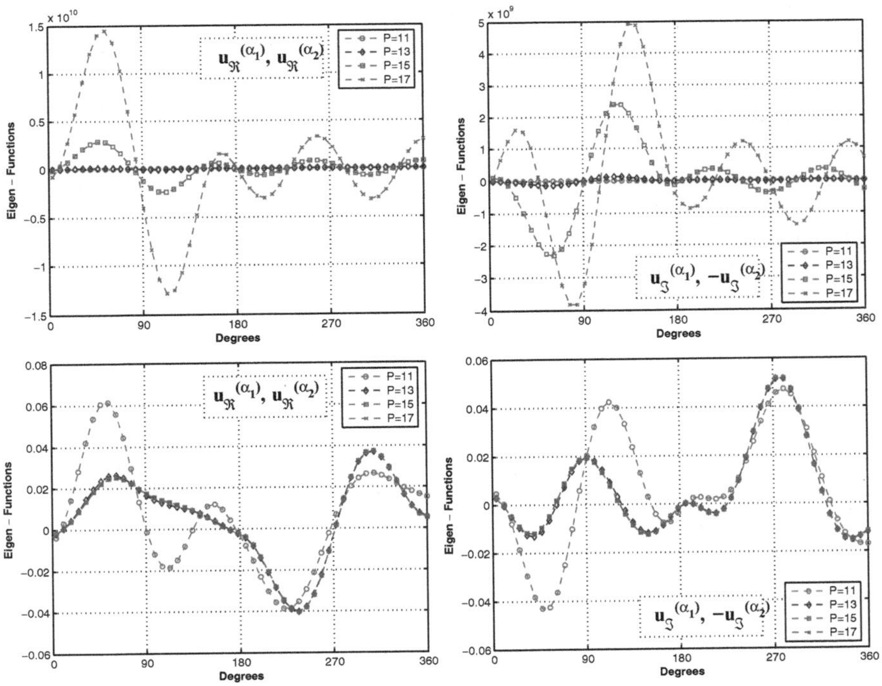

Fig. 13.23 The functions u_{r2} associated with $\tilde{\varphi}_2^{(\alpha_{1,2})}$, $\alpha_{1,2} = 0.5 \pm 0.0510612$, computed using 8 elements without (Top) and with (Bottom) the scalar product condition.

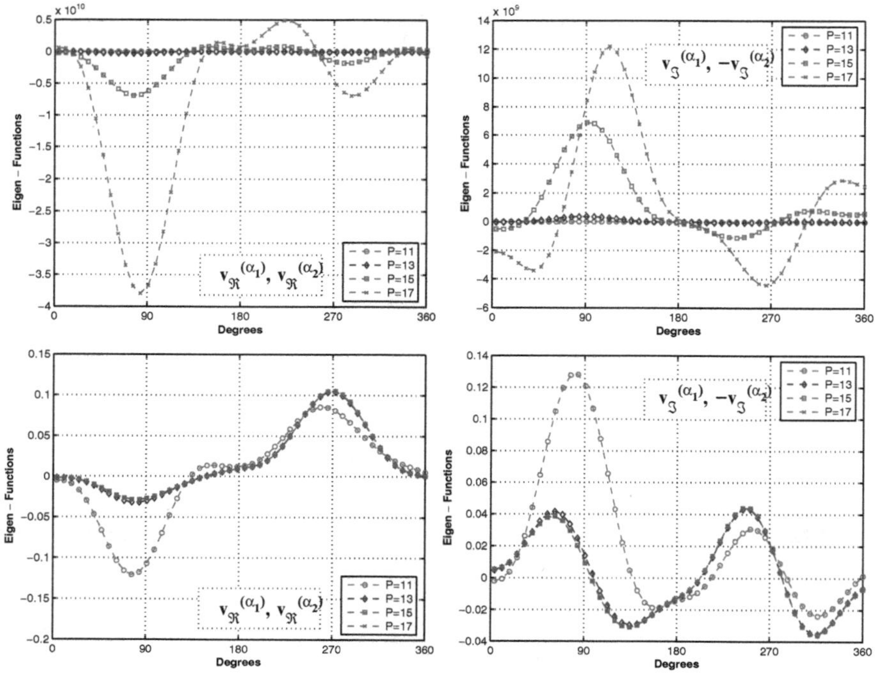

Fig. 13.24 The functions $u_{\theta 2}$ associated with $\tilde{\varphi}_2^{(\alpha_{1,2})}$, $\alpha_{1,2} = 0.5 \pm 0.0510612$, computed using eight elements without (top) and with (bottom) the scalar product condition.

It is visible from Figures 13.23, 13.24 that the functions computed without any additional condition are scattered, whereas the functions computed using the scalar product condition converge.

13.3.4 Extracting Complex ESIFs by the QDFM

Here we extend the QDFM to the case of complex ESIFs and multimaterial interfaces. The only change is that in Theorem 13.1, one has to replace α_1 and α_2 by $\Re(\alpha_1)$ and $\Re(\alpha_2)$:

$$J[R](\tilde{\boldsymbol{u}}, \mathbf{K}_m^{(\alpha_i)}[B]) = \int_I A_i(x_3)\, B(x_3)\, dx_3 + \mathcal{O}(R^{\Re(\alpha_1)-\Re(\alpha_i)+m+1}), \quad \text{as} \quad R \to 0. \tag{13.118}$$

Here $\Re(\alpha_1)$ is the real part of the smallest of all positive eigenvalues α_i. In the case of complex eigenvalues, $A_i(x_3) = A_i^{\Re}(x_3) + \imath A_i^{\Im}(x_3)$ and $B(x_3) = B^{\Re}(x_3) + \imath B^{\Im}(x_3)$. Choosing $m = 2$, we have in (13.118) $\mathcal{O}(R^{\Re(\alpha_1)-\Re(\alpha_i)+3})$. If $A_i(x_3)$ is a polynomial of degree N, it is expanded as a linear combination of Jacobi polynomials. For the functions $BJ(x_3)$, we choose them to satisfy $\partial_3^j BJ(x_3) = 0$ for $j = 0, 1, 2, 3$:

$$BJ^{\Re(k)}(x_3) = BJ^{\Im(k)}(x_3) = (1 - x_3^2)^4 \, \frac{J_4^{(k)}(x_3)}{h_k}, \quad h_k = \frac{2^9(k+4)!(k+4)!}{(2k+9)(k+8)!}. \tag{13.119}$$

If we take an approximation of $A_i(x_3)$ as a polynomial of N th order, being a linear combination of Jacobi polynomials,

$$A_i(x_3) = A_i^{\Re}(x_3) + \imath A_i^{\Im}(x_3), \quad \begin{cases} A_i^{\Re}(x_3) = \tilde{a}_0^{\Re} J_4^{(0)}(x_3) + \tilde{a}_1^{\Re} J_4^{(1)}(x_3) + \cdots + \tilde{a}_N^{\Re} J_4^{(N)}(x_3), \\ A_i^{\Im}(x_3) = \tilde{a}_0^{\Im} J_4^{(0)}(x_3) + \tilde{a}_1^{\Im} J_4^{(1)}(x_3) + \cdots + \tilde{a}_N^{\Im} J_4^{(N)}(x_3), \end{cases} \tag{13.120}$$

then we can obtain the coefficients $\tilde{a}_i$ directly by applying Theorem 13.1 with different extraction polynomials:

$$\int_{-1}^{1} A_i(x_3) BJ^{(k)}(x_3)\, dx_3 = J[R]^{\Re(k)} + \imath J[R]^{\Im(k)}, \qquad k = 0, 1, \ldots, N, \tag{13.121}$$

where

$$\begin{aligned} J[R]^{\Re(k)} &= \int_{-1}^{1} (A_i^{\Re}(x_3) BJ^{\Re(k)}(x_3) - A_i^{\Im}(x_3) BJ^{\Im(k)}(x_3)) dx_3, \\ J[R]^{\Im(k)} &= \int_{-1}^{1} (A_i^{\Re}(x_3) BJ^{\Im(k)}(x_3) + A_i^{\Im}(x_3) BJ^{\Re(k)}(x_3)) dx_3, \end{aligned} \tag{13.122}$$

and

$$\tilde{a}_k^{\Re} = \frac{J[R]^{\Re(k)} + J[R]^{\Im(k)}}{2}, \qquad \tilde{a}_k^{\Im} = \frac{-J[R]^{\Re(k)} + J[R]^{\Im(k)}}{2}. \tag{13.123}$$

In view of (13.118), the $J[R]$ integral evaluated for the quasidual functions $K_m^{(\alpha_i)}[B^{(k)}]$, $k = 0, 1, \ldots, N$, provides approximations of the coefficients $\tilde{a}_k$. Note that the polynomial degree is the superscript k. Of course, in general, $A_i(x_3)$ is an unknown function, and we find a projection of it only into spaces of polynomials. It is expected that as we increase the polynomial space, the approximation is progressively better.

To demonstrate the accuracy of the proposed methods, two example problems are considered. The first is a crack at a bimaterial interface between two isotropic materials for which semianalytical solutions are known. Thus the accuracy of the numerical results can be evaluated.

The second example problem is a crack in a compact test specimen at an interface of two anisotropic materials. Although the loading is perpendicular to the crack face, because of the anisotropy of the materials, all three modes are excited.

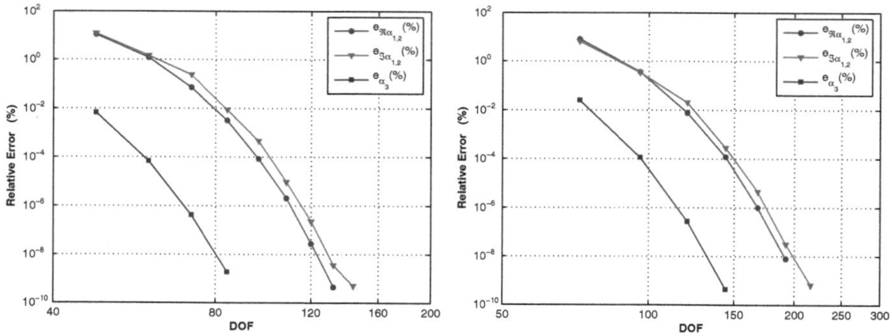

Fig. 13.25 Relative error (percentage) in eigenvalues α_1^{FE}, α_2^{FE}, α_3^{FE}, for example A, computed by two (left) and four (right) elements.

13.3.5 Numerical Example: A Crack at the Interface of Two Isotropic Materials

Consider a bimaterial interface that is composed of two homogeneous materials (Figure 13.21). The two materials are isotropic, both having Poisson ratio of $\nu = 0.3$, the Young's modulus of the upper material (Material #1) is $E = 10$ and of the lower material (Material #2) is $E = 1$. This example was chosen to present the performance of the method for cases of complex eigenvalues. The exact first three eigenvalues for this example problem, as reported in [202], are

$$\alpha_{1,2} = 0.5 \pm \imath 0.07581177769, \qquad \alpha_3 = 0.5. \tag{13.124}$$

The relative error as a percentage in the first three eigenvalues computed using two and four elements is shown in Figure 13.25. For the first complex eigenvalue, the relative error is split into real and imaginary parts:

$$e_{\Re\alpha_{1,2}} = 100\frac{\Re(\alpha_{1,2}) - \Re(\alpha_{1,2}^{\mathrm{FE}})}{\Re(\alpha_{1,2})}, \qquad e_{\Im\alpha_{1,2}} = 100\frac{\Im(\alpha_{1,2}) - \Im(\alpha_{1,2}^{\mathrm{FE}})}{\Im(\alpha_{1,2})}. \tag{13.125}$$

The eigenfunctions, duals, and shadows associated with the first three eigenvalues are presented in Figures 13.26, 13.27, and 13.28 computed using four elements, $p = 6$.

Obtaining the eigenpairs and shadows for the first three eigenvalues, we choose the ESIF to be, for example, a polynomial of order 2. Thus, the solution is

$$\begin{aligned}\tilde{\boldsymbol{u}} = {} & A_1(x_3)r^{\alpha_1}\tilde{\boldsymbol{\varphi}}_0^{(\alpha_1)}(\theta) + \partial_3 A_1(x_3)r^{\alpha_1+1}\tilde{\boldsymbol{\varphi}}_1^{(\alpha_1)}(\theta) + \partial_3^2 A_1(x_3)r^{\alpha_1+2}\tilde{\boldsymbol{\varphi}}_2^{(\alpha_1)}(\theta)\\ & + A_2(x_3)r^{\alpha_2}\tilde{\boldsymbol{\varphi}}_0^{(\alpha_2)}(\theta) + \partial_3 A_2(x_3)r^{\alpha_2+1}\tilde{\boldsymbol{\varphi}}_1^{(\alpha_2)}(\theta) + \partial_3^2 A_2(x_3)r^{\alpha_2+2}\tilde{\boldsymbol{\varphi}}_2^{(\alpha_2)}(\theta)\\ & + A_3(x_3)r^{\alpha_3}\tilde{\boldsymbol{\varphi}}_0^{(\alpha_3)}(\theta) + \partial_3 A_3(x_3)r^{\alpha_3+1}\tilde{\boldsymbol{\varphi}}_1^{(\alpha_3)}(\theta) + \partial_3^2 A_3(x_3)r^{\alpha_3+2}\tilde{\boldsymbol{\varphi}}_2^{(\alpha_3)}(\theta).\end{aligned} \tag{13.126}$$

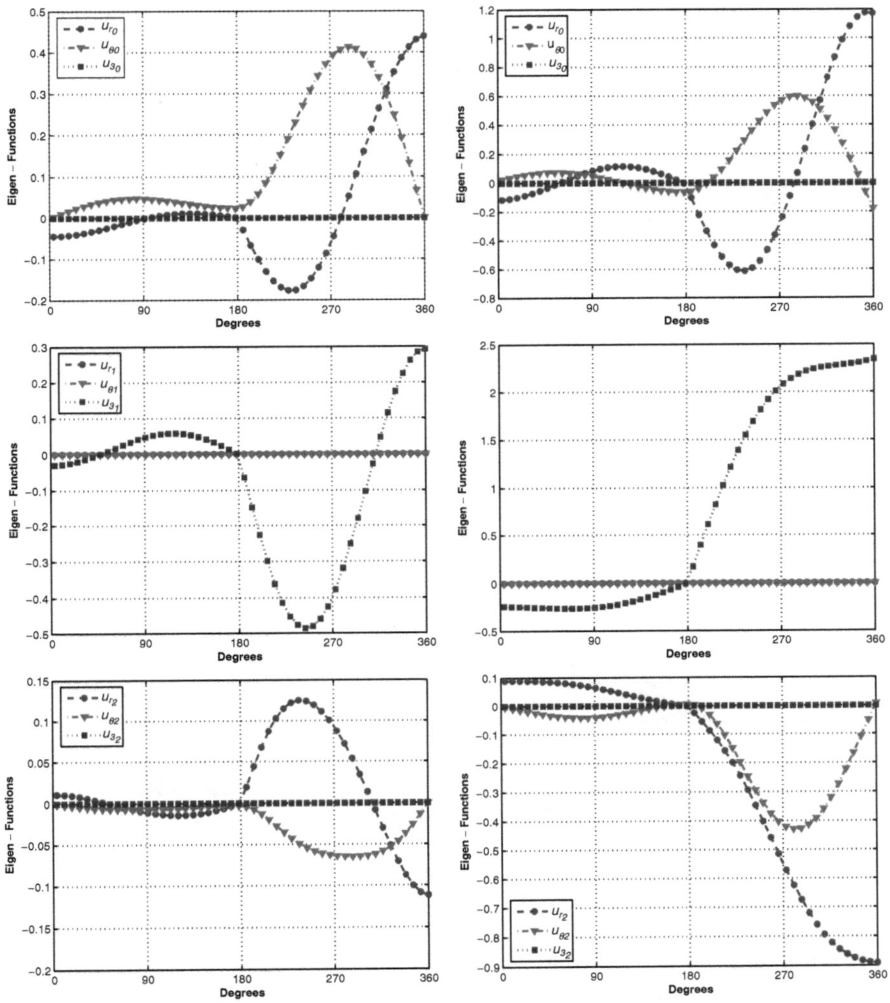

Fig. 13.26 The real part of the eigenfunctions (left) and dual eigenfunctions (right) associated with $\alpha_{1,2} = 0.5 \pm \imath 0.075812$, computed by four elements, $p = 6$.

Note that the eigenpairs and shadows are obtained numerically, and therefore (13.126) represents an approximation of the exact solution only.

For example, consider the following exact ESIFs (polynomials of order 3):

$$A_{1,2}^{Ex}(x_3) = (3 + 4x_3 + 5x_3^2) \pm i(2 + 3x_3 + 4x_3^2), \quad A_3^{Ex}(x_3) = 5 + 4x_3 + 2x_3^2. \tag{13.127}$$

If we prescribe on a traction-free cracked domain Dirichlet boundary conditions according to (13.126)-(13.127), the exact solution at each r, θ, x_3 is as in

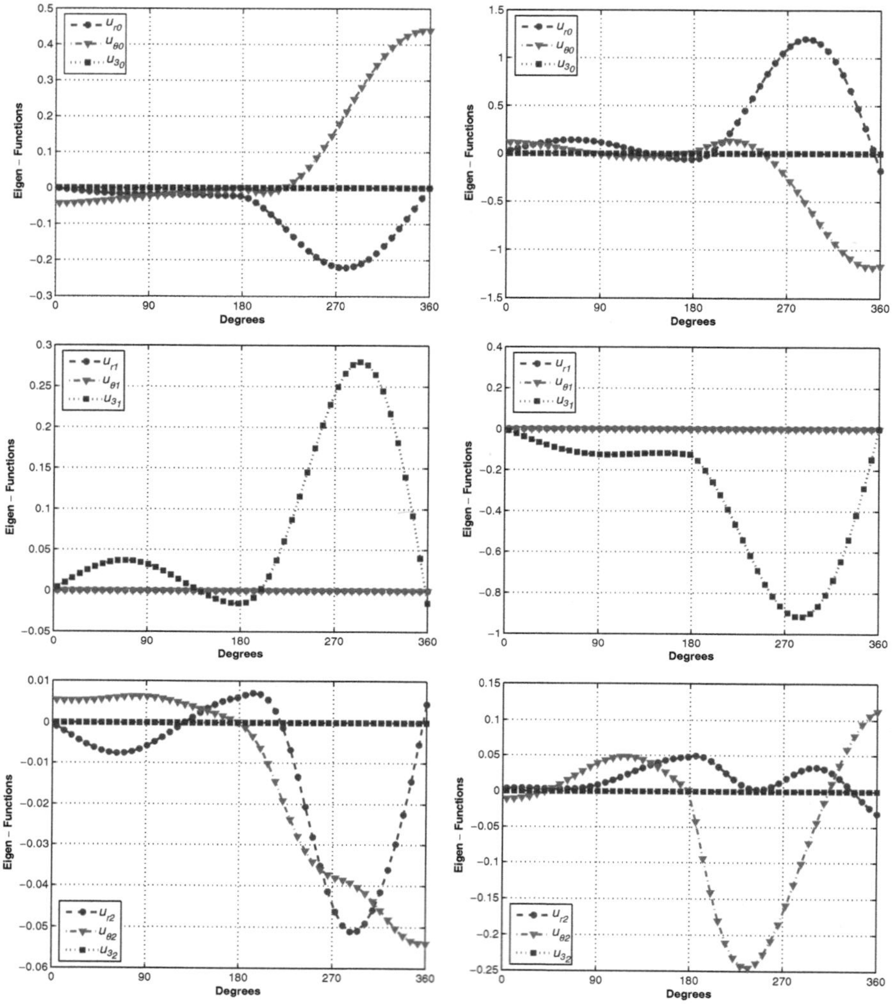

Fig. 13.27 The image part of the eigenfunctions (left) and dual eigenfunctions (right) associated with $\alpha_{1,2} = 0.5 \pm \imath 0.075812$, computed by four elements, $p = 6$.

(13.126)-(13.127). Consider a 3-D domain as shown in Figure 13.1 with $\omega = 2\pi$. The domain is discretized using a p -FE mesh, with geometric progression toward the singular edge with a factor of 0.15, having four layers of elements. In the x_3 direction, a uniform discretization using five elements has been adopted. In Figure 13.29 we present the mesh used for the cracked domain and the convergence rate of the relative error in the energy norm.

We specify on the entire boundary $\partial\Omega$, Dirichlet boundary conditions according to (13.126). Therefore the exact solution at any point $\boldsymbol{x} \equiv (r, \theta, x_3)$ should be (13.126).

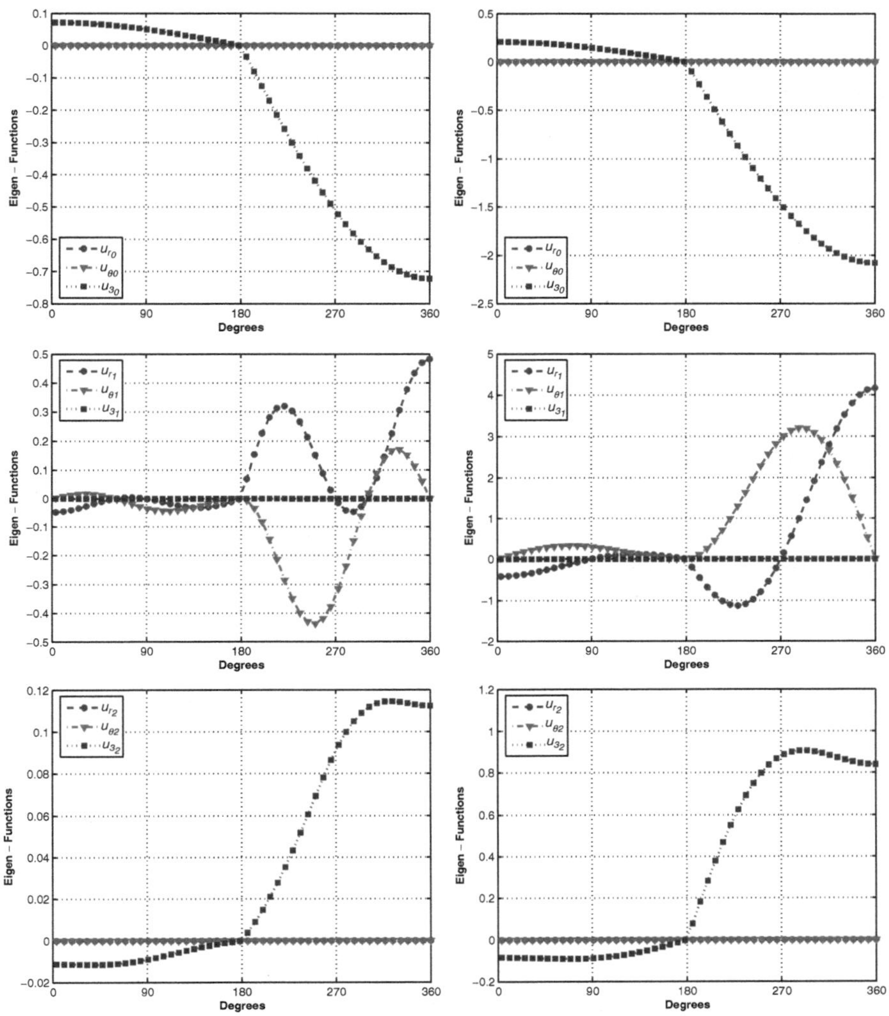

Fig. 13.28 Eigenfunctions (left) and dual eigenfunctions (right) associated with $\alpha_3 = 0.5$, computed by four elements, $p = 6$.

When $J[R]$ is computed with the quasidual function $\mathbf{K}_2^{(\alpha_i)}$ and $BJ^{(k)}(x_3)$, we expect to obtain, according to (13.118), (13.120) and (13.123), the coefficients $\tilde{a}_k^{\Re(\alpha_i)}, \tilde{a}_k^{\Im(\alpha_i)}$ for complex eigenvalues or $\tilde{a}_k^{(\alpha_i)}$ for real eigenvalues. The ESIF is then easily represented by a linear combination of the Jacobi polynomials in (13.120). We extract the ESIFs at $R = 0.05$ using the *numerically computed* dual eigenpairs and their shadows with $K_2^{(\alpha_i)}[B^{(k)}]$.

We present the relative error as a percentage of the extracted $A_{1,2}^{\Re}(x_3)$, $A_{1,2}^{\Im}(x_3)$, $A_3(x_3)$ of order $3, 4, 5$ in Figure 13.30.

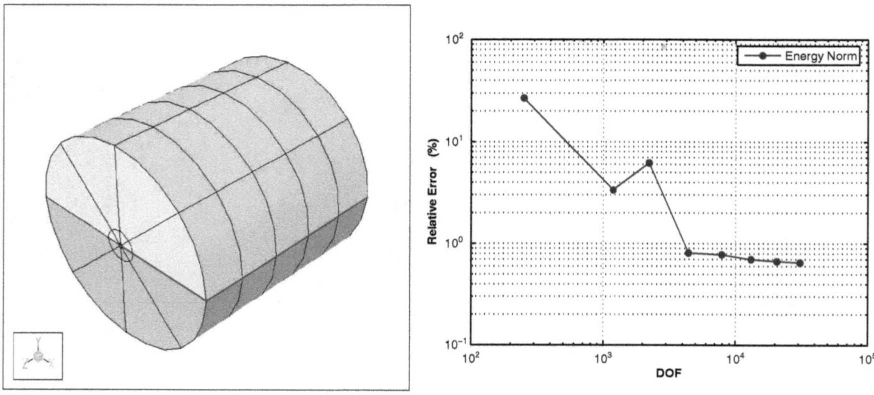

Fig. 13.29 (Left): The p -FEM model, having 160 elements. (Right): Convergence rate of the relative error in the energy norm.

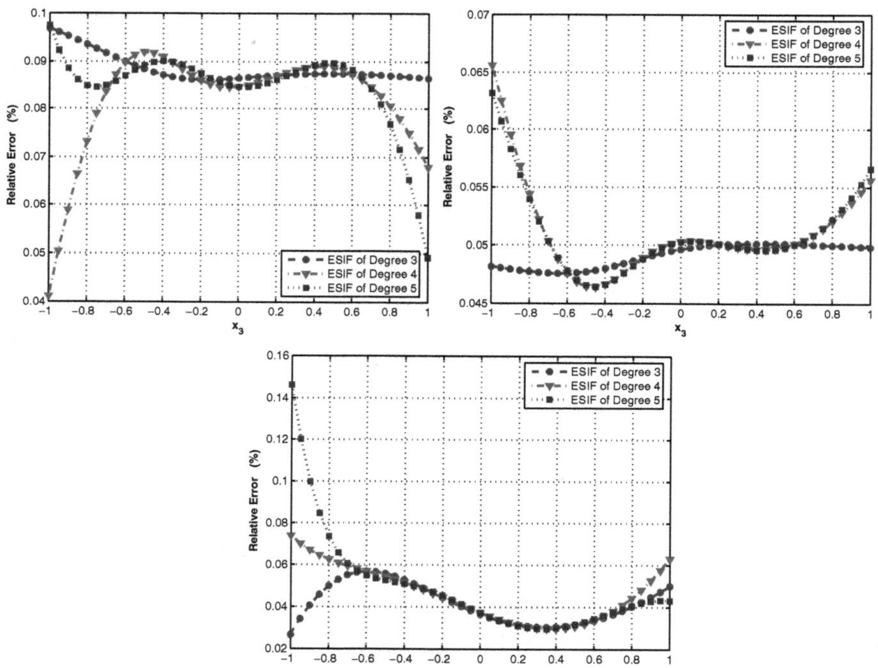

Fig. 13.30 Relative error of extracted ESIF. Eigenfunctions computed using $p = 6$ and four element model, ESIF computed using $BJ^{(k)}$ with $k = 3, 4, 5$ and $R = 0.05$.

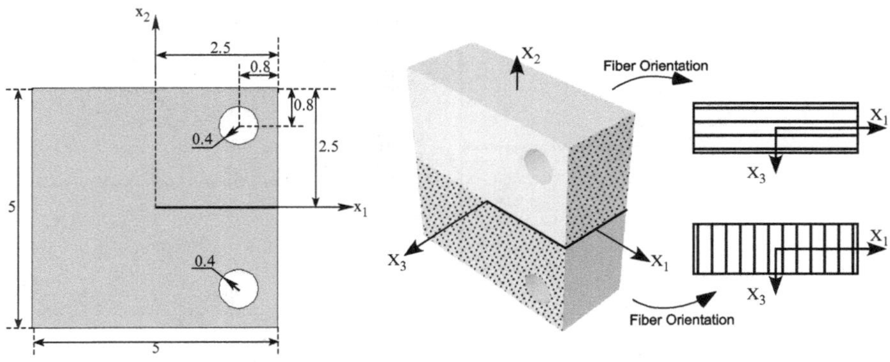

Fig. 13.31 Dimensions of CTS. The thickness of the specimen is 2 ranging over $-1 < x_3 < 1$.

Observe that the relative error of the extracted ESIFs is less than 0.2% . These results indicate that the method is accurate and efficient, and may be applied to realistic engineering problems for which analytical solutions are unavailable, as addressed in the next subsection.

13.3.6 Numerical Example: CTS, Crack at the Interface of Two Anisotropic Materials

Consider the classical compact tension specimen (CTS) of a constant thickness 2 ($-1 < x_3 < 1$), shown in Figure 13.31. The CTS's faces are traction-free and it is loaded by bearing loads at the tearing holes having an equivalent force of 100 Newton in the x_2 direction as seen in Figure 13.17. Although the loading is independent of x_3, because of the vertex singularities at $x_3 = \pm 1$ we anticipate a variation in the ESIFs as the vertices are approached. The domain is discretized using a p -FE mesh with geometric progression toward the singular edge with a factor of 0.15 where the smallest layer in the vicinity of the edge is at $r = 0.15^2$. In the x_3 direction we also used a mesh graded in a geometric progression close to the vertex singularity at $x_3 = \pm 1$. The smallest layer in the vicinity of the vertex is $-1 < x_3 < -1 + 0.15^2$, $1 < x_3 < 1 - 0.15^2$: see Figure 13.17.

The CTS is made of two orthotropic materials, both made of the same high-modulus graphite-epoxy system (13.116) with different fiber orientations. The orientation of fibers of the upper material is along the x_1 direction whereas the orientation at the lower material is along the x_3 direction.

The first three eigenvalues for this example problem, computed using an eight elements model and $p = 15$, are given in (13.117). The eigenfunctions, duals, and shadows associated with the first three eigenvalues are presented in Figures 13.32, 13.33, and 13.34 computed by eight elements, $p = 15$.

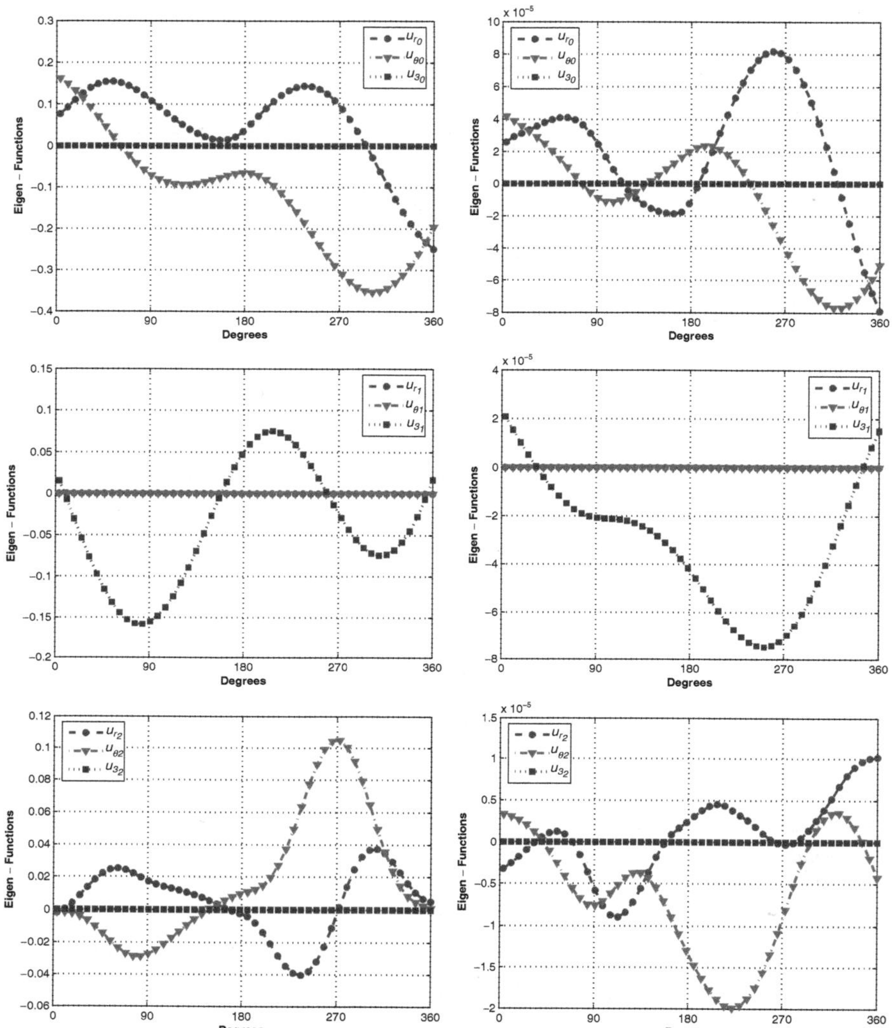

Fig. 13.32 Real part of the eigenfunctions (left) and dual eigenfunctions (right) associated with $\alpha_{1,2} = 0.5 \pm \imath 0.0510612$, computed by eight elements, $p = 15$.

We extract the ESIFs $A_1, A_2,$ and A_3 by increasing the polynomial order of approximation: 3, 5, 7, 9 and 11 at $R = 0.05$ (there was no noticeable difference between the ESIFs extracted at $R = 0.05$ and at $R = 0.1$). The extracted ESIFs are presented in Figure 13.35.

One may observe the good convergence of the ESIFs as the order of the extraction polynomial is increased. Although the ESIFs are influenced by the vertex singularity at $x_3 = \pm 1$, as we increase their polynomial order, the extracted ESIFs converge

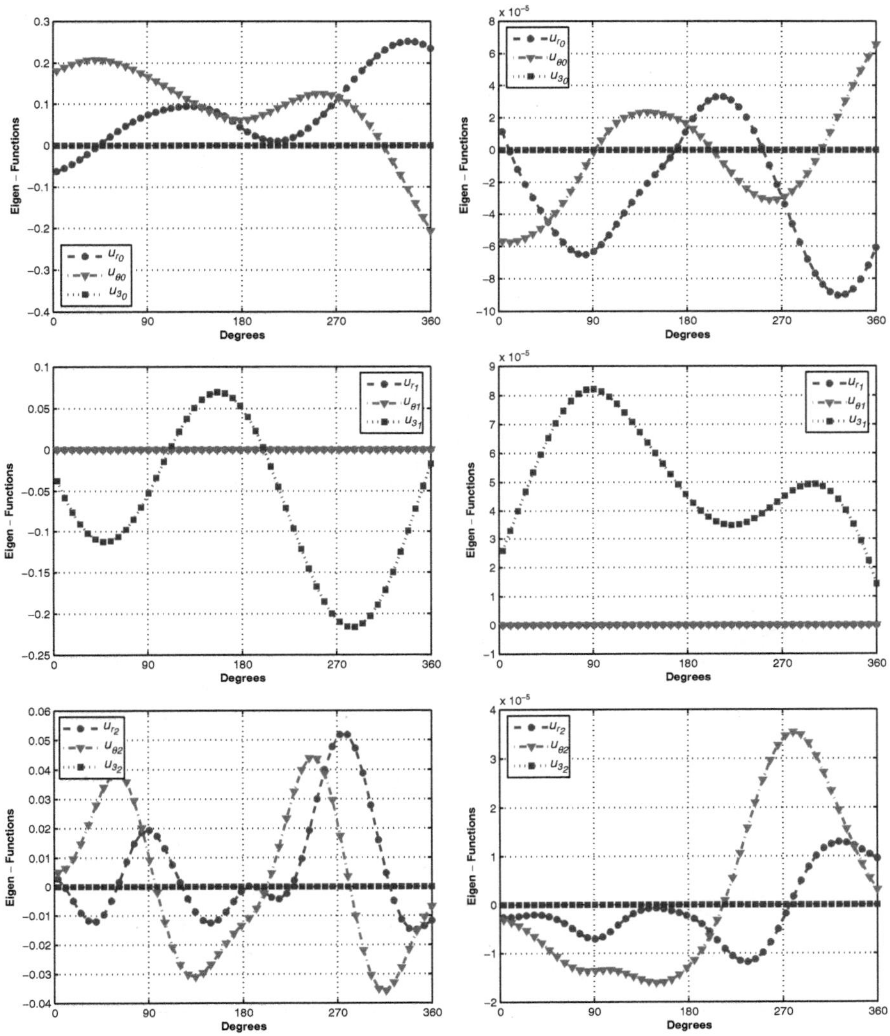

Fig. 13.33 Imaginary part of the eigenfunctions (left) and dual eigenfunctions (right) associated with $\alpha_{1,2} = 0.5 \pm \imath 0.0510612$, computed by eight elements, $p = 15$.

closer to the vertices and provide a better approximation. This example demonstrates the efficiency and accuracy of the ESIF extraction method, and its excellent results also in the close vicinity of the vertices.

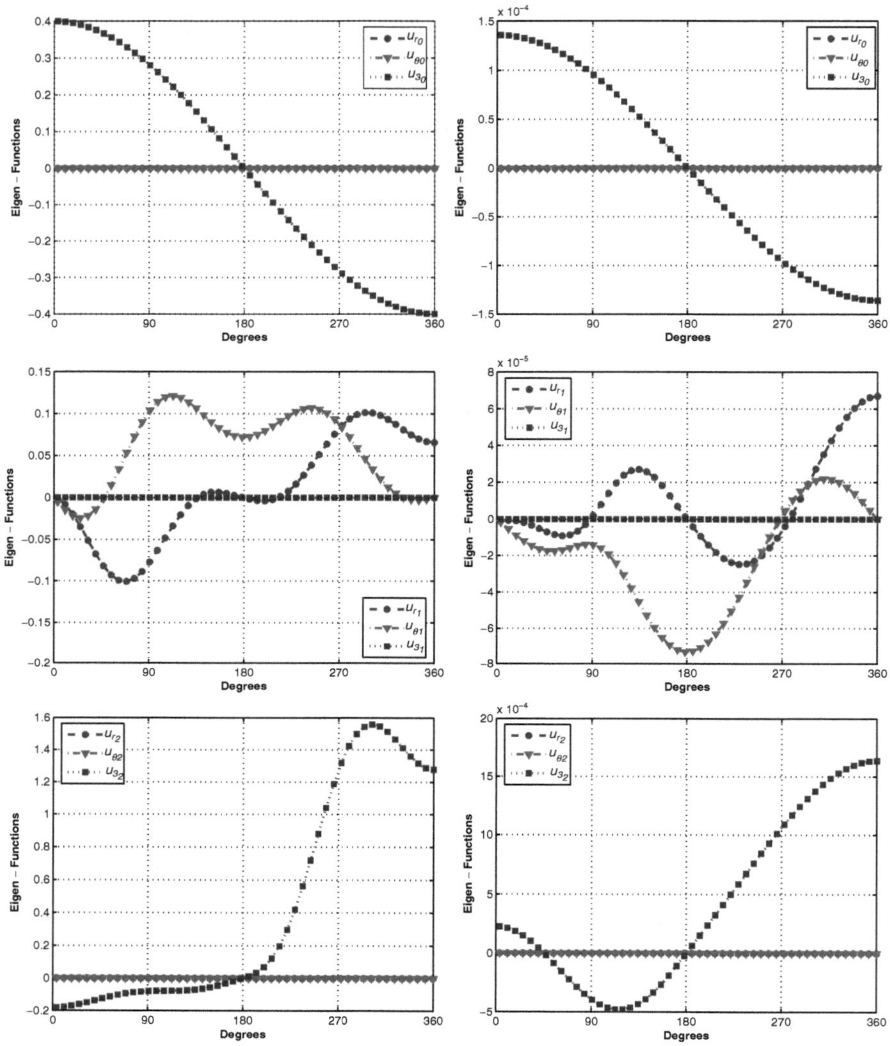

Fig. 13.34 Eigenfunctions (left) and dual eigenfunctions (right) associated with $\alpha_3 = 0.5$, computed by eight elements, $p = 15$.

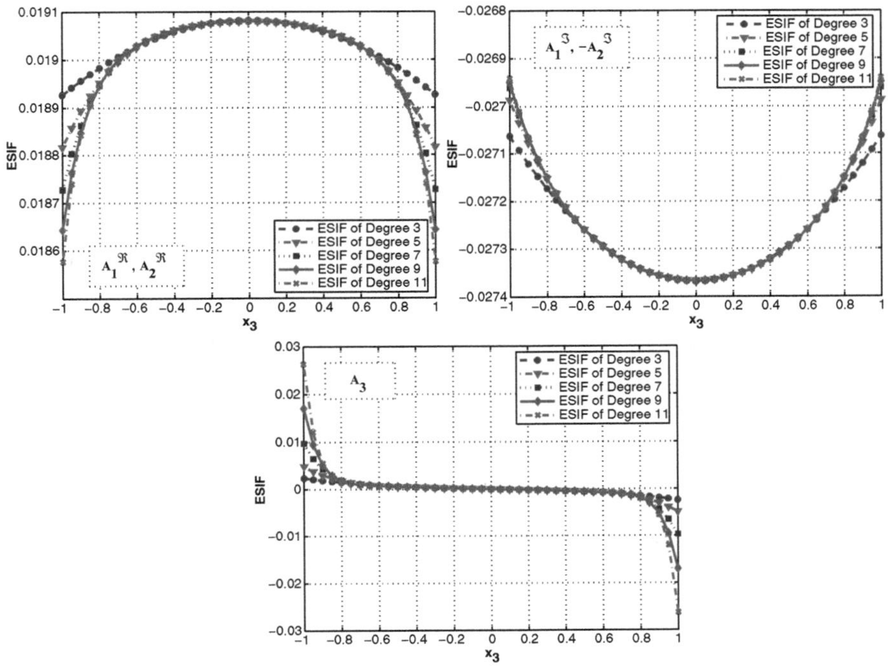

Fig. 13.35 ESIFs extracted using $BJ^{(k)}$ with $k = 3, 5, 7, 9, 11$ for the CTS problem.

Chapter 14
Remarks on Circular Edges and Open Questions

This last chapter is devoted to our latest results on circular edges, and some open questions that are the aim and scope of future research. In daily practice, in reality, most edges are curved in three-dimensional domains and therefore these are of utmost engineering interest. Here we concentrate on circular edges (a "penny-shaped crack" being a special renowned case) in a 3-D domain, and derive explicitly a singular series expansion in the vicinity of such an edge for the simplest scalar elliptic operator, the Laplace operator. The displacements and stress fields associated with the elasticity system are provided in a recent paper [208].

14.1 Circular Singular Edges in 3-D Domains: The Laplace Equation

The first three singular terms for the solution of the Laplace equation in the vicinity of a circular edge with homogeneous Dirichlet boundary conditions was analyzed from a theoretical viewpoint already in [186]. The first two terms in the Neumann case are provided in [16] when the edge is the boundary of a smooth plane crack surface. Here, we provide a systematic analysis of the explicit mathematical description of the solution in the vicinity of a circular edge, and formulas for the computation of all terms in the series expansion.

Let us consider as a model a domain generated by rotating the 2-D plane Ω having a reentrant corner with an opening $\omega \in (0, 2\pi]$ (the case of a crack, $\omega = 2\pi$, is included) along the x_3 axis, as shown in Figure 14.1. The cylindrical coordinate system r, θ, x_3 and the coordinate system attached to the circular edge ρ, φ, θ is shown in Figure 14.1.

It is important to emphasize that the domain's geometry does not need to be axisymmetric, nor the boundary conditions away from the singular edge, but only the generated circular singular edge. An example of several different circular singular edges to which the analysis in this manuscript is applicable are shown in

Z. Yosibash, *Singularities in Elliptic Boundary Value Problems and Elasticity and Their Connection with Failure Initiation*, Interdisciplinary Applied Mathematics 37, DOI 10.1007/978-1-4614-1508-4_14,

Fig. 14.1 Model domain of interest Ω and the coordinate systems.

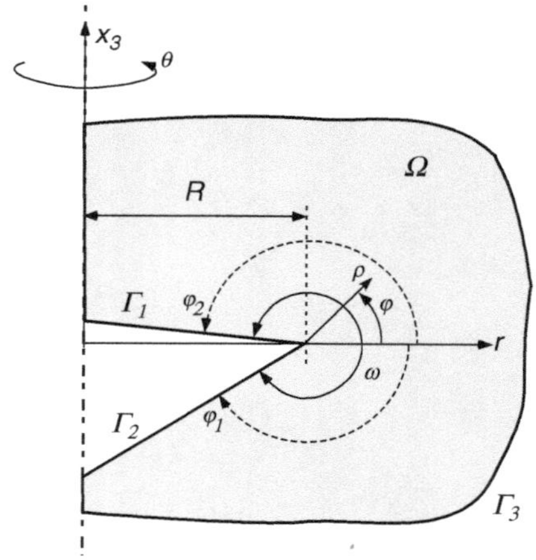

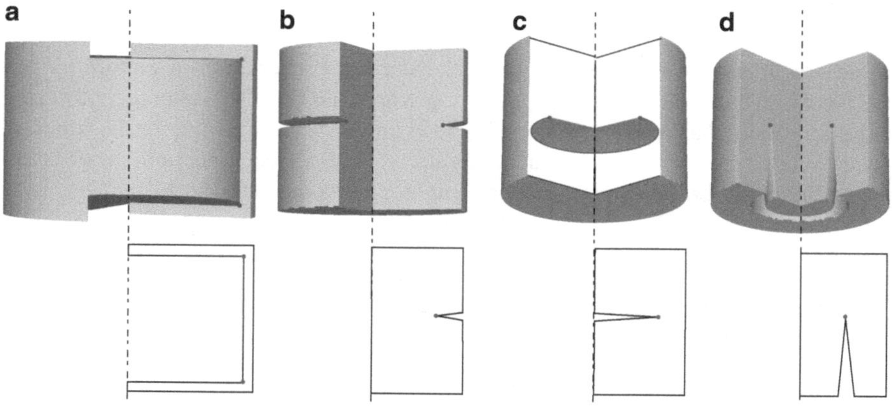

Fig. 14.2 Different types of singular circular edges (only a sector is plotted, so the circular edge is clearly visible. The domain in (c) includes the renowned "penny-shaped" crack.

Figure 14.2. For example, the lower singular edge in Figure 14.2(a) is determined by $\varphi \in (-\pi, \pi/2)$, and the outer circular crack in Figure 14.2(b) is determined by $\varphi \in (0, 2\pi)$, whereas the penny-shaped crack in (c) is determined by $\varphi \in (-\pi, \pi)$. Finally the reentrant corner with the solid angle ω in Figure 14.2 (d) is determined by $\varphi \in ((\pi - \omega)/2, (\pi + \omega)/2)$.

We are interested in solutions $\tau(\boldsymbol{x})$ of the equation

$$\triangle^{3D}\tau \stackrel{\text{def}}{=} \left(\partial_{rr} + \frac{1}{r}\partial_r + \frac{1}{r^2}\partial_{\theta\theta} + \partial_{33}\right)\tau = 0. \tag{14.1}$$

Homogeneous Dirichlet or Neumann boundary conditions are considered on $\Gamma_1 \times [0, 2\pi]$ and $\Gamma_2 \times [0, 2\pi]$.

The solution in the vicinity of the edge is of interest, so we perform a change of coordinates as follows:

$$r = \rho \cos\varphi + R, \quad x_3 = \rho \sin\varphi. \tag{14.2}$$

The Laplace operator in the new coordinates is given by

$$\Delta^{3D} = \partial_{\rho\rho} + \frac{1}{\rho}\partial_\rho + \frac{1}{\rho^2}\partial_{\varphi\varphi} + \frac{1}{r}\left[\cos\varphi\partial_\rho - \frac{1}{\rho}\sin\varphi\partial_\varphi\right] + \frac{1}{r^2}\partial_{\theta\theta}, \tag{14.3}$$

An asymptotic solution close to $\rho/R \ll 1$ is sought. First, for simplicity, let us assume that the solution in an axisymmetric domain is sought, i.e., it is independent of θ.

14.1.1 Axisymmetric Case, $\partial_{\theta\theta}\tau \equiv 0$

If both the domain and boundary conditions are independent of θ, then the last term in (14.3) vanishes, and the Laplace operator for $\rho/R \ll 1$ reads

$$\Delta^{\text{Axi}} = \partial_{\rho\rho} + \frac{1}{\rho}\partial_\rho + \frac{1}{\rho^2}\partial_{\varphi\varphi} + \frac{1}{r}\left[\cos\varphi\partial_\rho - \frac{1}{\rho}\sin\varphi\partial_\varphi\right]. \tag{14.4}$$

Remark 14.1 *Since $r \to \infty$ as $R \to \infty$, one may observe that $\Delta^{Axi} \overset{R\to\infty}{\longrightarrow} \Delta^{2D}$.*

Axisymmetric solutions τ of (14.1) are equivalently the solutions of $\frac{r}{R}\Delta^{\text{Axi}}$ $\tau = 0$, i.e.,

$$(1+\frac{\rho}{R}\cos\varphi)\left[\partial_{\rho\rho} + \frac{1}{\rho}\partial_\rho + \frac{1}{\rho^2}\partial_{\varphi\varphi}\right]\tau + \frac{1}{R}\left[\cos\varphi\partial_\rho - \frac{1}{\rho}\sin\varphi\partial_\varphi\right]\tau = 0. \tag{14.5}$$

Multiplying by ρ^2, we find another equivalent equation,

$$\left[(\rho\partial_\rho)^2 + \partial_{\varphi\varphi}\right]\tau + \frac{\rho}{R}\left[\cos\varphi(\rho\partial_\rho) - \sin\varphi\partial_\varphi + \cos\varphi\big((\rho\partial_\rho)^2 + \partial_{\varphi\varphi}\big)\right]\tau = 0. \tag{14.6}$$

As already recognized in a previous chapter, the solution in the vicinity of the singular point in the 2-D cross-section Ω can be obtained as an asymptotic series defined by eigenpairs of a one-dimensional boundary value problem on the interval $\varphi \in (\varphi_1, \varphi_1 + \omega)$. We denoted this eigenpair by α and $\phi_0(\varphi)$. Then it is conceivable (to be shown in the sequel) that for the axisymmetric case, a solution is formed as an asymptotic series of the form

$$\tau = A\rho^\alpha \sum_{i=0}^{\infty}\left(\frac{\rho}{R}\right)^i \phi_i(\varphi). \tag{14.7}$$

Boundary Conditions: To satisfy the homogeneous boundary conditions, the series representation has to satisfy the following constraints on $\varphi = \varphi_1$ and $\varphi = \varphi_2 = \varphi_1 + \omega$:

$$\phi_i(\varphi = \varphi_1, \varphi_2) = 0 \text{ in the Dirichlet case,} \tag{14.8}$$

$$\phi_i'(\varphi = \varphi_1, \varphi_2) = 0 \text{ in the Neumann case.} \tag{14.9}$$

Substitute (14.7) in (14.6) to obtain

$$\begin{aligned} A\Big\{ &\left[\alpha^2\phi_0 + \phi_0''\right] \\ &+\frac{\rho}{R}\left[\left((\alpha+1)^2\phi_1 + \phi_1''\right) + \alpha\cos\varphi\phi_0 - \sin\varphi\phi_0' + \cos\varphi\left(\alpha^2\phi_0 + \phi_0''\right)\right] \\ &+\frac{\rho^2}{R^2}\left[\left((\alpha+2)^2\phi_2+\phi_2''\right) + (\alpha+1)\cos\varphi\phi_1 - \sin\varphi\phi_1' + \cos\varphi\left((\alpha+1)^2\phi_1+\phi_1''\right)\right] \\ &+\frac{\rho^3}{R^3}\left[\left((\alpha+3)^2\phi_3+\phi_3''\right) + (\alpha+2)\cos\varphi\phi_2 - \sin\varphi\phi_2' + \cos\varphi\left((\alpha+2)^2\phi_2+\phi_2''\right)\right] \\ &+\cdots\Big\} = 0. \end{aligned} \tag{14.10}$$

To satisfy the above equation for any A and ρ, the following relationships must hold:

$$\alpha^2\phi_0 + \phi_0'' = 0, \tag{14.11}$$

$$(\alpha+1)^2\phi_1 + \phi_1'' = -\left(\alpha\cos\varphi\phi_0 - \sin\varphi\phi_0'\right) - \cos\varphi\left(\alpha^2\phi_0 + \phi_0''\right), \tag{14.12}$$

$$\begin{aligned} (\alpha+2)^2\phi_2 + \phi_2'' &= -\left((\alpha+1)\cos\varphi\phi_1 - \sin\varphi\phi_1'\right) - \cos\varphi\left((\alpha+1)^2\phi_1 + \phi_1''\right), \\ (\alpha+3)^2\phi_3 + \phi_3'' &= -\left((\alpha+2)\cos\varphi\phi_2 - \sin\varphi\phi_2'\right) - \cos\varphi\left((\alpha+2)^2\phi_2 + \phi_2''\right), \\ &\dots \end{aligned}$$

Substituting the RHS of equation (14.11) in (14.12), one obtains

$$\alpha^2\phi_0 + \phi_0'' = 0, \quad \varphi_1 < \varphi < \varphi_2, \tag{14.13}$$

$$(\alpha+1)^2\phi_1 + \phi_1'' = -\left(\alpha\cos\varphi\phi_0 - \sin\varphi\phi_0'\right), \quad \varphi_1 < \varphi < \varphi_2, \tag{14.14}$$

$$\begin{aligned} (\alpha+i)^2\phi_i + \phi_i'' &= -\left[(\alpha+i)(\alpha+i-1)\cos\varphi\phi_{i-1} - \sin\varphi\phi_{i-1}' + \cos\varphi\phi_{i-1}''\right], \\ &\qquad i \geq 2, \quad \varphi_1 < \varphi < \varphi_2. \end{aligned} \tag{14.15}$$

These equations have to be completed by the boundary conditions (14.8) or (14.9). Note the following:

- The equation (14.13) with BCs (14.8) or (14.9) is the one-dimensional eigenvalue problem corresponding to the 2-D problem over Ω (see, e.g., (1.7)), with eigenvalue α and the primal eigenfunction ϕ_0.

Table 14.1 First four eigenpairs for a crack with homogeneous Neumann BCs.

k	0	1	2	3	4	$\cdots$
α_k	0	$\frac{1}{2}$	1	$\frac{3}{2}$	2	$\cdots$
$\phi_{k,0}$	1	$\sin\frac{\varphi}{2}$	$\cos\varphi$	$\sin\frac{3\varphi}{2}$	$\cos 2\varphi$	$\cdots$

- A recursive system of ordinary differential equations is obtained; once ϕ_0 is computed from (14.13) it can be inserted in (14.14) to obtain the shadow associated with the edge's curvature ϕ_1. Then these can be inserted in (14.15) to obtain the second shadow associated with the edge's curvature ϕ_2, etc.
- Only particular solutions in (14.14) and (14.15) are required.

Because (14.7) corresponds to only one representative eigenpair, the complete solution should be a sum over all eigenpairs $\alpha_k, \phi_{k,i}$. Thus it is a double sum series:

$$\tau = \sum_k A_k \rho^{\alpha_k} \sum_{i=0}^{\infty} \left(\frac{\rho}{R}\right)^i \phi_{k,i}(\varphi). \tag{14.16}$$

Remark 14.2 *For each primal eigenfunction and shadow $\phi_{k,i}(\varphi)$, the first index k represents the eigenvalue α_k to which this eigenfunction is associated, whereas the second index $i \geq 1$ represents the rank of the curvature shadow terms. Here $\alpha_k = \frac{k\pi}{\omega}$, where $k = 0, 1, 2, \ldots$ for homogeneous Neumann BCs, and $k = 1, 2, 3, \ldots$ for homogeneous Dirichlet BCs.*

Remark 14.3 *If $\alpha + 1$ is not an eigenvalue of equation* (14.13) *with BCs* (14.8) *or* (14.9)*, there exists a unique solution ϕ_1 to equation* (14.14)*. On the other hand, if $\alpha + 1$ is itself an eigenvalue, then it can happen either that* (14.14) *has no solution (then the ansatz* (14.7) *has to be completed with logarithmic terms), or* (14.14) *has infinitely many solutions. The same situation holds for equation* (14.15)*, depending on whether $\alpha + i$ is an eigenvalue or not.*

In the special case of a crack, we have $\alpha_k = \frac{k}{2}$, and therefore resonances (i.e., $\alpha + i$ is an eigenvalue) always occur. Nevertheless, as proved in [43], logarithmic terms never appear: Equations (14.14) *and* (14.15) *with Dirichlet or Neumann BCs are always solvable. An* orthogonality condition *against the eigenvector makes the solution unique; see* (14.18)*.*

14.1.1.1 A Specific Example Problem: Penny-Shaped Crack with Axisymmetric Loading and Homogeneous Neumann BCs

As an example problem, consider the penny-shaped crack shown in Figure 14.2(c), $\varphi_1 = -\pi$, $\omega = 2\pi$ ($\varphi_2 = \pi$), in an axisymmetric domain. For the crack in a 2-D cross-section with homogeneous Neumann BCs, 2-D eigenpairs are known; see Table 14.1: They are obtained by solving (14.13) complemented by BCs (14.9).

Equations (14.13)-(14.15) can be solved (cf. Remark 14.3) for $\alpha_k = 0, 1/2, 1, 3/2$, obtaining $\phi_{0,i}$, $\phi_{1,i}$, $\phi_{2,i}$, $\phi_{3,i}$. They yield the following series solution for a penny-shaped crack with homogeneous Neumann BCs:

$$\begin{aligned}
\tau = &A_0 \\
&+A_1\rho^{\frac{1}{2}}\left[\sin\frac{\varphi}{2} + \left(\frac{\rho}{R}\right)\frac{1}{4}\sin\frac{\varphi}{2} + \left(\frac{\rho}{R}\right)^2\left(\frac{1}{12}\sin\frac{\varphi}{2} - \frac{3}{32}\sin\frac{3\varphi}{2}\right)\right. \\
&\qquad\left. + \left(\frac{\rho}{R}\right)^3\left(\frac{1}{16}\sin\frac{\varphi}{2} - \frac{1}{30}\sin\frac{3\varphi}{2} + \frac{5}{128}\sin\frac{5\varphi}{2}\right) + \cdots\right] \\
&+A_2\rho\left[\cos\varphi - \left(\frac{\rho}{R}\right)\frac{1}{4} + \left(\frac{\rho}{R}\right)^2\frac{3}{16}\cos\varphi - \left(\frac{\rho}{R}\right)^3\left(\frac{9}{128} + \frac{5}{64}\cos 2\varphi\right) + \cdots\right] \\
&+A_3\rho^{\frac{3}{2}}\left[\sin\frac{3\varphi}{2} - \left(\frac{\rho}{R}\right)\frac{1}{4}\sin\frac{\varphi}{2} - \left(\frac{\rho}{R}\right)^2\frac{1}{32}\left(3\sin\frac{\varphi}{2} - \frac{16}{5}\sin\frac{3\varphi}{2}\right)\right. \\
&\qquad\left. + \left(\frac{\rho}{R}\right)^3\left(-\frac{3}{40}\sin\frac{\varphi}{2} + \frac{5}{128}\sin\frac{3\varphi}{2} - \frac{3}{70}\sin\frac{5\varphi}{2}\right) + \cdots\right] \\
&+\cdots.
\end{aligned} \tag{14.17}$$

It is worthwhile to notice that we enforced the following orthogonality conditions on the shadow terms,

$$\int_{\varphi_1=-\pi}^{\varphi_2=\pi} \phi_{k,i}(\varphi)\,\phi_{k+i,0}(\varphi)\,d\varphi = 0, \quad k = 0,1,2,3 \text{ and } i = 1,2,3, \tag{14.18}$$

making them unique.

One may notice that for $R \to \infty$ (the crack edge curvature tends to zero) only the first terms are nonzero so the solution (14.17) reduces to the 2-D solution

$$\tau \overset{R\to\infty}{\longrightarrow} A_0 + A_1\rho^{1/2}\sin\frac{\varphi}{2} + A_2\rho\cos\varphi + A_3\rho^{3/2}\sin\frac{3\varphi}{2} + \cdots.$$

Remark 14.4 *The eigenfunctions and shadows associated with A_0 and A_2 above are* polynomials *in local Cartesian variables $x_1 = \rho\cos\varphi$ and $x_2 = \rho\sin\varphi$. This can be predicted by the general theory [49, 95].*

To verify the correctness of the solution (14.17), we consider a torus with inner radius $r_1 = 1.5$ and outer radius $r_2 = 2.5$ having a circular crack with the tip at $R = 2$, see Figure 14.3.

Taking $A_1 = 1$ and $A_k = 0, k \neq 1$, (notice that the outer boundary of the torus is $\rho_{out} = 1/2$ and $\rho/R = 1/4$ in the considered example), we prescribed on the outer surface of the torus Dirichlet boundary conditions according to (14.17):

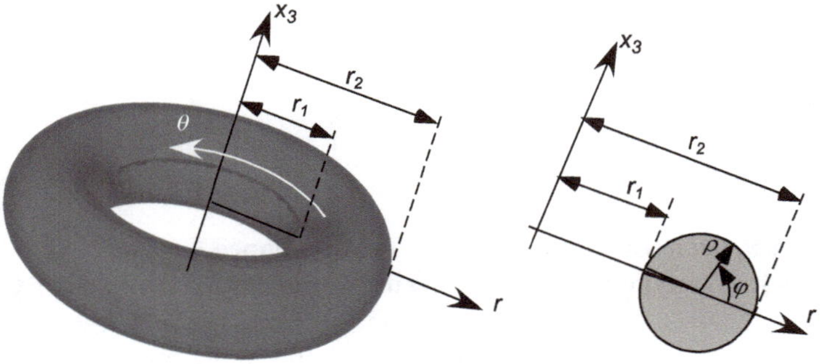

Fig. 14.3 The axisymmetric domain of interest (torus).

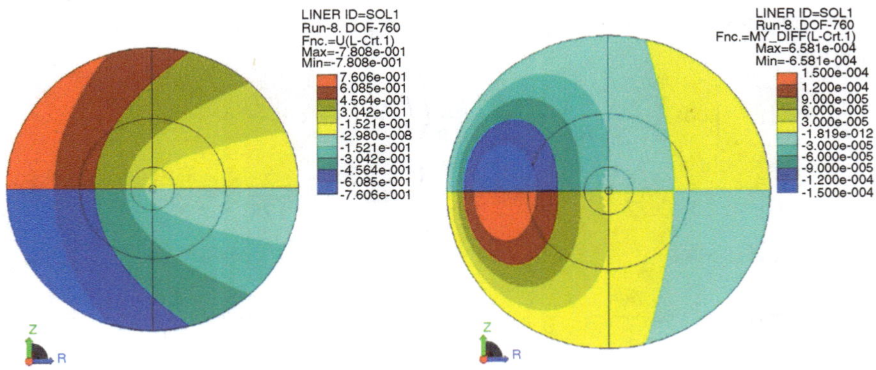

Fig. 14.4 Solution (left) and error (right) for the axisymmetric Laplacian with homogeneous Neumann BCs with $\rho_{\text{out}} = 1/2$, $\rho/R = 1/4$, and $A_1 = 1$. Series up to $(\rho/R)^3$. The axis of symmetry is to the left of the shown domain with the crack from the center of the circle to the left.

$$\tau = \sqrt{\frac{1}{2}}\left[\sin\frac{\varphi}{2} + \left(\frac{1}{4}\right)\frac{1}{4}\sin\frac{\varphi}{2} + \left(\frac{1}{4}\right)^2\left(\frac{1}{12}\sin\frac{\varphi}{2} - \frac{3}{32}\sin\frac{3\varphi}{2}\right)\right.$$
$$\left. + \left(\frac{1}{4}\right)^3\left(\frac{1}{16}\sin\frac{\varphi}{2} - \frac{1}{30}\sin\frac{3\varphi}{2} + \frac{5}{128}\sin\frac{5\varphi}{2}\right)\right],$$

with homogeneous Neumann boundary conditions on the crack face. Because the problem is axisymmetric, we construct a *two-dimensional axisymmetric* finite element (FE) model and solve the Laplace equation over the axisymmetric cross-section using a high-order FE analysis. In Figure 14.4, left, the finite element solution (τ^{FE}) at polynomial level $p = 8$ is shown, whereas in Figure 14.4, right, the difference between the analytical and FE solutions is shown $\tau - \tau^{\text{FE}}$. As may be observed $\tau - \tau^{\text{FE}}$ is three and a half orders of magnitude smaller compared to τ, indicating the correctness of the derived analytical solution. If only terms up to $(\rho/R)^2$ are applied to the boundary of the domain, then the error $\tau - \tau^{\text{FE}}$ increases by one order of magnitude, as expected.

Table 14.2 First three eigenpairs for a crack with homogeneous Dirichlet BCs.

k	α_k	$\phi_{k,0}(\varphi)$
1	$\frac{1}{2}$	$\cos\frac{\varphi}{2}$
2	1	$\sin\varphi$
3	$\frac{3}{2}$	$\cos\frac{3\varphi}{2}$

14.1.1.2 A Specific Example Problem: Penny-Shaped Crack with Axisymmetric Loading and Homogeneous Dirichlet BCs

As in Section 14.1.1.1 we present here the first terms in the asymptotic series solution for a penny-shaped crack, $\varphi_1 = -\pi$, $\omega = 2\pi$ in an axisymmetric domain with homogeneous Dirichlet BCs (14.8). Now the eigenpairs are as shown in Table 14.2.

We obtain the following expression for the first terms in the asymptotic series solution:

$$\begin{aligned}
\tau = A_1\rho^{\frac{1}{2}}\Bigg[&\cos\frac{\varphi}{2} - \left(\frac{\rho}{R}\right)\frac{1}{4}\cos\frac{\varphi}{2} + \left(\frac{\rho}{R}\right)^2\left(\frac{1}{12}\cos\frac{\varphi}{2} + \frac{3}{32}\cos\frac{3\varphi}{2}\right)\\
&+\left(\frac{\rho}{R}\right)^3\left(-\frac{1}{16}\cos\frac{\varphi}{2} - \frac{1}{30}\cos\frac{3\varphi}{2} - \frac{5}{128}\cos\frac{5\varphi}{2}\right) + \cdots\Bigg]\\
+A_2\rho\sin\varphi&\\
+A_3\rho^{\frac{3}{2}}\Bigg[&\cos\frac{3\varphi}{2} - \left(\frac{\rho}{R}\right)\frac{1}{4}\cos\frac{\varphi}{2} + \left(\frac{\rho}{R}\right)^2\left(\frac{3}{32}\cos\frac{\varphi}{2} + \frac{1}{10}\cos\frac{3\varphi}{2}\right)\\
&+\left(\frac{\rho}{R}\right)^3\left(-\frac{3}{40}\cos\frac{\varphi}{2} - \frac{5}{128}\cos\frac{3\varphi}{2} - \frac{3}{70}\cos\frac{5\varphi}{2}\right) + \cdots\Bigg]\\
+\cdots.&
\end{aligned}$$

Here we still enforce the orthogonality conditions (14.18) in order to have uniqueness.

The first terms in the asymptotic solution for the same specific problem are provided in [186, p. 293] (the third term in this reference is erroneous).

14.1.1.3 A Specific Example Problem: Circumferential Crack with Axisymmetric Loading and Homogeneous Neumann BCs

As in Section 14.1.1.1 we present here the first terms in the asymptotic series solution for a circumferential crack (see Figure 14.2(d)), $\varphi_1 = -\frac{\pi}{2}$, $\omega = 2\pi$ in an axisymmetric domain with homogeneous Neumann BCs (14.9), still taking the orthogonality condition (14.18) into account:

$$\begin{aligned}
\tau = A_0 + A_1\rho^{\frac{1}{2}}\Bigg[&\left(\sin\frac{\varphi}{2} - \cos\frac{\varphi}{2}\right)\\
&+\left(\frac{\rho}{R}\right)\left(\frac{1}{4}\left(\sin\frac{\varphi}{2} + \cos\frac{\varphi}{2}\right) + \frac{1}{12}\left(\sin\frac{3\varphi}{2} - \cos\frac{3\varphi}{2}\right)\right) + \cdots\Bigg].
\end{aligned} \tag{14.19}$$

14.1.2 General Case

If no axisymmetric assumption is imposed on the data (only the edge is circular), then the full Laplace operator Δ^{3D} in (14.3) has to be considered. As for (14.5)-(14.6), we find that solutions τ of (14.1) are equivalently the solutions of $(\frac{r}{R})^2\rho^2\Delta^{3D}\tau = 0$, i.e.,

$$(1+\frac{\rho}{R}\cos\varphi)^2\left[(\rho\partial_\rho)^2+\partial_{\varphi\varphi}\right]\tau + \frac{\rho}{R}(1+\frac{\rho}{R}\cos\varphi)\Big[\cos\varphi(\rho\partial_\rho)-\sin\varphi\partial_\varphi\Big]\tau + \left(\frac{\rho}{R}\right)^2\partial_{\theta\theta}\tau = 0. \quad (14.20)$$

To condense formulas, let us introduce the operators

$$m_0(\rho\partial_\rho;\partial_\varphi) = (\rho\partial_\rho)^2+\partial_{\varphi\varphi}, \quad m_{01}(\rho\partial_\rho;\partial_\varphi) = \cos\varphi(\rho\partial_\rho)-\sin\varphi\partial_\varphi. \quad (14.21)$$

Then equation (14.20) is equivalent to

$$m_0\tau+\frac{\rho}{R}\Big[2\cos\varphi m_0+m_{01}\Big]\tau+\left(\frac{\rho}{R}\right)^2\Big[\cos^2\varphi m_0+\cos\varphi m_{01}+\partial_{\theta\theta}\Big]\tau = 0. \quad (14.22)$$

In the general case, for a circular edge the following form of expansion series is appropriate:

$$\tau = \sum_{\ell=0,2,4,\dots}\sum_{k=0}\partial_\theta^\ell A_k(\theta)\,\rho^{\alpha_k}\left(\frac{\rho}{R}\right)^\ell\sum_{i=0}^{\infty}\left(\frac{\rho}{R}\right)^i\phi_{\ell,k,i}(\varphi). \quad (14.23)$$

Remark 14.5 *Observe that $\phi_{0,k,i} = \phi_{k,i}$ (associated with the curvature for an axisymmetric case), so these are known for the axisymmetric analysis.*

Comparing this asymptotic expansion to the one along a straight edge given in Chapter 10, e.g. (10.52) , one notices one extra sum, implying that for each primal eigenfunction there are two levels of shadow functions: one set is associated with the derivatives of A_k (the index ℓ), and the other set is associated with the "curvature terms," i.e., the powers ρ/R (index i).

The splitting in (14.22) provides an elegant and convenient way to formulate the series expansion of the solution. Introducing the definition for a general term in the expansion (14.23),

$$\Phi_{\ell,k,i} \stackrel{\text{def}}{=} \rho^{\alpha_k}\left(\frac{\rho}{R}\right)^{\ell+i}\phi_{\ell,k,i}(\varphi), \quad (14.24)$$

we observe that

$$m_0(\rho\partial_\rho;\partial_\varphi)\Phi_{\ell,k,i}(\rho,\varphi) = \rho^{\alpha_k}\left(\frac{\rho}{R}\right)^{\ell+i}m_0(\alpha_k+\ell+i;\partial_\varphi)\phi_{\ell,k,i}(\varphi), \quad (14.25)$$

$$m_{01}(\rho\partial_\rho;\partial_\varphi)\Phi_{\ell,k,i}(\rho,\varphi) = \rho^{\alpha_k}\left(\frac{\rho}{R}\right)^{\ell+i}m_{01}(\alpha_k+\ell+i;\partial_\varphi)\phi_{\ell,k,i}(\varphi).$$

Substituting (14.23) into (14.22), one deduces

$$
\begin{aligned}
0 = A_k(\theta) \times \Big\{ & m_0(\alpha_k)\phi_{0,k,0} \\
& + \Big(\frac{\rho}{R}\Big)\Big[m_0(\alpha_k+1)\phi_{0,k,1} + \big(2\cos\varphi m_0(\alpha_k) + m_{01}(\alpha_k)\big)\phi_{0,k,0}\Big] \\
& + \Big(\frac{\rho}{R}\Big)^2\Big[m_0(\alpha_k+2)\phi_{0,k,2} + \big(2\cos\varphi m_0(\alpha_k+1) + m_{01}(\alpha_k+1)\big)\phi_{0,k,1} \\
& \qquad + \big(\cos^2\varphi m_0(\alpha_k) + \cos\varphi m_{01}(\alpha_k)\big)\phi_{0,k,0}\Big] + \cdots\Big\} \\
+A_k''(\theta) \times \Big\{ & \Big(\frac{\rho}{R}\Big)^2\Big[m_0(\alpha_k+2)\phi_{2,k,0} + \phi_{0,k,0}\Big] \\
& + \Big(\frac{\rho}{R}\Big)^3\Big[m_0(\alpha_k+3)\phi_{2,k,1} + \big(2\cos\varphi m_0(\alpha_k+2) \\
& \qquad + m_{01}(\alpha_k+2)\big)\phi_{2,k,0} + \phi_{0,k,1}\Big] \\
& + \Big(\frac{\rho}{R}\Big)^4\Big[m_0(\alpha_k+4)\phi_{2,k,2} + \big(2\cos\varphi m_0(\alpha_k+3) \\
& \qquad + m_{01}(\alpha_k+3)\big)\phi_{2,k,1} + \big(\cos^2\varphi m_0(\alpha_k+2) \\
& \qquad + \cos\varphi m_{01}(\alpha_k+2)\big)\phi_{2,k,0} + \phi_{0,k,2}\Big] + \cdots\Big\}. \qquad (14.26)
\end{aligned}
$$

Equation (14.26) has to hold for any $(\rho/R)^i$ and for any $\partial_\theta^\ell A_k$, resulting in the following recursive set of ordinary differential equations for the determination of the eigenfunctions and shadows $\phi_{\ell,k,i}(\varphi)$:

$$
\begin{aligned}
m_0(\alpha_k+\ell+i) \quad \phi_{\ell,k,i} & \qquad\qquad (14.27) \\
= -\big(2\cos\varphi m_0(\alpha_k+\ell+i-1) + m_{01}(\alpha_k+\ell+i-1)\big)\phi_{\ell,k,i-1} & \\
-\big(\cos^2\varphi m_0(\alpha_k+\ell+i-2) + \cos\varphi m_{01}(\alpha_k+\ell+i-2)\big)\phi_{\ell,k,i-2} & \\
-\phi_{\ell-2,k,i}, \qquad \text{for} \quad \ell = 0,2,4,6,\ldots, \quad \text{and} \quad i \geq 0. &
\end{aligned}
$$

Here, by convention, the ϕ's with negative indices are zero.

Equations (14.27) for $\ell = 0$ are equivalent to equations (14.13)-(14.15) associated with the axisymmetric case, and for $\ell = 2,4,6,\ldots$ results in (14.28) associated with the nonaxisymmetric case:

$\ell = 0$

equations (14.13)-(14.15) for the axisymmetric case hold;

$\ell = 2,4,6,\ldots, \qquad i \geq 0,$

$$\begin{aligned}
&(\alpha_k + i + \ell)^2 \phi_{\ell,k,i} + \phi''_{\ell,k,i} \\
&\quad = -(\ell + i + \alpha_k - 1)\,[2(\ell + i + \alpha_k) - 1] \cos\varphi \phi_{\ell,k,(i-1)} \\
&\qquad + \sin\varphi \phi'_{\ell,k,(i-1)} - 2\cos\varphi \phi''_{\ell,k,(i-1)} \\
&\qquad -(\ell + \alpha_k + i - 2)(\ell + \alpha_k + i - 1)\cos^2\varphi \phi_{\ell,k,(i-2)} \\
&\qquad \phi_{(\ell-2),k,i}\,. \qquad (14.28)
\end{aligned}$$

Equations (14.28) are complemented by the homogeneous Dirichlet or Neumann boundary conditions:

$$\phi_{\ell,k,i}(\varphi) = 0 \quad (\varphi = \varphi_1, \varphi_1 + \omega), \quad \text{in case of Drichlet BCs,} \qquad (14.29)$$

$$(\phi_{\ell,k,i})'(\varphi) = 0 \quad (\varphi = \varphi_1, \varphi_1 + \omega), \quad \text{in case of Neumann BCs.} \qquad (14.30)$$

14.1.2.1 A Specific Example Problem: Penny-Shaped Crack for a Nonaxisymmetric Loading and Homogeneous Neumann BCs

Again we consider as an example problem a penny-shaped crack $\varphi_1 = -\pi$, $\omega = 2\pi$, however, the loading may be nonaxisymmetric, as well as the outer boundary of the 3-D domain of interest. The eigenfunctions and a part of the shadow functions $\phi_{0,k,i}(\varphi)$ have been provided by (14.17). As an example, the solution of $\phi_{2,1,0}(\varphi)$, $(\ell = 2, k = 1, i = 0)$ may be obtained from (14.28) for $k = 1$, $\alpha_1 = 1/2$, and $i = 0$, $\ell = 2$. All ϕ's with negative indices in the RHS vanish except one

$$\left(\frac{1}{2} + 0 + 2\right)^2 \phi_{2,1,0} + \phi''_{2,1,0} = -\phi_{0,1,0}, \qquad (14.31)$$

and the homogeneous Neumann BCs read

$$\phi'_{2,1,0}(\varphi = \pm\pi) = 0.$$

From (14.17), $\phi_{0,1,0} = \sin\frac{\varphi}{2}$, and thus the solution of (14.31) can be taken as the particular solution alone:

$$\phi_{2,1,0} = -\frac{1}{6}\sin\frac{\varphi}{2}. \qquad (14.32)$$

Once $\phi_{2,1,0}$ is available, one may proceed to the computation of $\phi_{2,1,1}(\varphi)$ $(\ell = 2, k = 1, i = 1)$ obtained from (14.28), for $k = 1$, $\alpha_1 = 1/2$, and $i = 1$, $\ell = 2$:

$$\begin{aligned}
\left(\frac{1}{2} + 1 + 2\right)^2 \phi_{2,1,1} + \phi''_{2,1,1} &= -\left(2 + 1 + \frac{1}{2} - 1\right)\left[2\left(2 + 1 + \frac{1}{2}\right) - 1\right]\cos\varphi \phi_{2,1,0} \\
&\quad + \sin\varphi \phi'_{2,1,0} - 2\cos\varphi \phi''_{0,1,0} - \phi_{0,1,1}. \qquad (14.33)
\end{aligned}$$

Substituting $\phi_{2,1,0}$ from (14.32) and $\phi_{0,1,1} = \frac{1}{4}\sin\frac{\varphi}{2}$ from (14.17), the particular solution to (14.33) that satisfies the homogeneous Neumann BCs is

$$\phi_{2,1,1} = -\frac{1}{8}\sin\frac{\varphi}{2} + \frac{7}{60}\sin\frac{3\varphi}{2}. \tag{14.34}$$

This procedure may be continued, to finally obtain the terms in the series expansion:

$$\begin{aligned}
\tau = {} & A_0(\theta) \\
& + A_0''(\theta)\left(\frac{\rho}{R}\right)^2\left[-\frac{1}{4} + \left(\frac{\rho}{R}\right)\frac{5}{16}\cos\varphi - \left(\frac{\rho}{R}\right)^2\left(\frac{19}{128} + \frac{11}{64}\cos 2\varphi\right) + \cdots\right] + \cdots \\
& + A_1(\theta)\rho^{\frac{1}{2}}\left[\sin\frac{\varphi}{2} + \left(\frac{\rho}{R}\right)\frac{1}{4}\sin\frac{\varphi}{2} + \left(\frac{\rho}{R}\right)^2\left(\frac{1}{12}\sin\frac{\varphi}{2} - \frac{3}{32}\sin\frac{3\varphi}{2}\right)\right. \\
& \qquad \left. + \left(\frac{\rho}{R}\right)^3\left(\frac{1}{16}\sin\frac{\varphi}{2} - \frac{1}{30}\sin\frac{3\varphi}{2} + \frac{5}{128}\sin\frac{5\varphi}{2}\right) + \cdots\right] \\
& + A_1''(\theta)\rho^{\frac{1}{2}}\left(\frac{\rho}{R}\right)^2\left[-\frac{1}{6}\sin\frac{\varphi}{2} + \left(-\frac{1}{8}\sin\frac{\varphi}{2} + \frac{7}{60}\sin\frac{3\varphi}{2}\right)\left(\frac{\rho}{R}\right) + \cdots\right] + \cdots .
\end{aligned} \tag{14.35}$$

Again, the factors corresponding to A_0 (and all terms of even order) and their derivatives are polynomial in (x_1, x_2).

Remark 14.6 *In the vicinity of a crack with a straight edge along the x_3 axis, the solution admits the expansion*

$$\begin{aligned}
\tau = {} & A_0(z_3) + A_0''(z_3)r^2\left(-\frac{1}{4}\right) + \cdots \\
& + A_1(z_3)r^{\frac{1}{2}}\sin\frac{\varphi}{2} + A_1''(z_3)r^{\frac{5}{2}}\left(-\frac{1}{6}\sin\frac{\varphi}{2}\right) + \cdots .
\end{aligned} \tag{14.36}$$

One may observe that (14.36) *is composed of the same leading terms associated with $i = 0$ as in the expansion* (14.35), *as expected.*

14.1.2.2 A Specific Example Problem: Penny-Shaped Crack for a Nonaxisymmetric Loading and Homogeneous Dirichlet BCs

As an example problem, the first terms of the asymptotic solution for a penny-shaped crack $\varphi_1 = -\pi$, $\omega = 2\pi$, with homogeneous Dirichlet boundary conditions are provided:

$$\tau = A_1(\theta)\rho^{\frac{1}{2}}\left[\cos\frac{\varphi}{2} - \frac{1}{4}\cos\frac{\varphi}{2}\left(\frac{\rho}{R}\right) + \left(\frac{1}{12}\cos\frac{\varphi}{2} + \frac{3}{32}\cos\frac{3\varphi}{2}\right)\left(\frac{\rho}{R}\right)^2\right.$$

$$+\left(-\frac{1}{16}\cos\frac{\varphi}{2}-\frac{1}{30}\cos\frac{3\varphi}{2}-\frac{5}{128}\cos\frac{5\varphi}{2}\right)\left(\frac{\rho}{R}\right)^3+\cdots\Bigg]$$

$$+\ A_1''(\theta)\rho^{\frac{1}{2}}\left(\frac{\rho}{R}\right)^2\left[-\frac{1}{6}\cos\frac{\varphi}{2}+\left(\frac{1}{8}\cos\frac{\varphi}{2}+\frac{7}{60}\cos\frac{3\varphi}{2}\right)\left(\frac{\rho}{R}\right)+\cdots\right]$$

$$+\cdots. \tag{14.37}$$

14.1.2.3 A Specific Example Problem: Hollow Cylinder with Nonaxisymmetric Loading and Homogeneous Neumann BCs

Consider the circular edge having a solid angle of $3\pi/2$ as the upper corner in Figure 14.2(a) with homogeneous Neumann BCs. In this case $\varphi \in (-\frac{\pi}{2}, \pi)$, $\alpha_0 = 0$, and $\alpha_1 = 2/3$, $\alpha_4 = \alpha_5 = \alpha_6 = 1\ldots$, and the first few terms in the asymptotic solution are given by

$$\tau = A_0 + A_0''\left(\frac{\rho}{R}\right)^2\left(-\frac{1}{4}\right) \tag{14.38}$$

$$+A_1\rho^{2/3}\Bigg[\sin\frac{2\varphi}{3}-\frac{1}{\sqrt{3}}\cos\frac{2\varphi}{3}$$

$$+\left(\frac{\rho}{R}\right)\frac{1}{60}\left(5\sqrt{3}\cos\frac{\varphi}{3}-\sqrt{3}\cos\frac{5\varphi}{3}+15\sin\frac{\varphi}{3}+3\sin\frac{5\varphi}{3}\right)$$

$$+\left(\frac{\rho}{R}\right)^2\frac{1}{160}\left(12\sin\frac{2\varphi}{3}-4\sqrt{3}\cos\frac{2\varphi}{3}-15\sin\frac{4\varphi}{3}-5\sqrt{3}\cos\frac{4\varphi}{3}\right)\Bigg]$$

$$+A_1''\rho^{2/3}\left(\frac{\rho}{R}\right)^2\left[\frac{1}{20}\left(\sqrt{3}\cos\frac{2\varphi}{3}-3\sin\frac{2\varphi}{3}\right)\right]+\cdots.$$

14.1.2.4 A Specific Example Problem: Exterior Circular Crack wich Nonaxisymmetric Loading and Homogeneous Neumann BCs

Consider the circular external crack as in Figure 14.2(b) with homogeneous Neumann BCs. In this case, $\varphi \in (0, 2\pi)$, and the first few terms in the asymptotic solution are given by

$$\tau = A_0 + A_0''\left(\frac{\rho}{R}\right)^2\left(-\frac{1}{4}\right) \tag{14.39}$$

$$+A_1\rho^{1/2}\left[\cos\frac{\varphi}{2}-\left(\frac{\rho}{R}\right)\frac{1}{4}\cos\frac{\varphi}{2}+\left(\frac{\rho}{R}\right)^2\left(\frac{1}{12}\cos\frac{\varphi}{2}+\frac{3}{32}\cos\frac{3\varphi}{2}\right)\right]$$

$$+A_1''\rho^{1/2}\left(\frac{\rho}{R}\right)^2\left[-\frac{1}{6}\cos\frac{\varphi}{2}\right]+\cdots.$$

14.2 Circular Singular Edges in 3-D Domains: The Elasticity System

For the sake of completeness, the stress tensor in the vicinity of a nonaxisymmetric penny-shaped traction-free crack is provided here explicitly. It involves, of course, the same typical expansion as for the Laplacian, and we list here terms up to and including order $\sqrt{\rho}$. The derivation is lengthy, and the interested reader may find the details in [208]:

$$\begin{Bmatrix} \sigma_{\rho\rho} \\ \sigma_{\theta\theta} \\ \sigma_{\varphi\varphi} \\ \sigma_{\rho\theta} \\ \sigma_{\rho\varphi} \\ \sigma_{\theta\varphi} \end{Bmatrix} = \left(\frac{1}{4}\right) \frac{K_{\mathrm{I}}(\theta)}{\sqrt{2\pi\rho}} \left[\begin{pmatrix} 5\cos\frac{\varphi}{2} - \cos\frac{3\varphi}{2} \\ \frac{4\lambda}{\lambda+\mu}\cos\frac{\varphi}{2} \\ 3\cos\frac{\varphi}{2} + \cos\frac{3\varphi}{2} \\ 0 \\ \sin\frac{\varphi}{2} + \sin\frac{3\varphi}{2} \\ 0 \end{pmatrix} \right.$$

$$\left. + \left(\frac{\rho}{R}\right) \begin{pmatrix} \frac{5\lambda+13\mu}{4(\lambda+\mu)}\cos\frac{\varphi}{2} - \frac{\lambda+9\mu}{4(\lambda+\mu)}\cos\frac{3\varphi}{2} \\ \frac{2(2\lambda+\mu)(\lambda+5\mu)}{(\lambda+\mu)^2}\cos\frac{\varphi}{2} - \frac{3\lambda+2\mu}{\lambda+\mu}\cos\frac{3\varphi}{2} \\ \frac{3(\lambda+9\mu)}{4(\lambda+\mu)}\cos\frac{\varphi}{2} + \frac{\lambda+9\mu}{4(\lambda+\mu)}\cos\frac{3\varphi}{2} \\ 0 \\ -\frac{\lambda-7\mu}{4(\lambda+\mu)}\sin\frac{\varphi}{2} - \frac{\lambda-7\mu}{4(\lambda+\mu)}\sin\frac{3\varphi}{2} \\ 0 \end{pmatrix} + \cdots \right]$$

$$+ \left(\frac{-1}{4}\right) \frac{K_{\mathrm{I}}'(\theta)}{\sqrt{2\pi\rho}} \left(\frac{\rho}{R}\right) \begin{pmatrix} 0 \\ 0 \\ 0 \\ \frac{2(\lambda-\mu)}{\lambda+\mu}\cos\frac{\varphi}{2} - \frac{2(\lambda+3\mu)}{\lambda+\mu}\cos\frac{3\varphi}{2} \\ 0 \\ \frac{2(\lambda+3\mu)}{\lambda+\mu}\sin\frac{\varphi}{2} + \frac{2(\lambda+3\mu)}{\lambda+\mu}\sin\frac{3\varphi}{2} \end{pmatrix} + \cdots$$

$$+\frac{3}{4}\frac{K_{\mathrm{II}}(\theta)}{\sqrt{2\pi\rho}}\left[\begin{pmatrix} -\frac{5}{3}\sin\frac{\varphi}{2}+\sin\frac{3\varphi}{2} \\ -\frac{4\lambda}{3(\lambda+\mu)}\sin\frac{\varphi}{2} \\ -\sin\frac{\varphi}{2}-\sin\frac{3\varphi}{2} \\ 0 \\ \frac{1}{3}\left(\cos\frac{\varphi}{2}+3\cos\frac{3\varphi}{2}\right) \\ 0 \end{pmatrix}\right.$$

$$+\left(\frac{\rho}{R}\right)\left.\begin{pmatrix} -\frac{51\lambda+107\mu}{60(\lambda+\mu)}\sin\frac{\varphi}{2}+\frac{\lambda+9\mu}{12(\lambda+\mu)}\sin\frac{3\varphi}{2} \\ \frac{2\left(34\lambda^2+83\lambda\mu+45\mu^2\right)}{15(\lambda+\mu)^2}\sin\frac{\varphi}{2}+\frac{3\lambda+2\mu}{3(\lambda+\mu)}\sin\frac{3\varphi}{2} \\ -\frac{\lambda+9\mu}{12(\lambda+\mu)}\sin\frac{\varphi}{2}-\frac{\lambda+9\mu}{12(\lambda+\mu)}\sin\frac{3\varphi}{2} \\ 0 \\ -\frac{23\lambda+31\mu}{60(\lambda+\mu)}cos\frac{\varphi}{2}+\frac{-\lambda+7\mu}{12(\lambda+\mu)}\cos\frac{3\varphi}{2} \\ 0 \end{pmatrix}+\cdots\right]$$

$$+\frac{3}{4}\frac{K'_{\mathrm{II}}(\theta)}{\sqrt{2\pi\rho}}\left(\frac{\rho}{R}\right)\begin{pmatrix} 0 \\ 0 \\ 0 \\ \frac{2(\lambda-\mu)}{3(\lambda+\mu)}\sin\frac{\varphi}{2}+\sin\frac{3\varphi}{2} \\ 0 \\ -\frac{2(\lambda+3\mu)}{3(\lambda+\mu)}\cos\frac{\varphi}{2}+\cos\frac{3\varphi}{2} \end{pmatrix}+\cdots$$

$$+\frac{K_{\mathrm{III}}(\theta)}{\sqrt{2\pi\rho}}\left[\begin{pmatrix} 0 \\ 0 \\ 0 \\ \sin\frac{\varphi}{2} \\ 0 \\ \cos\frac{\varphi}{2} \end{pmatrix}+\left(\frac{\rho}{R}\right)\begin{pmatrix} 0 \\ 0 \\ 0 \\ \frac{7}{4}\sin\frac{\varphi}{2}-\sin\frac{3\varphi}{2} \\ 0 \\ \frac{5}{4}\cos\frac{\varphi}{2}-\cos\frac{3\varphi}{2} \end{pmatrix}+\cdots\right]$$

$$
+\frac{K'_{\mathrm{III}}(\theta)}{\sqrt{2\pi\rho}}\left(\frac{\rho}{R}\right)\begin{pmatrix} -\frac{4}{5}\sin\frac{\varphi}{2} \\ \frac{4}{5}\frac{7\lambda+5\mu}{\lambda+\mu}\sin\frac{\varphi}{2} \\ 0 \\ 0 \\ -\frac{2}{5}\cos\frac{\varphi}{2} \\ 0 \end{pmatrix}+\cdots
$$

Remark 14.7 *For $R \to \infty$ the states of the stresses tend to a plane-strain state. Indeed, by computing $\nu(\sigma_{\rho\rho}+\sigma_{\varphi\varphi})$, one obtains $\nu\left[2\frac{K_I}{\sqrt{2\pi\rho}}\cos\frac{\varphi}{2}-2\frac{K_{II}}{\sqrt{2\pi\rho}}\sin\frac{\varphi}{2}\right]$, which equals to $\sigma_{\theta\theta}$. This is exactly the connection $\sigma_{33}=\nu(\sigma_{11}+\sigma_{22})$ according to a plane-strain situation.*

Remark 14.8 *The primal eigenstresses $\sigma_{\rho\rho}$ and $\sigma_{\varphi\varphi}$ do not depend on the material properties for traction-free boundary conditions on crack faces. However, their shadows do depend on the material properties.*

Remark 14.9 *Comparing the terms associated with the first derivatives of the SIFs ($K'_I(\theta), K'_{II}(\theta), K'_{III}(\theta)$) in [103] with* (14.40)*, one notices that these are identical for $K'_I(\theta)$ and $K'_{III}(\theta)$. The term that multiplies $K'_{II}(\theta)$ in [103] appears in* (14.40)*, but in our expression there are another two expressions proportional to* $\cos\frac{3\varphi}{2}$ *and* $\sin\frac{3\varphi}{2}$ *that are absent in [103].*

14.3 Further Theoretical and Practical Applications

The field of fracture mechanics and its relation to singular solutions has attracted an extensive amount of research during the past half century, and there are still many open questions on singular solutions of elliptic boundary value problems and their connection to failure initiation and propagation in structures of engineering relevance. Some of the unsolved questions that are the topic of ongoing research are as follows:

Among the many failure laws proposed to predict failure initiation at sharp V-notch tips in brittle, 2-D-like structures, which one best correlates to experimental observations (i.e., which one is a valid one)? Is it possible to use experimental "material properties" such as the fracture toughness and strength alone to formulate such failure laws, or should other material properties be defined and measured for V-notched configurations? Such questions also relate to failure laws in materials that are anisotropic (like sapphire) and composite materials.

Because in practice no V-notch is sharp and all contain a small radius at the notch tip, what is the influence of such notch tip radii on the failure initiation law? Can one extend the brittle elastic failure initiation laws to hard metals having a small yielding zone at the V-notch tip?

On top of the above questions, and related to them, the size effect is also a major concern especially because asymptotic analysis is used, i.e., to which structure scale is the asymptotic analysis valid, and can one apply the same solutions and failure laws to micron structures associated with the electronic industry (structures having dimensions at the μmeter level)? Answers to such topics are extremely important from the engineering point of view, especially in the growing area of the electronic device industry.

Surprisingly enough, in realistic 3-D structures no failure law exists even for a crack configuration that is able to predict crack propagation in a general direction. Furthermore, because both edge and vertex singularities exist, the questions whether failure begins along the edge or at the vertex, and what parameters trigger that failure are still unsolved.

Since in realistic 3-D structures cracks usually have curved edges, there is a need to explicitly formulate the asymptotic solution in the vicinity of a curved edge. This has also to be extended to sharp V-notch curved edges.

These and many other important and unsolved problems associated with singular solutions of elliptic partial differential equations and their connection to failure initiation will most probably be the topic of further research in the next half century, with many exciting and beneficial results to engineering practice.

Appendix A
Definition of Sobolev, Energy, and Statically Admissible Spaces and Associated Norms

To define the family of functions that are said to belong to a Sobolev space, we first need some preliminary definitions which are relatively simple and well known. We address both the heat conduction and elasticity systems in two or three dimensions. Let $\tau(\boldsymbol{x})$ and $\chi(\boldsymbol{x})$ be Lebesgue measurable functions (temperature fields), $\boldsymbol{u}$ and $\boldsymbol{v}$ displacement vectors, and $\boldsymbol{\sigma}(\boldsymbol{x})$ and $\boldsymbol{\varsigma}(\boldsymbol{x})$ stress vectors in a two or three-dimensional domain $\Omega \ni \boldsymbol{x}$.

Definition A.1. The L^2 **inner product** is defined for
temperature fields:

$$(\tau, \chi)_{L^2(\Omega)} \stackrel{\text{def}}{=} \int_\Omega \tau \chi d\boldsymbol{x};$$

displacement vectors:

$$(\boldsymbol{u}, \boldsymbol{v})_{L^2(\Omega)} \stackrel{\text{def}}{=} \int_\Omega u_i v_i d\boldsymbol{x};$$

stress tensors:

$$(\underset{\approx}{\boldsymbol{\sigma}}, \underset{\approx}{\boldsymbol{\varsigma}})_{L^2(\Omega)} \stackrel{\text{def}}{=} \int_\Omega \sigma_{ij} \varsigma_{ij} d\boldsymbol{x}.$$

Note that $i, j = 1, 2$ for a two-dimensional domain and $i, j = 1, 2, 3$ for a three-dimensional domain.

Definition A.2. The $L^2(\Omega)$ **norm** is defined
for temperature field:

$$\|\tau\|_{L^2(\Omega)} \stackrel{\text{def}}{=} \sqrt{\left(\int_\Omega \tau^2 d\boldsymbol{x}\right)};$$

Z. Yosibash, *Singularities in Elliptic Boundary Value Problems and Elasticity and Their Connection with Failure Initiation*, Interdisciplinary Applied Mathematics 37,
DOI 10.1007/978-1-4614-1508-4_15, © Springer Science+Business Media, LLC 2012

for a displacement vector:

$$\|\boldsymbol{u}\|_{L^2(\Omega)} \stackrel{\text{def}}{=} \sqrt{\left(\int_\Omega u_i u_i \, d\boldsymbol{x}\right)};$$

and for a stress tensor:

$$\|\underset{\approx}{\boldsymbol{\sigma}}\|_{L^2(\Omega)} \stackrel{\text{def}}{=} \sqrt{\left(\int_\Omega \sigma_{ij}\sigma_{ij} \, d\boldsymbol{x}\right)}.$$

With this definition we say that $f \in L^2(\Omega)$ if $\|f\|_{L^2(\Omega)} < \infty$, where f stands either for τ, $\boldsymbol{u}$, or $\underset{\approx}{\boldsymbol{\sigma}}$.

We now define a *locally integrable* function:

Definition A.3. Given a domain Ω and a locally compact subdomain $\Omega_R \subset \Omega$, the set of **locally integrable** functions is denoted by

$$L^1_{loc}(\Omega) \stackrel{\text{def}}{=} \left\{ f \mid \int_{\Omega_R} f d\boldsymbol{x} < \infty. \right\}$$

For a vector with d elements, a displacement vector in d-dimensions for example, we use the notation $\boldsymbol{u} \in \left(L^1_{loc}(\Omega)\right)^d$. Functions in $L^1_{loc}(\Omega)$ can behave arbitrarily badly near the boundary $\partial\Omega$. In order to define the Sobolev space, we need to generalize the concept of derivative of functions that may not be differentiable at a point, but are nevertheless continuous.

Definition A.4. We say that a given function $f \in L^1_{loc}(\Omega)$ has a **weak derivative**, denoted by $D_w f$, if there exists a function $g \in L^1_{loc}(\Omega)$ such that

$$\int_\Omega g(x)\phi(x)dx = -\int_\Omega f(x)\phi'(x)dx \qquad \forall \phi \in C^\infty_o.$$

If such a g exists then $D_w f = g$.

An example of a function that has a weak derivative at point $x = 0$, although it does not have a classical derivative, is $f(x) = 1 - |x|$ with $D_w f = \begin{cases} 1, & x < 0, \\ -1, & x > 0, \end{cases}$ (see [32, Chapter 1]).

We do not use the Sobolev space for stresses. Therefore definitions A.5-A.7 address only temperature and displacement vectors.

Definition A.5. The H^1 **inner product** (Sobolev inner product) is defined as follows:

For temperature fields let $\tau, \chi \in L^1_{loc}(\Omega)$, and suppose that the weak derivatives $D_w\tau$ and $D_w\chi$ exist. Then

$$(\tau, \chi)_{H^1(\Omega)} \stackrel{\text{def}}{=} \int_\Omega \partial_i\tau \, \partial_i\chi d\boldsymbol{x} + \int_\Omega \tau\chi d\boldsymbol{x} = (\mathbf{grad}\tau, \, \mathbf{grad}\chi)_{L^2(\Omega)} + (\tau, \, \chi)_{L^2(\Omega)} \, .$$

For a displacement vector (d dimensions): Let $\boldsymbol{u} \in \left(L^1_{loc}(\Omega)\right)^d$, and suppose that the weak derivatives $D_w u_i$ exist for $i = 1, \ldots, d$. Then

$$(\boldsymbol{u}, \boldsymbol{v})_{H^1(\Omega)} \stackrel{\text{def}}{=} \int_\Omega \partial_i u_j \; \partial_i v_j \, d\boldsymbol{x} + \int_\Omega u_i v_i \, d\boldsymbol{x},$$

$i, j = 1, \ldots, d.$

The weak derivatives are needed to ensure the existance of the integration of the first term in the above definition for functions that may not have classical derivatives at distinct points in one dimension, along a curve in two dimensions, or along surfaces in three dimensions.

Based on the $H^1(\Omega)$ inner product we may define the H^1 norm:

Definition A.6. The H^1 norm, called the **Sobolev norm**, is defined as follows:

For temperature:

$$\|\tau\|_{H^1(\Omega)} \stackrel{\text{def}}{=} \sqrt{(\tau, \tau)_{H^1(\Omega)}};$$

for a displacement vector:

$$\|\boldsymbol{u}\|_{H^1(\Omega)} \stackrel{\text{def}}{=} \sqrt{(\boldsymbol{u}, \boldsymbol{u})_{H^1(\Omega)}}.$$

Another useful definition that will be used in the following is the **Sobolev seminorm** :

Definition A.7. For temperature:

$$|\tau|_{H^1(\Omega)} \stackrel{\text{def}}{=} \sqrt{\int_\Omega \sum_{i=1}^{d} (\partial_i \tau)^2 d\boldsymbol{x}};$$

for displacement vectors:

$$|\boldsymbol{u}|_{H^1(\Omega)} \stackrel{\text{def}}{=} \sqrt{\int_\Omega \sum_{i,j=1}^{d} (\partial_i u_j)^2 d\boldsymbol{x}}.$$

Based on the Sobolev inner product we may define the **Sobolev space** for scalar functions:

Definition A.8. $H^1(\Omega) \stackrel{\text{def}}{=} \left\{\tau \in L^1_{loc}(\Omega) \mid \|\tau\|_{H^1(\Omega)} < \infty\right\}.$

An analogous definition holds for vector functions as for the displacement vectors. The space H^1_o contains all functions that belong to a Sobolev space, with an additional constraint that the value of the function is zero on the boundary of the domain on which Dirichlet boundary conditions are prescribed:

Definition A.9. $H_o^1(\Omega) \stackrel{\text{def}}{=} \{\tau \mid \tau \in H^1(\Omega),\ \ \tau = 0 \text{ on } \partial\Omega_D\}$.

We now proceed to the definition of the *energy space* and *energy norm*. These are required for the weak formulation of the heat conduction and elasticity problems.

Definition A.10. The **energy inner product** is defined as follows:
For temperature:

$$(\tau, \chi)_{\mathcal{E}(\Omega)} = \mathcal{B}(\tau, \chi) \stackrel{\text{def}}{=} \int_\Omega (\mathbf{grad}\tau)^T [k] \mathbf{grad}\chi \, d\boldsymbol{x} = \int_\Omega k_{ij}\ \partial_i \tau\ \partial_j \chi \, d\boldsymbol{x};$$

for elasticity:

$$(\boldsymbol{u}, \boldsymbol{v})_{\mathcal{E}(\Omega)} = \mathcal{B}(\boldsymbol{u}, \boldsymbol{v}) \stackrel{\text{def}}{=} \int_\Omega ([D]\boldsymbol{u})^T\ [E][D]\boldsymbol{v} d\boldsymbol{x},$$

where $[k]$ is the matrix of heat conduction coefficients defined in Chapter 1, $[D]$ is the differential matrix given for a three-dimensional domain in (1.41), and $[E]$ is the material matrix. For an elastic isotropic material,

$$\mathcal{B}(\boldsymbol{u}, \boldsymbol{v}) = \int_\Omega 2\mu \varepsilon_{ij}(\boldsymbol{u})\varepsilon_{ij}(\boldsymbol{v}) + \lambda\ \partial_i u_i\ \partial_j u_j \ d\boldsymbol{x}.$$

We associate the **energy norm** with the energy inner product.

Definition A.11.

$$\|\tau\|_{\mathcal{E}(\Omega)} \stackrel{\text{def}}{=} \sqrt{\frac{1}{2}\mathcal{B}(\tau, \tau)},$$

$$\|\boldsymbol{u}\|_{\mathcal{E}(\Omega)} \stackrel{\text{def}}{=} \sqrt{\frac{1}{2}\mathcal{B}(\boldsymbol{u}, \boldsymbol{u})}.$$

Note that the energy norm is *equivalent* to the H^1 seminorm for the two elliptic problems of interest, i.e.,

$$C_1 | \bullet |_{H^1} \geq \| \bullet \|_{\mathcal{E}} \geq C_2 | \bullet |_{H^1},$$

where the constants $C_1 \geq C_2$ are positive, and C_2 may become infinite for an elastic material as it becomes incompressible ($\lambda \to \infty$, i.e. $\nu \to 1/2$). The **energy space** contains all functions that have a bounded energy inner product:

Definition A.12. For a temperature field:

$$\mathcal{E}(\Omega) \stackrel{\text{def}}{=} \{\tau \mid \mathcal{B}(\tau, \tau) < \infty\};$$

for elasticity:

$$\mathcal{E}(\Omega) \stackrel{\text{def}}{=} \{\boldsymbol{u} \mid \mathcal{B}(\boldsymbol{u}, \boldsymbol{u}) < \infty\}.$$

We also have to define $\mathcal{E}_o(\Omega)$, which are all functions in the energy norm having a zero value along a part (or the whole) of the boundary. In the elasticity case, one of the components of the displacement field may be zero, for example $u_1 = 0$, on a part of the boundary, denoted by $(\partial\Omega_D)_a$, and $u_3 = 0$ on another part, denoted by $(\partial\Omega_D)_b$. For this particular case we define

$$\mathcal{E}_o(\Omega) = \{\boldsymbol{u} \mid \boldsymbol{u} \in \mathcal{E}(\Omega), \quad u_1 = 0 \text{ on } (\partial\Omega_D)_a, \quad u_3 = 0 \text{ on } (\partial\Omega_D)_b\}.$$

For the complementary weak form, we need to define the **statically admissible space**:

Definition A.13. For a temperature field:

$$\mathcal{E}_c(\Omega) \stackrel{\text{def}}{=} \left\{\boldsymbol{q} \mid \int_\Omega \boldsymbol{q} \cdot [k]^{-1} \boldsymbol{q}\, d\boldsymbol{x} < \infty, \ \nabla \cdot \boldsymbol{q} = Q \ \text{ in } \Omega\right\};$$

for elasticity:

$$\mathcal{E}_c(\Omega) \stackrel{\text{def}}{=} \left\{\boldsymbol{\sigma} \mid \int_\Omega \boldsymbol{\sigma}^T [E]^{-1} \boldsymbol{\sigma}\, d\boldsymbol{x} < \infty, \ -\partial_i \sigma_{ji} = f_j \ \text{ in } \Omega, \ i = 1, \dots, d\right\},$$

where Q is the heat source in (1.34) and f_j are the elements of the body force vector in (1.44).

Appendix B
Analytic Solution to 2-D Scalar Elliptic Problems in Anisotropic Domains

We consider here the exact solution for an elliptic problem with piecewise constant coefficients, representing an anisotropic 2-D domain. Analytical methods are applied to compute the eigenpairs by transformation of the coordinate system.

We restrict our discussion to the Dirichlet boundary conditions, and we locate the Cartesian coordinate system such that the x_1 axis coincides with one of the straight boundaries intersecting at the singular point at the origin. The problem for which we seek the solution is

$$-k_{\gamma\delta}\frac{\partial^2\tau}{\partial x_\gamma \partial x_\delta} = 0, \qquad \gamma, \delta = 1, 2, \qquad \text{in } \Omega, \tag{B.1}$$

$$\tau = 0, \qquad \text{on } \Gamma_\ell, \quad \ell = 1, 2, \tag{B.2}$$

where $k_{\gamma\delta} = k_{\delta\gamma}$ and $k_{\gamma\delta}$ satisfy the ellipticity restriction, i.e., $k_{11}k_{22} - k_{12}^2 > 0$. The domain and notation are presented in Figure B.1.

By performing a linear transformation of the coordinates of the form

$$\xi = \xi(x_1, x_2), \qquad \eta = \eta(x_1, x_2), \tag{B.3}$$

we would like to transform the scalar operator in (B.1) to the Laplacian in the coordinates ξ, η. First we find the required transformation. Of course, this transformation will also transform the domain to a different geometry in the neighborhood of O.

We define

$$\xi_{,\gamma} \stackrel{\text{def}}{=} \frac{\partial \xi}{\partial x_\gamma} \quad \text{and} \quad \xi_{,\gamma\delta} \stackrel{\text{def}}{=} \frac{\partial^2 \xi}{\partial x_\gamma \partial x_\delta},$$

$$\eta_{,\gamma} \stackrel{\text{def}}{=} \frac{\partial \eta}{\partial x_\gamma} \quad \text{and} \quad \eta_{,\gamma\delta} \stackrel{\text{def}}{=} \frac{\partial^2 \eta}{\partial x_\gamma \partial x_\delta}.$$

Z. Yosibash, *Singularities in Elliptic Boundary Value Problems and Elasticity and Their Connection with Failure Initiation*, Interdisciplinary Applied Mathematics 37,
DOI 10.1007/978-1-4614-1508-4_16,

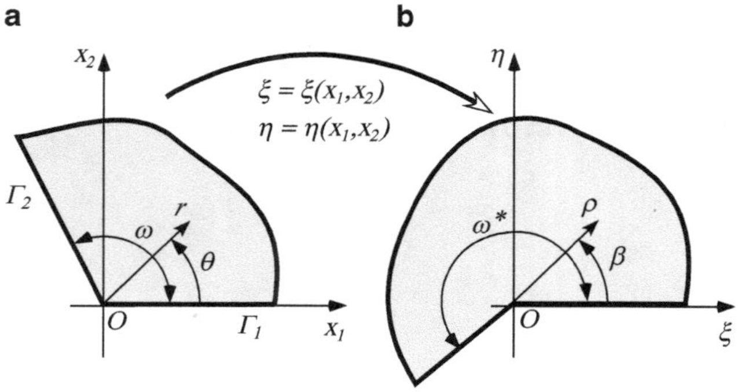

Fig. B.1 The domain of interest and notation **(a)** and the domains after coordinates transformation in **(b)**.

With this notation, the differential operator in (B.1) is represented as

$$
\begin{aligned}
k_{11}\frac{\partial^2\tau}{\partial x_1^2} + 2k_{12}\frac{\partial^2\tau}{\partial x_1\partial x_2} + k_{22}\frac{\partial^2\tau}{\partial x_2^2}
&= \left[k_{11}\eta_{,1}^2 + 2k_{12}\eta_{,1}\eta_{,2} + k_{22}\eta_{,2}^2\right]\frac{\partial^2\tau}{\partial\eta^2} \\
&+ \left[k_{11}\xi_{,1}^2 + 2k_{12}\xi_{,1}\xi_{,2} + k_{22}\xi_{,2}^2\right]\frac{\partial^2\tau}{\partial\xi^2} \\
&+ \left[k_{11}\eta_{,1}\xi_{,1} + k_{22}\eta_{,2}\xi_{,2} + k_{12}(\eta_{,1}\xi_{,2} + \eta_{,2}\xi_{,1})\right]\frac{\partial^2\tau}{\partial\xi\partial\eta}.
\end{aligned} \tag{B.4}
$$

To obtain the Laplace operator in terms of the new coordinates, the following restrictions have to be satisfied:

$$k_{11}\eta_{,1}^2 + 2k_{12}\eta_{,1}\eta_{,2} + k_{22}\eta_{,2}^2 = k_{11}\xi_{,1}^2 + 2k_{12}\xi_{,1}\xi_{,2} + k_{22}\xi_{,2}^2, \tag{B.5}$$

$$k_{11}\eta_{,1}\xi_{,1} + k_{22}\eta_{,2}\xi_{,2} + k_{12}(\eta_{,1}\xi_{,2} + \eta_{,2}\xi_{,1}) = 0. \tag{B.6}$$

We have two equations for the four unknowns $\xi_{,1}, \xi_{,2}, \eta_{,1}, \eta_{,2}$. To preserve the x_1-axis (i.e., that the x_1-axis will be in the same direction as the ξ axis), we choose $\eta_{,1} = 0$. In addition, we require that $\xi_{,1} = 1$. Thus, applying the two constraints on (B.5) and (B.6), the following system is obtained:

$$
\begin{aligned}
\xi_{,1} &= 1, \\
\xi_{,2} &= -\frac{k_{12}}{k_{22}}, \\
\eta_{,1} &= 0, \\
\eta_{,2} &= \frac{\sqrt{k_{11}k_{22} - k_{12}^2}}{k_{22}}.
\end{aligned} \tag{B.7}
$$

The system of equations (B.7) is satisfied by choosing

$$\xi = x_1 - \frac{k_{12}}{k_{22}} x_2, \tag{B.8}$$

$$\eta = \frac{\sqrt{k_{11}k_{22} - k_{12}^2}}{k_{22}} x_2. \tag{B.9}$$

The coordinate transformation (B.9) transforms (B.1) into the Laplace problem in (ξ, η) :

$$\frac{\partial^2 \tau}{\partial \xi^2} + \frac{\partial^2 \tau}{\partial \eta^2} = 0 \text{ on the mapped domain.} \tag{B.10}$$

The transformation (B.8)-(B.9) is linear, so that straight lines remain straight and the boundaries Γ_1 and Γ_2 remain straight. The line $x_2 = 0$ (the x_1-axis) is mapped into $\eta = 0$; thus the Γ_1 boundary is along the ξ-axis. Let us demonstrate that in the new coordinate system, the corner opening angle changes. The slope of a straight line in the ξ, η coordinate system is obtained from (B.8)-(B.9):

$$\frac{\eta}{\xi} = \frac{\sqrt{k_{11}k_{22} - k_{12}^2}\frac{x_2}{x_1}}{k_{22} - k_{12}\frac{x_2}{x_1}}. \tag{B.11}$$

The equation that describes Γ_2 is given by $\frac{x_2}{x_1} = \tan\omega$, so that on inserting this ratio in (B.11), we obtain the slope of the line Γ_2 in the ξ, η plane:

$$\frac{\eta}{\xi} = \tan\omega^* = \frac{\sqrt{k_{11}k_{22} - k_{12}^2}\tan\omega}{k_{22} - k_{12}\tan\omega}. \tag{B.12}$$

We have so far transformed the problem to a new domain presented in Figure B.1(b), having an opening angle ω^*, such that in the mapped domain we have to solve the "usual" Laplace equation with Dirichlet boundary conditions on the two straight boundaries intersecting at O. For the new domain, the eigenvalues are known:

$$\alpha_i = i\pi/\omega^* = i\pi \left[\arctan\left(\frac{\sqrt{k_{11}k_{22} - k_{12}^2}\tan\omega}{k_{22} - k_{12}\tan\omega} \right) \right]^{-1}. \tag{B.13}$$

And the ith term in the "singular solution" in the ξ, η domain is

$$\rho^{\alpha_i} \sin(\alpha_i \beta). \tag{B.14}$$

However, we are interested in the x_1, x_2 plane, so we have to transform the solution.

Using the connection $\rho^2 = \xi^2 + \eta^2$, $\tan\beta = \eta/\xi$, and inserting (B.8)-(B.9) instead of ξ and η, one obtains

$$\rho^2 = x_1^2 + \frac{k_{11}}{k_{22}} x_2^2 - 2\frac{k_{12}}{k_{22}} x_1 x_2, \tag{B.15}$$

$$\tan\beta = \frac{\sqrt{k_{11}k_{22} - k_{12}^2}\frac{x_2}{x_1}}{k_{22} - k_{12}\frac{x_2}{x_1}}. \tag{B.16}$$

Take $x_1 = r\cos\theta$ and $x_2 = r\sin\theta$, then equations (B.15)-(B.16) become

$$\rho = r\sqrt{\left(\frac{k_{11}}{k_{22}}\sin^2\theta + \cos^2\theta - \frac{k_{12}}{k_{22}}\sin 2\theta\right)}, \tag{B.17}$$

$$\beta = \arctan\left[\frac{\sqrt{k_{11}k_{22} - k_{12}^2}\sin\theta}{k_{22}\cos\theta - k_{12}\sin\theta}\right]. \tag{B.18}$$

Finally, inserting (B.17)-(B.18) in (B.14) yields the solution in the x_1, x_2 plane in the vicinity of the singular point:

$$\begin{aligned} \tau &= \sum_i A_i r^{\alpha_i}\left(\frac{k_{11}}{k_{22}}\sin^2\theta + \cos^2\theta - \frac{k_{12}}{k_{22}}\sin 2\theta\right)^{\alpha_i/2} \\ &\quad \times \sin\left[\alpha_i \arctan\left(\frac{\sqrt{k_{11}k_{22} - k_{12}^2}\sin\theta}{k_{22}\cos\theta - k_{12}\sin\theta}\right)\right] \end{aligned} \tag{B.19}$$

$$\text{with} \qquad \alpha_i = i\pi\left[\arctan\left(\frac{\sqrt{k_{11}k_{22} - k_{12}^2}\tan\omega}{k_{22} - k_{12}\tan\omega}\right)\right]^{-1}. \tag{B.20}$$

B.1 Analytic Solution to a 2-D Scalar Elliptic Problem in an Anisotropic Bimaterial Domain

We derive here the analytic solution for an elliptic problem with piecewise constant coefficients, that is associated with an anisotropic bimaterial 2-D domain as shown in Figure B.2 having homogeneous Neumann boundary conditions.

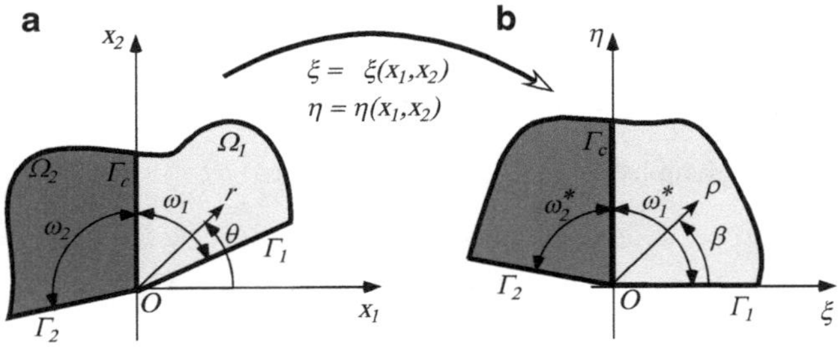

Fig. B.2 The domain of interest and notation **(a)** and the domains after coordinate transformation in **(b)**.

We locate the Cartesian coordinate system such that the x_2-axis coincides with the interface boundary denoted by Γ_c:

$$-k^{(\ell)}_{\gamma\delta}\frac{\partial^2\tau}{\partial x_\gamma \partial x_\delta} = 0 \qquad \text{in } \Omega_\ell, \quad \gamma,\delta = 1,2, \quad \ell = 1,2, \tag{B.21}$$

$$k^{(\ell)}_{\gamma\delta}\frac{\partial\tau}{\partial x_\gamma} n_\delta = 0 \qquad \text{on } \Gamma_\ell, \quad \ell = 1,2. \tag{B.22}$$

On the interface boundary we have the following continuity conditions:

$$\tau^{(1)}(\pi/2) = \tau^{(2)}(\pi/2), \tag{B.23}$$

$$\left[k^{(1)}_{\gamma\delta}\frac{\partial\tau}{\partial x_\gamma} n_\delta\right]_{\pi/2} = \left[k^{(2)}_{\gamma\delta}\frac{\partial\tau}{\partial x_\gamma} n_\delta\right]_{\pi/2}. \tag{B.24}$$

As in the case for a single domain, we perform a linear transformation in each of the subdomains to obtain the Laplace problem in subdomains that are geometrically different from the original ones, but the Neumann boundary conditions and the continuity conditions are preserved.

Consider (B.21) in Ω_ℓ (we drop the domain index, keeping in mind that we discuss *one* subdomain at a time) we perform the transformation (see also [109][chapter V])

$$\xi = \left(k_{11}^{-1}\sqrt{k_{11}k_{22} - k_{12}^2}\right) x_1, \tag{B.25}$$

$$\eta = -k_{12}k_{11}^{-1}x_1 + x_2. \tag{B.26}$$

This transformation preserves the x_2-axis (which becomes now η-axis), so that the general elliptic operator takes the form of the Laplacian in (ξ,η). The boundary Γ_1 given in the plane x_1, x_2 by the equation $x_2 = \tan(\pi/2-\omega_1)x_1$ remains a straight line given by

$$\eta = \tan(\pi/2 - \omega_1^*)\xi, \tag{B.27}$$

and the boundary Γ_2 transforms to the boundary given by

$$\eta = \tan(\pi/2 + \omega_2^*)\xi, \tag{B.28}$$

with the angles ω_ℓ^* given by

$$\omega_1^* = \pi/2 - \arctan\left(\frac{k_{11}^{(1)}\tan(\pi/2 - \omega_1) - k_{12}^{(1)}}{\sqrt{k_{11}^1 k_{22}^{(1)} - (k_{12}^{(1)})^2}}\right), \tag{B.29}$$

$$\omega_2^* = \pi/2 + \arctan\left(\frac{k_{11}^{(2)}\tan(\pi/2 + \omega_2) - k_{12}^{(2)}}{\sqrt{k_{11}^2 k_{22}^{(2)} - (k_{12}^{(2)})^2}}\right). \tag{B.30}$$

B.1.1 Treatment of the Boundary Conditions

The Neumann boundary conditions (B.24) are expressed in each subdomain:

$$(k_{11}n_1 + k_{12}n_2)\frac{\partial\tau}{\partial x_1} + (k_{21}n_1 + k_{22}n_2)\frac{\partial\tau}{\partial x_2} = 0, \quad (x) \in \Gamma_\ell,\ \ell = 1,2. \tag{B.31}$$

Using the coordinate transformation presented in (B.25)-(B.26), the boundary conditions on the transformed domains become

$$ck_{11}^{-1}\left[cn_2\frac{\partial}{\partial\eta} + (k_{11}n_1 + k_{12}n_2)\frac{\partial}{\partial\xi}\right]\tau = 0, \quad (\xi,\eta) \in \Gamma_\ell,\ \ell = 1,2, \tag{B.32}$$

where $c \overset{\text{def}}{=} \sqrt{k_{11}k_{22} - k_{12}^2}$.

It is more natural to represent the boundary conditions in terms of polar coordinates:

$$ck_{11}^{-1}\left\{[(c\sin\beta + k_{12}\cos\beta)n_2 + k_{11}(\cos\beta)n_1]\frac{\partial u}{\partial\rho}\right.$$
$$\left. + [(c\cos\beta - k_{12}\sin\beta)n_2 - k_{11}(\sin\beta)n_1]\frac{1}{\rho}\frac{\partial u}{\partial\beta}\right\} = 0, (\rho,\beta) \in \Gamma_\ell,\ \ell = 1,2. \tag{B.33}$$

In each subdomain the solution to the Laplace equation is

$$\tau^{(1)} = \rho^\alpha\,(A\sin\alpha\beta + B\cos\alpha\beta), \tag{B.34}$$

$$\tau^{(2)} = \rho^\alpha\,(C\sin\alpha\beta + D\cos\alpha\beta). \tag{B.35}$$

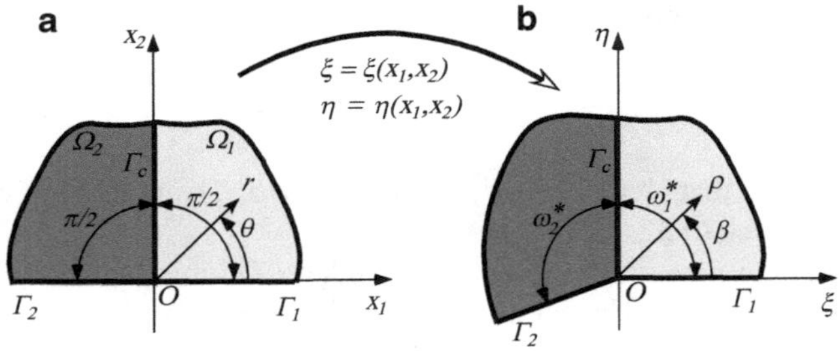

Fig. B.3 The example problem **(a)** and the domains after coordinate transformation in **(b)**.

Applying the boundary conditions (B.33), together with the continuity conditions for the interface boundary on (B.34) and (B.35), one obtains a system of four homogeneous equations for A, B, C, and D. For a nontrivial solution, the determinant of the coefficient matrix has to be zero. This condition provides an explicit equation from which the eigenvalues α_i can be determined (see also [109]):

$$\sqrt{k_{11}^{(1)}k_{22}^{(1)} - (k_{12}^{(1)})^2}\tan\alpha\omega_1^* = -\sqrt{k_{11}^{(2)}k_{22}^{(2)} - (k_{12}^{(2)})^2}\tan\alpha\omega_2^*. \tag{B.36}$$

For each α_i that is a solution to (B.36), the solutions (B.34) and (B.35) are determined up to a constant multiplier. The method is demonstrated in the following.

B.1.2 An Example

We compute the first five eigenvalues and eigenfunctions to the example problem shown in Figure B.3 with the following coefficients:

$$\begin{aligned} k_{11}^{(1)} &= k_{11}^{(2)} = k_{22}^{(1)} = k_{22}^{(2)} = 1.0, \\ k_{12}^{(1)} &= 0.0, \quad k_{12}^{(2)} = -0.75. \end{aligned} \tag{B.37}$$

Inserting the $k_{\gamma\delta}^{(\ell)}$ into (B.29), we obtain $\omega_1^* = \pi/2$ and $\omega_2^* = \pi/2 + \arctan(3/\sqrt{7})$.

Boundary Conditions

- Boundary conditions on Γ_1 : On this boundary $n_1 = 0, n_2 = -1$, and $\beta = 0$, so that the boundary condition (B.33) is simplified to

$$\left(-k_{12}^{(1)}\frac{\partial}{\partial\rho} - c^{(1)}\frac{1}{\rho}\frac{\partial}{\partial\beta}\right)\tau^{(1)}|_{\beta=0} = 0.$$

Inserting now $\tau^{(1)}$ from (B.34), we obtain

$$\alpha k_{12}^{(1)} B + c^{(1)} A = 0,$$

but $k_{12}^{(1)} = 0$, so that $A = 0$ and $\tau^{(1)} = B\rho^{\alpha} \cos \alpha\beta$. The number of equations is therefore reduced to 3.

- Boundary conditions on the interface boundary: On this boundary $n_1 = \pm 1$, $n_2 = 0$, and $\beta = \pi/2$. The condition of continuity of τ becomes

$$\tau^{(1)} = \tau^{(2)} \quad \Rightarrow \quad B \cos \frac{\alpha\pi}{2} = C \sin \frac{\alpha\pi}{2} + D \cos \frac{\alpha\pi}{2}, \tag{B.38}$$

and the condition of continuity of $k_{\gamma\delta} \frac{\partial \tau}{\partial x_\gamma} n_\delta$ becomes

$$B c^{(1)} \sin \frac{\alpha\pi}{2} = -C c^{(2)} \cos \frac{\alpha\pi}{2} + D c^{(2)} \sin \frac{\alpha\pi}{2}. \tag{B.39}$$

- Boundary condition on Γ_2 : On this boundary $n_1 = 0, n_2 = -1$, and $\beta = \pi/2 + \omega_2^*$, so the boundary condition after some algebraic manipulations becomes

$$\begin{aligned} &\Big\{C \left[c^{(2)} \cos\left((\alpha-1)(\pi/2+\omega_2^*)\right) + k_{12}^{(2)} \sin\left((\alpha-1)(\pi/2+\omega_2^*)\right)\right] \\ &\quad + D\left[-c^{(2)} \sin\left((\alpha-1)(\pi/2+\omega_2^*)\right) + k_{12}^{(2)} \cos\left((\alpha-1)(\pi/2+\omega_2^*)\right)\right]\Big\} = 0. \end{aligned} \tag{B.40}$$

Equations (B.38)-(B.40) form a homogeneous system of equations for the unknown vector $\{A\ B\ C\}^T$. To obtain a nontrivial solution, the determinant of the matrix that multiplies $\{A\ B\ C\}^T$ has to be zero, which is equivalent to (B.36):

$$\tan(\alpha\pi/2) = -\frac{\sqrt{7}}{4} \tan\left((\pi/2 + \arctan(3/\sqrt{7}))\alpha\right),$$

for which the first five nonzero roots are

$$\alpha_1 = 0.7567822, \quad \alpha_2 = 1.6253436, \quad \alpha_3 = 2.3148250, \quad \alpha_4 = 3.1720534,$$
$$\alpha_5 = 3.9476365.$$

The results were obtained using Mathematica and are accurate to the seventh digit, as shown.

Once the α_γ are found, the eigenfunctions $s_\gamma(\theta)$ can be computed. For example, $s_1(\theta)$, which corresponds to $\alpha_1 = 0.7567822$, is computed as follows: In Ω_1, we have $\tau_1^{(1)}(\rho, \beta) = B\rho^{\alpha_1} \cos \alpha_1 \beta$, but since Ω_1 is isotropic ($\rho = r$, $\beta = \theta$)

$$\tau_1^{(1)}(r, \theta) = B r^{0.7567822} \cos(0.7567822\, \theta), \quad 0 \le \theta \le \pi/2. \tag{B.41}$$

In Ω_2 the eigenfunction is

$$\tau_1^{(2)}(\rho,\beta) = B\rho^{0.7567822}\,[-0.1770724\sin(0.7567822\beta) + 1.4407124\cos(0.7567822\beta)]\,. \tag{B.42}$$

Expressing (ρ,β) in terms of (r,θ), we obtain

$$\rho = r\sqrt{k_{11}^{-1}k_{22}\cos^2\theta - k_{11}^{-1}k_{12}\sin 2\theta + \sin^2\theta},$$
$$\tan\beta = -k_{12}/c + \frac{k_{11}}{c}\frac{x_2}{x_1},$$

which for Ω_2 becomes

$$\rho = r\sqrt{1 + \frac{3}{4}\sin 2\theta},$$
$$\beta = \arctan\left[\frac{3}{\sqrt{7}} + \frac{4}{\sqrt{7}}\tan\theta\right].$$

Substituting the above relationships into (B.42), the first eigenpair in Ω_2 is obtained

$$\begin{aligned}\tau_1^{(2)}(r,\theta) = {} & Br^{0.7567822}\left(1 + \frac{3}{4}\sin 2\theta\right)^{0.37839} \\ & \times\left\{-0.1770724\sin\left[0.7567822\arctan\left(\frac{3}{\sqrt{7}} + \frac{4}{\sqrt{7}}\tan\theta\right)\right]\right. \\ & \left. +1.4407124\cos\left[0.7567822\arctan\left(\frac{3}{\sqrt{7}} + \frac{4}{\sqrt{7}}\tan\theta\right)\right]\right\}, \\ & \pi/2 \le \theta \le \pi.\end{aligned} \tag{B.43}$$

One may check that if we substitute $\theta = \pi/2$ in (B.43), we obtain $\tau_1^{(2)}(r,\pi/2) = 0.3728193Br^{0.7567822}$, which is exactly the expression obtained after substituting $\theta = \pi/2$ in (B.41).

Problem B.1. Consider the same problem over the domain in Figure B.3, given by equation (B.21), having heat transfer coefficients as in (B.37), with the same interface continuity conditions as in (B.23)-(B.24), except that homogeneous Dirichlet boundary conditions are prescribed:

$$\tau = 0 \qquad \text{on } \Gamma_\ell, \quad \ell = 1,2, \tag{B.44}$$

instead of (B.22).

Show that for this case the characteristic equation for determining the eigenvalues is

$$\frac{-\sqrt{7}}{4}\tan(\alpha\pi/2) = \tan\left(\alpha(\pi/2 + \arctan(3/\sqrt{7}))\right), \tag{B.45}$$

for which the first nonzero roots are

$$\alpha_1 = 0.8202618, \quad \alpha_2 = 1.5254221, \quad \alpha_3 = 2.4080637, \quad \alpha_4 = 3.1256247,$$
$$\alpha_5 = 3.9275639.$$

and that

$$\tau_1^{(1)}(r,\theta) = Ar^{0.8202618}\sin(0.8202618\,\theta), \;\; 0 \le \theta \le \pi/2, \tag{B.46}$$

$$\begin{aligned}\tau_1^{(2)}(r,\theta) = {} & Ar^{0.8202618}\left(1 + \frac{3}{4}\sin 2\theta\right)^{0.410130} \\ & \times \left\{1.03973\sin\left[0.8202618\arctan\left(\frac{3}{\sqrt{7}} + \frac{4}{\sqrt{7}}\tan\theta\right)\right]\right. \\ & \left. - 0.136956\cos\left[0.8202618\arctan\left(\frac{3}{\sqrt{7}} + \frac{4}{\sqrt{7}}\tan\theta\right)\right]\right\}, \;\; \pi/2 \\ & \le \theta \le \pi. \end{aligned} \tag{B.47}$$

Appendix C
Asymptotic Solution at the Intersection of Circular Edges in a 2-D Domain

Throughout the book we have considered only cases in which the boundaries intersecting at the singular points were straight lines. We show in this appendix by a simple example problem that the leading singular term corresponding to a domain with curved boundaries that intersect at a specific angle are the same as if the boundaries were straight lines intersecting at the same angle. For this purpose we use complex analysis.

Let us denote a complex number in the x_1, x_2 plane by z, i.e., $z = x_1 + \imath x_2$, where $\imath \stackrel{\text{def}}{=} \sqrt{-1}$. Let us consider the Laplace equation in a domain having curved boundaries that intersect at the origin. As a concrete example, consider the two arcs having a 90-degree corner at the origin created by the intersection of two arcs as shown in the left of Figure C.1. These arcs are part of two circles of radii r_1 and r_2 with centers at $-r_1$ and $-\imath r_2$ respectively.

The second point of intersection of the two circles is denoted by a:

$$a = -\frac{2r_1 r_2}{r_1^2 + r_2^2}(r_2 + \imath r_1) = a_{\Re} + \imath a_{\Im} = |a| e^{\imath \theta^*}, \tag{C.1}$$

where $|a| = \sqrt{a_{\Re}^2 + a_{\Im}^2}$ and $\theta^* = \arctan \frac{a_{\Im}}{a_{\Re}}$. We wish to obtain the solution of

$$\nabla^2 \tau(x_1, x_2) = 0, \quad (x_1, x_2) \in \Omega_z, \tag{C.2}$$

in a neighborhood of the origin, where Ω_z is the domain shown in the left of Figure C.1. Let us assume either Dirichlet or Neumann homogeneous boundary conditions on the two arc boundaries.

We recall the following properties from complex analysis (see, e.g., [39, pp. 183-187]):

- A function $\tau(x_1, x_2)$ is called *harmonic* at a point z if it satisfies the Laplace equation at that point. So, the function τ we are seeking is of course a harmonic function.

Z. Yosibash, *Singularities in Elliptic Boundary Value Problems and Elasticity and Their Connection with Failure Initiation*, Interdisciplinary Applied Mathematics 37,
DOI 10.1007/978-1-4614-1508-4_17, © Springer Science+Business Media, LLC 2012

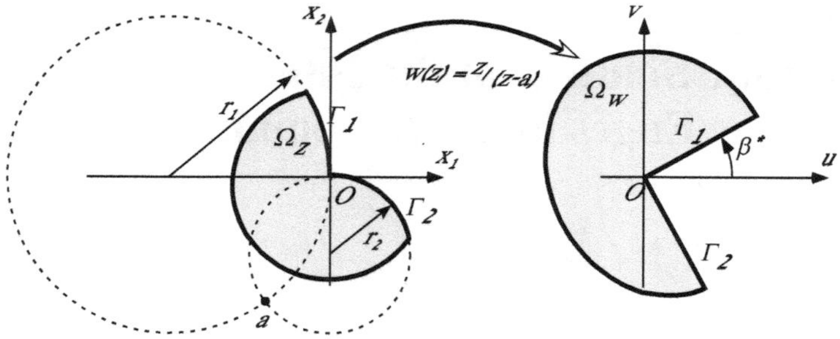

Fig. C.1 The domain with curved boundaries intersecting at a point.

- A mapping $w(z) = u + \imath v$ is called a *conformal mapping* at a point z if it is analytic ($w'(z) = dw/dz$ exists), and $w'(z) \neq 0$.
- A conformal mapping has the very important property that two curves in the z plane that intersect at a given angle ω, are mapped into two curves in the w plane that intersect at the same angle ω.
- If $\tau(x_1, x_2) = \tau(z)$ is a harmonic function in a domain Ω_z and we let $w(z)$ to be a conformal mapping that maps Ω_z into a different domain Ω_w in the w plane, then the function $\tau(w(z)) = \tau(u, v)$ is harmonic in Ω_w.
- Homogeneous Dirichlet and/or Neumann boundary conditions on the boundaries of a domain Ω_z remain homogeneous Dirichlet and/or Neumann boundary conditions on the boundaries of the domain Ω_w mapped by a conformal mapping.

In view of the above properties, we are interested in finding a conformal mapping that will map the domain Ω_z having circular boundaries into a domain Ω_w having straight boundaries. We again recall some known properties from complex analysis:

- A *bilinear mapping* $w(z) = \frac{c_1 z + c_2}{c_3 z + c_4}$ is conformal at any point except at $z = c_4/c_3$, $c_3 \neq 0$.
- A bilinear mapping maps lines or circles in z plane to circles or lines in the w plane.
- If a point on a circle in the z plane is mapped to ∞ in the w plane, then the circle is mapped into a line.

We introduce the bilinear mapping

$$w(z) = \frac{z}{z - a}, \tag{C.3}$$

which maps the two arcs Γ_1 and Γ_2 in the z plane into straight lines in the w plane, so that they intersect at the origin at the same angle of 90 degrees (see the right part of Figure C.1).

Problem C.1. It is known that the arcs Γ_1 and Γ_2 in the z plane are mapped into straight lines in the w plane. Show that the mapped straight lines have angles of $\arctan(\frac{-a_{\Re}}{-a_{\Im}})$ and $\arctan(\frac{a_{\Im}}{-a_{\Re}})$ with the u–axis.

Hint: Look at the mappings of the imaginary and real axes x_2 and x_1, i.e., take $z = \imath\varepsilon$ and $z = \varepsilon$, and determine their mappings as $\varepsilon \to 0$.

The straight boundary Γ_1 in the w plane forms an angle of

$$\beta^* = \arctan\left(\frac{-a_{\Re}}{-a_{\Im}}\right) \tag{C.4}$$

with the u–axis. We therefore have in the w plane the Laplace equation in the transformed domain Ω_w with homogeneous Dirichlet or Neumann boundary conditions on the straight lines, for which we know the solution to be of the form

$$\tau(\rho, \beta) \sim \rho^{\alpha} s^{+}(\beta), \tag{C.5}$$

where ρ, β are polar coordinates in the w plane. Let us assume that homogeneous Dirichlet boundary conditions are prescribed, so the solution can be explicitly given as

$$\begin{aligned} \tau(\rho, \beta) &= \sum_k A_k \rho^{\alpha_k} \sin[\alpha_k(\beta - \beta^*)) = \Im\left\{\sum_k A_k \rho^{\alpha_k} e^{\imath\alpha_k(\beta-\beta^*)}\right\}, \\ \alpha_k &= 2k/3, \ \ k = 1, 2, \ldots, \end{aligned} \tag{C.6}$$

where A_k are real coefficients.

Noting the representation of any complex number in the w plane

$$w = \rho e^{\imath\beta},$$

one has that

$$\rho^{\alpha_k} e^{\imath\alpha_k(\beta-\beta^*)} = w^{\alpha_k} e^{-\imath\alpha_k\beta^*}. \tag{C.7}$$

Inserting (C.7) in (C.6), one obtains

$$\tau(\rho, \beta) = \Im\left\{\sum_k A_k w^{\alpha_k} e^{-\imath\alpha_k\beta^*}\right\}. \tag{C.8}$$

But w is given in terms of z in (C.3), and thus

$$\tau = \Im\left\{\sum_k A_k \left(\frac{z}{(z-a)}\right)^{\alpha_k} e^{-\imath\alpha_k\beta^*}\right\}. \tag{C.9}$$

Consider first the expression $\left(\frac{z}{(z-a)}\right)^{\alpha_k}$. To evaluate it in the vicinity of the singular point, where $|z/a| \ll 1$, we observe that

$$\frac{z}{(z-a)} = (-1)\frac{z}{(a-z)} = (-1)\frac{z}{a}\left(\frac{1}{1-z/a}\right). \tag{C.10}$$

The last expression in (C.10) may be expanded in a series for $|z/a| \ll 1$:

$$\frac{1}{1-z/a} = 1 + \frac{z}{a} + \left(\frac{z}{a}\right)^2 + \left(\frac{z}{a}\right)^3 + \cdots. \tag{C.11}$$

Combining (C.10) with (C.11), we obtain

$$\frac{z}{(z-a)} = (-1)\frac{z}{a}\left[1 + \frac{z}{a} + \left(\frac{z}{a}\right)^2 + \left(\frac{z}{a}\right)^3 + \cdots\right] \tag{C.12}$$

and

$$\left(\frac{z}{z-a}\right)^{\alpha_k} = (-1)^{\alpha_k}\left(\frac{z}{a}\right)^{\alpha_k}\left[1 + \frac{z}{a} + \left(\frac{z}{a}\right)^2 + \left(\frac{z}{a}\right)^3 + \cdots\right]^{\alpha_k}. \tag{C.13}$$

Reuse the argument that in the vicinity of the singular point $|z/a| \ll 1$ to obtain

$$\left[1 + \frac{z}{a} + \left(\frac{z}{a}\right)^2 + \left(\frac{z}{a}\right)^3 + \cdots\right]^{\alpha_k} = \left[1 + \alpha_k\left(\frac{z}{a}\right) + \frac{\alpha_k^2 + \alpha_k}{2}\left(\frac{z}{a}\right)^2 + \mathcal{O}\left(\frac{z}{a}\right)^3\right], \tag{C.14}$$

and after substituting in (C.13), we finally get

$$\begin{aligned}\left(\frac{z}{z-a}\right)^{\alpha_k} &= (-1)^{\alpha_k}\left(\frac{z}{a}\right)^{\alpha_k}\left[1 + \alpha_k\frac{z}{a} + \frac{1}{2}(\alpha_k^2 + \alpha_k)\left(\frac{z}{a}\right)^2 + \mathcal{O}\left(\frac{z}{a}\right)^3\right] \\ &= (-1)^{\alpha_k}\left[\left(\frac{z}{a}\right)^{\alpha_k} + \alpha_k\left(\frac{z}{a}\right)^{\alpha_k+1} + \mathcal{O}\left(\frac{z}{a}\right)^{\alpha_k+2}\right].\end{aligned} \tag{C.15}$$

Note that $(-1)^{\alpha_k} = e^{\imath\alpha_k\pi}$ and that $z = re^{\imath\theta}$, and furthermore, $a = |a|e^{\imath\theta^*}$, so that (C.15) becomes

$$\begin{aligned}\left(\frac{z}{z-a}\right)^{\alpha_k} = e^{\imath\alpha_k\pi}\Bigg[&|a|^{-\alpha_k}r^{\alpha_k}e^{\imath\alpha_k(\theta-\theta^*)} + \alpha_k|a|^{-(\alpha_k+1)}r^{\alpha_k+1}e^{\imath(\alpha_k+1)(\theta-\theta^*)} \\ &+ \frac{\alpha_k^2 + \alpha_k}{2}|a|^{-(\alpha_k+2)}r^{\alpha_k+2}e^{\imath(\alpha_k+2)(\theta-\theta^*)} + \mathcal{O}(r^{\alpha_k+3})\Bigg].\end{aligned} \tag{C.16}$$

Insert now (C.16) in (C.9) to obtain

$$\tau = \Im\left\{\sum_k A_k e^{\imath\alpha_k(\pi-\beta^*)}\left[|a|^{-\alpha_k} r^{\alpha_k} e^{\imath\alpha_k(\theta-\theta^*)} + \alpha_k|a|^{-(\alpha_k+1)} r^{\alpha_k+1} e^{\imath(\alpha_k+1)(\theta-\theta^*)}\right.\right.$$
$$\left.\left. + \frac{\alpha_k^2+\alpha_k}{2}|a|^{-(\alpha_k+2)} r^{\alpha_k+2} e^{\imath(\alpha_k+2)(\theta-\theta^*)} + \mathcal{O}(r^{\alpha_k+3})\right]\right\}. \tag{C.17}$$

If we define $A_{k0} \stackrel{\text{def}}{=} A_k|a|^{-\alpha_k}$, $A_{k1} \stackrel{\text{def}}{=} A_k\alpha_k|a|^{-(\alpha_k+1)}$, $A_{k1} \stackrel{\text{def}}{=} A_k\frac{\alpha_k^2+\alpha_k}{2}|a|^{-(\alpha_k+2)}$, etc: then (C.17) becomes

$$\tau = \Im\left\{\sum_k A_{k0} r^{\alpha_k} e^{\imath\alpha_k(\theta-\theta^*+\pi-\beta^*)} + A_{k1} r^{\alpha_k+1} e^{\imath[(\alpha_k+1)(\theta-\theta^*)+\alpha_k(\pi-\beta^*)]}\right.$$
$$\left. + A_{k2} r^{\alpha_k+2} e^{\imath[(\alpha_k+2)(\theta-\theta^*)+\alpha_k(\pi-\beta^*)]} + \mathcal{O}(r^{\alpha_k+3})\right\}. \tag{C.18}$$

And finally, we obtain the solution in Ω_z in terms of r, θ :

$$\begin{aligned}\tau(r,\theta) = \sum_k & A_{k0} r^{\alpha_k}\sin[\alpha_k(\theta-\theta^*+\pi-\beta^*)]\\ & + A_{k1} r^{\alpha_k+1}\sin[(\alpha_k+1)(\theta-\theta^*)+\alpha_k(\pi-\beta^*)]\\ & + A_{k2} r^{\alpha_k+2}\sin[(\alpha_k+2)(\theta-\theta^*)+\alpha_k(\pi-\beta^*)]\\ & + \mathcal{O}(r^{\alpha_k+3}).\end{aligned} \tag{C.19}$$

We may conclude that having curved boundaries in the vicinity of the singular point does not affect the leading singularity, but adds additional terms with higher exponents, and the solution in general for these cases is given by

$$\tau = \sum_k \sum_{j=1,2,\cdots} A_{kj} r^{\alpha_k+j} s_{kj}^{+}(\theta). \tag{C.20}$$

Let us consider a concrete example, where the two arcs are on circles having the same radius $r_1 = r_2 = R$. According to (C.1), $a = -R - \imath R$ and $\theta^* = \arctan\frac{-R}{-R} = 5\pi/4$, and using (C.4), $\beta^* = \pi/4$. Inserting these values into (C.19), the solution for the concrete example is

$$\begin{aligned}\tau(r,\theta) = & A_{10} r^{2/3}\sin\left[2\left(\theta-\frac{5\pi}{4}\right)/3+\frac{\pi}{2}\right]\\ & + A_{11} r^{5/3}\sin\left[5\left(\theta-\frac{5\pi}{4}\right)/3+\frac{\pi}{2}\right]\end{aligned} \tag{C.21}$$

$$
\begin{aligned}
&+ A_{12} r^{8/3} \sin\left[8\left(\theta - \frac{5\pi}{4}\right)/3 + \frac{\pi}{2}\right] \\
&+ \cdots \\
&+ A_{20} r^{4/3} \sin\left[4\left(\theta - \frac{5\pi}{4}\right)/3 + \pi\right] \\
&+ A_{21} r^{7/3} \sin\left[7\left(\theta - \frac{5\pi}{4}\right)/3 + \pi\right] \\
&+ A_{22} r^{10/3} \sin\left[10\left(\theta - \frac{5\pi}{4}\right)/3 + \pi\right] \\
&+ \cdots .
\end{aligned}
$$

Appendix D
Proof that Eigenvalues of the Scalar Anisotropic Elliptic BVP with Constant Coefficients Are Real

Consider the domain Ω shown in Figure D.1, which has a single corner at point P. Consider the scalar anisotropic strongly elliptic problem with constant coefficients, formulated in Cartesian coordinates:

$$\mathcal{L}(\tau) \equiv -\nabla_{\boldsymbol{x}}^T \cdot ([k]\nabla_{\boldsymbol{x}}\tau) = 0, \quad \boldsymbol{x} \in \Omega, \tag{D.1}$$

$$B(\tau) = 0, \quad \boldsymbol{x} \in \Gamma_\ell, \quad \ell = 1, 2, \tag{D.2}$$

where $\nabla_{\boldsymbol{x}}^T = \left(\frac{\partial}{\partial x_1}, \frac{\partial}{\partial x_2}\right)$ and $[k]$ is the matrix containing the constant coefficients:

$$[k] = \begin{pmatrix} k_{11} & k_{12} \\ k_{21} & k_{22} \end{pmatrix}, \quad k_{12} = k_{21}.$$

The matrix $[k]$ is positive definite, so that $[k]^{-1}$ exists and is positive definite and $[k]^{\frac{1}{2}}$, $[k]^{-\frac{1}{2}}$ are well defined.

We consider three types of boundary conditions:

$$B(\tau) = \text{trace}(\mathcal{L}(\tau)), \qquad \text{Dirichlet boundary condition}, \tag{D.3}$$

$$B(\tau) = \boldsymbol{n}_x^T \cdot [k]\nabla_{\boldsymbol{x}}\tau, \qquad \text{Neumann boundary condition}, \tag{D.4}$$

$$B(\tau) = \boldsymbol{n}_x^T[k]\nabla_{\boldsymbol{x}}\tau + c \text{ ktrace}(\mathcal{L}(u)), \quad \text{Newton boundary condition}. \tag{D.5}$$

Here $\boldsymbol{n}_x$ is the outward normal vector to the boundary in the $\boldsymbol{x}$ coordinate system and c is a constant.

Theorem D.1. *The solution to the above elliptic problem in the vicinity of the corner O can be expanded in a series of the form*

$$u = r^\alpha s\left(\frac{\boldsymbol{x}}{|\boldsymbol{x}|}\right), \qquad \text{with } \alpha \in \mathbb{R}, \tag{D.6}$$

where r is the distance from the vertex O to a point in the domain.

Z. Yosibash, *Singularities in Elliptic Boundary Value Problems and Elasticity and Their Connection with Failure Initiation*, Interdisciplinary Applied Mathematics 37,
DOI 10.1007/978-1-4614-1508-4_18, © Springer Science+Business Media, LLC 2012

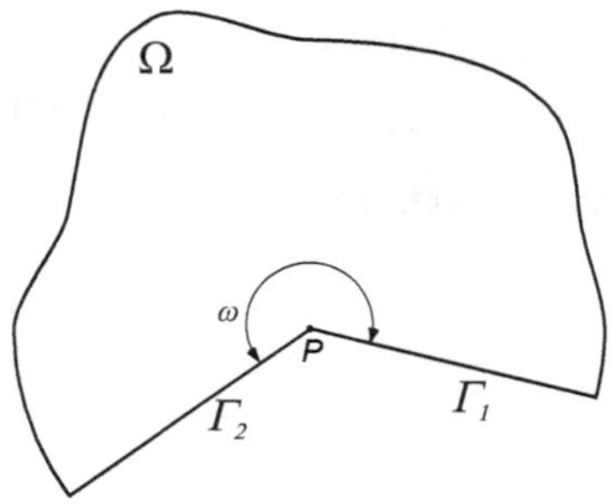

Fig. D.1 The domain and notation.

Proof. Define a linear coordinate transformation

$$y = [M]x, \text{ so that } [M] = [k]^{-\frac{1}{2}}. \tag{D.7}$$

We easily see that $\nabla_y = [M]\nabla_x$, so the operator $\mathcal{L}$ in the new coordinates becomes

$$\tilde{\mathcal{L}} = -\nabla_y^T[M]^T\cdot([k][M]\nabla_y) = -\nabla_y^T([k]^{-\frac{1}{2}})^T\cdot\left([k][k]^{-\frac{1}{2}}\nabla_y\right) = -\nabla_y^T\cdot\nabla_y = -\nabla_y{}^2,$$

which is exactly the Laplace operator in the new coordinate system.

Let us now examine the change in the boundary conditions due to the linear coordinate transformation. For the Dirichlet boundary conditions we retain the same $B(u)$, since the trace of the operator is now the trace of the new Laplace operator. In case of Neumann boundary conditions we now have

$$\tilde{B} = \boldsymbol{n}_x^T\cdot[k]\nabla_x = \boldsymbol{n}_y^T[M]^T\cdot[k][M]\nabla_y = \boldsymbol{n}_y^T\cdot([k]^{-\frac{1}{2}})^T[k][k]^{-\frac{1}{2}}\nabla_y = \boldsymbol{n}_y^T\cdot\nabla_y,$$

which is the usual Neumann boundary condition for the Laplace operator. The Newton boundary conditions change as a result of the change of variables as follows:

$$\tilde{B} = \boldsymbol{n}_x^T\cdot[k]\nabla_x + c\ \text{trace}\mathcal{L} = \boldsymbol{n}_y^T\cdot\nabla_y + c\ \text{trace}\tilde{\mathcal{L}},$$

which is exactly the Newton boundary condition for $\tilde{\mathcal{L}}$.

So far, we have shown that the scalar elliptic anisotropic operator with the various boundary conditions (D.3)-(D.5) can be transformed to the Laplace operator over a different domain with the usual Dirichlet, Neumann, and Newton boundary conditions.

It is well known that for the newly formulated Laplace problem

$$\tilde{\mathcal{L}}(\tau) = 0, \quad \boldsymbol{y} \in \tilde{\Omega}, \qquad \tilde{B}(\tau) = 0, \quad \boldsymbol{y} \in \tilde{\Gamma}_\ell,$$

where $\tilde{\Omega}$ and $\tilde{\Gamma}_\ell$ are the transformed domain and the boundaries, the solution in the vicinity of the vertex O (which under the linear transformation remain a vertex) can be expanded in a series of the form

$$\tau = \rho^{\tilde{\alpha}} \tilde{s}\left(\frac{y}{|y|}\right), \qquad \tilde{\alpha} \in \mathbb{R} \tag{D.8}$$

when $\rho \ll 1$ is the distance from the vertex to a point in the domain, where $\tilde{s}$ is an analytic function of the polar coordinate $\tilde{\theta}$.

Consider

$$\rho^2 = \boldsymbol{y}^T \cdot \boldsymbol{y} = \boldsymbol{x}^T [M]^T \cdot [M]\boldsymbol{x} = \boldsymbol{x}^T ([k]^{-\frac{1}{2}})^T \cdot [k]^{-\frac{1}{2}} \boldsymbol{x} = \boldsymbol{x}^T \cdot [k]^{-1}$$

$$\boldsymbol{x} = |\boldsymbol{x}|^2 \left(\frac{\boldsymbol{x}}{|\boldsymbol{x}|}\right)^T \cdot [k]^{-1} \left(\frac{\boldsymbol{x}}{|\boldsymbol{x}|}\right),$$

so that we get

$$\rho^2 = r^2 G_k \left(\frac{\boldsymbol{x}}{|\boldsymbol{x}|}\right). \tag{D.9}$$

Here $G_k\left(\frac{x}{|x|}\right)$ is an analytic function of $\left(\frac{x}{|x|}\right)$, since $[k]^{-1}$ is positive definite and symmetric. Substituting (D.9) into (D.8), we have

$$\tau = r^{\tilde{\alpha}} \left[G_k \left(\frac{\boldsymbol{x}}{|\boldsymbol{x}|}\right)\right]^{\frac{\tilde{\alpha}}{2}} \tilde{s}\left(\frac{\boldsymbol{y}}{|\boldsymbol{y}|}\right). \tag{D.10}$$

Notice that

$$\left(\frac{\boldsymbol{y}}{|\boldsymbol{y}|}\right) = \left(\frac{[M]\boldsymbol{x}}{|[M]\boldsymbol{x}|}\right) = \left(\frac{[k]^{-\frac{1}{2}}\boldsymbol{x}}{\sqrt{\boldsymbol{x}^T[M]^T \cdot [M]\boldsymbol{x}}}\right) = \left(\frac{[k]^{-\frac{1}{2}}\boldsymbol{x}}{\sqrt{\boldsymbol{x}^T \cdot [k]^{-1}\boldsymbol{x}}}\right),$$

and substituting into (D.10) we have

$$\tau = r^{\tilde{\alpha}} \left\{\left[G_k \left(\frac{\boldsymbol{x}}{|\boldsymbol{x}|}\right)\right]^{\frac{\tilde{\alpha}}{2}} \tilde{s}\left(\frac{[k]^{-\frac{1}{2}}|\boldsymbol{x}|}{\sqrt{\boldsymbol{x}^T[k]^{-1}\boldsymbol{x}}}\frac{\boldsymbol{x}}{|\boldsymbol{x}|}\right)\right\} \Rightarrow \tau = r^{\tilde{\alpha}} \tilde{\tilde{s}}\left(\frac{\boldsymbol{x}}{|\boldsymbol{x}|}\right). \qquad \blacksquare$$

The last result shows that the eigenvalues of our problem are all real.

In [115] a general proof is provided that the eigenpairs for the scalar elliptic problem are real in open and closed isotropic multimaterial corners and also for an open anisotropic multimaterial corners. The only possible case of complex eigenvalues in a scalar elliptic problem is the multi-material internal corner with at least one of the materials being anisotropic. For example, consider the internal singular point at the bimaterial interface shown in Figure 3.11. In Ω_1, occupying the sector $0 \leq \theta \leq \pi/2$, the governing equation is the Laplace equation $[k] = [I]$, whereas in Ω_2, occupying the sector $\pi/2 \leq \theta \leq 2\pi$, the scalar elliptic equation holds with $k_{11} = 10$, $k_{22} = 1$, $k_{12} = k_{21} = 0$. Continuity of the solution and the fluxes is assumed at the materials interface. Then, the singularity exponents between 0 and 1 are given by $\alpha_{1,2} = 0.8816020381 \pm \imath 0.3230787589$ (see [115]).

Appendix E
A Path-Independent Integral and Orthogonality of Eigenfunctions for General Scalar Elliptic Equations in 2-D Domains

Let us consider the general scalar elliptic PDE in two dimensions:

$$-\partial_\beta \left(k_{\beta\gamma}\partial_\gamma \tau\right) = 0 \quad \text{in } \Omega, \qquad \beta, \gamma = 1, 2, \tag{E.1}$$

which can be more conveniently represented as

$$-\nabla \cdot ([k]\nabla\tau) = 0 \quad \text{in } \Omega, \tag{E.2}$$

where Ω is the light gray domain in the vicinity of the singular point bounded by the four boundaries Γ_1-Γ_4 as shown in Figure E.1.

Lemma E.1. *Let χ and τ satisfy* (E.2). *Then*

$$\oint_{\partial\Omega} ([k]\nabla\tau \cdot \boldsymbol{n})\chi - ([k]\nabla\chi \cdot \boldsymbol{n})\tau \, d\Gamma = 0. \tag{E.3}$$

Proof. Multiplying (E.1) by a function χ and integrating over Ω, one obtains

$$-\iint_\Omega \partial_\beta \left(k_{\beta\gamma}\partial_\gamma \tau\right) \chi \, dx_1 dx_2 = 0, \tag{E.4}$$

which in expanded form is explicitly given as

$$\iint_\Omega \left[\frac{\partial}{\partial x_1}\left(k_{11}\frac{\partial\tau}{\partial x_1} + k_{12}\frac{\partial\tau}{\partial x_2}\right)\chi + \frac{\partial}{\partial x_2}\left(k_{21}\frac{\partial\tau}{\partial x_1} + k_{22}\frac{\partial\tau}{\partial x_2}\right)\chi\right] dx_1 dx_2 = 0. \tag{E.5}$$

For convenience let us define

$$\begin{aligned} -q_1 &= k_{11}\frac{\partial\tau}{\partial x_1} + k_{12}\frac{\partial\tau}{\partial x_2}, \qquad & -g_1 &= k_{11}\tfrac{\partial\chi}{\partial x_1} + k_{12}\tfrac{\partial\chi}{\partial x_2}, \\ q_2 &= k_{21}\frac{\partial\tau}{\partial x_1} + k_{22}\frac{\partial\tau}{\partial x_2}, \qquad & g_2 &= k_{21}\tfrac{\partial\chi}{\partial x_1} + k_{22}\tfrac{\partial\chi}{\partial x_2}. \end{aligned} \tag{E.6}$$

Z. Yosibash, *Singularities in Elliptic Boundary Value Problems and Elasticity and Their Connection with Failure Initiation*, Interdisciplinary Applied Mathematics 37,
DOI 10.1007/978-1-4614-1508-4_19,

Fig. E.1 Domain of interest in the vicinity of a reentrant corner.

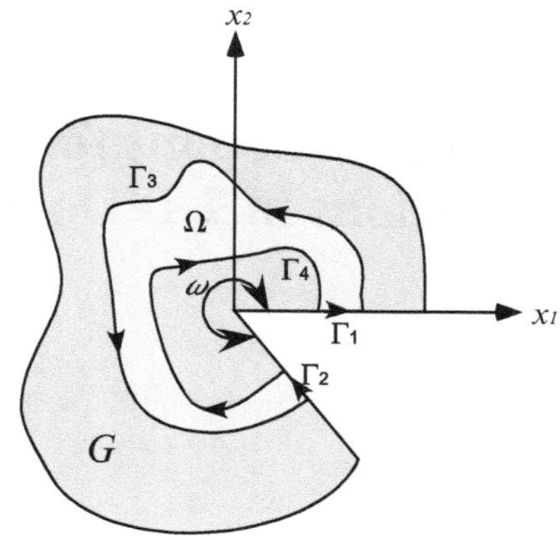

With this notation, (E.5) becomes

$$-\iint_{\Omega} \frac{\partial q_1}{\partial x_1} \chi \, dx_1 dx_2 + \iint_{\Omega} \frac{\partial q_2}{\partial x_2} \chi \, dx_1 dx_2 = 0. \tag{E.7}$$

Employing Green's theorem, the first and second integrals in (E.7) become

$$-\iint_{\Omega} \frac{\partial q_1}{\partial x_1} \chi \, dx_1 dx_2 = -\oint_{\partial\Omega} q_1 \chi \, dx_2 + \iint_{\Omega} q_1 \frac{\partial \chi}{\partial x_1} \, dx_1 dx_2, \tag{E.8}$$

$$\iint_{\Omega} \frac{\partial q_2}{\partial x_2} \chi \, dx_1 dx_2 = \oint_{\partial\Omega} q_2 \chi \, dx_1 - \iint_{\Omega} q_2 \frac{\partial \chi}{\partial x_2} \, dx_1 dx_2. \tag{E.9}$$

Substituting (E.8) and (E.9) in (E.7), one obtains

$$\oint_{\partial\Omega} -q_1 \chi \, dx_2 + q_2 \chi \, dx_1 - \iint_{\Omega} \left(-q_1 \frac{\partial \chi}{\partial x_1} + q_2 \frac{\partial \chi}{\partial x_2} \right) dx_1 dx_2 = 0. \tag{E.10}$$

Back-substituting q_1 and q_2 given in (E.6) into the second term of (E.10), then rearranging terms, and observing that in the first term, $-q_1 \; dx_2 + q_2 \; dx_1 = -k_{\beta\gamma} \frac{\partial \tau}{\partial x_\beta} n_\gamma d\Gamma \stackrel{\text{def}}{=} -q_n d\Gamma$, (E.10) becomes

$$-\oint_{\partial\Omega} q_n \chi d\Gamma - \iint_{\Omega} \left[\left(k_{11} \frac{\partial \chi}{\partial x_1} + k_{12} \frac{\partial \chi}{\partial x_2} \right) \frac{\partial \tau}{\partial x_1} + \left(k_{21} \frac{\partial \chi}{\partial x_1} + k_{22} \frac{\partial \chi}{\partial x_2} \right) \frac{\partial \tau}{\partial x_2} \right] dx_1 dx_2 = 0, \tag{E.11}$$

which after using the definitions in (E.6), we obtain

$$-\oint_{\partial\Omega} q_n \chi d\Gamma - \iint_{\Omega} \left(-g_1 \frac{\partial \tau}{\partial x_1} + g_2 \frac{\partial \tau}{\partial x_2}\right) dx_1 dx_2 = 0. \tag{E.12}$$

Employing Green's theorem on the second integral of (E.12), one gets

$$\iint_{\Omega} g_1 \frac{\partial \tau}{\partial x_1}\, dx_1 dx_2 = \oint_{\partial\Omega} g_1 \tau\, dx_2 - \iint_{\Omega} \frac{\partial g_1}{\partial x_1} \tau\, dx_1 dx_2, \tag{E.13}$$

$$-\iint_{\Omega} g_2 \frac{\partial \tau}{\partial x_2}\, dx_1 dx_2 = -\oint_{\partial\Omega} g_2 \tau\, dx_1 + \iint_{\Omega} \frac{\partial g_2}{\partial x_2} \tau\, dx_1 dx_2. \tag{E.14}$$

We insert (E.13) and (E.14) into (E.12), then resubstitute the definitions of g_1 and g_2 and use their connection to g_n to obtain

$$-\oint_{\partial\Omega} q_n \chi d\Gamma + \oint_{\partial\Omega} g_n \tau d\Gamma + \iint_{\Omega} \tau \frac{\partial}{\partial x_i}\left(k_{ij} \frac{\partial \chi}{\partial x_j}\right) dx_i dx_j = 0. \tag{E.15}$$

If χ is so chosen to be a solution of (E.1), then the last integral in (E.15) vanishes, and we obtain

$$\oint_{\partial\Omega} k_{\beta\gamma} \frac{\partial \tau}{\partial x_\beta} n_\gamma \chi d\Gamma - \oint_{\partial\Omega} k_{\beta\gamma} \frac{\partial \chi}{\partial x_\beta} n_\gamma \tau d\Gamma = 0. \quad \blacksquare \tag{E.16}$$

Let us denote by Γ a generic path starting at any point along Γ_1 and terminating at any point along Γ_2.

Definition E.1. We define the path integral I_Γ as

$$I_\Gamma \stackrel{\text{def}}{=} \int_\Gamma \left[\left([k]\nabla\tau \cdot \mathbf{n}\right)\chi - \left([k]\nabla\chi \cdot \mathbf{n}\right)\tau\right] d\Gamma = \int_\Gamma \left[k_{\beta\gamma} \frac{\partial \tau}{\partial x_\beta} n_\gamma \chi - k_{\beta\gamma} \frac{\partial \chi}{\partial x_\beta} n_\gamma \tau\right] d\Gamma, \tag{E.17}$$

which expressed in polar coordinates becomes

$$I_\Gamma \stackrel{\text{def}}{=} \int_\Gamma \left\{\left(\frac{\partial \tau}{\partial r}\chi - \tau \frac{\partial \chi}{\partial r}\right)\left(k_{11}\cos^2\theta + k_{12}\sin 2\theta + k_{22}\sin^2\theta\right) + \frac{1}{r}\left(\frac{\partial \tau}{\partial \theta}\chi - \tau \frac{\partial \chi}{\partial \theta}\right)\left[\frac{(k_{22}-k_{11})}{2}\sin 2\theta + k_{12}\cos 2\theta\right]\right\} d\Gamma. \tag{E.18}$$

Then the following lemma is immediately obtained:

Lemma E.2. *Let τ and χ satisfy* (E.2) *with Dirichlet or Neumann homogeneous boundary conditions on Γ_1 and Γ_2 (i.e., τ and $\chi = 0$, or $k_{\beta\gamma}\frac{\partial \tau}{\partial x_\beta} n_\gamma$ and $k_{\beta\gamma}\frac{\partial \chi}{\partial x_\beta} n_\gamma = 0$). Then I_Γ is path-independent.*

Proof. The proof follows immediately from Lemma E.1. Starting with (E.16), and noticing that on both Γ_1 and Γ_2 homogeneous boundary conditions are prescribed, one is left with

$$\int_{\Gamma_3}\left[k_{\beta\gamma}\frac{\partial\tau}{\partial x_\beta}n_\gamma\chi - k_{\beta\gamma}\frac{\partial\chi}{\partial x_\beta}n_\gamma\tau\right]d\Gamma + \int_{\Gamma_4}\left[k_{\beta\gamma}\frac{\partial\tau}{\partial x_\beta}n_\gamma\chi - k_{\beta\gamma}\frac{\partial\chi}{\partial x_\beta}n_\gamma\tau\right]d\Gamma = 0. \tag{E.19}$$

According to the definition in (E.17), equation (E.16) states that

$$I_{\Gamma_3} + I_{\Gamma_4} = 0 \Rightarrow I_{\Gamma_3} = -I_{\Gamma_4}. \tag{E.20}$$

Since the integration for Γ_3 is in the opposite direction to Γ_4, (E.20) states that I_{Γ_3} is equal to I_{Γ_4} (when integrating in the same direction), and both are arbitrary paths staring on Γ_1 and ending on Γ_2. This means that I_Γ is path-independent. ■

Orthogonality of the Primal and Dual eigenfunctions

Here we prove that the primal and dual eigenfunctions of a scalar elliptic problem are orthogonal. Let us choose a circular path (having a given radius R) around the singular point, Γ_R for example, as shown in Figure 1.9.

Taking as τ the ith eigenpair and as χ the jth dual eigenpair, both functions satisfy the conditions of Lemmas E.1-E.2, and inserting them in the definition of I_{Γ_R} according to (E.18), one obtains

$$I_{\Gamma_R} = R^{\alpha_i-\alpha_j}\int_0^\omega \Big\{(\alpha_i+\alpha_j)s_i^+(\theta)s_j^-(\theta)\left[k_{11}\cos^2\theta + k_{12}\sin 2\theta + k_{22}\sin^2\theta\right]$$
$$+\Big((s_i^+)'(\theta)s_j^-(\theta) - (s_j^-)'(\theta)s_i^+(\theta)\Big)\left[\frac{(k_{22}-k_{11})}{2}\sin 2\theta + k_{12}\cos 2\theta\right]\Big\}\,d\theta,$$

with no summation on i and j.

If $\alpha_i \neq \alpha_j$, then the following condition must hold so that I_{Γ_R} will not be R-dependent (otherwise it would violate Lemma E.2):

$$\int_0^\omega \Big\{(\alpha_i+\alpha_j)s_i^+(\theta)s_j^-(\theta)\left[k_{11}\cos^2\theta + k_{12}\sin 2\theta + k_{22}\sin^2\theta\right] +$$
$$+\Big((s_i^+)'(\theta)s_j^-(\theta) - (s_j^-)'(\theta)s_i^+(\theta)\Big)$$
$$\times\left[\frac{(k_{22}-k_{11})}{2}\sin 2\theta + k_{12}\cos 2\theta\right]\Big\}\,d\theta = 0, \quad \alpha_i \neq \alpha_j,$$

with no summation on i and j. (E.21)

For $\alpha_i = \alpha_j$, one obtains

$$I_\Gamma^{(\alpha_i)} = \int_0^\omega \{2\alpha_i s_i^+(\theta) s_i^-(\theta) \left[k_{11}\cos^2\theta + k_{12}\sin 2\theta + k_{22}\sin^2\theta\right]$$

$$+ \left((s_i^+)'(\theta) s_i^-(\theta) - (s_i^-)'(\theta) s_i^+(\theta)\right) \left[\frac{(k_{22}-k_{11})}{2}\sin 2\theta + k_{12}\cos 2\theta\right]\Big\} \, d\theta,$$

with no summation on i,

which simplifies in the case of the Laplace equation (where $s_i^+ \equiv s_i^-$) to

$$I_\Gamma^{(\alpha_i)} = 2\alpha_i \int_0^\omega (s_i^+)^2(\theta) d\theta, \qquad \text{no summation on } i. \tag{E.22}$$

Appendix F
Energy Release Rate (ERR) Method, its Connection to the J-integral and Extraction of SIFs

F.1 Derivation of the ERR

The energy release rate in LEFM was given its name by Irwin in 1956 [86], but was already used in 1920 by Griffith and later shown to be equal to the J-integral by Cherepanov and Rice. In this chapter the various forms of the ERR and their connections are provided.

F.1.1 The Energy Argument [94]

From basic energy arguments, for a given crack of length a to propagate in an elastic domain, an energy rate inequality has to hold:

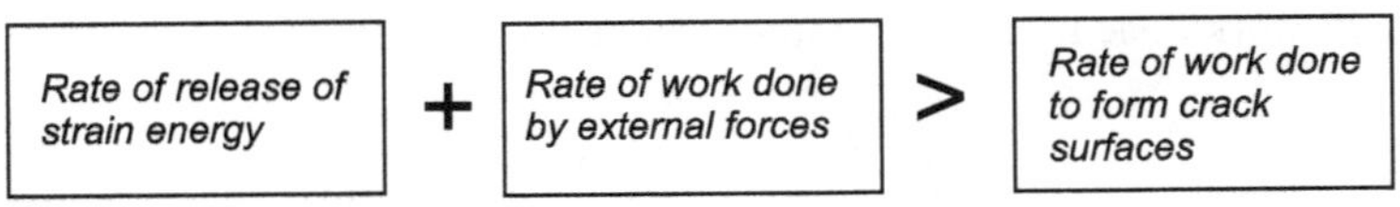

If we call the sum of the two left-hand-side blocks the *energy release rate* (ERR) $\mathcal{G}$, which is the rate at which energy is supplied to initiate a crack growth, then

$$\mathcal{G} \stackrel{\text{def}}{=} -\frac{d\mathcal{U}}{da} + \frac{dW}{da}. \tag{F.1}$$

In a 2-D domain Ω, the strain energy is given by

$$\mathcal{U} \stackrel{\text{def}}{=} \frac{1}{2}\int_{\Omega} \sigma_{\gamma\delta}\varepsilon_{\gamma\delta}\, d\Omega, \qquad \gamma, \delta = 1, 2, \tag{F.2}$$

Z. Yosibash, *Singularities in Elliptic Boundary Value Problems and Elasticity and Their Connection with Failure Initiation*, Interdisciplinary Applied Mathematics 37,
DOI 10.1007/978-1-4614-1508-4_20, © Springer Science+Business Media, LLC 2012

and using Clapeyron's theorem (in the absence of body forces), see for example [167, p. 86], we obtain

$$\mathcal{U} = \frac{1}{2}\int_{\Omega} \sigma_{\gamma\delta}\varepsilon_{\gamma\delta}\, d\Omega = \frac{1}{2}\int_{\partial\Omega} T_\gamma u_\gamma d\Gamma. \tag{F.3}$$

For the rate of change of the strain energy with respect to the crack extension, one obtains

$$\frac{d\mathcal{U}}{da} = \frac{1}{2}\int_{\partial\Omega}\left(\frac{\partial T_\gamma}{\partial a}u_\gamma + T_\gamma\frac{\partial u_\gamma}{\partial a}\right) d\Gamma. \tag{F.4}$$

Consider next the rate of work done by the external forces, assuming that no body forces exist when the crack increases from a to $a + \Delta a$. This equals the tractions at $a + \Delta a$ times the change in displacements from a to $a + \Delta a$. Expanding the displacements on the boundary using a Taylor series, one obtains

$$\Delta u_\gamma = u_\gamma(a + \Delta a) - u_\gamma(a) = \Delta a \left.\frac{\partial u_\gamma}{\partial a}\right|_a + \mathcal{O}(\Delta a)^2. \tag{F.5}$$

Similarly, the tractions on the boundary at $a + \Delta a$ are

$$T_\gamma(a + \Delta a) = T_\gamma(a) + \Delta a \left.\frac{\partial T_\gamma}{\partial a}\right|_a + \mathcal{O}(\Delta a)^2. \tag{F.6}$$

Therefore the change in energy due to external forces is

$$\Delta W = \int_{\partial\Omega}\left(T_\gamma(a)\Delta a \left.\frac{\partial u_\gamma}{\partial a}\right|_a + \mathcal{O}(\Delta a)^2\right) d\Gamma. \tag{F.7}$$

The rate of work done by external forces can be computed now as

$$\frac{dW}{da} = \lim_{\Delta a\to 0}\frac{\Delta W}{\Delta a} = \lim_{\Delta a\to 0}\int_{\partial\Omega}\left(T_\gamma(a)\left.\frac{\partial u_\gamma}{\partial a}\right|_a + \mathcal{O}(\Delta a)\right) d\Gamma = \int_{\partial\Omega} T_\gamma\frac{\partial u_\gamma}{\partial a} d\Gamma. \tag{F.8}$$

Finally, we substitute (F.8) and (F.4) into (F.1) to obtain

$$\mathcal{G} = \frac{1}{2}\int_{\partial\Omega}\left(T_\gamma\frac{\partial u_\gamma}{\partial a} - u_\gamma\frac{\partial T_\gamma}{\partial a}\right) d\Gamma. \tag{F.9}$$

F.1.2 The Potential Energy Argument [94]

The departure point is the potential energy Π, which is a functional defined over *all displacement functions that satisfy the essential (displacements) boundary conditions*

$$\Pi \stackrel{\text{def}}{=} \frac{1}{2}\int_{\Omega} \sigma_{\gamma\delta}\varepsilon_{\gamma\delta}\, d\Omega - \int_{\partial\Omega_T} T_\gamma u_\gamma d\Gamma, \tag{F.10}$$

where $\partial\Omega_T$ is the part of the boundary on which traction boundary conditions are prescribed. We use again Clapeyron's theorem for the first term in (F.10) to obtain

$$\Pi = \frac{1}{2}\int_{\partial\Omega} T_\gamma u_\gamma d\Gamma - \int_{\partial\Omega_T} T_\gamma u_\gamma d\Gamma. \tag{F.11}$$

Since $\partial\Omega = \partial\Omega_T \cup \partial\Omega_u$, (F.11) becomes

$$\begin{aligned} \Pi &= \frac{1}{2}\int_{\partial\Omega_T} T_\gamma u_\gamma d\Gamma + \frac{1}{2}\int_{\partial\Omega_u} T_\gamma u_\gamma d\Gamma - \int_{\partial\Omega_T} T_\gamma u_\gamma d\Gamma \\ &= \frac{1}{2}\int_{\partial\Omega_u} T_\gamma u_\gamma d\Gamma - \frac{1}{2}\int_{\partial\Omega_T} T_\gamma u_\gamma d\Gamma. \end{aligned} \tag{F.12}$$

Before we consider the expression $\frac{\partial\Pi}{\partial a}$, we observe that on $\partial\Omega_u$, the expression $\frac{\partial u_i}{\partial a}$ equals 0 because the displacements are prescribed functions independent of a, and on $\partial\Omega_T$, the expression $\frac{\partial T_\gamma}{\partial a}$ is 0 because the tractions are prescribed functions independent of a. Therefore,

$$\frac{\partial\Pi}{\partial a} = \frac{1}{2}\int_{\partial\Omega_u} u_\gamma \frac{\partial T_\gamma}{\partial a} d\Gamma - \frac{1}{2}\int_{\partial\Omega_T} T_\gamma \frac{\partial u_\gamma}{\partial a} d\Gamma. \tag{F.13}$$

Since $\frac{\partial u_\gamma}{\partial a} = 0$ on $\partial\Omega_u$, we can subtract from the RHS of (F.13) the term

$$\frac{1}{2}\int_{\partial\Omega_u} T_\gamma \frac{\partial u_\gamma}{\partial a} d\Gamma,$$

and since $\frac{\partial T_\gamma}{\partial a} = 0$ on $\partial\Omega_T$,, we can also add the term

$$\frac{1}{2}\int_{\partial\Omega_T} u_\gamma \frac{\partial T_\gamma}{\partial a} d\Gamma.$$

Thus, (F.13) becomes

$$\frac{\partial\Pi}{\partial a} = \frac{1}{2}\int_{\partial\Omega} u_\gamma \frac{\partial T_\gamma}{\partial a} d\Gamma - \frac{1}{2}\int_{\partial\Omega} T_\gamma \frac{\partial u_\gamma}{\partial a} d\Gamma = \frac{1}{2}\int_{\partial\Omega} \left(u_\gamma \frac{\partial T_\gamma}{\partial a} - T_\gamma \frac{\partial u_\gamma}{\partial a} \right) d\Gamma. \tag{F.14}$$

Comparing (F.14) with (F.9) one notices that

$$\mathcal{G} = -\frac{\partial\Pi}{\partial a} \tag{F.15}$$

F.2 Griffith's Energy Criterion [70, 71]

The first paper addressing the fracture mechanics criterion of an infinite two-dimensional domain weakened by a crack (obtained as the limit of an ellipse when the minor axis tends to zero) was proposed by A. Griffith in 1920 [70]. This first publication was corrected four years later in [71], and we provide here the original derivation including the correction, which is not derived in [71], but in much later papers such as [94, 161].

When tractions are prescribed over the entire boundary, $\partial\Omega = \partial\Omega_T$, the potential energy in (F.11) becomes

$$\Pi = \frac{1}{2}\int_{\partial\Omega} T_\gamma u_\gamma d\Gamma - \int_{\partial\Omega} T_\gamma u_\gamma d\Gamma = -\frac{1}{2}\int_{\partial\Omega} T_\gamma u_\gamma d\Gamma = -\mathcal{U}. \tag{F.16}$$

Therefore, the ERR criterion in (F.15) becomes

$$\mathcal{G} = \frac{\partial \mathcal{U}}{\partial a} = \frac{1}{2}\frac{\partial}{\partial a}\int_{\partial\Omega} T_\gamma u_\gamma d\Gamma \quad \text{when } T_\gamma \text{ are prescribed on all } \partial\Omega. \tag{F.17}$$

Because Griffith considered an infinite domain containing an elliptical hole with the large radius denoted by a, loaded by stresses on its entire infinitely long boundary, see the left part of Figure F.1, the strain energy is infinite $\mathcal{U}(a) = \infty$. In this case, one may consider the strain energy of the same infinitely large plate that is free of holes called $\mathcal{U}(0)$; obviously, this configuration is independent of a. Although $\mathcal{U}(0)$ is also infinite, the difference

$$\Delta\mathcal{U}(a) = \mathcal{U}(a) - \mathcal{U}(0) \tag{F.18}$$

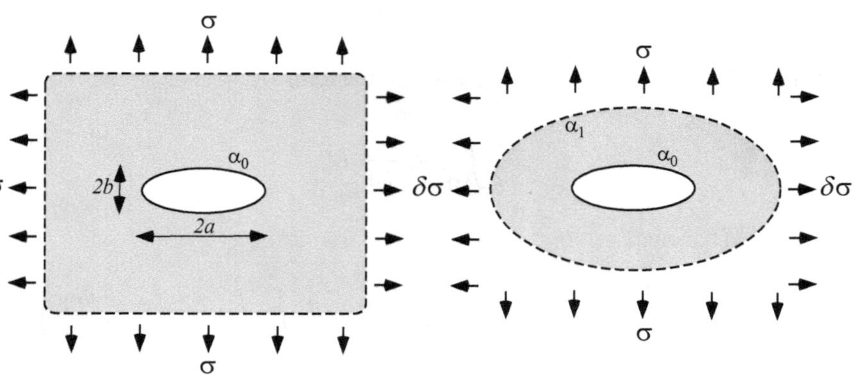

Fig. F.1 An ellipse in an infinite domain (left) and two concentric ellipses (right) subject to remote stress (in Griffith's works $\delta = 1$).

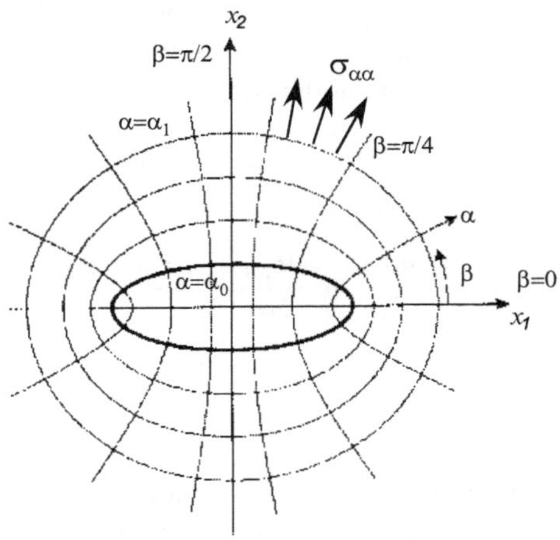

Fig. F.2 Elliptical coordinates and notation.

is a finite value. So, we may insert in (F.17) instead of $\mathcal{U}(a)$ the expression obtained from (F.11), $\mathcal{U}(a) = \Delta\mathcal{U}(a) + \mathcal{U}(0)$, to obtain

$$\mathcal{G} = \frac{\partial(\Delta\mathcal{U}(a) + \mathcal{U}(0))}{\partial a} = \frac{\partial\Delta\mathcal{U}(a)}{\partial a}. \tag{F.19}$$

Thus, Griffith in his works considered the release rate of the *strain energy difference* without fully justifying its use, under the assumption that the tractions at infinity should be prescribed and independent of the ellipse's dimensions. As will be shown in the sequel, his first work [70] is erroneous because the tractions computed at any given path surrounding the elliptical hole, although vanishing as the path increased to infinity, yet are dependent on the elliprse's dimensions and contribute a finite strain energy. His second work corrects this error.

Griffith used Inglis's analysis [83]; Inglis obtained the stress field in an infinite plate containing an elliptical hole. This permitted crack-like geometries to be treated by making the minor axis of the ellipse small. For the elliptical hole, it is convenient to work in elliptical coordinates α and β as shown in Figure F.2. *Note that in this section,* α_0 *and* α_1 *denote elliptical coordinates of the inner and outer ellipses and not the eigenvalues.* The Cartesian coordinates are expressed by

$$x_1 = c\cosh\alpha\cos\beta, \qquad x_2 = c\sinh\alpha\sin\beta, \tag{F.20}$$

where c is a constant. If β is eliminated in (F.20), one obtains

$$\frac{x_1^2}{c\cosh^2\alpha} + \frac{x_2^2}{c\sinh^2\alpha} = 1. \tag{F.21}$$

If we set the boundary of the ellipse to be α_0, then it is evident from (F.21) that the major and minor axes of the elliptical hole are:

$$a = c\cosh\alpha_0, \qquad b = c\sinh\alpha_0. \tag{F.22}$$

A crack of length $2a$ is obtained at the limit when $b = c \sinh \alpha_0 \to 0$. Thus $\alpha_0 = 0$. In this case, $a = c \cosh \alpha_0$, so $c = a$. In an elliptical coordinate system, the equilibrium equations in the absence of body forces are given by

$$\frac{\partial \sigma_{\alpha\alpha}}{\partial \alpha} + \frac{\partial \sigma_{\alpha\beta}}{\partial \beta} + \frac{c^2 \sin 2\beta}{h_\alpha^2} \sigma_{\alpha\beta} + \frac{c^2 \sinh 2\alpha}{2h_\alpha^2} (\sigma_{\alpha\alpha} - \sigma_{\beta\beta}) = 0, \quad (F.23)$$

$$\frac{\partial \sigma_{\beta\beta}}{\partial \beta} + \frac{\partial \sigma_{\alpha\beta}}{\partial \alpha} + \frac{c^2 \sinh 2\alpha}{h_\alpha^2} \sigma_{\alpha\beta} - \frac{c^2 \sin 2\beta}{2h_\alpha^2} (\sigma_{\alpha\alpha} - \sigma_{\beta\beta}) = 0, \quad (F.24)$$

where

$$2h_\alpha^2 = c^2 (\cosh 2\alpha - \cos 2\beta). \quad (F.25)$$

The solution to the above equations for an infinite plate having an elliptical traction-free hole, and loaded at infinity by a constant normal stress σ, as shown on the left side of Figure F.1, is provided in [83]. The stress component $\sigma_{\alpha\alpha}$ and displacement component u_α along an elliptical curve characterized by α, $u_\beta = 0$, are given by

$$\sigma_{\alpha\alpha} = \sigma \frac{\sinh 2\alpha (\cosh 2\alpha - \cosh 2\alpha_0)}{(\cosh 2\alpha - cos 2\beta)^2}, \quad (F.26)$$

$$\frac{u_\alpha}{h_\alpha} = \frac{a^2 \sigma}{8\mu} \left[(\kappa - 1) \cosh 2\alpha - (\kappa + 1) \cos 2\beta + 2 \cosh 2\alpha_0 \right], \quad (F.27)$$

where κ is the Kolosov constant that is $(3 - \nu)/(1 + \nu)$ for plane-stress and $3 - 4\nu$ for plane-strain. Of course, as $\alpha \to \infty$, the stress $\sigma_{\alpha\alpha}$ has to approach σ. We may expand the hyperbolic terms in (F.26) for large α's, then divide both nominator and denominator by $e^{4\alpha}/4$ to obtain

$$\sigma_{\alpha\alpha} = \sigma \frac{1 - 2e^{-2\alpha} \cosh 2\alpha_0}{(\cosh 2\alpha - \cos 2\beta)^2} + \mathcal{O}(e^{-6\alpha}). \quad (F.28)$$

Using the binomial expansion $(1 - x)^{-2} = 1 + 2x + 3x^2 + \mathcal{O}(x^3)$ for $|x| < 1$ for the denominator, one obtains

$$\sigma_{\alpha\alpha} = \sigma \left[1 + 2e^{-2\alpha} (2 \cos 2\beta - \cosh 2\alpha_0) \right] + \mathcal{O}(e^{-4\alpha}). \quad (F.29)$$

It can be noticed that $\sigma_{\alpha\alpha}$ at any given α depends on the elliptical hole via α_0. The term having the α_0 dependence vanishes as $\alpha \to \infty$ (in an infinite domain). Nevertheless, at any given α, it has a contribution to the strain energy when the stresses are multiplied by the displacement term u_α (as will be shown). This is the error in Griffith's first paper in 1920, because if one wishes to use (F.19), the tractions on the entire boundary have to be *independent of the elliptical hole*.

We use now the stresses and displacements in (F.29) and (F.27) to show the erroneous result of Griffith in his 1920 paper. The strain energy inside an elliptical subplate characterized by the coordinate $\alpha_1 \to \infty$ shown on the right side of Figure F.1 (according to (F.3)) gives

$$\begin{aligned} \mathcal{U}_{\alpha_1} &= \frac{1}{2}\int_0^{2\pi} \left(\sigma_{\alpha\alpha}\frac{u_\alpha}{h_\alpha}\right)|_{\alpha_1}\, d\beta \\ &= \frac{\pi a^2\sigma^2}{8\mu}\left[\frac{\kappa-1}{2}e^{2\alpha_1} + (3-\kappa)\cosh 2\alpha_0\right] + \mathcal{O}(e^{-2\alpha_1}). \end{aligned} \tag{F.30}$$

Let us investigate the first two terms in the strain energy:

$$\mathcal{U}_{\alpha_1} = \frac{\pi\sigma^2(\kappa-1)}{16\mu}a^2 e^{2\alpha_1} + \frac{\pi\sigma^2(3-\kappa)}{8\mu}a^2\cosh 2\alpha_0 + \mathcal{O}(e^{-2\alpha_1}). \tag{F.31}$$

For a crack, the first term in (F.31) tends to infinity as $\alpha_1 \to \infty$, the second term is constant ($\alpha_0 = 0$), and the remainder tends to zero. To use (F.19) we have to compute the strain energy of an infinite plate with no crack, which is the limit of (F.31) as $a \to 0$ and $\alpha_1 \to \infty$. In this case only the first expression in (F.31) remains, and tends to infinity because $e^{2\alpha_1} \to \infty$ much faster than $a^2 \to 0$. Thus, we finally obtain

$$\begin{aligned} \Delta\mathcal{U} &= \lim_{\alpha_1\to\infty,\alpha_0\to 0}[\mathcal{U}_{\alpha_1,\alpha_0}(a) - \mathcal{U}_{\alpha_1,\alpha_0}(0)] \\ &= \lim_{\alpha_0\to 0}\frac{\pi\sigma^2(3-\kappa)}{8\mu}a^2\cosh 2\alpha_0 = \frac{\pi\sigma^2 a^2}{8\mu}(3-\kappa). \end{aligned} \tag{F.32}$$

This is the erroneous value reported in [70]. In a closing footnote in the same paper Griffith writes that the method used to calculate the strain energy of the cracked plate is in error, but only in his 1924 paper [71] does he explain

> ... but in the solution there given (1920 paper) the calculation of the strain energy was erroneous, in that the expressions used for the stresses gave values at infinity differing from the postulated uniform stress at infinity by an amount which, though infinitesimal, yet made a finite contribution to the energy when integrated round the boundary. This difficulty has been overcome by slightly modifying the expressions for the stresses, so as to make this contribution to the energy vanish...

The details of the calculations were not provided by Griffith, and only more than 40 years later in [161] was a method for computation of this stress adjustment provided. We use here Keating and Sinclair's [94] adjustment based on [161] to provide the correction to Griffith's 1920 paper.

The problem in [70] has been emphasized by Sih and Liebowitz [161] on a model problem of a circular hole in an infinite plate for which an analytical solution for stresses and displacements exists for an infinite plate as well as a concentric annulus. For a circular hole in an infinite plate under constant tension at infinity (as on the left of Figure F.1 when $b = a$, $\delta = 1$), the stresses and displacements at a given circle of radius r are computed. These stresses depend on the hole radius a. Computing the strain energy difference (compared to the plate without a hole) and taking the limit as $r \to \infty$, they obtained

$$\Delta \mathcal{U} = \frac{\pi \sigma^2 a^2}{4\mu}(3 - \kappa). \tag{F.33}$$

This result (for a circular hole) is exactly twice that of Griffith's erroneous result of 1920 for a crack. When the analysis has been redone for a concentric annulus, where on the outer boundary *precise constant stresses* $\sigma_{rr} = \sigma$ are prescribed (as in the right of Figure F.1 when $b = a$, and $\sigma_{rr} = \sigma$ on the outer boundary), then taking the limit as the outer boundary tends to infinity, they obtained:

$$\Delta \mathcal{U} = \frac{\pi \sigma^2 a^2}{4\mu}(1 + \kappa) \tag{F.34}$$

This result (for a circular hole) is exactly twice as compared to Griffith's correct result of 1924 for a crack.

This example problem in [161] clearly demonstrates that the stresses obtained from the analysis of an infinite plate, which depend on the hole's dimensions (and may vanish as the radius tends to infinity), provide a different strain energy as compared to the case in which the stresses on the boundary of interest are indeed independent of the hole's dimensions.

As a remedy, they show that a modification to the stresses obtained as the solution to the infinite plate problem (which depend on the hole dimension) can be incorporated so to eliminate this dependence (this modification applies to the displacements also, but it does not eliminate their dependence on the hole dimension). The modification ensures that the strain energy at the infinite plate limit computed from the modified stresses results in the same strain energy computed accurately using annular concentric holes with constant stresses on the outer boundary.

Returning to Griffith's work, observe that if σ is replaced by $\sigma(1+2e^{-2\alpha_1}\cosh\alpha_0)$, then at the limit when $\alpha_1 \to \infty$ both result in the same constant stress at infinity. But in this case the stress in (F.29) for $\alpha_1 \gg 1$ reads

$$\begin{aligned} \sigma_{\alpha\alpha} &= \sigma(1 + 2e^{-2\alpha_1}\cosh\alpha_0)\left[1 + 2e^{-2\alpha_1}(2\cos 2\beta - \cosh 2\alpha_0)\right] + \mathcal{O}(e^{-4\alpha_1}) \\ &= \sigma(1 + 4e^{-2\alpha_1}\cos 2\beta) + \mathcal{O}(e^{-4\alpha_1}), \end{aligned} \tag{F.35}$$

and they are clearly *independent of* α_0 and thus do not contribute a finite energy as $\alpha_1 \to \infty$. Substituting $\sigma(1 + 2e^{-2\alpha_1}\cosh\alpha_0)$ instead of σ in (F.27), then at $\alpha = \alpha_1$ one obtains

$$\frac{u_\alpha}{h_\alpha} = \frac{a^2\sigma}{8\mu}\left[\frac{1}{2}(\kappa - 1)e^{2\alpha_1} - (\kappa + 1)\cos 2\beta + 2\cosh 2\alpha_0\right] + \mathcal{O}(e^{-2\alpha_1}). \tag{F.36}$$

Substituting (F.35) and (F.36) in the expression of the strain energy as in (F.30), one obtains

$$\begin{aligned}\mathcal{U}_{\alpha_1} &= \frac{1}{2}\int_0^{2\pi}\left(\sigma_{\alpha\alpha}\frac{u_\alpha}{h_\alpha}\right)|_{\alpha_1}\,d\beta \\ &= \frac{\pi a^2\sigma^2}{8\mu}\left[\frac{\kappa-1}{2}e^{2\alpha_1}+(\kappa+1)\cosh 2\alpha_0\right]+\mathcal{O}(e^{-2\alpha_1}),\end{aligned} \tag{F.37}$$

which yields the correct expression for the change in strain energy as given by Griffith in [71]:

$$\Delta\mathcal{U}(a) = \frac{\pi\sigma^2 a^2}{8\mu}(\kappa+1). \tag{F.38}$$

The change in strain energy obtained in (F.38) is for creating *two crack tips* (generating an embedded crack in an infinite domain). If the surface energy density is denoted by γ, then for the creation of a crack of dimension $2a$ having upper and lower faces, the surface energy required is $4a\gamma$. Accordingly, Griffith's original criterion requires that the change in the strain energy for the crack creation satisfy the inequality

$$\Delta\mathcal{U}(a) \geq 4a\gamma \quad\rightarrow\quad \frac{\partial\Delta\mathcal{U}(a)}{\partial a} \geq 4\gamma. \tag{F.39}$$

Because (F.38) represents the change in strain energy for two crack tips created simultaneously, the energy release rate for a single crack tip is

$$\mathcal{G} = \frac{1}{2}\frac{\partial\Delta\mathcal{U}(a)}{\partial a}, \tag{F.40}$$

and on substituting this in (F.39), we finally obtain

$$\mathcal{G} = \frac{1}{2}\frac{\partial\Delta\mathcal{U}(a)}{\partial a} = \frac{a\pi\sigma^2}{8\mu}(\kappa+1) \geq 2\gamma. \tag{F.41}$$

Or in other words, fracture does not occur if

$$\sigma\sqrt{a} \leq \sqrt{\frac{8E\gamma}{\pi(\nu+1)(\kappa+1)}}. \tag{F.42}$$

We now provide briefly, without any proof, the energy release rate for a plate with a central crack of length $2a$ loaded uniaxially by a stress at infinity in the x_2 direction, i.e., $\sigma_{22}(x_2\rightarrow\infty)=\sigma$:

$$\mathcal{G} = \frac{1}{2}\frac{\partial\Delta\mathcal{U}(a)}{\partial a} = \begin{cases}\frac{1}{2}\frac{2\pi a\sigma^2}{E} & \text{plane-stress} \\ \frac{1}{2}\frac{2(1-\nu^2)\pi a\sigma^2}{E} & \text{plane-strain}\end{cases} \geq 2\gamma \tag{F.43}$$

so that fracture does not occur if

$$\sigma\sqrt{a} \leq \begin{cases} \sqrt{\frac{2E\gamma}{\pi a}} & \text{plane-stress} \\ \sqrt{\frac{2E\gamma}{\pi a(1-\nu^2)}} & \text{plane-strain} \end{cases} \tag{F.44}$$

F.3 Relations Between the ERR and the SIFs

F.3.1 Symmetric (Mode I) Loading

Consider the state of stress at the crack tip for a plate in uniaxial tension in the x_2 direction. In the coordinate system centered on the crack tip, as shown in Figure F.3, neglecting terms of higher order,

$$\sigma_{22} = \frac{K_{\mathrm{I}}}{\sqrt{2\pi x_1}}, \qquad 0 \leq x_1. \tag{F.45}$$

The displacement of the crack face is

$$u_2 = \frac{(1+\nu)(\kappa+1)}{E}\frac{K_{\mathrm{I}}}{\sqrt{2\pi}}\sqrt{-x_1'}, \qquad x_1' \leq 0. \tag{F.46}$$

Assume now that the crack length increases by a small amount Δa, as shown in Figure F.3. In this case $x_1' = x_1 - \Delta a$:

$$u_2 = \frac{(1+\nu)(\kappa+1)}{E}\frac{K_{\mathrm{I}}}{\sqrt{2\pi}}\sqrt{\Delta a - x_1}, \qquad 0 \leq x_1 \leq \Delta a. \tag{F.47}$$

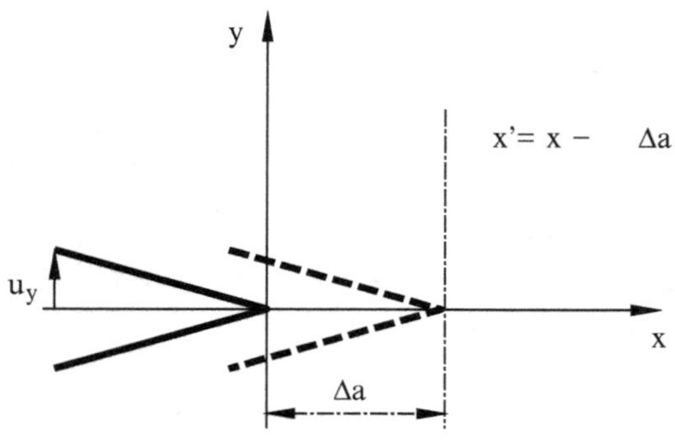

Fig. F.3 Notation.

The work required to return the crack to its original length, that is, to close the length increment Δa, is

$$\Delta\mathcal{U} = 2\int_0^{\Delta a} \frac{1}{2}\sigma_{22}u_2\,dx_1 = \frac{(1+\nu)(\kappa+1)}{E}\frac{K_{\mathrm{I}}^2}{2\pi}\int_0^{\Delta a}\sqrt{\frac{\Delta a - x_1}{x_1}}\,dx_1. \quad \text{(F.48)}$$

The integral expression is readily evaluated. Letting $x_1 = z\Delta a$, we obtain

$$\int_0^{\Delta a}\sqrt{\frac{\Delta a - x}{x}}\,dx = \Delta a\int_0^1\sqrt{\frac{1-z}{z}}\,dz = \frac{\pi}{2}\,\Delta a.$$

Hence

$$\Delta\mathcal{U} = \frac{(1+\nu)(\kappa+1)}{4E}K_{\mathrm{I}}^2\Delta a. \quad \text{(F.49)}$$

In order to restore the crack to its initial length, energy equal to $\Delta\mathcal{U}$ had to be imparted to the elastic body. This is the energy expended in crack growth, called Griffith's surface energy (see, for example, [90]). The potential energy had to decrease by the same amount when the crack increment occurred. Hence

$$\mathcal{G} = -\lim_{\Delta a\to 0}\frac{\Delta\Pi}{\Delta a} = -\frac{\partial\Pi}{\partial a} = \frac{(1+\nu)(\kappa+1)}{4E}K_{\mathrm{I}}^2. \quad \text{(F.50)}$$

Problem F.1. Explain the reasons for writing

$$\Delta\mathcal{U} = 2\int_0^{\Delta a}\frac{1}{2}\sigma_{22}u_2\,dx_1$$

in (F.48).

F.3.2 Antisymmetric (Mode II) Loading

When the loading is purely antisymmetric, then the relationship between the energy release rate and the stress intensity factor is analogous to the symmetric case. However, instead of equations (F.45)-(F.46), we have

$$\sigma_{12} = \frac{K_{\mathrm{II}}}{\sqrt{2\pi x_1}}, \qquad 0 \le x_1, \quad \text{(F.51)}$$

and

$$u_1 = \frac{(1+\nu)(\kappa+1)}{E}\frac{K_{\mathrm{II}}}{\sqrt{2\pi}}\sqrt{-x_1}, \qquad x_1 \le 0. \quad \text{(F.52)}$$

The derivation of the relationship between $\mathcal{G}$ and K_{II} under the assumption of perfectly antisymmetric loading, is left to the reader in the following exercise.

Problem F.2. Make a sketch, analogous to Figure F.3, for the case of perfectly antisymmetric loading and show that

$$\mathcal{G} = \frac{(1+\nu)(\kappa+1)}{4E} K_{\text{II}}^2. \tag{F.53}$$

F.3.3 Combined (Mode I and Mode II) Loading

In view of the fact that the solutions corresponding to mode I and mode II loadings are *energy orthogonal*, we have

$$\Pi(\mathbf{u}_{\text{I}} + \mathbf{u}_{\text{II}}) = \Pi(\mathbf{u}_{\text{I}}) + \Pi(\mathbf{u}_{\text{II}}), \tag{F.54}$$

where $\mathbf{u}_{\text{I}}$ and $\mathbf{u}_{\text{II}}$ are the solutions of the mode I and mode II loadings, respectively. Therefore, in the case of combined loading, we have

$$\mathcal{G} = \frac{(1+\nu)(\kappa+1)}{4E} (K_{\text{I}}^2 + K_{\text{II}}^2). \tag{F.55}$$

Consequently, the stress intensity factors are related to $\mathcal{G}$, which is computable, by

$$(K_{\text{I}}^2 + K_{\text{II}}^2) = \begin{cases} E\mathcal{G} & \text{for plane-stress,} \\ \frac{E\mathcal{G}}{1-\nu^2} & \text{for plane-strain.} \end{cases} \tag{F.56}$$

F.3.4 Computation of $\mathcal{G}$ by the Stiffness Derivative Method

For the computation of the stress intensity factors, a variation of the energy release rate method is used when finite element methods are employed. This method is called the *stiffness derivative method*, and was first proposed by Parks [140], and we describe it here. Let us assume that we have computed the finite element linear elastic solution in a domain having a crack of length a, and denote it by $\boldsymbol{b}_a$. The potential energy (in the absence of body forces) is

$$\Pi(a) = \frac{1}{2}\boldsymbol{b}_a^T[K(a)]\boldsymbol{b}_a - \boldsymbol{b}_a^T\boldsymbol{r}_a,$$

where $[K] = [K(a)]$ is the stiffness matrix and $\boldsymbol{r} = \boldsymbol{r}_a$ is the load vector. Following a virtual crack extension Δa, the new crack length becomes $(a + \Delta a)$. Because of the extension of the crack length, a potential energy difference $\Delta\Pi$ will occur, which is the difference between the potential energy of the cracked domain and the potential energy of the virtually cracked domain. The load vector $\boldsymbol{r}$ does not change, because the loading is independent of the crack length. The new potential energy for a crack of length $a + \Delta a$ is:

$$\Pi(a + \Delta a) = \frac{1}{2}\boldsymbol{b}_{a+\Delta a}^T [K]_{a+\Delta a} \boldsymbol{b}_{a+\Delta a} - \boldsymbol{r}^T \boldsymbol{b}_{a+\Delta a}. \tag{F.57}$$

The stiffness matrix $[K]_{a+\Delta a}$ may be split into two matrices, where the first is the stiffness matrix of a cracked domain containing a crack of length a, and the second matrix is the remainder between the two stiffness matrices:

$$[K]_{a+\Delta a} = [K]_a + \Delta[K]. \tag{F.58}$$

The displacement vector $\boldsymbol{b}_{a+\Delta a}$ may also be split:

$$\boldsymbol{b}_{a+\Delta a} = \boldsymbol{b}_a + \Delta \boldsymbol{b}. \tag{F.59}$$

Substituting equations (F.58) and (F.59) into (F.57) results in

$$\begin{aligned}
\Pi(a + \Delta a) &= \frac{1}{2}(\boldsymbol{b}_a + \Delta\boldsymbol{b})^T([K]_a + \Delta[K])(\boldsymbol{b}_a + \Delta\boldsymbol{b}) - \boldsymbol{r}^T(\boldsymbol{b}_a + \Delta\boldsymbol{b}) \\
&= \Pi(a) + \frac{1}{2}\boldsymbol{b}_a^T[K]_a\Delta\boldsymbol{b} + \frac{1}{2}\Delta\boldsymbol{b}^T[K]_a\boldsymbol{b}_a \\
&\quad + \frac{1}{2}\Delta\boldsymbol{b}^T[K]_a\Delta\boldsymbol{b} + \frac{1}{2}\boldsymbol{b}_a^T\Delta[K]\boldsymbol{b}_a + \frac{1}{2}\boldsymbol{b}_a^T\Delta[K]\Delta\boldsymbol{b} \\
&\quad + \frac{1}{2}\Delta\boldsymbol{b}^T\Delta[K]\boldsymbol{b}_a + \frac{1}{2}\Delta\boldsymbol{b}^T\Delta[K]\Delta\boldsymbol{b} - \boldsymbol{r}^T\Delta\boldsymbol{b}.
\end{aligned} \tag{F.60}$$

The potential energy is a scalar and therefore each of the terms in (F.60) is transposable. Let us transpose the two terms in (F.60)

$$\begin{aligned}
\frac{1}{2}\boldsymbol{b}_a^T[K]_a\Delta\boldsymbol{b} &= \frac{1}{2}\left(\boldsymbol{b}_a^T[K]_a\Delta\boldsymbol{b}\right)^T = \frac{1}{2}\Delta\boldsymbol{b}^T[K]_a\boldsymbol{b}_a, \\
\boldsymbol{r}^T\Delta\boldsymbol{b} &= \left(\boldsymbol{r}^T\Delta\boldsymbol{b}\right)^T = \Delta\boldsymbol{b}^T\boldsymbol{r},
\end{aligned}$$

and therefore the potential energy $\Pi(a + \Delta a)$ is reduced to

$$\begin{aligned}
\Pi(a + \Delta a) &= \Pi(a) + \frac{1}{2}\Delta\boldsymbol{b}^T[K]_a\Delta\boldsymbol{b} + \frac{1}{2}\boldsymbol{b}_a^T\Delta[K]\boldsymbol{b}_a \\
&\quad + \frac{1}{2}\boldsymbol{b}_a^T\Delta[K]\Delta\boldsymbol{b} + \frac{1}{2}\Delta\boldsymbol{b}^T\Delta[K]\boldsymbol{b}_a + \frac{1}{2}\Delta\boldsymbol{b}^T\Delta[K]\Delta\boldsymbol{b} \\
&\quad + \Delta\boldsymbol{b}^T\underbrace{([K]_a\boldsymbol{b}_a - \boldsymbol{r})}_{\text{this is zero}}.
\end{aligned} \tag{F.61}$$

The last term in (F.61) vanishes, because at a the equilibrium condition holds. Substituting (F.61) into the energy release rate equation (F.15) and taking the limit as Δa tends to zero, one obtains

$$\begin{aligned}\lim_{\Delta a\to 0}\frac{\Delta\Pi}{\Delta a} &= \lim_{\Delta a\to 0}\frac{\Pi(a+\Delta a)-\Pi(a)}{\Delta a}\\ &= \lim_{\Delta a\to 0}\frac{\frac{1}{2}\Delta\boldsymbol{b}^T[K]_a\Delta\boldsymbol{b}}{\Delta a}+\frac{\frac{1}{2}\boldsymbol{b}_a^T\Delta[K]\boldsymbol{b}_a}{\Delta a}+\frac{\frac{1}{2}\boldsymbol{b}_a^T\Delta[K]\Delta\boldsymbol{b}}{\Delta a}\\ &\quad+\frac{\frac{1}{2}\Delta\boldsymbol{b}^T\Delta[K]\boldsymbol{b}_a}{\Delta a}+\frac{\frac{1}{2}\Delta\boldsymbol{b}^T\Delta[K]\Delta\boldsymbol{b}}{\Delta a}.\end{aligned} \tag{F.62}$$

When $\Delta a\to 0$ also $\Delta\boldsymbol{b}\to 0$ and $\Delta[K]\to 0$, and therefore (F.62) becomes

$$\lim_{\Delta a\to 0}\frac{\Delta\Pi}{\Delta a}=\lim_{\Delta a\to 0}\frac{\frac{1}{2}\boldsymbol{b}_a^T\Delta[K]\boldsymbol{b}_a}{\Delta a}\overset{\text{def}}{=}\frac{1}{2}\boldsymbol{b}^T\frac{\partial[K]}{\partial a}\boldsymbol{b}. \tag{F.63}$$

Thus

$$\mathcal{G}=-\frac{1}{2}\boldsymbol{b}^T\frac{\partial[K]}{\partial a}\boldsymbol{b}. \tag{F.64}$$

In finite element computations, $\partial[K]/\partial a$ is usually approximated by finite differences:

$$\frac{\partial[K]}{\partial a}\approx\frac{[K(a+\Delta a)]-[K(a-\Delta a)]}{2\Delta a}. \tag{F.65}$$

This involves recomputation of the stiffness matrices for only those elements that have a vertex on the crack tip.

Since $\mathcal{G}$ is related to a combination of the stress intensity factors of the two modes, the possibility of computing each of them is restricted to cases in which only mode I or mode II alone exists: When only mode I is present,

$$-\frac{1}{2}\boldsymbol{b}^T\frac{[K]_{a+\Delta a}-[K]_a}{\Delta a}\boldsymbol{b}=\mathcal{G}=\frac{K_{\mathrm{I}}^2}{E^*}. \tag{F.66}$$

The stiffness derivative method enhancement for the computation of SIFs for mixed-mode crack problems (extraction of K_{I} and K_{II} independently) was addressed by Shumin and Xing in [160]. They prescribe the solution of the stresses in the vicinity of the singular point as

$$\begin{aligned}\sigma_{11} &= \frac{K_{\mathrm{I}}}{\sqrt{2\pi r}}\cos\frac{\theta}{2}\left(1-\sin\frac{\theta}{2}\sin\frac{3\theta}{2}\right)-\frac{K_{\mathrm{II}}}{\sqrt{2\pi r}}\sin\frac{\theta}{2}\left(2+\cos\frac{\theta}{2}\cos\frac{3\theta}{2}\right)\\ &\quad+\text{remainder},\end{aligned}$$

$$\sigma_{22} = \frac{K_{\mathrm{I}}}{\sqrt{2\pi r}} \cos\frac{\theta}{2}\left(1+\sin\frac{\theta}{2}\sin\frac{3\theta}{2}\right) + \frac{K_{\mathrm{II}}}{\sqrt{2\pi r}} \sin\frac{\theta}{2}\cos\frac{\theta}{2}\cos\frac{3\theta}{2}$$
$$+\text{remainder},$$
$$\sigma_{12} = \frac{K_{\mathrm{I}}}{\sqrt{2\pi r}} \cos\frac{\theta}{2}\sin\frac{\theta}{2}\cos\frac{3\theta}{2} + \frac{K_{\mathrm{II}}}{\sqrt{2\pi r}} \cos\frac{\theta}{2}\left(1-\sin\frac{\theta}{2}\sin\frac{3\theta}{2}\right)$$
$$+\text{remainder}, \tag{F.67}$$

where the remainder terms are of higher order than $1/\sqrt{r}$. The normal stresses σ_{11} and σ_{22} related to mode I, are even whereas the normal stresses related to mode II are odd. The shear stress σ_{12} related to mode I is odd, whereas the shear stress related to mode II is even. Therefore, a subregion loaded by symmetric forces will describe mode I and a subregion loaded by antisymmetric forces will describe mode II.

Accordingly, an arbitrary subregion may be chosen where the loads along the boundary of the subregion are resolved into two types, one symmetric and the other antisymmetric. Based on each type of load, a stiffness matrix is constructed ($[K]^{\text{sym}}$, $[K]^{\text{asy}}$), and the displacement vectors can be found ($\{\mathbf{u}^{\text{sym}}\}$, $\{\mathbf{u}^{\text{asy}}\}$) and the stress intensity factors K_{I}, K_{II} are extracted separately by

$$-\frac{1}{2}\{\mathbf{u}^{\text{sym}}\}^T \frac{[K]^{\text{sym}}_{a+\Delta a} - [K]^{\text{sym}}_a}{\Delta a}\{\mathbf{u}^{\text{sym}}\} = \mathcal{G}^{\text{sym}} = \frac{K_{\mathrm{I}}^2}{E*},$$
$$-\frac{1}{2}\{\mathbf{u}^{\text{asy}}\}^T \frac{[K]^{\text{asy}}_{a+\Delta a} - [K]^{\text{asy}}_a}{\Delta a}\{\mathbf{u}^{\text{asy}}\} = \mathcal{G}^{\text{asy}} = \frac{K_{\mathrm{II}}^2}{E*}, \tag{F.68}$$

where the displacement components are extracted from the finite element solution.

The stiffness derivative method for stress intensity factors extraction was extended by Wu in [194] for 2-D bimaterial cracked domains. The method involves a finite element model of the bimaterial domain and a subregion selection for calculating $\mathcal{G}$. The procedure of load separation into symmetric and antisymmetric sets of loads for the bimaterial subdomain is prescribed by Wu, and numerical examples are presented.

A formulation for the stiffness derivative method for anisotropic 2-D mixed-mode cracked domains is presented by Hamoush and Salami in [78]. The extraction procedure involves two considered independent equilibrium states with field variables (1) and (2). Each of the equilibrium states is separated into a symmetric and antisymmetric set of loads, and therefore four sets of energy release rates are computed out of which the stress intensity $K_{\mathrm{I}}^{(1)}$, $K_{\mathrm{II}}^{(1)}$, $K_{\mathrm{I}}^{(2)}$, $K_{\mathrm{II}}^{(2)}$ are extracted. On the other hand, a superposition of the two equilibrium states is considered where two more calculations (symmetric and antisymmetric) of $\mathcal{G}$ are obtained. Hamoush and Salami show that the energy release rate of the symmetric and antisymmetric sets of the superposition is a linear combination of the two sets of equilibrium states and their stress intensity factors. Therefore, the stress intensity factors of the superposition equilibrium state, $K_{\mathrm{I}}^{(1,2)}$ and $K_{\mathrm{II}}^{(1,2)}$, may be extracted.

F.3.5 The Stiffness Derivative Method for 3-D Domains

The calculation of the energy release rate using the stiffness derivative method for a 3-D cracked domain was presented by Banks-Sills and Sherman in [18]. The computation of the energy release rate in the vicinity of the edge is of the form

$$\int_0^L \mathcal{G}(s)\delta a(s)d\Gamma = -\frac{1}{2}\boldsymbol{b}^T \Delta[K]\boldsymbol{b}, \tag{F.69}$$

where $\mathcal{G}(s)$ is Griffith's energy release rate at point s along the crack front, $\delta a(s)$ is the change in crack length, L is the length of the crack, and $\boldsymbol{b}$ is the vector of displacement found from finite element analysis along the nodes of the model. The change in the stiffness matrix, $\Delta[K]$, for a given virtual crack increased by δa is of the form

$$\Delta[K] \approx [K]_{a+\delta a} - [K]_a. \tag{F.70}$$

Banks-Sills and Sherman suggest the types of elements and number of nodes needed for accurate extraction of energy release rates along either straight edges or curved edges.

Although a method for the energy release rate of a 3-D cracked domain was established using the stiffness derivative method for the entire edge, the extraction method of stress-intensity functions using the stiffness derivative method is reduced to a pointwise method along the edge. The extraction method of the stress-intensity factors (or stress-intensity functions in the 3-D case) involves comparing the energy release rate computed by a numerical method (finite element method) with stress-intensity factors that are not directly connected. Because the energy release rate $\mathcal{G}$ of the 3-D cracked domains are based on pointwise calculations, extracting the stress-intensity factors at a specific point along the edge each time, the 3-D stiffness derivative method is based on the pointwise method as well. Moreover, because calculation of $\mathcal{G}$ is based on either a plane-strain or plane-stress assumption, the stiffness derivative method carries the same assumptions.

F.4 The J-Integral and its Relation to ERR

One of the well-known integrals in fracture mechanics is the so-called J-integral. It was introduced for two-dimensional domains by Cherepanov in 1967 [38] and independently by Rice in 1968 [147]. Therefore it is also known as the Cherepanov-Rice integral. Consider a crack along the x_1-axis with area Ω in the vicinity of the crack tip, bounded by the closed path Γ^*, as shown in Figure F.4. Notice that the x_1-axis here is in the direction of the crack propagation, not as in the conventional notation. Consider also a path Γ, which may be any path beginning on one face and

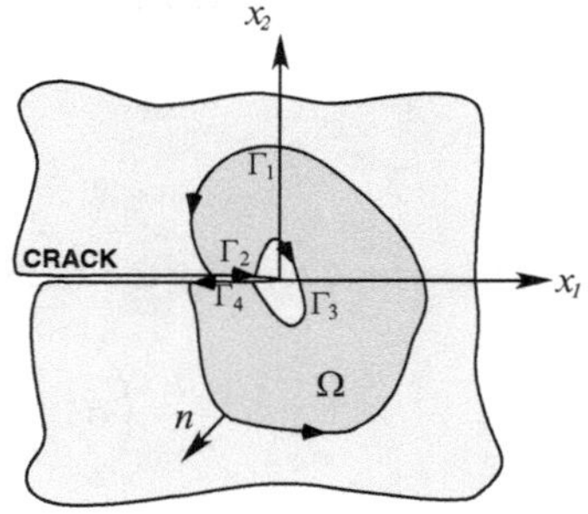

Fig. F.4 Two-dimensional domain with a crack. Γ is any curve surrounding the crack.

terminating on the other face of the crack, surrounding the crack tip. The J-integral is defined as follows:

$$J \stackrel{\text{def}}{=} \int_{\Gamma} \left(\mathcal{U} dx_2 - \boldsymbol{T}^T \frac{\partial \boldsymbol{u}}{\partial x_1} d\Gamma \right), \tag{F.71}$$

where $\mathcal{U}$ is the strain-energy density ($\mathcal{U} = \mathcal{U} = \int_0^{\varepsilon} \sigma_{\gamma\delta} d\varepsilon_{\gamma\delta}$), which equals $\frac{1}{2}\sigma_{\gamma\delta}\varepsilon_{\gamma\delta}$ for elastic domains.

Theorem F.1. *J is path-independent.*

Proof. Let us choose a closed path such as $\Gamma^* = \Gamma_1 + \Gamma_2 + \Gamma_3 + \Gamma_4$, enclosing an area $\Omega(\Gamma^*)$. Consider the following two expressions:

$$\oint_{\Gamma^*} \mathcal{U} dx_2 = \oint_{\Gamma^*} \mathcal{U} n_1 d\Gamma \underset{\text{Green's thm.}}{=} \iint_{\Omega(\Gamma^*)} \frac{\partial \mathcal{U}}{\partial x_1} d\Omega, \tag{F.72}$$

$$\begin{aligned} \oint_{\Gamma^*} \boldsymbol{T}^T \frac{\partial \boldsymbol{u}}{\partial x_1} d\Gamma &\underset{\text{Cauchy lemma}}{=} \oint_{\Gamma^*} (\sigma_{\gamma\delta} n_\delta) \frac{\partial u_\gamma}{\partial x_1} d\Gamma = \oint_{\Gamma^*} \left(\sigma_{\gamma\delta} \frac{\partial u_\gamma}{\partial x_1} \right) n_\delta d\Gamma \\ &\underset{\text{Green's thm.}}{=} \iint_{\Omega(\Gamma^*)} \frac{\partial}{\partial x_\delta} \left(\sigma_{\gamma\delta} \frac{\partial u_\gamma}{\partial x_1} \right) d\Omega. \end{aligned} \tag{F.73}$$

Subtracting the above two equations, i.e., (F.72)-(F.73), one obtains

$$\oint_{\Gamma^*} \left(\mathcal{U} dx_2 - \boldsymbol{T}^T \frac{\partial \boldsymbol{u}}{\partial x_1} d\Gamma \right) = \int_{A(\Gamma^*)} \left[\frac{\partial \mathcal{U}}{\partial x_1} - \frac{\partial}{\partial x_\delta} \left(\sigma_{\gamma\delta} \frac{\partial u_\gamma}{\partial x_1} \right) \right] dx_1 dx_2. \tag{F.74}$$

We examine now each of the two terms forming the RHS integrand in (F.74). Applying the chain rule to the first term of the RHS integrand, one obtains

$$\frac{\partial \mathcal{U}}{\partial x_1} = \frac{\partial \mathcal{U}}{\partial \varepsilon_{\gamma\delta}} \cdot \frac{\partial \varepsilon_{\gamma\delta}}{\partial x_1} = \sigma_{\gamma\delta} \frac{\partial \varepsilon_{\gamma\delta}}{\partial x_1} = \frac{1}{2} \sigma_{\gamma\delta} \frac{\partial}{\partial x_1} \left(\frac{\partial u_\gamma}{\partial x_\delta} + \frac{\partial u_\delta}{\partial x_\gamma} \right). \tag{F.75}$$

Relying on the symmetric relation of the stress tensor $\sigma_{\gamma\delta} = \sigma_{\delta\gamma}$, (F.76) reduces to

$$\frac{\partial \mathcal{U}}{\partial x_1} = \sigma_{\gamma\delta}\frac{\partial^2 u_\gamma}{\partial x_1 \partial x_\delta}. \tag{F.76}$$

The second term on the RHS of the integrand in (F.74) is

$$\frac{\partial}{\partial x_\delta}\left(\sigma_{\gamma\delta}\frac{\partial u_\gamma}{\partial x_1}\right) = \frac{\partial \sigma_{\gamma\delta}}{\partial x_\delta}\frac{\partial u_\gamma}{\partial x_1} + \sigma_{\gamma\delta}\frac{\partial^2 u_\gamma}{\partial x_1 \partial x_\delta}. \tag{F.77}$$

In view of the equilibrium equation *without body forces*, which reads $\frac{\partial \sigma_{\gamma\delta}}{\partial x_j} = 0$, (F.77) reduces to

$$\frac{\partial}{\partial x_\delta}\left(\sigma_{\gamma\delta}\frac{\partial u_\gamma}{\partial x_1}\right) = \sigma_{\gamma\delta}\frac{\partial^2 u_\gamma}{\partial x_1 \partial x_\delta}. \tag{F.78}$$

Substituting (F.76) and (F.78) in (F.74), we notice that the area integral vanishes:

$$\int_{\Gamma^*}\left(\mathcal{U}dx_2 - \boldsymbol{T}^T\frac{\partial \boldsymbol{u}}{\partial x_1}d\Gamma\right) = 0. \tag{F.79}$$

The path integrals over Γ_2 and Γ_4 vanish because $dx_2 = 0$, and either $\boldsymbol{T} = 0$ or $\boldsymbol{u} = 0$, so that $\frac{\partial \boldsymbol{u}}{\partial x_1} = 0$ on both paths. By changing the direction of integration on Γ_3, (F.79) is simplified:

$$\int_{\Gamma_1}\left(\mathcal{U}dx_2 - \boldsymbol{T}^T\frac{\partial \boldsymbol{u}}{\partial x_1}d\Gamma\right) = \int_{\Gamma_3}\left(\mathcal{U}dx_2 - \boldsymbol{T}^T\frac{\partial \boldsymbol{u}}{\partial x_1}d\Gamma\right) \equiv J. \tag{F.80}$$

The paths Γ_1 and Γ_3 are randomly selected, so the right-hand side as well as the left-hand side of (F.80) may be considered an invariant where Γ is a path in the domain that starts at one edge and ends at the other edge of the crack. ■

Remark F.1. In the proof of path-independence of the J-integral, we rely on the equilibrium equation without body forces and the kinematic conditions of small strain. However, no restriction was made on a linear constitutive model. Therefore, J is path-independent for small-deformation nonlinear elasticity or the deformation theory of plasticity.

The same formulation for the J-integral was provided by Cherepanov in [38]. Cherepanov prescribed the energy-release rate of a 2-D cracked domain and showed that by choosing a circular path around the crack tip, the energy-release rate is independent of the radius of the circular path. Although Cherepanov considered a 2-D domain, he allowed the existence of the stress component σ_{33}, i.e., the method is applicable to 3-D problems under a plane-strain assumption.

For linear elastic materials, the J-integral can be shown to be equal to the energy release rate $\mathcal{G}$. To derive this connection mathematically, one needs to apply the

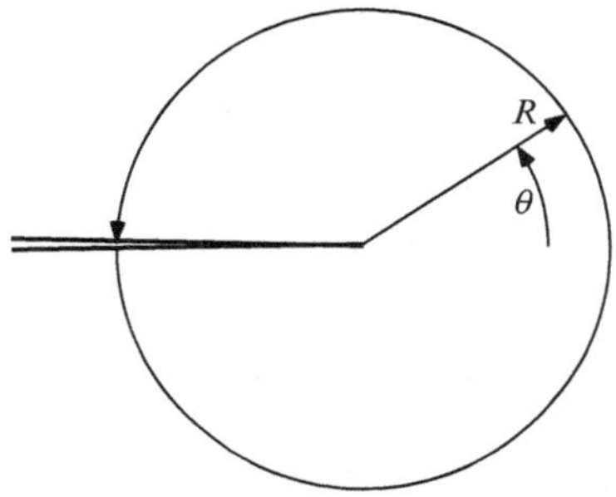

Fig. F.5 Path for J-integral computation.

divergence theorem on the potential energy variation to obtain an expression for the J-integral. Because of the crack tip singularity, a straightforward application of the divergence theorem is flawed, and a mathematically rigorous derivation addressing the effect of this singularity can be found, for example, in [88]. Because it is being technical and tedious, we do not give this proof here, but use only the outcome. Since the J-integral is equal to the ERR, and we have shown the relation between the ERR and the stress intensity factors K_{I} and K_{II}, one has

$$J = \mathcal{G} = \frac{K_{\text{I}}^2 + K_{\text{II}}^2}{E^*}, \qquad E^* = \begin{cases} E & \text{for plane-stress,} \\ \frac{E}{1-\nu^2} & \text{for plane-strain,} \end{cases} \tag{F.81}$$

and therefore the computation of the J-integral allows one to determine the stress-intensity factors. Some studies have focused on simplifying the extraction method. One of the studies as presented by Wu [193] extracts the stress-intensity factors based on the J-integral. The extraction simplification method relies on the relations between the stress, strain, and displacement components in a tangential set of coordinates.

Theorem F.2. *Under the assumption of plane-strain and mode I loading only, for an isotropic material with TF/TF boundary conditions,* $J = \frac{K_I^2}{E/(1-\nu^2)}$.

Proof. Consider a circular path of radius R in the vicinity of the crack tip, as shown in Figure F.5. In order to compute the J-integral, we first consider the first term in the J-integral:

$$\int \mathcal{U} dx_2.$$

Along the circular path one has $dx_2 = R\cos\theta d\theta$, and thus

$$\int \mathcal{U} dx_2 = \int_{-\pi}^{\pi} \mathcal{U}|_{r=R}\, R\cos\theta d\theta. \tag{F.82}$$

The strain-energy density for an isotropic elastic material in the plane state ($\sigma_{13} = \sigma_{23} = 0$) is

$$\mathcal{U} = \frac{-\nu}{2E}\left(\sigma_{11} + \sigma_{22} + \sigma_{33}\right)^2 + \frac{1+\nu}{2E}\left(\sigma_{11}^2 + \sigma_{22}^2 + \sigma_{33}^2 + 2\sigma_{12}^2\right). \tag{F.83}$$

For plane-strain, $\sigma_{33} = \nu(\sigma_{11} + \sigma_{22})$, whereas for plane-stress, $\sigma_{33} = 0$. The stress tensor in the vicinity of the crack tip is expressed in terms of K_{I} and θ at a given radius R by the expressions in Table 5.3. Substituting these expressions in (F.83) and then in (F.82), we obtain

$$\int_{-\pi}^{\pi} \mathcal{U}|_{r=R}\, R\cos\theta d\theta = \frac{(1-2\nu)(1+\nu)K_{\mathrm{I}}^2}{4E}. \tag{F.84}$$

The second term in the expression for the J-integral is given as follows:

$$\int \boldsymbol{T}^T \frac{\partial \boldsymbol{u}}{\partial x_1} d\Gamma = \int_{-\pi}^{\pi} \left[(\sigma_{11}\cos\theta + \sigma_{12}\sin\theta)\frac{\partial u_1}{\partial x_1} + (\sigma_{12}\cos\theta + \sigma_{22}\sin\theta)\frac{\partial u_2}{\partial x_1}\right]_{r=R} R d\theta. \tag{F.85}$$

The stress tensor and the displacements are expressed in terms of K_{I} and θ at a given radius R by the expressions in Table 5.3. Using the chain rule,

$$\frac{\partial}{\partial x_1} = \cos\theta\frac{\partial}{\partial r} - \frac{1}{r}\sin\theta\frac{\partial}{\partial\theta},$$

one obtains

$$\int \boldsymbol{T}^T \frac{\partial \boldsymbol{u}}{\partial x_1} d\Gamma = \frac{(1+\nu)(-3+2\nu)K_{\mathrm{I}}^2}{4E}. \tag{F.86}$$

Combining (F.84) and (F.86), we finally obtain

$$J = \frac{K_{\mathrm{I}}^2}{E/(1-\nu^2)}. \qquad \blacksquare$$

Problem F.3. Following the same steps as in the proof of Theorem F.2, show that under the assumption of plane-strain with mode I and mode II loading, for an isotropic material with TF/TF boundary conditions, $J = \frac{K_{\mathrm{I}}^2 + K_{\mathrm{II}}^2}{E/(1-\nu^2)}$.

References

1. *STRESS CHECK Master Guide - V-7*, Engineering Software Research and Development, Inc., 7750 Clayton Road, Suite 204, St. Louis, MO 63117, www.esrd.com, 2004.
2. M. Abramowitz and A. Stegun, *Handbook of mathematical functions with formulas, graphs and mathematical tables*, Nat. Bureau of Standards, Applied Mathematics Series, 1964.
3. G. Amar and Z. Yosibash, *p-FEM for formulating an elastic criterion for predicting mechanical failures at 2-D singular points*, in p-FEM2000, May 31-June 2: Summaries of Papers., St. Louis, MO, USA, 2000, Washington Univ, p. 15.
4. E. Anderson, Z. Bai, C. Bischof, J. Demmel, J. Dongarra, J. Du Croz, A. Greenbaum, S. Hammarling, A. McKenney, S. Ostrouchov, and S. D., *LAPACK Users' Guide - Release 2.0*, SIAM, 1994.
5. T. Anderson, *Fracture Mechancis Fundamentals and Application*, CRC Press, 2005.
6. B. Andersson, U. Falk, I. Babuška, and T. Von-Petersdorff, *Reliable stress and fracture mechanics analysis of complex components using a h-p version of FEM*, Int. Jour. Numer. Meth. Engrg., 38 (1995), pp. 2135–2163.
7. T. Apel, V. Mehrmann, and D. Watkins, *Structured eigenvalue method for computation of corner singularities in 3D anisotropic elastic structures*, Computer Meth. Appl. Mech. Engrg., 191 (2002), pp. 4459–4473.
8. I. Babuška and A. Aziz, *Survey lectures on the mathematical foundations of the finite element method*, in The mathematical foundations of the finite element method with applications to partial differential equations, A. Aziz, ed., Academic Press, New-York, NY, USA, 1972, pp. 3–343.
9. I. Babuška and B. Guo, *Regularity of the solution of elliptic problems with piecewise analytic data. part I: Boundary value problems for linear elliptic equation of second order*, SIAM J. Math. Anal., 19 (1988), pp. 172–203.
10. ———, *Regularity of the solution of elliptic problems with piecewise analytic data. part II: The trace spaces and application to the boundary value problems with nonhomogeneous boundary conditions*, SIAM J. Math. Anal., 20 (1989), pp. 763–781.
11. ———, *Approximation properties of the h-p version of the finite element method*, Computer Meth. Appl. Mech. Engrg., 133 (1996), pp. 319–346.
12. I. Babuška and A. Miller, *The post-processing approach in the finite element method - Part 2: The calculation of stress intensity factors*, Int. Jour. Numer. Meth. Engrg., 20 (1984), pp. 1111–1129.
13. I. Babuška and H.-S. Oh, *The p-version of the finite element method for domains with corners and for infinite domains*, Numer. Methods PDEs, 6 (1990), pp. 371–392.

Z. Yosibash, *Singularities in Elliptic Boundary Value Problems and Elasticity and Their Connection with Failure Initiation*, Interdisciplinary Applied Mathematics 37, DOI 10.1007/978-1-4614-1508-4,

14. I. BABUŠKA AND M. SURI, *The p and h-p versions of the finite element method, basic principles and properties*, SIAM review, 36 (1994), pp. 578–632.
15. I. BABUŠKA, T. VON-PETERSDORFF, AND B. ANDERSSON, *Numerical treatment of vertex singularities and intensity factors for mixed boundary value problems for the Laplace equation in R^3*, SIAM Jour. Numer. Anal., 31 (1994), pp. 1265–1288.
16. M. BACH, S. A. NAZAROV, AND W. L. WENDLAND, *Propagation of a penny shaped crack under the Irwin criterion*, in Analysis, numerics and applications of differential and integral equations (Stuttgart, 1996), vol. 379 of Pitman Res. Notes Math. Ser., Longman, Harlow, 1998, pp. 17–21.
17. L. BANKS-SILLS AND C. ISHBIR, *A conservative integral for bimaterial notches subjected to thermal stresses*, Int. Jour. Numer. Meth. Engrg., 60 (2004), pp. 1075–1102.
18. L. BANKS-SILLS AND D. SHERMAN, *On the computation of stress intensity factors for 3-dimensional geometries by means of the stiffness derivative and J-integral methods*, International Journal of Fracture, 53 (1992), pp. 1–20.
19. L. BANKS-SILLS, P. WAWRZYNEK, B. CARTER, A. INGRAFFEA, AND I. HERSHKOVITZ, *Methods for calculating stress intensity factors in anisotropic materials: Part II Arbitrary geometry*, Engrg. Frac. Mech., 74 (2007), pp. 1293–1307.
20. R. BARSOUM, *Application of the finite element iterative method to the eigenvalue problem of a crack between dissimilar media*, Int. Jour. Numer. Meth. Engrg., 26 (1988), pp. 541–554.
21. ———, *Theoretical basis of the finite element iterative method for the eigenvalue problem in stationary cracks*, Int. Jour. Numer. Meth. Engrg., 26 (1988), pp. 531–539.
22. Z. P. BAŽANT AND L. F. ESTENSSORO, *Surface singularity and crack propagation*, Int. Jour. Solids and Structures, 15 (1979), pp. 405–426.
23. Z. P. BAŽANT AND L. M. KEER, *Singularities of elastic stresses and of harmonic functions at conical notches or inclusions*, Int. Jour. Solids and Structures, 10 (1974), pp. 957–964.
24. A. BEAGLES AND A.-M. SÄNDIG, *Singularities of rotationally symmetric solutions of boundary value problems for the Lamé equations*, ZAMM - Z. Angew. Math. Mech., 71 (1991), pp. 423–431.
25. A. BEAGLES AND J. WHITEMAN, *General conical singularities in three-dimensional Poisson problems*, Math. Meth. Appl. Sci., 11 (1989), pp. 215–235.
26. C. BERNARDI, M. DAUGE, AND I. MADAY, *Spectral methods for axisymmetric domains*, Gauthier-Villars - NH, Paris, France, 1999.
27. H. BLUM AND M. DOBROWOLSKI, *On finite element methods for elliptic equations on domains with corners*, Computing, 28 (1982), pp. 53–63.
28. D. B. BOGY, *On the problem of edge-bonded elastic quarter-planes loaded at the boundary*, Int. Jour. Solids and Structures, 6 (1970), pp. 1287–1313.
29. D. BRAESS, *Finite elements. Theory, fast solvers, and applications in solid mechanics*, Cambridge University Press, London, UK, 2nd ed., 2001.
30. O. BRANDT, K. PLOOG, B. R., AND M. HOHENSTEIN, *Breakdown of continuum elasticity theory in the limit of monoatomic films*, Physical Review Letters, 68 (1992), pp. 1339–1342.
31. S. C. BRENNER, *Multigrid methods for the computation of singular solutions and stress intensity factors I: Corner singularities*, Mathematics of Computations, 68 (1999), pp. 559–583.
32. S. C. BRENNER AND R. L. SCOTT, *The mathematical theory of finite element methods*, Springer-Verlag, Berlin,Heidelberg,New York, 1994.
33. W. CARPENTER, *Calculation of fracture machanics parameters for general corner*, Int. Jour. Fracture, 24 (1984), pp. 45–58.
34. ———, *Insensitivity of the reciprocal work contour integral method to higher order eigenvectors*, Int. Jour. Fracture, 73 (1995), pp. 93–108.
35. W. CARPENTER AND C. BYERS, *A path independent integral for computing stress intensities for V-notched cracks in a bi-material*, Int. Jour. Fracture, 35 (1987), pp. 245–268.
36. D.-H. CHEN, *Analysis of singular stress field around the inclusion corner tip*, Engrg. Frac. Mech., 49 (1994), pp. 533–546.

37. D.-H. CHEN, *Logarithmic singular stress field in a semi-infinite plate consisting of two edge-bonded wedges subjected to surface tractions*, Int. Jour. Fracture, 75 (1996), pp. 357–378.
38. G. CHEREPANOV, *Crack propagation in continuous media*, Journal of Applied Mathematics and Mechanics, 31 (1967), pp. 503–512.
39. R. CHURCHILL, *Complex variables and applications*, McGraw-Hill, 1960.
40. M. COSTABEL AND M. DAUGE, *General edge asymptotics of solution of second order elliptic boundary value problems I & II*, Proc. Royal Soc. Edinburgh, 123A (1993), pp. 109–184.
41. ———, *Computation of corner singularities in linear elasticity*, in Boundary value problems and integral equtions in nonsmooth domains, M. Costabel, M. Dauge, and S. Nicaise, eds., Marcel Dekker, New York, Basel, Hong-Kong, 1995, pp. 59–68.
42. ———, *Crack singularities for general elliptic systems*, Math. Nach., 235 (2002), pp. 29–49.
43. M. COSTABEL, M. DAUGE, AND R. DUDUCHAVA, *Asymptotics without logarithmic terms for crack problems*, Communication in PDEs, 28 (2003), pp. 869–926.
44. M. COSTABEL, M. DAUGE, AND Y. LAFRANCHE, *Fast semi-analytic computation of elastic edge singularities*, Computer Meth. Appl. Mech. Engrg., 190 (2001), pp. 2111–2134.
45. M. COSTABEL, M. DAUGE, AND S. NICAISE, *Corner Singularities and Analytic Regularity for Linear Elliptic Systems*. In preparation.
46. M. COSTABEL, M. DAUGE, AND Z. YOSIBASH, *A quasidual function method for extracting edge stress intensity functions*, SIAM Jour. Math. Anal., 35 (2004), pp. 1177–1202.
47. M. COSTABEL AND E. STEPHAN, *Curvature terms in asymptotic expansions for solutions of boundary integral equations on curved polygons*, Jour. Integral Equations, 5 (1983), pp. 353–371.
48. B. COTTERELL AND J. RICE, *On a slightly curved or kinked crack*, Int. Jour. Fracture, 16 (1980), pp. 155–169.
49. M. DAUGE, *Elliptic boundary value problems in corner domains - smoothness and asymptotics of solutions*, Lecture notes in Mathematics 1341, Springer-Verlag, Heidelberg, 1988.
50. M. DAUGE AND M. POGU, *Existence et régularité de la fonction potentiel pour des écoulements subcritiques s'établissant autour d'un corps à singularité conique*, Annales Facultè des Sciences de Toulouse, 9 (1988), pp. 213 – 242.
51. A. DE AZA, J. CHEVALIER, G. FANTOZZI, M. SCHEHEL, AND R. TORRECILLAS, *Crack growth resistence of alumina, zirconia and zirconia toughened alumina ceramics for joint prostheses*, Biomaterials, 23 (2002), pp. 937–945.
52. J. P. DEMPSEY, *Power-logarithmic stress singularities at bi-material corners and interface cracks*, Jour. Adhesion Sci. Technol., 9 (1995), pp. 253–265.
53. J. P. DEMPSEY AND G. B. SINCLAIR, *On the stress singularities in the plane elasticity of the composite edge*, Jour. of Elasticity, 9 (1979), pp. 373–391.
54. ———, *On the singular behavior at the vertex of a bi-material wedge*, Jour. of Elasticity, 11 (1981), pp. 317–327.
55. X. DENG, *General crack-tip fields for stationary and steadily growing interface cracks in anisotropic bimaterials*, Trans. ASME, Jour. Appl. Mech., 60 (1993), pp. 183–189.
56. A. DIMITROV, H. ANDRA, AND E. SCHNACK, *Efficient computation of order and mode of corner singularities in 3d-elasticity*, Int. J. Num. Meth. Engrg., 52 (2001), pp. 805–827.
57. R. DUDUCHAVA AND W. WENDLAND, *The Wiener-Hopf method for systems of pseudodifferential equations with an application to crack problems*, Integral Equations Operator Theory, 23 (1995), pp. 294–335.
58. M. L. DUNN, W. SUWITO, AND S. CUNNINGHAM, *Fracture initiation at sharp notches: Correlation using critical stress intensities*, Int. Jour. Solids and Structures, 34 (1997), pp. 3873–3883.
59. M. L. DUNN, W. SUWITO, S. CUNNINGHAM, AND C. W. MAY, *Fracture initiation at sharp notches under mode I, mode II, and mild mixed mode loading*, Int. Jour. Fracture, 84 (1997), pp. 367–381.
60. A. ENGLAND, *A crack between dissimilar media*, Trans. ASME, Jour. Appl. Mech., 32 (1965), pp. 400–402.

61. F. ERDOGAN, *Stress distribution in a nonhomogeneous elastic plane with cracks*, Trans. ASME, Jour. Appl. Mech., 30 (1963), pp. 232–236.
62. ———, *Stress distribution in bonded dissimilar materials with cracks*, Trans. ASME, Jour. Appl. Mech., 32 (1965), pp. 418–423.
63. M. FRANKLE, D. MUNZ, AND Y. YANG, *Stress singularities in a bimaterial joint with inhomogeneous temperature distribution*, Int. Jour. Solids and Structures, 33 (1996), pp. 2039–2054.
64. F. GHAHREMANI, *A numerical variational method for extracting 3D singularities*, Int. Jour. Solids and Structures, 27 (1991), pp. 1371–1386.
65. D. GILBARG AND N. S. TRUDINGER, *Elliptic partial differential equations of second order*, Springer-Verlag, Berlin, Heidelberg, New-York, 1977.
66. F. GOMEZ AND M. ELICES, *Fracture of componenets with sharp v-shaped notches*, Engrg. Frac. Mech., 70 (2003), pp. 1913–1927.
67. F. GOMEZ, M. ELICES, AND J. PLANAS, *The cohesive crack concept: application to pmma at* $-60°$ *c*, Engrg. Frac. Mech., 72 (2005), pp. 1268–1285.
68. W. J. GORDON AND C. A. HALL, *Transfinite element methods: Blending functions interpolation over arbitrary curved element domains*, Numer. Math., 21 (1973), pp. 109–129.
69. R. GREGORY, *Green's functions, bi-linear forms, and completeness of the eigenfunctions for the elastostatic strip and wedge*, Jour. of Elasticity, 9 (1979), pp. 283–309.
70. A. A. GRIFFITH, *The phenomena of rupture and flow in solids*, Philosophical Tran. Roy. Soc. London, Ser A, 221 (1920), pp. 163–198.
71. ———, *The theory of rupture*, in Proc. 1st Int. Congr. Appl. Mech., Delfth, 1924, pp. 55–63.
72. P. GRISVARD, *Elliptic problems in nonsmooth domains*, Pitman Publishing, England, 1985.
73. ———, *Singularities in boundary value problems*, Masson, France, 1992.
74. L. GU AND T. BELYTSCHKO, *A numerical study of stress singularities in a two-material wedge*, Int. Jour. Solids and Structures, 31 (1994), pp. 865–889.
75. B. GUO, *The h-p version of the fininte element method for solving boundary value problems in polyhedral domains*, in Boundary value problems and integral equtions in nonsmooth domains, M. Costabel, M. Dauge, and S. Nicaise, eds., Marcel Dekker, New York, Basel, Hong-Kong, 1995, pp. 101–120.
76. B. GUO AND H.-S. OH, *The method of auxiliary mapping for the finite element solutions of elliptic partial differential equations on nonsmooth domains in* R^3. http://home.cc.umanitoba.ca/g̃uo//mapping-3d.pdf, 1996.
77. S. E. HALLSTROM, *A generalised fracture mechanics approach to fracture initiated at corners*, PhD thesis, Dept. of Aeronautics, Royal Iinstitute of Technology, Stockholm, Sweden, 1997.
78. S. HAMOUSH AND M. SALAMI, *A stiffness derivative technique to determine mixed-mode stress intensity factors of rectilinear anisotropic solids*, Engineering Fracture Mechanics, 44 (1993), pp. 297–305.
79. R. HARTRANFT AND G. SIH, *The use of eigenfunction expansions in the general solution of three-dimensional crack problems*, Jour. Math. Mech., 19 (1967), pp. 123–138.
80. T. HATTORI, *A stress singularity parameter approach for evaluating the adhesive strength of single-lap joints*, Int. Jour. Japanese Soc. Mech. Eng., Ser. I, 34 (1991), pp. 326–331.
81. T. HATTORI, S. SAKATA, AND G. MURAKAMI, *A stress singularity parameter approach for evaluating the interfacial reliability of plastic encapsulated LSI devices*, Jour. Electronic Packaging, 111 (1989), pp. 243–248.
82. J. HUTCHINSON, M. MEAR, AND J. RICE, *Crack paralleling an interface between dissimilar materials*, Trans. ASME, Jour. Appl. Mech., 54 (1987), pp. 828–832.
83. C. IGLIS, *Stresses in a plate due to the presence of cracks and sharp corners*, Transactions of the Institute of Naval Architects, 55 (1913), pp. 219–241.
84. T. IKEDA, M. NAGAI, K. YAMANAGA, AND N. MIYAZAKI, *Stress intensity factor analyses of interface cracks between dissimilar anisotropic materials using the finite element method*, Engrg. Frac. Mech., 73 (2006), pp. 2067–2079.

85. K. IKEGAMI, *Some topics of mechanical problems in electronic packaging*, Advances in Electronic Packaging, ASME-EEP, 1 (1992), pp. 567–573.
86. G. R. IRWIN, *Onset of fast crack propagation in high strength steel and aluminum alloys*, in Sagamore Research Conference Proceedings - Vol 2, 1956, pp. 289–305.
87. ———, *Analysis of stresses and strains near the end of a crack transversing a plate*, Trans. ASME, Jour. Appl. Mech., 24 (1957), pp. 361–364.
88. Z.-H. JIN AND C. SUN, *On J-integral and potential energy variation*, Int. Jour. Fracture, 126 (2004), pp. L19–L24.
89. A. KALANDIIA, *Remarks on the singularity of elastic solutions near corners*, Jour. Appl. Math. Mech., 33 (1969), pp. 127–131.
90. M. F. KANNINEN AND C. H. POPELAR, *Advanced Fracture Mechanics*, Oxford University Press, New York, NY, USA, 1985.
91. E. KARNIADAKIS AND S. SHERWIN, *Spectral/hp element methods for CFD*, Oxford University Press, New York, 1999.
92. ———, *Spectral/hp element methods for CFD - 2nd edition*, Oxford University Press, New York, 2006.
93. S. KARP AND F. J. KARAL, *The elastic field behavior in the neighborhood of a crack of arbitraty angle*, Comm. on Pure & Appl. Math., 15 (1962), pp. 413–421.
94. R. KEATING AND G. SINCLAIR, *On the fundamental energy argument of elastic fracture mechanics*, Int. Jour. Fracture, 74 (1995), pp. 43–61.
95. V. A. KONDRATIEV, *Boundary value problems for elliptic equations in domains with conical or angular points*, Transact. Moscow Math. Soc., 16 (1967), pp. 227–313.
96. V. KOZLOV AND V. MAZ'YA, *On stress singularities near the boundary of a polygonal crack*, Proc. Roy. Soc. Edinburgh Sect. A, 117 (1991), pp. 31–37.
97. V. KOZLOV, V. MAZ'YA, AND J. ROSSMANN, *Elliptic boundary value problems in domains with point singularities*, AMS Mathematical Surveys and Monographs, Vol. 52, Providence, RI, 1997.
98. ———, *Spectral problems associated with corner singularities of solutions to elliptic equations*, AMS Mathematical Surveys and Monographs, Vol. 85, Providence, RI, 2001.
99. A. KUFNER AND A.-M. SÄNDIG, *Some Applications of Weighted Sobolev Spaces*, Teubner: Leipzig, 1987.
100. P. LAURENT-GENGOUX AND D. NEVEU, *Calcul des singularités par le méthode des eléménts finis*, Mathematical Modelling and Numerical Analysis, 24 (1990), pp. 85–101.
101. P. LAZZARIN AND R. ZMABARDI, *A finite-volume-energy based approach to predict the static and fatigue behavior of components with sharp v-shaped notches*, Int. Jour. Fracture, 112 (2001), pp. 275–298.
102. N. LEBEDEV, *Special functions and their applications*, Prentice-Hall, 1965.
103. J. LEBLOND AND O. TORLAI, *The stress-field near the front of an arbitrarily shaped crack in a 3-dimensional elastic body*, Jour. of Elasticity, 29 (1992), pp. 97–131.
104. K. LEE AND Y. H. CHO, *Boundary element analysis of thermal stress intensity factors for cusp cracks*, Engrg. Frac. Mech., 37 (1990), pp. 787–798.
105. D. LEGUILLON, *Computation of 3D-singularities in elasticity*, in Boundary value problems and integral equations in nonsmooth domains - Lecture notes on pure and applied mathematics, Vol. 167, M. Costabel, M. Dauge, and S. Nicaise, eds., New York, 1995, Marcel Dekker, pp. 161–170.
106. ———, *A critetion for crack nucleation at a notch in homogeneous materials*, C.R. Acad. Sci. Paris, Ser IIb, 329 (2001), pp. 97–102.
107. ———, *Strength or toughness? A criterion for crack onset at a notch*, Eur. J. of Mechanics A/Solids, 21 (2002), pp. 61–72.
108. D. LEGUILLON AND S. MURER, *Crack deflection in a biaxial stress state*, Int. Jour. Fracture, 150 (2008), pp. 75–90.
109. D. LEGUILLON AND E. SANCHEZ-PALENCIA, *Computation of singular solutions in elliptic problems and elasticity*, John Wiley & Sons, New York, NY, 1987.

110. D. LEGUILLON AND Z. YOSIBASH, *Crack onset at a V-notch. Influence of the notch tip radius*, Int. Jour. Fracture, 122 (2003), pp. 1–21.
111. R. LEHMAN, *Developments near an analytic corner or solutions of elliptic partial differential equations*, Jour. Math. Mech., 8 (1959), pp. 727–760.
112. N. LIU AND J. ALTIERO, *A new boundary element method for the solution of plane steady-state thermoelastic fracture mechanics problems*, Appl. Math. Modelling, 16 (1992), pp. 618–628.
113. A. LOVE, *A treatise on the mathematical theory of elasticity*, Dover, New York, USA, 1944.
114. C.-C. MA, *Plane solution of thermal stresses for anisotropic bimaterial elastic wedges*, Jour. Thermal Stresses, 18 (1995), pp. 219–245.
115. V. MANTIC, F. PARIS, AND J. BERGER, *Singularities in 2D anisotropic potential problems in multi-material corners. Real variable approach*, Int. Jour. Solids and Structures, 40 (2003), pp. 5197–5218.
116. V. MAZ'YA AND J. ROSSMANN, *On a problem of Babuška (Stable asymptotics of the solution of the Dirichlet problem for elliptic equations of second order in domains with angular points)*, Mathematische Nachrichten, 155 (1992), pp. 199–220.
117. ———, *Elliptic Equations in Polyhedral Domains*, American Mathematical Society, Providence, 2010.
118. E. MAZZA AND J. DUAL, *Mechanical behavior of a μm-sized single crystal silicon structure with sharp notches*, Jour. Mech. Phy. Sol., 47 (1999), pp. 1795–1821.
119. M. MICHAEL AND R. HARTRANFT, *Thermal stress singularities in microelectronic*, in Proceedings - 41st Electronic Components & Technology Conference, Atlanta, GA, USA, IEEE, Piscataway, NJ, 1991, pp. 273–277.
120. T. MIYOSHI, M. SHIRATORI, H. OKUDA, AND N. TAKANO, *Stress singularities at the interface crack and the corner of edge-bonded dissimilar materials*, Advances in Electronic Packaging - ASME, (1992), pp. 551–557.
121. H. K. MOFFATT AND B. R. DUFFY, *Local similarity solutions and their limitations*, Jour. Fluid Mech., 96 (1980), pp. 299–313.
122. M. MOSKE, P. HO, D. MIKALSEN, J. CUOMO, AND R. ROSENBERG, *Measurement of thermal stress and stress relaxation in confined metal lines. 1. stress during thermal cycling*, Jour. App. Phys., 74 (1993), pp. 1716–1724.
123. A. MOVCHAN AND N. MOVCHAN, *Mathematical modelling of solids with nonregular boundaries*, CRC press, New York, NY, 1995.
124. Y. MURAKAMI, *Stress intensity factors handbook (two volumes)*, Pergamon Press, New York, NY, 1987.
125. N. I. MUSKHELISHVILI, *Some basic problems of the mathematical theory of elasticity*, P. Nordhoff, Groningen, Holland, 1953.
126. S. NAKANISHI, S. TANI, M. SUZUKI, AND N. SUMI, *Orthotropic rectangular plates with an eccentric crack and an inclined crack in steady state temperature fields*, Trans. Japan Soc. Mech. Engrs., 51 (1985), pp. 2094–2102. (In Japanese).
127. S. NAZAROV AND B. PLAMENEVSKY, *Elliptic problems in domains with piecewise smooth boundaries*, De Gruyter Expositions in Mathematics, 13, Berlin, New York, 1994.
128. J. NEČAS AND I. HLAVÁČEK, *Mathematical theory of elastic and elastico-plastic bodies: An Introduction*, Elsevier, Amsterdam-Oxford-New York, 1981.
129. T. NOSE AND T. FUJII, *Evaluation of fracture toughness for ceramic materials by a single edge pre-cracked-beam method*, Jour. Amer. Ceramic Soc., 71 (1988), pp. 328–333.
130. V. NOVOZHILOV, *On a necessary and sufficient criterion for brittle strength*, Jour. Appl. Math. Mech. (Translation of PMM), 33 (1969), pp. 212–222.
131. J. T. ODEN AND J. N. REDDY, *Variational methods in theoretical mechanics*, Springer-Verlag, New York, 1983.
132. H.-S. OH AND I. BABUŠKA, *P-version of the finite element method for the elliptic boundary value problems with interfaces*, Computer Meth. Appl. Mech. Engrg., 97 (1992), pp. 211–231.
133. M. OHRING, *The Materials Science of Thin Films*, Academic Press, 1992.

134. N. Omer, Z. Yosibash, M. Costabel, and M. Dauge, *Edge flux intensity functions in polyhedral domains and their extraction by a quasidual function method*, Int. Jour. Fracture, 129 (2004), pp. 97–130.
135. S. S. Pageau, P. F. Joseph, and S. B. J. Biggers, *Finite element analysis of anisotropic materials with singular inplane stress fields*, Int. Jour. Solids and Structures, 32 (1995), pp. 571–591.
136. S. S. Pageau, S. G. Kesavaram, S. J. Biggers, and P. F. Joseph, *Standardized complex and logarithmic eigensolutions for n-material wedges and junctions*, Int. Jour. Fracture, 77 (1997), pp. 51–76.
137. E. Pan and F. Yuan, *Boundary element analysis of three-dimensional cracks in anisotropic solids*, Int. Jour. Numer. Meth. Engrg., 48 (2000), pp. 211–237.
138. P. Papadakis, *Computational aspects of the determination of the stress intensity factors for two-dimensional elasticity*, PhD thesis, University of Maryland at College Park, College Park, Maryland, USA, Dec. 1988.
139. P. Papadakis and I. Babuška, *A numerical procedure for the determination of certain quantities related to stress intensity factors in two-dimensional elasticity*, Computer Meth. Appl. Mech. Engrg., 122 (1995), pp. 69–92.
140. D. M. Parks, *A stiffness derivative finite element technique for determination of crack tip stress intensity factors*, Int. Jour. Fracture, (1974), pp. 487–501.
141. D. Picard, D. Leguillon, and C. Putot, *A method to estimate the influence of the notch-root radius on the fracture toughness mesurements of ceramics*, Jour. Europ. Ceramic Soc., 26 (2005), pp. 1421–1427.
142. N. N. V. Prasad, M. H. Aliabadi, and D. P. Rooke, *The dual boundary element method for thermoelastic crack problems*, Int. Jour. Fracture, 66 (1994), pp. 255–272.
143. E. Priel, A. Bussiba, I. Gilad, and Z. Yosibash, *Mixed mode failure criteria for brittle elastic V-notched structures*, Int. Jour. Fracture, 144 (2007), pp. 247–265.
144. E. Priel, Z. Yosibash, and D. Leguillon, *Failure initiation at a blunt V-notch tip under mixed mode loading*, Int. Jour. Fracture, 149 (2008), pp. 143–173.
145. A. Ralston and P. Rabinowitz, *A first course in numerical analysis*, McGraw-Hill, New-York, USA, 2nd ed., 1977.
146. J. E. D. Reedy and T. R. Guess, *Butt tensile joint strength: interface corner stress intensity factor prediction*, Jour. Adhesion Science Technology, 9 (1995), pp. 237–251.
147. J. Rice, *A path independent integral and the approximate analysis of strain concentration by notches and cracks*, Journal of Applied Mechanics, 35 (1968), pp. 379–386.
148. J. Rice, *Limitations to the small-scale yielding approximation for crack-tip plasticity*, Jour. Mech. Phys. Solids, 22 (1974), pp. 17–26.
149. J. R. Rice, *Elastic fracture mechanics concepts for interfacial cracks*, Trans. ASME, Jour. Appl. Mech., 55 (1988), pp. 98–103.
150. J. R. Rice and G. C. Sih, *Plane problems of cracks in dissimilar media*, Trans. ASME, Jour. Appl. Mech., 32 (1965), pp. 418–423.
151. L. Rosa, J. Fernandes, and I. Duarte, *Subcritical crack growth in three engineering ceramics under biaxial conditions*, in ECF12 - Fracture from defects - Volume I, M. Brown, E. de los Rios, and K. Miller, eds., EMAS Publishing, 1998, pp. 509–514.
152. A. Rössle, *Spannungssingluraritäteen fur gekoppelte Strukturen in der Festkörpermechanik unter mechanisher und thermischer belastung*, master's thesis, Universität Stuttgart, Mathematisches Institut A, Stuttgart, Germany, January 1996.
153. ———, *Corner singularities and regularity of weak solutions for the two-dimensional Lamé equations on domains with angular corners*, Jour. Elasticity, 60 (2000), pp. 57–75.
154. A. Rössle and A.-M. Sändig, *Stress singularities in bonded elastic materials*, in The mathematics of finite element and applications, 1996, J. R. Whiteman, ed., John Wiley & Sons, New York, 1997, pp. 405–416.
155. A. I. Sauter and W. Nix, *Finite element calculations of thermal stresses in passivated and unpassivated lines bonded to substrates*, in Materials Research Society Symposium Proceedings, vol. 188, 1990, pp. 15–21.

156. E. J. SCHIERMEIER AND B. A. SZABÓ, *Numerical analysis of stress singularities in composite materials*, Engrg. Frac. Mech., 32 (1989), pp. 979–996.
157. C. SCHWAB, *p- and hp- Finite element methods*, Oxford Science Publications, UK, 1998.
158. A. SEWERYN, *Brittle fracture criterion for structures with sharp notches*, Eng. Frac. Mech., 47 (1994), pp. 673–681.
159. A. SEWERYN AND A. LUKASZEWICZ, *Verification of brittle fracture criteria for elements with v-shaped notches*, Eng. Frac. Mech., 69 (2002), pp. 1487–1510.
160. C. SHUMIN AND Z. XING, *Generalized stiffness derivative method for mixed-mode crack problems*, Mechanics Research Communications, 17 (1990), pp. 437–444.
161. G. SIH AND H. LIEBOWITZ, *On the Griffith energy criterion for brittle fracture*, Int. Jour. Solids and Structures, 3 (1967), pp. 1–22.
162. G. SIH AND B. MACDONALD, *Fracture mechanics applied to engineering problems - strain energy density fracture criterion*, Eng. Fracture Mechanics, 6 (1974), pp. 361–386.
163. G. C. SIH, *On the singular character of thermal stresses near a crack tip*, Trans. ASME, Jour. Appl. Mech., 29 (1962), pp. 587–589.
164. G. C. SIH AND J. R. RICE, *The bending of plates of dissimilar materials with cracks*, Trans. ASME, Jour. Appl. Mech., 31 (1964), pp. 477–482.
165. G. SINCLAIR, *Stress singularities in classical elasticity II: Asymptotic identification*, Appl. Mech. Rev., 57 (2004), pp. 385–439.
166. G. SINCLAIR, M. OKAJIMA, AND J. GRIFFIN, *Path independent integrals for computing stress intensity factors at sharp notches in elastic plates*, Int. Jour. Numer. Meth. Engrg., 20 (1984), pp. 999–1008.
167. I. S. SOKOLNIKOFF, *Mathematical Theory of Elasticity*, McGraw-Hill, New York, 1956.
168. J. E. STEINWALL AND H. JOHNSON, *Mechanical properties of thin film aluminum fibers: Grain size effects*, in Materials Research Society Symposium Proceedings - 188, M. Doerner, W. Oliver, G. Pharr, and F. Brotzen, eds., MRS, 1990, pp. 177–183.
169. M. W. STEKLOFF, *Sur les problèmes fondamentaux de la physique mathématique*, Ann. Sci. Ecole Norm. Sup., 19 (1902), pp. 455–490.
170. E. STEPHAN AND J. WHITEMAN, *Singularities of the Laplacian at corners and edges of three-dimensional domains and their treatment with finite element methods*, Math. Meth. Appl. Sci., 10 (1988), pp. 339–350.
171. M. STERN, E. BECKER, AND R. DUNHAM, *A contour integral computation of mixed-mode stress intensity factors*, Int. Jour. Fracture, 12 (1976), pp. 359–368.
172. M. STERN AND M. L. SONI, *On the computation of stress intensities at fixed-free corners*, Int. Jour. Solids and Structures, 12 (1976), pp. 331–337.
173. N. SUMI AND T. KATAYAMA, *Thermal stress singularities at tips of a Griffith crack in a finite rectangular plate*, Nuclear Eng. Desgn., 60 (1980), pp. 389–394.
174. Z. SUO, *Mechanics of interface fracture*, PhD thesis, Harvard University, Cambridge, Massachusetts, USA, May 1989.
175. W. SUWITO AND M. L. DUNN, *Fracture initiation at sharp notches in single crystal silicon*, Jour. Appl. Physics, 83 (1998), pp. 3574–3582.
176. B. A. SZABÓ AND R. L. ACTIS, *Finite element analysis in professional practice*, Computer Meth. Appl. Mech. Engrg., 133 (1996), pp. 209–228.
177. B. A. SZABÓ AND I. BABUŠKA, *Computation of the amplitude of stress singular terms for cracks and reentrant corners*, in Fracture Mechanics: Nineteenth Symposium, C. T. A., ed., ASTM STP 969, ASTM, Philadelphia, 1988, pp. 101–124.
178. ———, *Finite Element Analysis*, John Wiley & Sons, New York, 1991.
179. B. A. SZABÓ, I. BABUŠKA, AND B. K. CHAYAPATHY, *Stress computation for nearly incompressible materials by the p-version of the finite element method*, Int. Jour. Numer. Meth. Engrg., 28 (1989), pp. 2175–2190.
180. B. A. SZABÓ AND Z. YOSIBASH, *Numerical analysis of singularities in two-dimensions. Part 2: Computation of the generalized flux/stress intensity factors*, Int. Jour. Numer. Meth. Engrg., 39 (1996), pp. 409–434.
181. S. SZE, *VLSI Technology*, McGraw-Hill, 1983.

182. D. TAYLOR, *The theory of critical distances*, Elsevier Science, 2007.
183. C.-H. TSAI AND C.-C. MA, *Thermal weight function of cracked bodies subjected to thermal loading*, Engrg. Frac. Mech., 41 (1992), pp. 27–40.
184. K.-N. TU, J. W. MAYER, AND L. C. FELDMAN, *Electronic Thin Film Science for Electrical Engineers and Material Scientists*, Macmillan Publishing Company, 1992.
185. D. VASILOPOULOS, *On the determination of higher order terms of singular elastic stress fields near corners*, Numer. Math., 53 (1988), pp. 51–95.
186. T. VON PETERSDORFF AND E. STEPHAN, *Singularities of the solution of the Laplacian in domains with circular edges*, Applicable Analysis, 45 (1992), pp. 281 – 294.
187. S. WAN, M. DUNN, S. CUNNINGHAM, AND D. READ, *Elastic moduli, strength, and fracture initiation at sharp notches in etched single crystal silicon microstructures*, Jour. Appl. Phy., 85 (1999), pp. 3519–3534.
188. J. G. WILLIAMS, *Fracture Mechanics of Polymers*, Ellis Horwood, Chichester., 1984.
189. M. L. WILLIAMS, *Stress singularities resulting from various boundary conditions in angular corners of plates in extension*, Trans. ASME, Jour. Appl. Mech., 19 (1952), pp. 526–528.
190. ———, *On the stress distribution at the base of a stationary crack*, Trans. ASME, Jour. Appl. Mech., 24 (1957), pp. 109–114.
191. ———, *The stresses around a fault or crack in dissimilar media*, Bull. Seismological Society of America, 49 (1959), pp. 199–204.
192. S. WOLFRAM, *Mathematica 5*, Wolfram Research, Inc., 2004.
193. K. WU, *Representations of stress intensity factors by path-independent integrals*, Trans. ASME, Jour. Appl. Mech., 56 (1989), pp. 780–785.
194. Y. WU, *A new method for evaluation of stress intensities for interface cracks*, Engrg. Frac. Mech., 48 (1994), pp. 755–761.
195. A. YAKHOT, *A note on non-regular similarities*. Private Communication, 2003.
196. X. X. YANG AND Z. B. KUANG, *Contour integral method for stress intensity factors of interface crack*, Int. Jour. Fracture, 78 (1996), pp. 299–313.
197. X. YING, *A reliable root solver for automatic computation with application to stress analysis of a composite plane wedge*, PhD thesis, Washington University, St. Louis, Missouri, USA, Dec. 1986.
198. X. YING AND I. KATZ, *A uniform formulation for the calculation of stress singularities in the plane elasticity of a wedge composed of multiple isotropic materials*, Comput. Math. Applic., 14 (1987), pp. 437–458.
199. ———, *A reliable argument principle algorithm to find the number of zeros of an analytic function in a bounded domain*, Numerische Mathematik, 53 (1988), pp. 143–163.
200. Z. YOSIBASH, *Numerical analysis of singularities and first derivatives for elliptic boundary value problems in two-dimensions*, PhD thesis, Sever Institute of Technology, Washington University, St. Louis, Missouri, USA, Aug. 1994.
201. ———, *Numerical thermo-elastic analysis of singularities in two-dimensions*, Int. Jour. Fracture, 74 (1996), pp. 341–361.
202. ———, *Computing edge singularities in elastic anisotropic three-dimensional domains*, Int. Jour. Fracture, 86 (1997), pp. 221–245.
203. ———, *Numerical analysis on singular solutions of the Poisson equation in two-dimensions*, Comp. Mech., 20 (1997), pp. 320–330.
204. ———, *Thermal generalized stress intensity factors in 2-d domains*, Computer Meth. Appl. Mech. Engrg., 157 (1998), pp. 365–385.
205. Z. YOSIBASH, O. ADAN, R. SHNECK, AND H. ATLAS, *Thermo-mechanical failure criterion at the micron scale in electronic devices*, Int. Jour. Fracture, 122 (2003), pp. 47–64.
206. Z. YOSIBASH, A. BUSSIBA, AND I. GILAD, *Failure criteria for brittle elastic materials*, Int. Jour. Fracture, 125 (2004), pp. 307–333.
207. Z. YOSIBASH, E. PRIEL, AND D. LEGUILLON, *A failure criterion for brittle elastic materials under mixed-mode loading*, Int. Jour. Fracture, 141 (2006), pp. 291–312.

208. Z. YOSIBASH, S. SHANNON, M. DAUGE, AND M. COSTABEL, *Circular edge singularities for the Laplace equation and the elasticity system in 3-D domains*, Int. Jour. Fracture, 168 (2011), p. 31–52.
209. Z. YOSIBASH AND B. A. SZABÓ, *Convergence of stress maxima in finite element computations*, Communications Numer. Meth. Engrg., 10 (1994), pp. 683–697.
210. ———, *Numerical analysis of singularities in two-dimensions. Part 1: Computation of eigenpairs*, Int. Jour. Numer. Meth. Engrg., 38 (1995), pp. 2055–2082.
211. Z. YUE, H. XIAO, AND E. PAN, *Stress intensity factors of square crack inclined to interface of transversely isotropic bi-material*, Eng. Anal. with Boundary Elements, 31 (2007), pp. 50–65.
212. W. M. ZAJACZKOWSKI, *On the edge singularities in composite media*, in Proceedings of the Conference on Applications of Multiple Scaling in Mechanics, Ciarlet and Sanchez-Palencia, eds., Paris, 1987, Masson.
213. T. ZALTZMAN AND Z. YOSIBASH, *Vertex singularities associated with singular points for the 3-D Laplace equation*, Num. Meth. Partial-Diff. Eqs., 27 (2011), pp. 662–679.

Index

Z. Yosibash, *Singularities in Elliptic Boundary Value Problems and Elasticity and Their Connection with Failure Initiation*, Interdisciplinary Applied Mathematics 37,
DOI 10.1007/978-1-4614-1508-4, © Springer Science+Business Media, LLC 2012

FSC
www.fsc.org
MIX
Papier | Fördert
gute Waldnutzung
FSC® C083411

Kolibri 360 GmbH
Ferdinand-Jühlke-Straße 7
99095 Erfurt, Deutschland
produktsicherheit@kolibri360.de